# *Proceedings in Life Sciences*

# MICROBIAL ECOLOGY

Edited by
M. W. Loutit and J. A. R. Miles

With 173 Figures

Springer-Verlag
Berlin Heidelberg New York 1978

Associate Professor Margaret W. Loutit
Professor John A. R. Miles
Department of Microbiology
University of Otago
P.O.Box 56, Dunedin/New Zealand

ISBN 3-540-08974-8 Springer-Verlag Berlin Heidelberg New York
ISBN 0-387-08974-8 Springer-Verlag New York Heidelberg Berlin

Library of Congress Cataloging in Publication Data. Main entry under title: Microbial ecology. (Proceedings in life sciences) Bibliography: p. Includes index. 1. Microbial ecology. I. Loutit, M. W., 1929-. II. Miles, John Arthur Reginald, 1913-. QR100.M513. 576'.15. 78-14480.



Offsetprinting: Beltz Offsetdruck, Hemsbach/Bergstr., Bookbinding: Brühlsche Universitätsdruckerei, Gießen.

2131/3130-543210

# Preface

The suggestion for a symposium on microbial ecology was first put forward by Professor J.A.R. Miles in 1972. After gaining support from the New Zealand Microbiological Society and the Royal Society of New Zealand, a national committee with international representatives and a local committee were formed. Sponsorship was obtained from UNEP, UNESCO, ICRO and IOMS, IUBS and ICOME and the culmination was the First International Microbial Ecology Symposium in 1977 attended by over 400 scientists from 30 countries.

Certain facets of microbial ecology have been studied for over a century, but the recognition of microbial ecology as a discipline has come relatively recently. The National Committee decided that all aspects of microbial ecology should be discussed at the Symposium. The local organising committee, therefore, invited papers on the ecology of microorganisms and viruses associated with plants and animals, as well as microorganisms associated with soil and water and with general environmental problems.

Of the 240 papers presented, only a selection is published here. It is hoped that they will be of interest particularly to those who could not attend. The volume stands as a tribute to the foresight of John Miles in instigating the symposium and who retires this year from the Microbiology Department of the University of Otago where he has had a distinguished career. May his interest in microbial ecology long continue.

July, 1978 M.W. LOUTIT

# Contents

## Microbial Ecology - General



## Marine Microbial Ecology



## Freshwater Microbial Ecology

## Biodegradation in Aquatic Environments



## Microbial Ecology of Soils

## Microbial Ecology of Animals

## Microorganisms and the Gastro Intestinal Tract



## Microbial Insecticides

## Microbial Ecology of Plants

### The Rhizosphere



### Plant Diseases

## Rhizobium - Legume Symbiosis



## Environmental Problems Management and Control

## Microbiology of Food



*Requests for reprints should be addressed to this person

# Contributors

ALEXANDER, M., Department of Agronomy, Cornell University, Ithaca, New York 14853 / USA

ATLAS, R.M., Department of Biology, University of Louisville, Louisville, Kentucky 40208 / USA

BABICH, H., Biology Department, New York University, New York, New York 10003 / USA

BARTLETT, K., Department of Bacteriology and Biochemistry, P.O. Box 1700, Victoria, British Columbia V8W 2Y2 / Canada

BAZIN, M.J., Queen Elizabeth College, Campden Hill Road, London W8 / U.K.

BEAN, W.J., Jr., Laboratories of Virology, St. Jude Children's Research Hospital, P.O. Box 318, Memphis, Tennessee 38101 / USA

BEEVER, R.E., Plant Diseases Division, DSIR, Private Bag, Auckland / New Zealand

BERGERSEN, F.J., CSIRO Division of Plant Industry, Canberra 2601 / Australia

BETLACH, M.R., Department Crop and Soil Sciences and Cyclotron Laboratory, Michigan State University, E. Lansing, Mich. 48824 / USA

BETTELHEIM, K.A., National Health Institute, P.O. Box 7126, Wellington South / New Zealand

BLACKIE, M.J., Lincoln College, Canterbury / New Zealand

BROCKWELL, J., CSIRO Division of Plant Industry, Canberra, ACT 2601 / Australia

BURNS, D.J.W., Plant Diseases Division, DSIR, Private Bag, Auckland / New Zealand

CAPPENBERG, TH.E., Limnological Institute "Vijverhof", Nieuwersluis / The Netherlands

CASHION, P., University of New Brunswick, Fredericton, N.B. / Canada, E3B 5A3

CLARKE, R.T.J., Applied Biochemistry Division, D.S.I.R., Palmerston North / New Zealand

CLOSE, R.C., Lincoln College, Canterbury / New Zealand

COLE, A.L.J., Department of Botany, University of Canterbury, Christchurch / New Zealand

CORMIER, C., University of New Brunswick, Fredericton, N.B. / Canada, E3B 5A3

DAFT, M.J., Department of Biological Sciences, University of Dundee, Dundee, Scotland / U.K.

DAVIS, P.E., Soil Microbiology Department, Rothamsted Experimental Station, Harpenden, Herts AL5 2JQ / U.K.

DE LEY, J., Laboratory of Microbiology and Microbial Genetics, State University, B9000 Gent / Belgium

DELLA-LATTA, P., Biology Department, New York University, New York, New York 10003 / USA

DEPAUW-GILLET, M.C., Department of Botany, University of Liege, Sart Tilman, B-4000 Liege / Belgium

DIATLOFF, A., Plant Pathology Branch, Department of Primary Industries, Indooroopilly, Queensland 4068 / Australia

DICKSON, M.R., Biomedical E.M. Unit, University of New South Wales, Kensington, N.S.W. 2033 / Australia

DUMKE, D.S., Department of Soil Science and Agricultural Engineering, University of California, Riverside 92502 / USA

DYE, M., Soil Microbiology Department, Rothamsted Experimental Station, Harpenden, Herts AL5 2JQ / U.K.

ELLIOTT, D., Physiology Division, DSIR, Private Bag, Palmerston North / New Zealand

EREN, J., Rutgers University, New Brunswick, New Jersey 08903 / USA

ERVIN, J.O., Department of Soild Science and Agricultural Engineering, University of California, Riverside 92502 / USA

EUTICK, M.L., Department of Biochemistry, University of Sydney, Sydney, N.S.W. 2006 / Australia

FAHY, P., School of Microbiology, The University of New South Wales, Kensington, NSW 2033 / Australia

FEDEL, R., CSIRO, Division of Soils, Glen Osmond, South Australia 5064 / Australia

FILIP, Z., Department of Agricultural Microbiology, Justus Liebig University Giessen, 6300 Lahn-Giessen / FRG

FIRESTONE, M.K., Department Crop and Soil Sciences and Cyclotron Laboratory, Michigan State University, E. Lansing, Mich. 48824 / USA

FIRESTONE, R.B., Department Crop and Soil Sciences and Cyclotron Laboratory, Michigan State University, E. Lansing, Mich. 48824 / USA

FITZGERALD, J.M., Microbiology Department, Monash University, Vic., and Biochemistry Department, University of Georgia / USA

FLANAGAN, P.W., Institute of Arctic Biology, University of Alaska, Fairbanks, Alaska 99701 / USA

FLOODGATE, G.D., Marine Science Laboratories, University College of North Wales, Menai Bridge, Gwynedd LL59 5EH, North Wales / U.K.

FOSTER, R.C., CSIRO Division of Soils, Glen Osmond, South Australia

FRANKLIN, M., University of New Brunswick, Fredericton, N.B. / Canada, E3B 5A3

GARLAND, C.D., School of Microbiology, Biomedical E.M. Unit, University of New South Wales, Kensington, N.S.W. 2033 / Australia

GILES, K.L., Physiology Division, DSIR, Private Bag, Palmerston North / New Zealand

GILL, C.O., Meat Industry Research Institute of New Zealand, P.O. Box 617, Hamilton / New Zealand

GLENN, A.R., School of Environmental and Life Sciences, Murdoch University, Murdoch / Western Australia

GRANT, W.D., Cawthron Institute, P.O. Box 175, Nelson /New Zealand

GRASS, L., Soil and Water Conservation Research Division, ARS, USDA, Brawley, California 92227 / USA

GREENWOOD, R.M., Applied Biochemistry Division, DSIR, Private Bag, Palmerston North / New Zealand

HALSALL, D.M., Division of Plant Industry, CSIRO, P.O. Box 1600, Canberra City, ACT 2601 / Australia

HAUXHURST, J., Departments of Microbiology and Zoology, University of Georgia, Athens, Georgia 30602 and Skidaway Institute of Oceanography, Savannah, Georgia 31406 / USA

HEAP, H.A., New Zealand Dairy Research Institute, Private Bag, Palmerston North / New Zealand

HIGGINS, H.W., CSIRO, Division of Fisheries and Oceanography, P.O. Box 21, Cronulla, N.S.W. 2230 / Australia

HINSHAW, V.S., Laboratories of Virology, St. Jude Children's Research Hospital, P.O. Box 318, Memphis, Tennessee 38101 / USA

HOLDER-FRANKLIN, M.A., University of New Brunswick, Fredericton, N.B. / Canada, E3B 5A3

HUNGATE, R.E., Department of Bacteriology, University of California Davis, California 95616 / USA

HYSLOP, N.ST.G., P.O. Box 490, Manotick, Ontario KOA 2NO / Canada

JARVIS, A.W., New Zealand Dairy Research Institute, Private Bag, Palmerston North / New Zealand

JONGEJAN, E., Limnological Institute "Vijverhof", Nieuwersluis / The Netherlands

KALMAKOFF, J., Department of Microbiology, University of Otago Dunedin / New Zealand

KANEKO, T., Departments of Microbiology and Zoology, University of Georgia, Athens, Georgia 30602 and Skidaway Institute of Oceanography, Savannah, Georgia 31406 / USA

KAPER, J., Limnological Institute "Vijverhof", Nieuwersluis / The Netherlands

KAURI, T., Department of Microbial Ecology, Helgonav. 5, S-223, 62 Lund / Sweden

KELLAR, P.E., Freshwater Section, Ecology Division, DSIR., P.O. Box 415, Taupo / New Zealand

KELLY, W.J., Biochemistry Department, Lincoln College, Canterbury / New Zealand

KEMP, D.R., University of Hong Kong, Pokfulam Road / Hong Kong

KILPATRICK, D., Queen Elizabeth College, Campden Hill Road, London W8 / U.K.

KRICHEVSKY, M., Departments of Microbiology and Zoology, University of Georgia, Athens, Georgia 30602 and Skidaway Institute of Oceanography, Savannah, Georgia 31406 / USA

KUO, K.J., Microbiology Section, University of Nebraska, Lincoln, NE 68588 / USA

KURSTAK, E., Faculty of Medicine, Université de Montreal, Montreal, P.Q. / Canada

LAIRD, M., Research Unit on Vector Pathology, Memorial University of Newfoundland, St. John's, Newfoundland, A1C 5S7 / Canada

LAWRENCE, R.C., New Zealand Dairy Research Institute, Private Bag, Palmerston North / New Zealand

LEACH, J.E., Microbiology Section, University of Nebraska, Lincoln, NE 68588 / USA

LEE, A., School of Microbiology, University of New South Wales, Kensington, N.S.W. 2033 / Australia

LEE, J., Microbiology Department, Monash University, Vic., and Biochemistry Department, University of Georgia / USA

LEONARDOPOULOS, J., Department of Microbiology, Medical School, University of Athens, Athens, P.O. Box 1540 / Greece

LIE, T.A., Laboratory of Microbiology, Agricultural University, Wageningen / The Netherlands

LONGWORTH, J.F., DSIR Entomology Division, Auckland / New Zealand

LYNCH, J.M., Agricultural Research Council, Letcombe Laboratory, Wantage, Oxon OX12 9JT / U.K.

MacINNES, J.I., Department of Bacteriology and Biochemistry, P.O. Box 1700, Victoria, British Columbia V8W 2Y2 / Canada

MALAJCZUK, N., Division of Land Resources Management, CSIRO, Private Bag, Post Office Wembley / Western Australia

MARTIN, E.L., Microbiology Section, University of Nebraska, Lincoln, NE 68588 / USA

MARTIN, J.K., CSIRO Division of Soils, Glen Osmond / South Australia 5064

MARTIN, J.P., Department of Soil Science and Agricultural Engineering, University of California, Riverside 92502 / USA

MARTINEZ, M.R., Department of Botany, University of the Philippines at Los Banos College, Laguna 3720 / Philippines

McFETERS, G.A., Montana State University,Bozeman, MT 59715 / USA

McINTOSH, B.M., National Institute for Virology, Private Bag X4, Sandringham 2131 / South Africa

McMURTRY, C.M., Cawthron Institute, P.O. Box 175, Nelson / New Zealand

MEEK, B., Soil and Water Conservation Research Division, ARS, USDA, Brawley, California 92227 / USA

di MENNA, M.E., Ruakura Agricultural Research Centre, Private Bag, Hamilton / New Zealand

MICHITSCH, R., Biology Department, New York University, New York, New York 10003 / USA

MILHAM, P.J., Biological and Chemical Research Institute,P.M.B. 10, Rydalmere 2116, NSW / Australia

MISHUSTIN, E.N., Institute of Microbiology, the U.S.S.R. Academy of Sciences / U.S.S.R.

MISKIMIN, D.K., Food Science Department, Cook College, Rutgers, The State University New Brunswick, New Jersey 08903 / USA

MORIARTY, D.J.W., CSIRO, Division of Fisheries and Oceanography, Cleveland, 4163, Queensland / Australia

MORRISON, W.L., Biological and Chemical Research Institute, P.M.B. 10, Rydalmere 2116, NSW / Australia

NAYLOR, G.E., Applied Biochemistry Division, D.S.I.R., Palmerston North / New Zealand

NESBITT, H.J., School of Environmental and Life Sciences, Murdoch University, Murdoch / Western Australia

NEWHOOK, F.J., Department of Botany, University of Auckland, Private Bag, Auckland / New Zealand

NEWTON, K.G., Meat Industry Research Institute of New Zealand, P.O. Box 617, Hamilton / New Zealand

NICOLSON, T.H., Department of Biological Sciences, University of Dundee, Dundee DD1 4HN, Scotland / U.K.

NORMAN, A.G., University of Michigan, Ann Arbor, Michigan 48109 / USA

NUTMAN, P.S., Soil Microbiology Department, Rothamsted Experimental Station, Harpenden, Herts AL5 2JQ / U.K.

O'BRIEN, R.W., Department of Biochemistry, University of Sydney, Sydney, N.S.W. 2006 / Australia

OKPALA, E.U., Department of Plant/Soil Science, University of Nigeria, Nsukka / Nigeria

OKUK, A.B., Department of Plant/Soil Science, University of Nigeria, Nsukka / Nigeria

OLD, K.M., Department of Biological Sciences, University of Dundee, Dundee DD1 4HN, Scotland / U.K.

OLSON, B.H., Program in Social Ecology, University of California Irvine, California 92717 / USA

OLSON, S.B., Montana State University, Bozeman, MT 59715 / USA

OWEN, B.A., Queen Elizabeth College, Campden Hill Road, London W8 / U.K.

PAERL, H.W., Freshwater Section, Ecology Division, DSIR., P.O. Box 415, Taupo / New Zealand

PAPAVASSILIOU, J., Department of Microbiology, Medical School, University of Athens, Athens, P.O. Box 1540 / Greece

PARLE, J.N., Ruakura Agricultural Research Centre, Private Bag, Hamilton / New Zealand

PATEL, J.J., Physiology Division, DSIR, Private Bag, Palmerston North / New Zealand

PAVRI, K.M., Virus Research Centre, Poona / India

POWELL, C.Ll., Soil and Field Research Organisation, Ruakura Agricultural Research Centre, Hamilton / New Zealand

PRAMER, D., Rutgers University, New Brunswick, New Jersey 08903 / USA

PREVOST, B.A., Food Science Department, Cook College, Rutgers, The State University New Brunswick, New Jersey 08903 / USA

READ, D.J., Department of Botany, Sheffield University, Western Bank, Sheffield S10, 2TN / U.K.

REANNEY, D.C., Biochemistry Department, Lincoln College, Canterbury / New Zealand

REMACLE, J., Department of Botany, University of Liege, Sart Tilman, B-4000 Liege / Belgium

REYNDERS, L., I.U. Leuven, Faculteit van der Landbouwwetenschappen, Laboratorium voor Bodemvruchtbaarheid en Bodembiologie, Kard. Mercierlaan 92, 3030 Heverlee / Belgium

ROGAL, S., Physiology Division, DSIR, Private Bag, Palmerston North / New Zealand

ROSENZWEIG, W.D., Biology Department, New York University, New York, New York 10003 / USA

ROVIRA, A.D., CSIRO Division of Soils, Glen Osmond / South Australia

SAUNDERS, P.T., Queen Elizabeth College, Campden Hill Road, London W8 / U.K.

SAVAGE, D.C., Department of Microbiology, University of Illinois, Urbana, Illinous 61801 / USA

SCHIPPERS, B., Phytopathological Laboratory, 'Willie Commelin Scholten', Baarn 2670 / The Netherlands

SCHWINGHAMER, E.A., CSIRO Division of Plant Industry, Canberra ACT 2601 / Australia

SILVESTER. W.B., Botany Department, University of Auckland, Private Bag, Auckland/New Zealand. Present address: Department of Biological Sciences, University of Waikato, Hamilton / New Zealand

SKERMAN, T.M., Wallaceville Animal Research Centre, Ministry of Agriculture and Fisheries, Private Bag, Upper Hutt / New Zealand

SLAYTOR, M., Department of Biochemistry, University of Sydney, Sydney, N.S.W. 2006 / Australia

SMITH, A.M., Biological and Chemical Research Institute P.M.B. 10, Rydalmere 2116, NSW /Australia

SMITH, C.E.G., London School of Hygiene and Tropical Medicine, Keppel Street, London, WC1E 7HT / U.K.

SMITH, D.F., CSIRO, Division of Fisheries and Oceanography, P.O. Box 21, Cronulla, N.S.W. 2230 / Australia

SMITH, M.S., Department Crop and Soil Sciences and Cyclotron Laboratory, Michigan State University, E. Lansing, Mich. 48824 / USA

SMITH, S.E., Departments of Botany and Agricultural Biochemistry and Soil Science, University of Adelaide / South Australia

SOLBERG, M., Food Science Department, Cook College, Rutgers, The State University New Brunswick, New Jersey 08903 / USA

STANLEY, N.F., Department of Microbiology, University of W.A., Perth Medical Centre, Nedlands / Western Australia 6009

STARK, A.E., School of Community Medicine, University of New South Wales, Kensington, N.S.W. 2033 / Australia

STOTZKY, G., Biology Department, New York University, New York 10003 / USA

STUART, S.A., Montana State University, Bozeman, MT 59715 / USA

SWABY, R.J., CSIRO, Division of Soils, Glen Osmond / South Australia 5064

TAYLOR, J.B., Plant Diseases Division, DSIR, Private Bag, Auckland / New Zealand

TENG, P.S., Lincoln College, Canterbury / New Zealand

TIEDJE, J.M., Department Crop and Soil Sciences and Cyclotron Laboratory, Michigan State University, E. Lansing, Mich. 48824 / USA

TIMMERMANS, P.C.J.M., Laboratory of Microbiology, Agricultural University, Wageningen / The Netherlands

TIPPETT, J.T., Department of Botany, Monash University, Clayton, Victoria 3168 / Australia

TRUST, T.J., Department of Bacteriology and Biochemistry, P.O. Box 1700, Victoria, British Columbia V8W 2Y2 / Canada

van EGERAAT, A.W.S.M., Laboratory of Microbiology, Agricultural University, Wageningen / The Netherlands

van VUURDE, J.W.L., Phytopathological Laboratory, 'Willie Commelin Scholten', Baarn 2670 / The Netherlands

VENTURA, A.K., University of Miami School of Medicine, 1600 N.W. 10th Avenue, Miami, Florida 33136 / USA

VLASSAK, K., I.U. Leuven, Faculteit van der Landbouwwetenschappen, Laboratorium voor Bodemvruchtbaarheid en Bodembiologie, Kard. Mercierlaan 92, 3030 Heverlee / Belgium

WEBSTER, R.G., Laboratories of Virology, St. Jude Children's Research Hospital, P.O. Box 318, Memphis, Tennessee 38101 / USA

WESTE, G., Department of Botany, University of Melbourne, Parkville 3052 / Australia

WHITEHEAD, H.C.M., Physiology Division, DSIR, Private Bag, Palmerston North / New Zealand

WILLS, B.J., Department of Botany, University of Canterbury, Christchurch / New Zealand

WINARNO, R., Laboratory of Microbiology, Agricultural University, Wageningen / The Netherlands

WUEST, L., University of New Brunswick, Federicton, N.B. / Canada, E3B 5A3

# Microbial Ecology – General

# Microbial Ecology: What for Science, What for Society?

M. ALEXANDER

Much evidence exists to confirm the growing international activity in microbial ecology. The mere fact that such a distinguished body of scientists is convening to discuss the ecology of microorganisms is witness to the increasing importance of this field, and a growing number of individuals in many countries clearly are involved in research on the behavior of microorganisms in nature. Furthermore, recent years have witnessed the appearance of new journals or journals with new titles to reflect this greater activity, and many of the older journals also show the rising pace of activity. Needless to say, all of this reflects an increased allocation of research funds by society through its governments.

It is not difficult to propose reasons for the rising concern with microbial ecology. Without question, the general public has become more aware of the impacts of the surrounding environment on its health and welfare, and this greater awareness is not only of chemical and physical stresses on human populations but also of diverse facets of ecology. In addition, some of the greater activity may be attributed to the enactment of laws in certain countries that now require information on the metabolism of classes of chemicals in natural ecosystems or the susceptibility of microorganisms to potential pollutants; for example, regulations in the United States requiring information on the influence of pesticides on microbial processes and requirements in many countries for information on biodegradability and sometimes products of degradation of chemicals that may be or indeed are pollutants of waters and soils. Not only are the public at large and governmental agencies helping with the development of our field, but so too are ecologists concerned with higher organisms or ecosystems as they gain a greater appreciation of the role of microorganisms in primary productivity, mineralization, biogeochemistry and ecosystem function. Finally, the introduction of new methods and approaches has resulted in an ever-expanding base of information and opened new avenues for inquiry, so that microbiologists have begun to explore the function and structure of microbial ecosystems to a greater degree; thus, a variety of electron and fluorescence microscopic methods, procedures for measuring acetylene reduction as an indication of nitrogen fixation, means for labelling of particular populations to trace their fate in natural ecosystems, the availability of gas chromatography and mass spectrometry for establishing metabolic pathways in nature, and the existence of techniques for using $^{14}C$ and $^{32}P$ for characterizing processes in soil and water have all led to a greater wealth of knowledge.

This ever-increasing flow of information should lead to a reassessment of the broad directions of the research; that is, since so much is yet to be learned, it seems appropriate to view the expanding amount of data with a view to making suggestions on the general trends for future inquiry. However, too often, the growing stature and activity in a field go to the scientist's head and he develops the feeling that society owes him something. He defends his past and current activities as a pure quest for knowledge, a pursuit to satisfy man's inquisitive mind. Nevertheless, if the mere search for knowledge is the only justification for the greater research activity, one can justifiably argue for an equal sum of money to be devoted to the humanities. After all, if the purpose of the quest is solely to satisfy man's intellect, why should the quest be so well-funded just in the natural sciences? The plea would then be for equal funding for philosophy, the social sciences, art and literature. But in fact, which scientists are willing to transfer much of their money for research to the philosophers, social scientists and artists? Yet, if the function of basic science is simply to satisfy man's curiosity, then it is not unreasonable to argue that all legitimate areas of man's intellectual gropings should be supported to comparable extents.

In self defence, the basic scientist answers that his research is needed by society. It is important to man's health and welfare. If we look at ourselves and our colleagues, it is not unreasonable to ask how often is such a statement a true reflection of the scientist's feeling and how often is it really a camouflage for a degree of hypocrisy?

Some of us take a different approach and maintain that all scientific experimentation is important because it is rarely if ever sufficiently clear what are society's future needs in the applied and technological disciplines. Thus, inasmuch as coming trends in medicine, agriculture and industry are undefined and the kinds of problems society will create for itself are unknown, research should proceed along all avenues to provide adequate information to cope with the undefined problems of the future years. In this connection, microbial ecology serves as a fine example. Some of us might argue that information needs to be obtained on all microorganisms and all microbial processes in all habitats. Does not each fact have to be available, such a viewpoint maintains, to provide a basis for coping with society's forthcoming needs? Some of us thus hold that new information is obtained by examining each different soil, body of water, or other ecosystem that has not been heretofore tested. The consequence of such an approach is that all bacteria, fungi, algae and protozoa in each habitat should be catalogued, every process in all ecosystems should be characterized, and each microbial interaction in every habitat should be explored. In viewing an approach to microbial ecology of this sort, it is worth bearing in mind the distinction once nicely put by F.C. Steward, who said that it is essential to distinguish between REsearch and reSEARCH. Is the emphasis to be on repetition or on the quest? Should one collect essentially the same facts disguised under a somewhat different cloak or pursue a course which will provide really new information?

To return to the issues of the importance of science for its own sake and whether other areas of intellectual endeavor should not receive the same degree of support as basic science, it is not unreasonable to argue that the same degree of support does not denote the same extent of financial support. Much of scientific research is intrinsically expensive. Thus, if one compares the costs of a gas chromatograph with a paintbrush and canvas or with a writer's pen and pencil, it is immediately evident that more money is needed to maintain one scientist than a single painter or author. Even then, however, that kind of science or that type of art or literature requires a rigorous test: it must be <u>really</u> original, the minds involved must be society's best and the work must be truly pioneering.

Granting the need for such research to satisfy man's curiosity, let me propose another area of concern, "relevant basic science." To put this anomalous term into perspective, let us focus on three facts of great importance to society. The first fact: microorganisms are still <u>the</u> major cause of human death, either directly or indirectly. Millions of people die each year from communicable diseases often associated with malnutrition. No other kind of pollutant causes the high incidences of mortality or as much human misery as the microbial agent of communicable disease as it takes the lives of millions of people each year, particularly in the developing countries. Even in purely economic terms, the losses to society are greater because of microorganisms than any other pollutant.

The second fact: poor nutrition or undernutrition, typically together with communicable disease, takes an enormous toll of lives and causes immeasurable human misery in Latin America, Asia and Africa. Although statistics are not easy to find and those available are of dubious reliability, any visitor to a developing country readily sees the misery associated with an inadequate or insufficient supply of food.

The third fact: all people, whether in the technologically advanced or in the developing countries, live in a sea of chemicals. Most of these chemicals are natural, but a few are synthetic. Many of the compounds are safe, but some are dangerous. Epidemiologists are in general agreement that more than 50% of the cases of cancer result from environmental causes, and undoubtedly the natural and synthetic chemicals are responsible for many of these cancers as well as other maladies. Moreover, the waters and soils of many regions contain chemicals harmful to man, animals or plants. In addition, air has many toxicants and some of these atmospheric pollutants or natural substances result from microbial activity in soils and in waters.

What lessons can be learned in "relevant basic microbial ecology" from a consideration of these three facts? Consider first microorganisms as agents of communicable disease. In the developing countries, drugs are often unavailable or are so expensive that they can not be afforded. Needless to say, sanitation is poor or nonexistent. Even in countries where the water supply is treated with chlorine at the municipal treatment plant, the pressure in the water mains is often insufficient to keep the pipes filled; hence, polluted water enters through the omnipresent leaks in these pipes so that ultimately the consumer is drinking bacteriologically unsafe water. How can microbial ecology help? One might focus on just two aspects of the problem, namely survival and dissemination. Is enough known about the survival of the pathogens and their dispersal that it is possible to predict the best means and economic procedures to reduce survival and to suppress dissemination? If the answer to this question is no, then such work can be both basic and relevant, and this research on organisms of societal interest should be just as attractive to microbial ecologists as working on an organism whose survival or dispersal is totally unrelated to human problems. Thus, a study of the dispersal of *Salmonella*, *Shigella*, and *Vibrio*, rather than just chemotaxis of *E. coli*, would not only provide good science but would help overcome a major problem in developing countries. Such investigations are *not* applied but they are more useful to society for the foreseeable future, during which time millions will die in the developing countries because of diseases of the young and of individuals whose health is aggravated by malnutrition.

Consider how microorganisms and food production might be appropriate for "relevant basic microbial ecology." In addition to insect damage and weed competition, much of the world's potential food is destroyed by microorganisms acting either on the living plant or on the harvested commodity. Yet little is known of the ecology of the pathogenic or spoilage-inducing fungi and bacteria. Although pathologists have conducted some ecological studies, there is remarkably little known about the environmental factors affecting the distribution and survival of the host of fungi and several bacterial genera associated with the destruction of much of the food and feed supply of the world. Furthermore, the fund of information on the ecology of microorganisms causing diseases of insects and weed species is essentially zero; such information should help in their ultimate microbiological control.

Microorganisms have a direct and beneficial role in promoting food production. For example, soil inhabitants provide plants with nitrogen and phosphorus in a usable form and mycorrhizal fungi have great significance for plant growth. For microbial ecologists, the focus of interest is not the biochemistry, genetics, physiology or morphology of the responsible organisms but simply their ecology. Thus, of interest to us should be what limits the distribution of these microorganisms and why there is so little activity; that is, why is there so remarkably little nitrogen fixed or phosphorus released in most soils so that plants growing there suffer from nitrogen or phosphorus deficiencies? Ecologically important and of practical concern, too, would be finding means to enhance the survival or minimize the decline of these important organisms, questions that frequently can only be answered after the theoretical studies on survival and decline have been performed. The point at issue is not necessarily to do the applied research itself but only the basic studies so that others can conduct the final exploitation.

The third area to be considered in "relevant basic microbial ecology" is the interaction between microorganisms and chemical pollution. Many nontechnological topics await the ecologist here. Because synthetic chemicals are extremely important for society, they will be used and will appear in waters and soils. But it should be possible to balance the needs for chemical efficacy with structures that are biodegradable, so that the chemical will serve a useful function and yet will be destroyed readily by microbial communities once the function is no longer required. In this connection, the reasons for resistance of chemical structures to microbial attack must be defined, and establishing the mechanisms of chemical recalcitrance is a study as suitable for work in environmental metabolism as the decomposition of leaf litter. The difference is that the individuals ultimately interested in the results are those who suffer from pollution rather than one's professional colleagues as they read technical journals. Furthermore, the products of microbial degradation often are of considerable practical importance because these products may be persistent, more toxic than the original molecule or harmful to species that one would not have expected to be sensitive; hence, the metabolic pathway in natural ecosystems must be established, at least to the extent that the extracellular products are identified. This type of work employs the same approaches and faces the same problems that characterize studies of natural products in soils and waters. An illustration of the relevancy-irrelevancy overlap is evident in our recent finding that innocuous amines, which are quickly destroyed in nature, are converted to nitrosamines. This might seem like a mere curiosity that could represent basic research except for the fact that the nitrosamines are both persistent and carcinogenic. It is likewise critical to establish what environmental chemicals do at ambient concentrations to microbial processes that are important to ecosystem function and to the health and well-being of plant and animal communities. Thus, a compound such as $SO_2$ which might be expected to have little effect on microbial communities, is in fact a specific and highly active inhibitor of blue-green algae and processes they catalyze. Such investigations represent parts of basic studies of the interactions between microorganisms and chemicals, but the information obtained will help in developing an understanding of the changes in nature induced by man's actions.

Let me emphasize the view that society has the right to ask scientists in general and microbial ecologists in particular, "Why are you doing this or that type of research?" Individuals outside of the microbiology laboratory or away from the field site can legitimately say, "We pay your salary, we provide you with the equipment and the laboratories and we deserve something in return. We want more than technical journals we don't understand and occasional stories in the newspaper about which we care little." In response, microbial ecology as a discipline must and surely can provide answers, answers that will be useful in man's endeavor to control agents of communicable disease, improve food production and overcome chemical pollution. If microbial ecologists do not conduct the applied studies or are not willing to provide the relevant information for others to perform the applied research, then society surely has a right to call a halt to the growth of our field and to the extension of our quest for knowledge for its own sake.

And, mark my words, society surely will do so.

# Tropical Ecology and Human Health

N.F. STANLEY

This paper emphasises:
(a) the magnitude of the problem of tropical disease, (b) the additonal complexities to the host-parasite interaction introduced by human modifications of the environment and (c) the involvement of science and government in making more effective our research, understanding and control of specific diseases. I propose to use one group of microorganisms in one geographic area, arboviruses in North-West Australia to illustrate these points.

Some idea of the magnitude of the problem of tropical disease may be gained by realizing that there are about 2,000 million people on the planet with tropical disease and malnutrition. Lack of knowledge about host-parasite relationships affects our ability to remedy this situation, e.g. many virus diseases still remain uncontrolled and little understood, particularly in situations where the virus persists (e.g. in infections with hepatitis B, arboviruses, herpesviruses, slow viruses and C type RNA viruses).

We are faced with complex problems in human behavior as well as host-parasite interactions. We know, in theory, how to prevent such diseases as schistosomiasis, malaria, syphilis, gonhorrhoea and some forms of environmental cancer, but the prevention demands an altered life-style which millions of people are obviously reluctant to accept. Until this reluctance is overcome, what microbiological contributions can be made to add to our better understanding and control of particular diseases?

WHO is providing some answers by special programmes for research and training in tropical diseases. The objectives are "to develop through research, more effective control methods for tropical diseases, with the corollary objective of strengthening biomedical research in tropical areas". The major goals are "to obtain effective new vaccines, diagnostic tests, drugs and measures for vector control and to assist the tropical countries to improve their own research institutions and capabilities".

Initially six diseases have been selected for study, by the establishment of specific working groups for filariasis, leishmaniasis, leprosy, malaria, schistosomiasis and trypanosomiasis. Each of these diseases is a challenge, to the immunologists particularly. For example, in schistosomiasis, studies are required on concomitant immunity, cell-mediated events in liver damage, and antigenic configuration of pathogenic as opposed to benign species leading to diagnosis of carriers and better therapy. Malaria also offers a challenge. The importation of malaria cases from endemic areas continues to increase, but the immunological responses to parasitemia are not understood. Much more basic information is required if an effective vaccine is to be developed.

Other diseases require investigation. The immunology of the shock syndrome in dengue virus infection is still challenging. We need to obtain immunological diagnostic tests for filariasis and to understand the immunopathology leading to blindness. Although immunological studies with *Mycobacterium leprae* are under way, there is gross and serious ignorance of cholera immunology and pathogenesis.

In studies of these diseases we must also include the effect of nutrition. Malnutrition is a significant part of tropical diseases, and a great deal of work remains to be done on the immunology of malnutriton itself (2, 3, 4, 5). The Macy Foundation's definition of tropical medicine is pertinent in this respect. "Tropical medicine includes as its principal components, protozoan, helminth, bacterial and arthropod-borne virus infections, malnutrition and delivery of health care to the underprivileged".

In the title to the paper I have used the term "tropical ecology" and interpret it in relation to health as "the systematic study in tropical areas of biological and environmental factors that interact with human activities to promote or endanger the health of man, animals or crops". This means that in trying to solve health problems in the tropics we have to assume a multidisciplinary approach. We have to consider that in addition to all the factors customarily studied, the environment is being dramatically changed by the human activities of mining, irrigated agriculture, man-made lakes, tourism and defence. These may affect the host-parasite relationship. Such an approach is required in Northern Australia where as a prelude to possible increased human population densities changes are occurring and the situation is further complicated by the rapid movement of increasing numbers of people by air through this area between countries. In any study of tropical disease in this area then these factors have to be considered. Further, tropical diseases are by no means confined to the tropics and this increased movement of humans demands greater clinico-pathologic surveillance in temperate and highly urbanized communities. The recent establishment of the arbovirus disease 'dengue fever' in some Pacific Islands is believed to have resulted from the introduction by air of viremic humans from South-East Asia.

Spread of tropical disease is of considerable concern to Australia. Nearly 60% of the world's population is in Asia, immediately to our north. What is being done in Australia to cope with tropical ecology and human health? The immunological aspects of certain tropical diseases have stimulated the Australian Society for Immunology Council to create a Tropical Immunology Working Party which has resolved "to support the development of a multidisciplinary research and training programme in tropical diseases by continuing to encourage immunologists both to participate through visits to Western Pacific and South-East Asian centres and to orient their thoughts and research towards an understanding of tropical diseases" (1). Prior to this, the National Health and Medical Research Council had established a Working Party on "Tropical Ecology and Human Health" (11).

In Western Australia we have integrated our field studies on schistosomiasis, molluscs and water, with those on arboviruses, mosquitoes and vertebrate host reservoirs (10). <u>Schistosoma mansoni</u>, <u>haematobium</u> or <u>japonicum</u> have not been known to spread in Australia although apparently healthy individuals from Africa have lived in the temperate areas of the country, excreting ova for a large number of years. One aspect of the studies therefore concerns the identification of the major molluscs in northern tropical areas and a determination of their suitability as an intermediate host for the known pathogens. Although this disease, and malaria and dengue, are at present not spreading in Northern Australia, both dengue and malaria were endemic prior to the 1930's.

I wish to discuss a small part of Western Australia which is being dramatically changed by man-made lakes, irrigation and mining (12), and is therefore of interest in relation to comments already made.

The increased population densities associated with mining, irrigation, defence and tourism demand continual surveillance for tropical diseases such as malaria, schistosomiasis, dengue fever, arbovirus encephalitis, filariasis, etc. In our programme on this area we have considered the following: (a) the virus-mosquito-vertebrate host biocenose (6); (b) infections of humans, animals and birds (7); (c) the current distribution of mosquitoes, viruses and antibodies and the changing patterns associated with human activity (8); and (d) we have assessed the techniques employed in the studies and their limitations.

Arboviruses have evolved independently of man due to the fact that they replicate, without inducing serious illness, in both the arthropod vector and the vertebrate host reservoir. Virus detection and serological epidemiology have pin-pointed the geographic distribution, the antigenic types (>390), the arthropod vectors and the vertebrate host reservoirs. Man's involvement is usually accidental due to the movement of the virus to the human environment or to the development of human activity in the areas where the virus is enzootic. Some infections (yellow fever, Japanese B encephalitis, Murray Valley encephalitis (MVE), dengue), are clinically serious, but many are subclinical.

Following the construction of a man-made lake of considerable size, a diversion dam and an irrigation scheme in the Ord Valley of North-West Australia, we have followed patterns of arbovirus activity in this part of tropical Australia. Special attention has been focused on MVE, as this has caused outbreaks in South-East Australia, but never in South-West Australia. Our studies, together with those of Dr. Doherty and Dr. Marshall have confirmed that MVE is enzootic over Northern Australia and possibly New Guinea, with Culex annulirostris being the key arthropod vector. The major vertebrate host reservoirs have not been defined but birds are believed to be significantly involved in the natural biocenose. Cattle which were introduced to the Kimberley country of North West Australia in the last century now appear to be a significant vertebrate host reservoir.

In the Ord Valley, the largest town, Kununurra, is the focus for very large populations of birds, cattle and mosquitoes. Effective and continuous mosquito surveys have not been performed and consequently there are great gaps in our knowledge. Nevertheless, we are gaining some idea of the distribution of mosquitoes by season and their modification by man. Sixty-two species belonging to 8 genera have been identified in Western Australia. There are 28 tropical species, 24 non-tropical species and 10 species occur in both climates. The mosquito fauna of the entire area is dominated by C. annulirostris. Eighteen of the species are known disease vectors. Until recently, because of the use of insecticides, the irrigation areas have not been important as mosquito breeding areas.

Table 1. Virus isolations from Kimberley mosquitoes. Mosquito species from which viruses have been isolated. (May 1972 - November 1976)

| Mosquito species | Number of pools tested | Number of isolates | % |
|---|---|---|---|
| Culex annulirostris | 774 | 154 | 20 |
| Aedomyia catasticta | 111 | 37 | 33 |
| Aedes normanensis | 16 | 3 | 19 |
| Aedes tremulus | 6 | 1 | 17 |
| Culex fatigans | 96 | 2 | 2 |
| Total: | 1,003 | 197 | 19.6 |

One hundred and ninety-seven viruses have been isolated from 1,000 pools of mosquitoes (representing a 20% isolation rate). Flaviviruses were represented by MVE and Kunjin strains; the non-haemagglutinating viruses comprised Corriparta and Rhabdoviruses, some of which appear to be 'new' antigenic types.

Serological studies with haemagglutination inhibition (H-I) tests showed a high percentage with antibody to MVE virus in humans, birds and cattle. The data probably give a false picture because MVE virus shares antigens with closely related viruses such as Kunjin. Kunjin is the predominant isolate and the viruses isolated represent only those pathogenic for infant mice of one genetic constitution.

Table 2. Virus isolations from Kimberley mosquitoes: Current identification of 68 isolates from 3 vectors

| Vector | MVE | Kunjin | To be typed GpB | Sindbis | Koongol | Wongol | Untyped non-HA Koongol group | Corriparta | Rhabdovirus | Total |
|---|---|---|---|---|---|---|---|---|---|---|
| Culex annulirostris | 6 | 12 | 6 | 5 | 3 | 8 | 17 | 1 | 2 | 60 |
| Aedomyia catasticta | | | | | | | | 6 | 1 | 7 |
| [a]Aedes tremulus | | 1 | | | | | | | | 1 |
| Total | 6 | 13 | 6 | 5 | 3 | 8 | 17 | 7 | 3 | 68 |

[a] 2 mosquitoes only in this pool!

Table 3. Arbovirus antibody tests with human sera

| | Number | % positive by H-I tests with MVE | Sindbis | Ross River |
|---|---|---|---|---|
| Caucasian | | | | |
| Adults | 199 | 53 | 30 | 41 |
| Children | 94 | 24 | 14 | 21 |
| Aboriginal | | | | |
| Adults | 95 | 96 | 20 | 51 |
| Children | 53 | 77 | 36 | 68 |

It is obvious that to demonstrate a little more than the tip of this huge virus iceberg, an iceberg now being modified by man, we need at the very least, to: (a) use hosts additional to the new-born mouse for virus isolation, e.g. cell culture, chorioallantoic membrane of chick-embryos and intrathoracic inoculation of mosquitoes; (b) study the antigenic configuration of the flavivirus isolates; this may require polypeptide mapping; (c) determine the nature and extent of human infection with flaviviruses antigenically related to MVE: (d) ascertain the mechanisms of virus persistence in the natural vertebrate host reservoirs, particularly cattle and birds; and (e) continue monitoring the virus biocenose for obvious changes resulting from irrigation, bird movement, increased movement of humans, population density and the stabilization of the huge tropical lake.

Table 4. Arbovirus antibodies in cattle sera from the East Kimberley study site, the West Kimberley and South-West Australia[a]

| Site of cattle station | Number of sera tested | % positive by H-I tests with MVE | Sindbis | Ross River |
|---|---|---|---|---|
| Kununurra (study site area) | 316 | 80 | 4 | 2 |
| Kimberley (remote study site) | 564 | 37 | 4 | 2 |
| South-West of Australia | 200 | 1 | | 0.5 |

[a]Permission granted from Department of Agriculture for release of these results.

Table 5. Comparison of arbovirus antibody in sera of water birds and others

| | Total | Number positive by H-I tests with MVE | Sindbis | Ross River |
|---|---|---|---|---|
| Water birds | 170 | 96 | 33 | 5 |
| Others | 165 | 98 | 15 | 1 |

In conclusion, one could be justified in saying that man's use and control of water in both arid and tropical areas involves marked environmental changes which are not yet understood, but frequently associated in equatorial regions with increased water-associated disease. An increased human population density and the increased movement of humans demand increased and different surveillance now and in the future. Arbovirus ecology and epidemiology merely illustrate some of the complexities arising from man's modification of his environment in one site; but they also demonstrate the need, particularly with tropical diseases, for the planning and execution of national and international exercises. This involves many disciplines, but the key actors in the exercise must be the microbiologists and immunologists. They must be involved in the planning and integration of the research programmes where basic studies should proceed, *pari passu*, with practical exercises in developing countries, if we are to understand the ecology necessary for the control of tropical disease.

I finally quote Dr. Gordon Smith - "no major agricultural or engineering development should be permitted or funded without adequate provision for medical and other research to assess its health hazards, and for resources to control and remedy disease problems it may create" (9).

## A. References

1. Australian Society for Immunology Council, recommendation of the, December (1976).

2. Bell, R.G., Hazell, L.A.: Influence of dietary protein restriction on immune competence. I. Effect on the capacity of cells from various lymphoid organs to induce graft-vs.-host reactions. J. Exp. Med. 141, 127-137 (1975).

3. Bell, R.G., Hazell, L.A., Price, P.: Influence of dietary protein restriction on immune competence II. Effect on lymphoid tissue. Clin. Exp. Immunol. 26, 314-326 (1976).

4. Bell, R.G., Hazell, L.A., Sheridan, J.W.: The influence of dietary protein deficiency on haemopoietic cells in the mouse. Cell Tissue Kinet. 9, 305-311 (1976).

5. Bell, R.G., Turner, K.J., Gracey, M., Suharjono, Sunoto: Serum and small intestinal immunoglobulin levels in undernourished children. Amer. J. Clin. Nutr. 29, 392-397 (1976).

6. Liehne, C.G., Leivers, S., Stanley, N.F., Alpers, M.P., Paul, S., Liehne, P.F. S., Chan, K.H.: Ord River arboviruses isolations from mosquitoes. Aust. J. Exp. Biol. Med. Sci. 54, 499-504 (1976).

7. Liehne, C.G., Stanley, N.F., Alpers, M.P., Paul, S., Liehne, P.F.S., Chan, K.H.: Ord River arboviruses - serological epidemiology. Aust. J. Exp. Biol. Med. Sci. 54, 505-512 (1976).

8. Liehne, P.F.S., Stanley, N.F., Alpers, M.P., Liehne, C.G.: Ord River arboviruses - the study site and mosquitoes. Aust. J. Exp. Biol. Med. Sci. 54, 487-497 (1976).

9. Smith, C.E. Gordon: Changing patterns of disease in the tropics. In: Man-made Lakes and Human Health. Stanley, N.F., Alpers, M.P. (eds.). London-New York-San Francisco: Academic Press, pp. 345-362 (1975).

10. Stanley, N.F.: Ord River ecology. Search 3, 7-12 (1972).

11. Stanley, N.F.: Report to the Federal Minister for Health, Canberra, A.C.T., on Tropical Ecology and Human Health: A National Approach for Australian Studies, February (1976).

12. Stanley, N.F., Alpers, M.P. (eds.): Man-made Lakes and Human Health. London-New York-San Francisco: Academic Press, (1975).

# Atmospheric Pollution: Impacts on and Interactions with Microbial Ecology*

H. BABICH and G. STOTZKY

Microorganisms occupy a unique ecological position in their interactions with and impact on air pollution. Similar to plants and animals, including human beings, microbes are recipients of airborne contaminants and, in general, are sensitive to prolonged exposures to these contaminants. Microorganisms, however, are distinct, in that they: (a) are sources of substantial quantities of various particulate and gaseous atmospheric pollutants: (b) can utilize, either as an energy or a nutrient source, numerous air pollutants and, thereby, serve as sinks for many anthropogenic pollutants; and (c) when in the airborne state, are, themselves, air contaminants.

Microbes are primary sources of many gaseous atmospheric pollutants, including carbon monoxide (CO), hydrogen sulfide ($H_2S$), nitrogen oxides($NO_x$), ammonia ($NH_3$), methane ($CH_4$), and numerous other organics (e.g., ethylene ($C_2H_4$) and sulfur-containing volatiles, such as methyl mercaptan ($CH_3SH$), dimethyl sulfide ($(CH_3)_2S$), dimethyl disulfide ($((CH_3)_2S_2)$). Gaseous emissions from microbes, in combination with those from other biologic, as well as abiotic, sources, constitute the natural background levels to which are added emissions from anthropogenic sources. For example, sulfur compounds in the atmosphere are mainly (a) sulfur dioxide ($SO_2$), primarily from anthropogenic emissions, although small amounts of $SO_2$ are also released from volcanoes, and (b) $H_2S$, primarily from microbial reductions of sulfate in anaerobic environments and from microbial decomposition of sulfhydryl-containing amino acids in aerobic environments. Consequently, in localized regions, microbial production of $H_2S$ may release more sulfur into the atmosphere than that released as $SO_2$ from industrial activities. In addition, microbial decomposition of organic matter in soil results in the evolution of a variety of reduced sulfur-containing organics, such as $CH_3SH$, $(CH_3)_2S$, and $(CH_3)_2S_2$ (2, 3, 19, 24).

Microbes are also a source of volatile organometal compounds, as microbial methylation of mercury, selenium, tellurium, and arsenic results in the formation and evolution of mono- and dimethyl mercury, dimethyl tellurite, dimethyl selenide, and di- and trimethyl arsine, respectively. Methylation, in addition to changing the potential distribution of these metals in the biosphere, may influence the toxicity of these pollutants; for example, the microbial conversion of mercury results in methyl mercury, a potent neurotoxin (24, 27).

Gaseous compounds derived from microbial activity undergo a variety of catalytic and photochemical reactions in the atmosphere, with the subsequent formation of secondary products: dimethyl mercury is photolyzed by ultraviolet light in the atmosphere to yield elemental mercury, ethane, and $CH_4$ (27). Atmospheric sulfate salts are formed from $(CH_3)_2S$ and $H_2S$, albeit by different processes (2, 3).

---

* Because of the limitations in space and because of the general nature of this presentation, references are made primarily to review articles, which contain detailed references to the original papers.

Thus, microbes are also indirectly involved in the production of many secondary gaseous and particulate pollutants.

The continuous natural production of air pollutants without the subsequent accumulation of these materials implies the existence of natural scavenging processes, both biotic and abiotic, termed sinks. Atmospheric CO is removed by a variety of abiotic and biotic sinks: the former include reactions with hydroxyl radicals ($CO + OH \rightarrow CO_2 + H$) and with nitrous oxide ($CO + N_2O \rightarrow CO_2 + N_2$); the latter include utilization by anaerobic $CH_4$-producing bacteria (e.g., Methanosarcina bakerii, Methanobacterium formicum), fungi, blue-green algae, and green algae (2, 3, 9). The soil and phyllosphere microbiota also serve as major sinks for removal of gaseous hydrocarbons from the atmosphere, including those contained in automobile exhausts; aerobic microbial decomposition of $C_2H_4$ removes approximately $7 \times 10^6$ tons $yr^{-1}$ in the United States alone; nitrogen-fixing organisms are a potential sink for acetylene ($C_2H_2$), which is reduced to $C_2H_4$; and several bacteria (e.g., Mycobacterium methanicum, Methanomonas methanica), some fungi (e.g., Acremonium sp., Graphium sp.) and algae of the genus, Chlorella, are able to oxidize and utilize $CH_4$ (1, 3, 20, 24).

In addition to serving as both sources of and sinks for atmospheric contaminants, airborne microbes are, themselves, atmospheric pollutants. Aerial transmission of diseases of human beings (e.g., pulmonary anthrax, pneumococcal pneumonia, poliomyelitis, aspergillosis, histoplasmosis), of poultry (e.g., Marek's and Newcastle disease), of cattle (e.g., infectious bovine rhinotracheitis, foot and mouth disease) and of plants (e.g., powdery mildew of plum caused by Sphaerotheca pannosa, brown-rot blossom-blight of almond caused by Monilinia laxa) presents a constant hazard (3, 15).

Microorganisms and viruses, as well as plants and animals, are sensitive to many airborne contaminants, with toxicity being manifested on various levels of cellular, organismal, and population complexity. For example, the toxicity of $SO_2$ is a function of its high solubility in water and, consequently, to its solubility products (i.e., sulfurous acid ($H_2SO_3$), bisulfite ($HSO_3^-$), sulfite ($SO_3^{-2}$), protons ($H^+$)) and, subsequently, to their oxidation products (i.e., sulfuric acid ($H_2SO_4$), bisulfate ($HSO_4^-$), sulfate ($SO_4^{-2}$)). On the cellular level, $HSO_3^-$ has been shown to be mutagenic for bacteria and coliphage, to inhibit transforming activity of bacterial DNA, to inhibit translation of mRNA, and to alter transport across membranes (3).

On the organismal level, $SO_2$ has been shown to inhibit growth and respiration of bacteria, fungi, algae, and lichens; to retard germination of fungal spores; to decrease survival of airborne bacteria and animal viruses; and to inhibit nitrification rates in soil (3, 10). The responses of microbes to pollutants differ with the developmental forms (e.g., fungal hyphae vs. fungal spores) and processes (e.g., spore formation vs. spore germination) studied. For example, fumigation with 10 to 60 pphm ozone ($O_3$) for 4 hr $day^{-1}$ on 3 consecutive days reduced spore production by Colletotrichum lindemuthianum but stimulated spore production by Alternaria oleraceae; however, spore germination in both species was reduced. Mycelial growth of C. lindemuthianum, Alternaria citri, and Oospora citriauranti was inhibited, whereas that of Helminthosporium sativum was stimulated, by low levels of $O_3$ (2, 3). When spores of Trichoderma viride, Penicillium egyptiacum, Botrytis allii, and Colletotrichum lagenarium were fumigated with 10 pphm $O_3$ for 70 hr, sporulation of the resultant colonies of C. lagenarium was increased, whereas sporulation of the other species was inhibited (13).

On the population level, air pollutants can alter the interactions of microbes and viruses with plants and animals: e.g., the infectivity of bacteriophage S5 for

Serratia marcescens and of T1 for Escherichia coli was reduced on exposure of the phages to peroxyacetyl nitrate (PAN) and $HSO_3^-$, respectively. Simulated acidified rain (i.e., water amended with $H_2SO_4$, pH 3.5) increased the susceptibility of kidney bean plants to infection by Pseudomonas phaseolicola; conversely, infectivity of potato leaflets by zoospores of Phytophthora infestans was almost entirely eliminated after a prior 6 hr exposure of the zoospores to 60 ppm $SO_2$ in solution. In regions of high $SO_2$ pollution, various plant pathogenic fungi were either eliminated or reduced (e.g., Diplocarpon rosae, Venturia inaequalis, Cronartium ribicola, Melampsorella ceraatii); however, proliferation of some phytopathogenic fungi (e.g., Rhizosphaera kalkhoffii, Lophodermium abietis) has been reported in regions of high $SO_2$ pollution. Apparently, air pollutants have the potential to increase or decrease the resistance of the host organism to infection (3, 12, 21).

Because microorganisms are key components in the biogeochemical cycling of various chemicals, in the production of energy (chemosynthesis and photosynthesis), and in many food-chains, these phenomena may be adversely affected by air pollutants. For example, CO is a potent inhibitor of both symbiotic and nonsymbiotic nitrogen fixation; $SO_2$ and acid rain, presumably by lowering the pH of soil, inhibit ammonification, nitrosofication, and nitrification processes in soil: $SO_2$, $O_3$,PAN, and formaldehyde inhibit photosynthesis by several green algae; and accumulation of heavy metals in living plants and in plant litter inhibits their degradation by microorganisms (3).

The toxicity of an airborne pollutant to viruses and microbes is dependent on the physicochemical parameters of the environment into which the pollutant is deposited. For example, the toxicity of the particulate pollutant, cadmium (Cd), to microbes in culture and in soil was increased as the pH was increased, whereas the toxicity was reduced by the addition of the clay minerals, montmorillonite and kaolinite (4, 5, 6), organic matter, chelating agents, and competing cations, such as zinc or magnesium (7). The toxicity of the gaseous pollutant, $SO_2$, and of its solubility products, $HSO_3^-$ and $SO_3^{-2}$, to coliphage, bacteria, and fungi was increased as the pH was decreased (8). Other abiotic factors reported to influence $SO_2$ toxicity include temperature, water content, relative humidity, solar radiation (3), and concentration and type of clay minerals (10).

Microbes and plants also emit a variety of volatile organic compounds that not only contribute to the overall pollution of the atmosphere, but also influence various microbe-microbe and plant-microbe relations (11, 14, 23, 24). Because some anthropogenic emissions are similar to natural emissions (e.g., $C_2H_4$ is released from aerobic and anaerobic bacteria, fungi, and plants, as well as from industrial combustion processes), some anthropogenic pollutants might antagonize, enhance, or mimic natural emissions and, thereby, affect biotic phenomena that have resulted as adaptations to these natural emissions (3, 24).

Evolution of organic volatiles from one group of microbes ("producers") has been shown to influence the growth and development of other microbes ("responders"). Microbial phenomena such as bacteriostasis, sporostasis, fungistasis, and mycolysis may be controlled by organic volatiles released by microbes (3, 23, 24). For example, unidentified organic volatiles emanating from Enterobacter aerogenes, Bacillus cereus, Agrobacterium radiobacter, Micrococcus luteus, Proteus vulgaris, and S. marcescens inhibited sporulation of Zygorhynchus vuilleminii, Fusarium oxysporum, and Gelasinospora cerealis (16). Organic volatiles from E. coli and Nocardia corallina inhibited both growth and sporulation of T. viride and growth of Penicillium viridicatum, but stimulated sporulation in Z. vuilleminii (18). The degree of inhibition varied among the different bacterium-fungus combinations and could be altered by the presence of various abiotic scavengers, suggesting that several active volatile metabolites were involved. Furthermore, the degree of inhibition was concentration dependent, i.e., heavy bacterial development always inhibited growth and/or sporulation of all test fungi, whereas lighter bacterial growth either inhibited, did not alter, or stimulated growth and/or sporulation, depending on the bacterium-fungus combination (16, 18). Morphological abnormalities were also induced in fungi by exposure to the bacterial volatiles, e.g., shortening of conidiophores of Aspergillus giganteus, increased septation in T. viride and Z. vuilleminii, and hyphal distortion in Z. vuilleminii (17).

Organic volatiles are also released by various parts of living plants, including roots, leaves, flowers, bark, and germinating seeds. For example, germinating seeds of various gymnosperms and angiosperms emitted ethanol, and several also released ethylene, propylene, formic acid, acetone, propionaldehyde, acetaldehyde, formaldehyde, and methanol (25, 26). The volatiles from these seeds, especially acetaldehyde, were able to serve as sole carbon sources for the growth of bacteria (e.g., Pseudomonas fluorescens, B. cereus, Rhizobium japonicum) and fungi (T. viride, Penicillium vermiculatum, Mucor mucedo) (22).

Volatile, gaseous, and particulate air pollutants, derived from both natural and anthropogenic sources, affect not only the growth, reproduction, morphology, and other characteristics of microbes, but also their activity, ecology, and population dynamics in natural habitats. As all life in our biosphere is ultimately dependent on microbial activities, it is imperative that the potential effects of airborne pollutants on the microbiota, and the mediating influence of abiotic environmental factors, be clarified.

## A. References


1. Abeles, F.B., Craker, L.E., Forrence, L.E., Leather, G.R.: Fate of air pollutants: removal of ethylene, sulfur dioxide, and nitrogen dioxide by soil. Science 173, 914-916 (1971).
2. Babich, H., Stotzky, G.: Ecologic ramifications of air pollution. Soc. Auto. Eng. Trans. 81, 1955-1971 (1972).
3. Babich, H., Stotzky, G.: Air pollution and microbial ecology. Crit. Rev. Environ. Control 4, 353-421 (1974).
4. Babich, H., Stotzky, G.: Sensitivity of various bacteria, including actinomycetes, and fungi to cadmium and the influence of pH on sensitivity. Appl. Environ. Microbiol. 33, 681-695 (1977).
5. Babich, H., Stotzky, G.: Reductions in the toxicity of cadmium to microorganisms by clay minerals. Appl. Environ. Microbiol. 33, 696-705 (1977).
6. Babich, H., Stotzky, G.: Effect of cadmium on fungi and on interactions between fungi and bacteria in soil: influence of clay minerals and pH. Appl. Environ. Microbiol. 33, 1059-1066 (1977).
7. Babich, H., Stotzky, G.: Effects of cadmium on the biota: influence of environmental factors. In: Advances in Applied Microbiology, Vol. 22. Perlman, D. (ed.). New York: Academic Press, in press (1978).
8. Babich, H., Stotzky, G.: Influence of pH on inhibition of bacteria, fungi, and coliphages by bisulfite and sulfite. Environ. Res. in press (1978).
9. Bortner, M.H., Kummler, R.H., Jaffe, L.S.: Carbon monoxide in the earth's atmosphere. Water, Air, Soil Pollut. 3, 17-52 (1974).
10. Bozian, R.H., Stotzky, G.: Inhibition of nitrification in soil by $SO_2$, and the effect of kaolinite and montmorillonite on inhibition. Agron. Abstr. p. 135 (1976).

11. Fries, N.: Effects of volatile organic compounds on the growth and development of fungi. Trans. Br. Mycol. Soc. 60, 1-21 (1973).

12. Heagle, A.S.: Interactions between air pollutants and plant parasites. Ann. Rev. Phytopathol. 11, 365-388 (1973).

13. Hibben, C.R., Stotzky, G.: Effects of ozone on the germination of fungal spores. Can. J. Microbiol. 15, 1187-1196 (1969).

14. Hutchinson, S.A.: Biological activities of volatile fungal metabolites. Ann. Rev. Phytopathol. 11, 223-246 (1973).

15. Jacobson, A.R.: Viable particles in the air. In: Air Pollution, Vol. 1. Stern, A.C. (ed.). New York: Academic Press, pp. 95-119 (1968).

16. Moore-Landecker, E., Stotzky, G.: Inhibition of fungal growth and sporulation by volatile metabolites from bacteria. Can. J. Microbiol. 18, 957-962 (1972).

17. Moore-Landecker, E., Stotzky, G.: Morphological abnormalities of fungi induced by volatile microbial metabolites. Mycologia 65, 519-530 (1973).

18. Moore-Landecker, E., Stotzky, G.: Effects of concentration of volatile metabolites from bacteria and germinating seeds on fungi in the presence of selective absorbents. Can. J. Microbiol. 20, 97-103 (1974).

19. Rasmussen, K.H., Taheri, M., Kabel, R.L.: Global emissions and natural processes for removal of gaseous pollutants. Water, Air, Soil Pollut. 4, 33-64 (1975).

20. Rasmussen, R.A., Hutton, R.S.: Utilization of atmospheric organic volatiles as an energy source by microorganisms in the tropics. Chemosphere 1, 47-51 (1972).

21. Saunders, P.J.W.: Modification of the leaf surface and its environment by pollution. In: Ecology of Leaf Surface Micro-organisms. Preece, T.F. & Dickinson, C.H. (ed.). New York: Academic Press, pp. 81-89 (1971).

22. Schenck, S., Stotzky, G.: Effect on microorganisms of volatile compounds released from germinating seeds. Can. J. Microbiol. 21, 1622-1634 (1975).

23. Smith, A.M.: Ethylene in soil biology. Ann. Rev. Phytopathol. 14, 53-73 (1976).

24. Stotzky, G., Schenck, S.: Volatile organic compounds and microorganisms. Crit. Rev. Microbiol. 4, 333-382 (1976 ).

25. Stotzky, G., Schenck, S.: Observations on organic volatiles from germinating seeds and seedlings. Amer. J. Bot. 63, 798-805 (1976).

26. Vancura, V., Stotzky, G.: Gaseous and volatile exudates from germinating seeds and seedlings. Can. J. Bot. 54, 518-532 (1976).

27. Wood, J.M.: Biological cycles for toxic elements in the environment. Science 183, 1049-1052 (1974).

# The Correlation Between rRNA Similarities and Bacterial Ecology

J. DE LEY

## A. Introduction

There are a number of methods to detect relationships between bacteria which are closely related. These organisms belong usually within a well-established genus or within a well-established family; e.g. the Enterobacteriaceae. When taxa are less related there are almost no reliable methods to detect the nature and degree of kinship, particularly, phylogenetic relationship. On the other hand the phylogenetic dendrogram of higher organisms - in particular animals - is reasonably well known through a number of methods (for example, comparative anatomy, embryology, paleontology, stratigraphy, radioactive dating) which are not applicable to bacteria. For higher organisms it is also well known that there is a very good correlation between the above mentioned phylogenetic tree and the dendrogram comparing the amino acid sequences of cytochrome c (2, 7).

Determining the amino acid sequences of bacterial cytochrome c might lead to some indirect insight into the possible phylogenetic relationships of these organisms. This area is being explored (1) although bacterial cytochromes are much more complex and the methods are very time-consuming. It seemed to us therefore that a faster approach was urgently needed. We have explored the similarities in base sequences of ribosomal RNA (rRNA) cistrons by DNA:rRNA hybridizations. rDNA's are conservative genes (6, 8) because of their importance in the ubiquitous and very similar protein synthesis mechanisms in all living beings. In addition it is our experimental experience that to compare base sequence similarities is faster than to determine amino acid sequences.

## B. Methods

From a number of selected reference strains of bacteria we prepared $^{14}$C-labeled rRNA. From the same and many other well-described bacterial strains we prepared pure DNA, which was fixed on membrane filters (4). Each $^{14}$C-rRNA was hybridized with either homologous or heterologous DNA on a membrane. The DNA:rRNA hybrids on the filters were thermally denatured. Two parameters were determined: a) the amount of rRNA hybridized to DNA, expressed as % rRNA binding and b) $T_{m(e)}$, the mid-point temperature of the thermal elution curve of the hybrid, a measure of base similarity and mismatching. The methods used have been described extensively by De Ley and De Smedt (3).

## C. Results and Discussion

$^{14}$C-labeled rRNA was prepared from type and reference strains of e.g. Escherichia coli, Vibrio parahaemolyticus, Pseudomonas fluorescens, Ps. acidovorans, Ps. solanacearum, Alcaligenes faecalis, A. denitrificans, Chromobacterium violaceum, Janthinobacterium lividum, Agrobacterium tumefaciens, A. rhizogenes, Acetobacter aceti subsp. aceti, Gluconobacter oxydans subsp. oxydans, Zymomonas mobilis subsp.

mobilis, Azomonas agilis, Arthrobacter oxydans, etc. The homologous and heterologous similarities of rDNA versus each reference $^{14}C$-rRNA were represented in an rRNA similarity map, i.e. a plot of $T_{m(e)}$ versus % rRNA binding. Examples of these maps are given e.g. by De Smedt and De Ley (5). It appears that the % rRNA binding is usually smaller than 0.3% and that $T_{m(e)}$ is below 82° C in our experimental conditions. Strains belonging in a well-established taxon, usually but not always a genus, form a dense cluster in these maps; examples are Zymomonas, Acetobacter, Gluconobacter. When it is genotypically (by DNA:DNA hybridization) and phenotypically well established that genera are distinctly related in a family as e.g. in the Enterobacteriaceae and the Vibrionaceae, then the rRNA cistrons within each of these families show considerable similarities and the strains form a distinct cluster in the similarity maps. On the other hand there are a number of genera, consisting of two or more rRNA clusters. It is extremely interesting that each of these clusters consists of strains which share many phenotypical features. In other words there is a correlation between rRNA similarities and overall phenotypical similarities. Examples are: Alcaligenes consists of two rRNA clusters corresponding with A. faecalis and A. denitrificans (De Ley, Kersters and Segers, to be published); Chromobacterium consists of two rRNA clusters corresponding with C. violaceum and C. lividum (De Ley, Segers and Gillis, in press); Pseudomonas consists of at least three rRNA clusters corresponding with Ps. fluorescens, Ps. acidovorans and Ps. solanacearum (De Ley and De Vos, to be published). We shall see below that the similarity between rRNA features and phenotypic characteristics extends to a higher taxonomic level.

At a higher taxonomic level it appears from our data that there is a good correlation between $T_{m(e)}$ and the phenotypical features; this is quite to be expected since $T_{m(e)}$ is a measure of base sequence similarities between $^{14}C$-rRNA and and rDNA. There is also a distinct correlation between $T_{m(e)}$ and the ecological habitat of bacterial taxa. So far we discovered five large superfamilies of rRNA cistron similarities.

a) One rRNA superfamily contains the Enterobacteriaceae and Vibrionaceae. Each family forms a separate rRNA branch. The phenotypical and ecological similarities between both families is well-known. Many organisms from both families are animal pathogens of great epidemiological importance (Yersinia pestis, Salmonella, Shigella, Klebsiella, Vibrio comma, Aeromonas etc.).

b) The second rRNA superfamily consists of Agrobacterium, Rhizobium, Phyllobacterium, Acetobacter, Gluconobacter, Zymomonas and a few other genera. Most of these organisms live in, on or around plants. This may be partially due to the fact that these bacteria are unable to synthesize certain amino acids and/or growth factors, which are readily available in plant juices. Of course each genus developed its own set of separate features in the course of evolution, e.g. Agrobacterium developed the TIP principle and the 3-keto-sugar metabolism, Rhizobium acquired the ability to penetrate Leguminosae and to carry out symbiotic $N_2$ fixation, Acetobacter and Gluconobacter acquired many enzymes on the cytoplasmic membrane, some oxidizing ethanol to acetic acid, others sugars to sugar acids, still others oxidizing polyalcohols to reducing compounds, etc.; Zymomonas acquired the ability to ferment glucose to ethanol and $CO_2$ and prefers to live anaerobically.

c) A third rRNA family consists of Chromobacterium violaceum, Janthinobacterium lividum (De Ley, Segers, Gillis, in press), Ps.acidovorans, Ps. solanacearum, Alcaligenes and a few other taxa. All these organisms share a number of common ecological and phenotypic features: e.g. they are soil and water organisms, some are pathogenic for animals or plants, they are chemo-organotrophic, they frequently prefer organic acids as carbon and energy source, the metabolism is usually respiratory, growth factors are usually not required.

d) The fourth rRNA superfamily consists of the Pseudomonas fluorescens complex (Pseudomonas section I according to Bergey 8th Edition), the xanthomonads, Azotobacter and Azomonas. Most of these bacteria are soil and water inhabitants, the phytopathogenic xanthomonads are phylogenetically speaking probably a later branch from the Ps. fluorescens group. It is very significant that there is a great similarity between the amino acid sequences of cytochrome c of Ps. fluorescens section I and Azotobacter (1). There are thus two gene products pointing to a common phylogenetic origin of both taxa.

e) The only Gram-positive rRNA superfamily we have studied so far consists of Arthrobacter, the phytopathogenic corynebacteria, the saprophytic corynebacteria, Microbacterium, and a few other organisms. It is obvious that most of them are soil coryneforms. Their rRNA cistrons are quite different from those of Bacillus and of all the Gram-negative organisms we examined.

The results will be published in full elsewhere. Other taxa are being examined.

## D. References

1. Ambler, R.P.: Bacterial cytochromes c and molecular evolution. Syst. Zool. 22, 554-565 (1973).

2. Dayhoff, M.O.: Atlas of protein sequence and structure. Vol.5. Nat. Biomed. Res. Foundation, Silver Spring, Md. USA. (1972).

3. De Ley, J., De Smedt, J.: Improvements of the membrane filter method for DNA: rRNA hybridization. Antonie van Leeuwenhoek J. Microbiol. Serol. 41, 287-307 (1975).

4. De Ley, J., Tytgat, R.: Evaluation of membrane filter methods for DNA:DNA hybridization. Antonie van Leeuwenhoek J. Microbiol. Serol. 36, 461-476 (1970).

5. De Smedt, J., De Ley, J.: Intra- and inter-generic similarities of Agrobacterium rRNA cistrons. Int. J. Syst. Bacteriol. 27, 222-240 (1977).

6. Doi, R.H., Igarashi, R.T.: Conservation of ribosomal and messenger ribonucleic acid cistrons in Bacillus species. J. Bacteriol. 90, 384-390 (1965).

7. Margoliash, E., Fitch, W.M., Dickerson, R.E.: Molecular expression of evolutionary phenomena in the primary and tertiary structures of cytochrome c. In: Biochemical Evolution and the Origin of Life. Schoffeniels E. (ed.). Amsterdam-London: North Holland Publ. Co. pp. 52-95 (1971).

8. Moore, R.L., McCarthy, B.J.: Comparative study of ribosomal ribonucleic acid cistrons in Enterobacteria and Myxobacteria. J. Bacteriol. 94, 1066-1074 (1967).

# Predation by Slime Mould Amoebae

M.J. BAZIN, P.T. SAUNDERS, B.A. OWEN, and D. KILPATRICK

## A. Introduction

Dent et. al (3) grew amoebae of the cellular slime mould, Dictyostelium discoideum together with Escherichia coli in glucose-limited chemostat culture and compared their results to several mathematical models of microbial predation. Their data corresponded most closely to models predicting damped oscillations in the population densities of the two organisms. Significant deviation, however, was found between their observations and the predictions of the models they tested; for example, the period of the oscillations decreased markedly and the bacteria disappeared from the system at rates much faster than the dilution rate at times when the amoebae appeared to have almost stopped growing. We report here our attempts to explain this last feature and also outline an analysis of the system in terms of catastrophe theory.

Two hypotheses were proposed to explain the rapid disappearance of bacteria from the system at times when it appeared that little or no predation was taking place. First, it was proposed that at high densities the predator might compete with the prey for some limiting nutrient. Consequently glucose, the substrate limiting the growth of the bacteria, and the nitrogen source of the system were tested. The second hypothesis suggested that at high population densities the amoebae stopped growing and released digestive enzymes into the medium which caused lysis of much of the bacterial population. We tested this by estimating the activity of an amoebal exoenzyme with probable digestive properties, β-N-acetyl-glucosaminidase (β-NAG). No evidence for competition between the predator and its prey was found but some correlation between β-NAG activity per amoeba and the rapid decline in prey density was observed.

## B. Materials and Methods

Escherichia coli B/r and vegetative amoebae of Dictyostelium discoideum NC4 were grown in 200 ml chemostat cultures as described previously (3). The outflow of cultures was collected in a fraction collector maintained at $4^{0}$ to inhibit microbial growth and the number density of each organism in the fractions estimated using a Coulter Counter Model $Z_{BI}$.

In order to determine whether the amoebae competed with the bacteria for glucose, 1.0 ml of uniformly labelled $^{14}C$-glucose (specific activity 5.0 μC $mg^{-1}$) at a concentration of 1.0 mg $ml^{-1}$ was added to a chemostat culture at a time when both the amoebae and the bacteria were increasing in numbers and the amoebal density was relatively high i.e. just before a rapid decrease in the bacterial population. At timed intervals a 5.0 ml sample was removed and centrifuged at 700 g for 3 minutes. The supernatant (a) was retained. The pellet was resuspended in distilled water and centrifuged again. This procedure produced a final pellet containing most of the amoebae relatively free of bacteria plus a supernatant (b) which was combined with supernatant (a). The mixture of supernatants (a + b) was centrifuged twice at 9000 g to produce a pellet retaining most of the bacterial activity of the original culture. The bacterial and the amoebal pellets were each suspended in 1.0 ml of 0.1% Triton and combined with 15 ml of β-PBD Toluene: Triton in scintillation vials. The radioactivity in each vial was estimated by counting for 10 minutes in a Hewlett-Packard scintillation counter. The fate of glucose during a period in which both the bacteria and the amoebae were decreasing in density was determined by adding $^{14}C$-glucose at an appropriate time to another chemostat culture of the organisms and repeating the procedure.

The nitrogen source used was Vitamin-free Casamino Acids (Difco). The concentration of amino acids was measured throughout the course of one experiment by using supernatants of samples which had been centrifuged for 20 minutes at 900 g.

The activity of β-N-acetylglucosaminidase was estimated (2) using 0.1 ml of each fraction. The release of 4-methylumbelliferone from 0.4mM 4-methylumbelliferyl -N-acetyl-β-D- glucosaminide was measured flourimetrically.

## C. Results

Figure 1 shows the fate of $^{14}C$-glucose added before and after relative maxima in the population densities of the two organisms. During the phase when neither the prey nor the predator appeared to be growing, very little incorporation of glucose occurred. When both populations were growing, uptake by the bacteria was very rapid and was followed by slower incorporation into the amoebal fraction, presumably the result of the amoebae phagocytosing the radioactive bacteria.

The change in the numbers of the two organisms and the way in which the total amino acid concentration varied during a chemostat culture is shown in Figure 2. Although the amino acid concentration fluctuated, at no time did it appear to be reduced to growth-limiting levels.

The activity of β-NAG, which is shown in Fig. 2 also, reached maxima at times corresponding to maxima in the predator density while relative minima in enzyme activity occurred after those of the amoebal population.

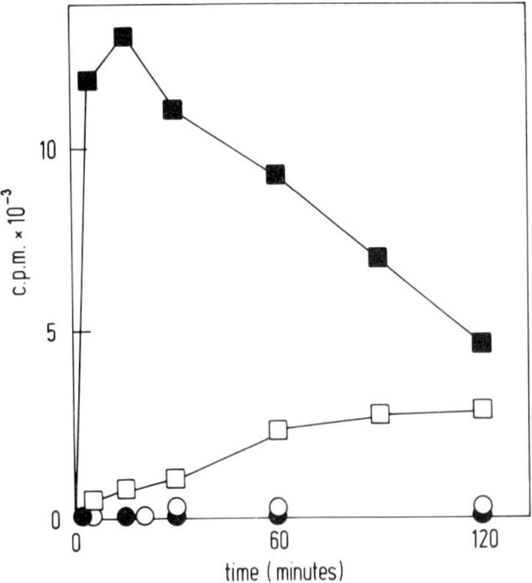


Fig. 1. Incorporation of $^{14}C$-glucose into E. coli and D. discoideum during rapid growth of populations (E. coli, closed squares; D. discoideum, open squares) and during a phase of decline in both populations (D. discoideum, closed circles; E. coli, open circles). Radioactive glucose was injected into chemostat cultures at times approximating the conditions illustrated in Fig. 2 after 76 h and 45 h of culture

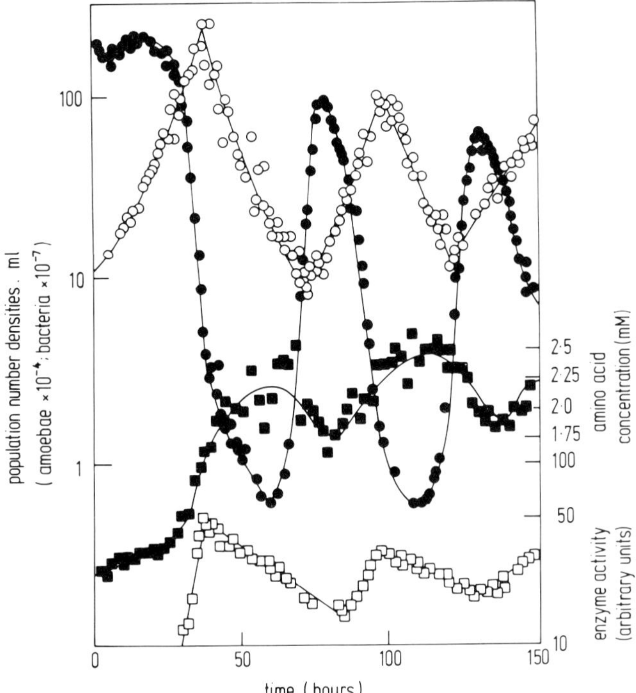


Fig. 2. Chemostat culture of E. coli and D. discoideum provided with 1.0 mg ml$^{-1}$ glucose at a dilution rate of 0.05 h$^{-1}$ showing the behaviour of the amoebal number density (open circles), bacterial number density (closed circles), total amino acid concentration (closed squares) and β-N-acetylglucosaminidase activity (open squares). Note: all variables except amino acid concentration, are plotted semilogarithmically so that the slopes of the lines generated by these data are equal to the specific rates of change of the variables concerned. The solid lines in the figure are provided for clarity and have been fitted by eye

## D. Discussion

The fate of radioactive tracer in the $^{14}$C-glucose experiments shows quite clearly that during growth phases the bacteria incorporate this sugar very rapidly and if any does enter the amoebae directly the amounts involved are insignificant. During periods of population decline neither organism appears to take up much of the label. Thus the hypothesis that competition between the prey and the predator for glucose causes the rapid decline of the bacterial population can be dismissed. The maintenance of relatively high levels of total amino acids in the culture argues against the idea of competition for a nitrogen source. On the other hand, the relatively high activity of β-NAG might indicate that this digestive enzyme is associated in some way with the rapid disappearance of bacteria from the system. This feature, however, does not seem to be a simple function of enzyme activity because at identical levels of activity the bacterial population may be increasing or decreasing. Rather, the controlling variable might be related to the ratio of enzyme activity to predator density as this quantity increases as amoebal numbers decrease.

Although these experimental results provide no definitive explanation for the behaviour of the system they do indicate that some features of the interaction between slime mould amoebae and bacteria cannot be explained solely in terms of the former preying upon the latter. As pointed out by Dent et al (3) the system is unlikely to be characterised by a set of simple ordinary differential equations of the type usually proposed to describe predator-prey interactions. Bazin (1) used a cusp catastrophe (4) to describe the behaviour of D. discoideum and E. coli in chemostat culture and obtained good qualitative agreement using prey density and culture age as control variables and predator specific growth rate as the state variable. We have extended this approach and found that a good quantitative fit of the data to a cusp catastrophe can be obtained. However, when the data is used to define the catastrophe set of a cusp catastrophe one of the "rules" of catastrophe theory is contravened. It appears, therefore, that interpretation of the results in these terms is incorrect. When we replace prey density as a control variable with the ratio of prey to predator present, behaviour appropriate to the conventions of catastrophe theory occurs. The results of this analysis therefore indicate that the dynamics of the interaction can be described in simpler terms if the average amount of prey available to each predator is used as a controlling variable rather than the total amount of prey present.

## E. Acknowledgements

This research was supported by the U.K. NERC. The authors are grateful to Ms Dimity Cox for provision of elasticated suspensory material.

## F. References

1. Bazin, M.J.: Predator-prey interactions between protozoa and bacteria. Ann. Appl. Biol. (In the press).

2. Dance, N., Price, D., Robinson, D. & Stirling, J.: β-galactosidase, β-glucosidase and N-acetyl-β-glucosaminidase in human kidney. Clin. Chim. Acta 24, 189-197 (1969).

3. Dent, V.E., Bazin, M.J. & Saunders, P.T.: Behaviour of Dictyostelium discoideum and Escherichia coli grown together in chemostat culture. Arch. Microbiol. 109, 187-194 (1976).

4. Zeeman, E.C.: Catastrophe Theory. Scientific American 234, 65-83 (1976).

# Marine Microbial Ecology

# Numerical Taxonomic Studies of Bacteria Isolated from Arctic and Subarctic Marine Environments

T. KANEKO, J. HAUXHURST, M. KRICHEVSKY, and R.M. ATLAS

## A. Introduction

Although 2/3 of the continental shelf of the United States is located around the state of Alaska, few microbiological studies have been conducted in this region. Using numerical taxonomic procedures, we have been characterizing bacterial communities overlying areas of the continental shelf of this region and studies so far have been conducted in the arctic Beaufort Sea and in the subarctic Gulf of Alaska. Microbial communities in surface ice, water, and sediment have been examined, and numerical taxonomic procedures used not only to determine the distribution of phenotypic groupings, but also to estimate the diversity of the bacterial populations. Further, the distribution of individual features in the bacterial populations has been determined and used as an indication of bacterial adaptation and function within particular ecosystems.

## B. Materials and Methods

I. Sampling and enumeration of bacterial populations. Samples were collected during the late summer and late winter in the Beaufort Sea and in summer in the northwest Gulf of Alaska and in the late winter in the northeast Gulf of Alaska. Ice samples were collected by scraping into sterile containers; water samples were collected with a Niskin sterile butterfly sampler and sediment samples with grab samplers.

Bacteria were enumerated by direct epifluorescent counting and by indirect plate counting. Plate count procedures were initiated within a few hours of collection and samples were maintained at 5°C during processing. Marine agar 2216 (Difco) was used for plating viable heterotrophic bacteria. After incubation at 4°C for 3 weeks for psychrophilic and psychrotrophic bacteria, and at 20°C for 2 weeks for mesophilic bacteria, colonies were counted with a binocular microscope and counts from triplicate plates averaged. For direct counting, samples were filtered through 0.2 μm Sartorius black filters, which were stained with acridine orange and viewed with an Olympus epifluorescent microscope with a BG-12 exciter filter.

II. Numerical taxonomic testing and analysis. Generally 20-25 isolates from each sample were selected at random from the countable plates for numerical taxonomic testing and approximately 2500 isolates were obtained for study. Pure cultures of the isolates were tested for over 300 characteristics including morphological, physiological, biochemical and nutritional features. Details of these procedures and results will be reported elsewhere.

Numerical taxonomic analyses were performed using single linkage sorting of Jaccard coefficients ($S_J$) (4). Phenotypic groupings were generally recognized at greater than or equal to 70% similarity. The Shannon diversity index ($\overline{H}$) was calculated from the cluster analyses for each sample (3). The formula $\overline{H} = \frac{3.3219}{N}(N \log N - \Sigma n_i \log n_i)$ was used where N = total numbers of isolates and $n_i$ = number of isolates in the $i^{th}$ cluster. The averages of the diversities from several samples collected in a geographic area were calculated to describe the overall diversity in a sample type, e.g., sediment in that area.

The percentage of the population from a sample possessing a particular feature was also determined and the averages of the feature frequencies from samples collected from a geographic area calculated.

## C. Results and Discussion

Both direct and viable counts of bacteria were higher in water samples from the Beaufort Sea than in samples from the Gulf of Alaska (Table 1) and could be due to differences in survival or predation rates. Numbers of bacteria in sediment were several orders of magnitude higher than in water samples. Estimations of numbers of organisms growing at 20°C were generally about the same as those growing at 4°C for both Beaufort Sea and Gulf of Alaska samples. Taxonomic testing showed that while a high percentage of organisms present in both locations could grow at temperatures of 4° - 20°C, different dominant phenotypic groupings occurred in the two sites.

Table 1. Counts of bacteria (numbers $ml^{-1}$ or $g^{-1}$)

| | Area | Season/Year | Direct Count | Viable Count 4°C | Viable Count 20°C |
|---|---|---|---|---|---|
| Ice | Beaufort Sea | Winter 76 | $9.9x10^4$ | $6.6x10^1$ | $4.5x10^1$ |
| Water | Beaufort Sea | Summer 75 | $8.2x10^5$ | $9.6x10^3$ | $7.3x10^3$ |
| | | Winter 76 | $1.8x10^5$ | $6.1x10^2$ | $1.1x10^3$ |
| | | Summer 76 | $5.2x10^5$ | $5.0x10^4$ | $2.7x10^4$ |
| | N.W. Gulf of Alaska | Summer 75 | $3.0x10^5$ | $1.0x10^2$ | $5.2x10^2$ |
| | N.E. Gulf of Alaska | Winter 76 | $1.4x10^5$ | $1.3x10^2$ | $2.4x10^2$ |
| Sediment | Beaufort Sea | Summer 75 | $6.2x10^7$ | $2.0x10^5$ | $2.9x10^5$ |
| | | Winter 76 | $3.7x10^8$ | $2.5x10^5$ | $6.9x10^4$ |
| | | Summer 76 | $2.1x10^9$ | $8.3x10^6$ | $1.0x10^7$ |
| | N.W. Gulf of Alaska | Summer 75 | - | $6.3x10^5$ | $8.0x10^5$ |
| | N.E. Gulf of Alaska | Winter 76 | $3.0x10^9$ | $9.9x10^5$ | $7.3x10^5$ |

The cluster analyses showed that organisms that were dominant in summer comprised only a small percentage of the population in winter. Organisms from the Gulf of Alaska were different from organisms from the Beaufort Sea. Within the Gulf of Alaska or Beaufort Sea, some phenotypic groupings were characteristic of a particular region, e.g., Prudhoe Bay; others were characteristic of sample type, e.g., sediment and some phenotypic clusters showed ubiquitous distribution.

The diversity of bacterial populations was higher in sediment than in water samples (Table 2) and there was greater diversity in water samples from the Gulf of Alaska compared with the Beaufort Sea. In the Beaufort Sea samples, greater diversity was apparent following incubation at 4°C compared with 20°C especially samples collected during winter. Water and ice samples from the Beaufort Sea collected in winter and incubated at 20°C had very low diversity supporting the claim that ecologic stress should result in low diversity (1).

Table 2. Diversity of bacterial populations expressed as a diversity index

| | Area | Season/Year | Diversity of Populations Enumerated at 4°C | 20°C |
|---|---|---|---|---|
| Ice | Beaufort Sea | Winter 76 | 3.0 | 1.4 |
| Water | Beaufort Sea | Summer 75 | 2.3 | 2.1 |
| | | Winter 76 | 2.6 | 1.1 |
| | | Summer 76 | 2.4 | - |
| | N.W. Gulf of Alaska | Summer 75 | 3.1 | 3.0 |
| | N.E. Gulf of Alaska | Winter 76 | 3.4 | 4.1 |
| Sediment | Beaufort Sea | Summer 75 | 3.5 | 3.5 |
| | | Winter 76 | 3.6 | 3.5 |
| | | Summer 76 | 4.0 | - |
| | N.W. Gulf of Alaska | Summer 75 | 3.7 | 2.7 |
| | N.E. Gulf of Alaska | Winter 76 | 4.2 | 3.4 |

Some of the characteristics of the bacterial populations isolated at 4°C are shown in Table 3. Most of the bacteria were Gram negative rods. Five-10 percent of the Gulf of Alaska isolates were Gram positive, whereas less than 0.5% of the Beaufort Sea isolates were Gram positive. A high percentage of isolates obtained from water samples collected in summer from the Beaufort Sea produced organic pigments but the percentage in winter was much lower. The orange pigmented bacteria appeared to belong mainly to the genus Flavobacterium, but at least ten phenotypic groupings were found. Pigmentation could have adaptive value during the continuous sunlight of the Arctic summer.

The average upper limit of growth temperature was higher for isolates from the Gulf of Alaska than from the Beaufort Sea. A greater percentage of isolates from the Beaufort Sea were true psychrophiles. During winter, for example, greater than 50% of the organisms isolated at 4°C from water and sediment in the Beaufort Sea were psychrophiles, i.e., could not grow at 20°C, compared to less than 20% in the Gulf of Alaska. Most isolates required NaCl for growth.

Table 3. Characteristics of microbial populations isolated at 4°C (% of populations)

| | SUMMER | | | | | WINTER | | | | |
|---|---|---|---|---|---|---|---|---|---|---|
| | Beaufort Sea | | | Gulf Ak. | | Beaufort Sea | | | Gulf Ak. | |
| | Ice | Water | Sed. | Water | Sed. | Ice | Water | Sed. | Water | Sed. |
| MORPHOLOGICAL: | | | | | | | | | | |
| rod shaped | - | 100 | 100 | 95 | 100 | 100 | 80 | 100 | 80 | 85 |
| gram negative | - | 100 | 100 | 90 | 100 | 100 | 100 | 100 | 94 | 95 |
| occur singly | - | 100 | 100 | 100 | 100 | 100 | 100 | 100 | 80 | 90 |
| motile | - | 0 | 50 | 60 | 35 | 20 | 40 | 50 | 30 | 35 |
| pigmented | - | 90 | 30 | 50 | 45 | 30 | 30 | 10 | 55 | 50 |
| PHYSIOLOGICAL: | | | | | | | | | | |
| Growth at 20°C | - | 90 | 50 | 90 | 70 | 80 | 50 | 40 | 85 | 80 |
| Growth at 37°C | - | 0 | 0 | 10 | 15 | 0 | 30 | 0 | 25 | 15 |
| Growth - 0% NaCl | - | 10 | 20 | 5 | 0 | 40 | 40 | 0 | 40 | 30 |
| Growth - 7.5% NaCl | - | 50 | 20 | 15 | 5 | 30 | 20 | 10 | 60 | 30 |
| Fac. anaerobe | - | 100 | 100 | 90 | 100 | 100 | 100 | 100 | 100 | 100 |
| ENZYMATIC: | | | | | | | | | | |
| catalase | - | 90 | 70 | 80 | 90 | 80 | 80 | 40 | 90 | 80 |
| oxidase | - | 0 | 10 | 75 | 90 | 30 | 20 | 30 | 65 | 85 |
| amylase | - | 10 | 30 | 55 | 50 | 20 | 20 | 60 | 30 | 45 |
| lipase | - | 20 | 60 | 55 | 60 | 60 | 50 | 70 | 80 | 85 |
| phosphatase | - | 90 | 90 | 85 | 70 | 40 | 30 | 30 | 95 | 95 |
| $NO_3^-$ reductase | - | 0 | 70 | 50 | 85 | 10 | 40 | 80 | - | - |
| NUTRITIONAL: | | | | | | | | | | |
| Glucose | - | 60 | 50 | 45 | 25 | 20 | 20 | 40 | 60 | 55 |
| Acetate | - | 20 | 20 | 45 | 25 | 0 | 40 | 10 | 10 | 10 |
| Lactate | - | 0 | 0 | 30 | 10 | 20 | 40 | 10 | 25 | 35 |
| Glycerol | - | 20 | 40 | 30 | 10 | 0 | 20 | 30 | 20 | 25 |
| Glutamate | - | 50 | 60 | 70 | 50 | 0 | 50 | 40 | 35 | 35 |
| Aspartate | - | 30 | 30 | 55 | 45 | 0 | 10 | 10 | 25 | 25 |

A high percentage of Gulf of Alaska isolates were oxidase positive, compared to a very low percentage of Beaufort Sea isolates. Almost 10% of Gulf of Alaska isolates could hydrolyse agar compared to 0% of the Beaufort Sea isolates. Algae containing agar are not normally found in the Beaufort Sea. A higher percentage of Gulf of Alaska isolates from summer samples showed amylase and lipase activity than Beaufort Sea isolates.

An interesting ecologic distribution for organisms capable of reducing nitrate was found. Nitrate reductase was characteristic of a higher percentage of organisms in sediment than in water. A very low percentage of isolates from the Beaufort Sea water could reduce nitrate. The ability to reduce nitrate is a key step in denitrification processes, which in marine environments occurs mostly in sediments (2).

Nutritional features of isolates were interesting; a very high percentage of isolates from the Beaufort Sea required growth factors. Many isolates required vitamins and amino acids. Only rarely could a single substrate be utilized by more than half the population. During summer, glucose and glutamate could be utilized by a relatively high percentage of the populations from both Gulf of Alaska and Beaufort Sea samples. The percentage of organisms capable of utilizing glucose was much lower during winter than summer in the Beaufort Sea. Isolates from Beaufort Sea ice samples showed very limited abilities to utilize individual substrates.

In summary, numerical taxonomic analyses can be used not only for taxonomy, but also ecologically to describe microbial communities. Bacterial diversity can be easily calculated from these analyses. Bacterial diversity was found to be lower in arctic than subarctic waters as might be predicted from ecologic stress. The occurrence of particular features in the microbial populations, such as temperature growth ranges and pigmentation also appear to demonstrate microbial adaptation to ecologic stress. Further, the distribution of particular features in bacterial populations, e.g., ability to reduce nitrate, correlates with the metabolic functions of microorganisms in particular ecosystems.

## D. Acknowledgment

This study was supported by the Bureau of Land Management through interagency agreement with the National Oceanic and Atmospheric Administration, under which a multi-year program responding to needs of petroleum development of the Alaskan continental shelf is managed by the Outer Continental Shelf Environmental Assessment Program (OCSEAP) office. We wish to thank Dr. R.R. Colwell for generously supplying the computer clustering program GTP2 used in this study.

## E. References

1. Odum, E.P.: Fundamentals of Ecology. Philadelphia: W.B. Saunders (1971).
2. Payne, W.J.: Reduction in nitrogenous oxides by microorganisms. Bacteriol. Rev. 37, 409-452 (1973).
3. Shannon, C.E., Weaver, W.: The Mathematical theory of communication. Urbana: Univ. of Ill. Press (1949).
4. Sokal, R., Sneath, P.H.A.: Principles of Numerical Taxonomy. San Francisco: W.H. Freeman (1963).

# Estimation of Bacterial Biomass in Water and Sediments Using Muramic Acid

D.J.W. MORIARTY

## A. Introduction

In many aquatic sediments, and the gut contents of animals which feed on them, techniques for estimating bacterial biomass do not give accurate results. This led me to develop a method for estimating biomass by using the relationship between muramic acid and carbon in bacteria (1-4). I have experienced a number of problems with both the assay procedure and the principle of the method and modifications are described which overcome these difficulties.

## B. Assay Procedure

I. Acid hydrolysis. After hydrolysis with 3M HCl in sealed tubes at $100^{o}C$ for 6 hours neutralisation with NaOH is possible, but HCl is better removed by evaporation *in vacuo*, thus reducing the final salt concentration. Most sediments can be dried before adding HCl but carbonate sediments must be degraded with 6M HCl until all $CO_2$ is driven off. A pH meter which reads negative pH can be used to adjust the HCl concentration to about 3 or 4M. Bacteria in water can be assayed after trapping the cells on 0.2 μm.Nuclepore polycarbonate filters which do not degrade during hydrolysis. Filters made of cellulose esters cannot be used because they release a substance which reacts with D-LDH.

II. Removal of cations. Divalent cations, especially $Ca^{++}$, inhibit bacterial luciferase and prevent extraction of D-lactic acid into ether. In siliceous sediments these cations can be removed by adjusting the pH to about 7.5 - 8.0 with NaOH and $Na_2HPO_4$ as described (1). Precipitates are removed by centrifugation. For coral reef sediments, $Na_2HPO_4$ (0.5M) should be added until no further $Ca^{++}$ is precipitated. This should be checked after centrifuging and adjusting the pH to 8.0 with NaOH. All samples are then freeze-dried.

III. Ether extraction. Dry samples are extracted with diethyl ether acidified with concentrated HCl (0.2% v/v) to remove D. lactate liberated by acid hydrolysis and other compounds such as glycollic acid. After drying to remove excess ether, samples are redissolved in water and pH adjusted to 7.5 - 8.0 (NaOH). Half of each sample is then frozen until required, for assay of residual D-lactate (or any compound which reacts with D-LDH (D-lactate dehyrogenase) to reduce (NAD).

IV. Alkaline hydrolysis. To liberate D-lactate from muramic acid the pH of the other half of the sample is adjusted to 12.5 and incubated at $35^{o}$ for 2 hours and the pH reduced to 9.0. The presence of muramic acid can be confirmed by running a thin layer chromatogram after the ether extract and assaying the region known to contain muramic acid (5). This procedure is necessary in samples containing much carbohydrate, as some carbohydrates can be degraded by alkali to D-lactate. An assay technique which does not use alkaline hydrolysis is desirable; gas-liquid chromatography (6) is being investigated.

V. Enzymatic assays. These are carried out as described (3), with the following modified reaction mixtures:- D-Lactate assay: Glutamate buffer 0.1M, pH 9.4-

0.5ml/assay; NAD, 33mg/ml in $H_2O$ - 10μl/assay; D-LDH, 5mg/ml - 2μl/assay; GPT, (glutamate-pyruvate transaminase) 10mg/ml (80 units/mg) - 1μl/assay. NADH assay: Phosphate buffer, 0.1M, pH 7.5 - 2ml/assay; FMN, 5mg/10ml $H_2O$ - 10μl/assay; Tetradecyl aldehyde, saturated in ethanol - 10μl/assay; Bacterial luciferase and bovine serum albumen, 1mg each/ml - 20μl/assay; D-lactate assay mixture - 0.5ml/assay. Three tubes per sample are prepared, one after acid hydrolysis, one after alkali hydrolysis and one after alkali hydrolysis plus 50 ng D-lactate as an internal standard. Activity of D-LDH and/or GPT may be less than specified by the manufacturer, perhaps due to high temperature during shipping, so an increase in incubation time to 60 minutes is sometimes necessary. Incomplete D-LDH reaction can be detected as an increase in the slope of a regression line for the lactate standards, when these are measured at intervals of about 30 min during storage in an ice bath.

An increase in intercept of a standard curve of lactate has been observed with some batches of D-LDH and GPT after placing the reaction tubes in an ice bath caused by contaminating glutamate dehydrogenase (GlDH), and can be stopped by briefly freezing and thawing the reaction tubes after the D-LDH incubation. Preferably, batches of D-LDH having no GlDH activity, and GPT with less than 0.0001% of GlDH should be used so that blank values for lactate will be less, thus giving greater sensitivity. Boehringer Mannheim will supply details of activity for each batch of enzyme they manufacture. To obtain precise results with bacterial luciferase, it is necessary to centrifuge for about 15 min at 3,000 x g and $0^oC$, to remove any solid material. Precise time intervals between adding luciferase and D-lactate mixture and starting the counting sequence are necessary.

## C. Principle of the Method

The method depends on measuring muramic acid (MA) and using an empirical formula to convert to biomass. It is assumed that all gram positive bacteria have a ratio of 44μg MA/mg C and all gram negative bacteria, 12μg MA/mg C (3). There is however, no sharp distinction in the ratio of muramic acid to carbon between these groups of bacteria, and these amounts have been determined for only a few bacteria. Further,if an arbitrary assumption is made about the proportions of gram negative and gram positive bacteria in natural populations, biomass estimates may be inaccurate, especially in anaerobic sediments when gram positive bacteria predominate. This problem has been overcome by combining muramic acid determinations with direct counts on portions of homogenised sediment. Using direct counts to give an independent estimate of biomass, muramic acid values can be used to calculate the proportion of gram positive bacteria (with thick cell walls) in a natural population. Some examples are given in Table 1, and more information will be published elsewhere. These estimates may not be accurate, because they are strongly dependent on the biomass of the cells. They show that a substantial proportion of the total population of bacteria in anaerobic zones of the sediment are gram positive and that there are none or few in open sea water. To obtain accurate estimates of bacterial biomass in sediments, it is necessary to use both direct counts and muramic acid values, as only about 10 to 30% of bacteria can be separated from the sediment by homogenising. In water which does not contain suspended sediment, direct counting is a much simpler procedure than muramic acid determination.

Table 1. Estimation of the proportion of gram positive bacteria with a thick peptidoglycan layer in marine populations from Moreton Bay, Qld. Direct counts were made using acridine orange. Bacteria were separated from sediments by homogenising. The sediment was strongly reducing at depths below 2 cm. Values for biomass per cell are only approximate, based on the diameters of the most common cells in each sample

| Sample | No./μgMA x $10^8$ | Biomass/cell μg x $10^{-7}$ | Gram positives % total population |
|---|---|---|---|
| Sea water | 15 | 0.5 | 0 |
| Sediment: | | | |
| 0 - 1 cm | 5 | 1.0 | 25 |
| 2 - 3 cm | 2.4 | 1.5 | 50 |
| 5 - 6 cm | 1.9 | 1.5 | 70 |

## D. References

1. Moriarty, D.J.W.: A method for estimating the biomass of bacteria in squatic sediments and its application in trophic studies. Oecologia (Berlin) 20, 219-229 (1975).
2. Moriarty, D.J.W.: Quantitative studies on bacteria and algae in the food of the mullet Mugil cephalus L. and the prawn Metapenaeus bennettae Racek & Dall. J. exp. mar. Biol. Ecol. 22, 131-143 (1976).
3. Moriarty, D.J.W.: Improved method using muramic acid to estimate biomass of bacteria in sediments. Oecologia (Berlin) 26, 317-323 (1977).
4. Moriarty, D.J.W.: Quantification of carbon, nitrogen and bacterial biomass in the food of some penaeid prawns. Aust. J. Mar. Freshwater Res., 28, 113-118 (1977).
5. Morrison, S.J., King, J.D., Bobbie, R.J., Bechtold, R.E., White, D.C.: Evidence for microfloral succession on allochthonous plant litter in Apalachicola Bay Florida, U.S.A. Mar. Biol. 41, 229-240 (1977).
6. Casagrande, D.J., Park, K.: Simple gas-liquid chromatographic technique for the analysis of muramic acid. J. Chromatogr. 135, 208-211 (1977).

# An Interspecies Regulatory Control of Dissolved Organic Carbon Production by Phytoplankton and Incorporation by Microheterotrophs

D.F. SMITH and H.W. HIGGINS

## A. Introduction

It has long been recognized that during photosynthesis phytoplankton release a portion of their fixed carbon to the surrounding water in the form of dissolved organic carbon (DOC). Since this material can be incorporated by microheterotrophs, the $^{14}CO_2$ method of Steeman-Nielsen (9) will measure both the $^{14}CO_2$ fixed by phytoplankton plus the $^{14}C$-labelled DOC which has been converted to particulate organic carbon (POC) by the microheterotroph populations. The fact that the measured radioactivity resides in particles of different size phytoplankton and microheterotrophs does not alter the quantity of primary production measured by the $^{14}CO_2$ method, but the distribution among size classes will determine the size of organisms which can subsequently make use of the primary production (6). Thus, to analyze the pathways by which primary production is utilised, the rates of DOC production by phytoplankton and incorporation by microheterotrophs must be estimated.

DOC production by phytoplankton has been extensively investigated, but the amounts of $^{14}C$-labelled DOC remaining in the medium after a few hours of incubation vary from $< 1\%$ to $>50\%$ of the total carbon fixed (1, 2, 3, 4, 5, 10, 11, 15). In these studies, however, neither the total quantity of DOC released nor the DOC production rate can be estimated; the amount of DOC remaining at the termination of the incubation has been measured, but not the DOC that has been respired or incorporated into heterotroph POC.

Ideally, in tracer experiments designed to measure POC and DOC production, and involving mixed populations of both phytoplankton and microheterotrophs, one should be able to introduce $^{14}C$-labelled DOC as well as dissolved inorganic carbon (DIC). Only recently has it become possible to employ one plankton sample to produce DOC with sufficiently high specific activity that a small volume of the DOC can be introduced as tracer to a second plankton sample. By conducting tracer kinetic experiments and employing highly radioactive DOC ($1.5 \times 10^5$ dpm/ml) to introduce label to some of the water samples, one can independently measure the rate of DOC production and incorporation by microheterotrophs (13).

## B. Materials and Methods

I. Preparation of C-labelled DIC and DOC. Commercial preparations of $Ba^{14}CO_3$ were acidified and the $^{14}CO_2$ collected in 1.0M NaOH: the pH of the solution was subsequently adjusted to pH = 8.5 by the addition of 1.0M HCl (14). This process ensures that the $DI^{14}C$ preparation added to a sample contains no radioactive contaminants which are non-volatile in an acidic solution.

Labelled DOC was prepared by the method of Wiebe and Smith (14) employing a portion of the natural phytoplankton population to which the labelled DOC would subsequently be introduced. In essence, this procedure involves the light-incubation of phytoplankton in seawater having the $DI^{12}C$ totally replaced by $DI^{14}C$.

II. Sample preparation and incubation. Water samples were collected from 2 m depth in South West Arm, a marine dominated estuary, located approximately 5 km, W. S.W. of the Division of Fisheries and Oceanography, Cronulla, N.S.W., Australia.

Water samples (5l) were stirred to ensure homogeneity during the dispensing of 100 ml volumes of the sample to 125 ml Erlenmeyer flasks. The flasks were incubated in an Aquatherm rotary shaker (New Brunswick Co., New Brunswick, N.J., U.S.A) at 180 rpm and at the ambient water temperature (17.4 - 18.2$^{o}$C). The flasks were illuminated by a bank of seven 40W fluorescent lights (ATLAS daylight) yielding an average illumination of 180 μ Einstein $m^{-2}$ $sec^{-1}$. Both dark and formalin-killed cell controls were run for each experiment. Incubations were terminated by filtering the contents of the flasks at recorded times.

Details of the experimental procedures employed to obtain the radioactivities (dpm) of the POC and DOC, the subsequent data analysis of the tracer kinetic curves, and the method employed to fractionate the particulate matter of seawater samples into discrete size classes are published elsewhere (8, 13, 14).

## C. Results and Discussion

I. Appearance of radioactivity in DOC during incubation with DI$^{14}$C. The curves shown in Figures 1 and 2 were obtained by incubating a marine water sample and a non-axenic culture of *Isochrysis galbani* in the light with DI$^{14}$C. The curves obtained from these experiments result from an increasing specific activity of a DOC pool of constant size (7). The plateau regions of the curves occur because the DOC pool size is constant and because the DOC specific activity has become equal to that of the administered DI$^{14}$C.

Analogous experiments were conducted with log phase *I. galbani* cultures which had been grown for three days in medium F containing antibiotics (Rifampcin or penicillin G). These cultures showed a continual increase in DOC radioactivity (Figures 3, 4) indicating that DOC was being produced at a rate greater than that

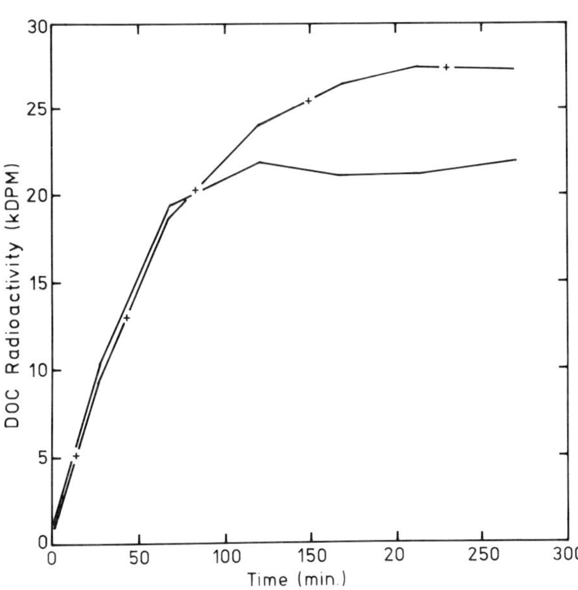


Fig. 1. Appearance of radioactivity in DOC excreted by phytoplankton. Duplicate seawater samples, 0.2 m M in DIC, were light incubated in the presence of DI$^{14}$C (1.46 × $10^{9}$ dpm/flask). Each 100 ml sample was filtered at a noted time and the radioisotope content of the sample DOC plotted against the time of filtration

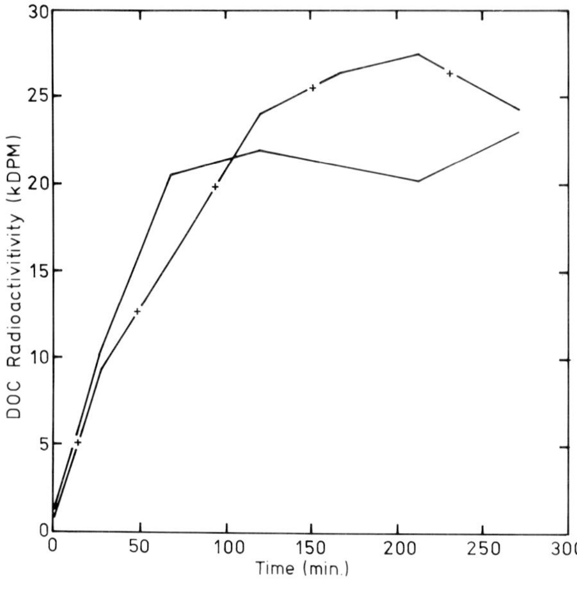


Fig. 2. Appearance of radioactivity in DOC excreted by a non-axenic culture of *Isochrysis galbani*. Duplicate samples were incubated in the light with $DI^{14}C$ ($3.33 \times 10^9$ dpm/flask) and the radioisotope content of the DOC plotted vs time. The algal cell densities were obtained by haemocytometer counts and bacterial cell densities by plate counts. In flask 1, (———) algal cell density = $9.0 \times 10^4$ and bacterial cell density = $4.0 \times 10^5$, cells/ml; flask 2 (—×—×—) algal cell density = $9.0 \times 10^4$ and bacterial cell density = $4.8 \times 10^5$ cells/ml

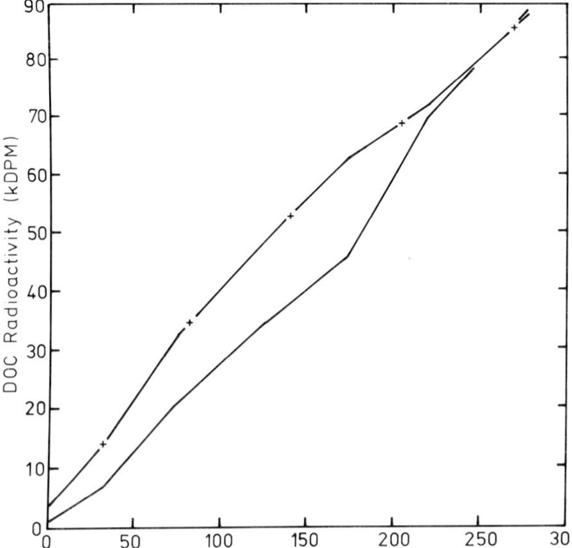


Fig. 3. Appearance of radioactivity in DOC excreted by an *I. galbani* culture pre-grown in Rifampcin (50 μg/ml). Pregrowth in antibiotic resulted in a decreased ratio of bacterial to algal cells. Flask 1 (———) algal cell density = $9.1 \times 10^4$, bacterial cell density = $1.2 \times 10^5$, cells/ml; flask 2 (—×—×—) algal cell density = $8.8 \times 10^4$, bacterial cell density = $1.1 \times 10^5$, cells/ml

at which it was incorporated into POC. The control cultures, grown without antibiotics, yielded tracer kinetic curves (Figure 2) which indicate that the DOC pool was in steady state during the experiment.

II. DOC production and uptake by discrete size classes of particles in seawater. The rates of DOC production and incorporation by discrete size classes of particles were independently obtained by incubations where tracer was added as $DI^{14}C$ or $DO^{14}C$ (Figs. 5, 6). Each particle size class was suspended in membrane-filtered (0.45 μm pore size) seawater at the original particle size class concent-

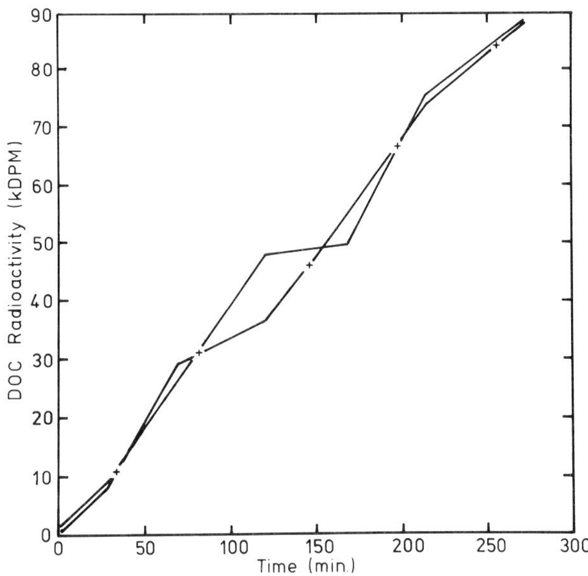


Fig. 4. Appearance of radioactivity in DOC excreted by an *I. galbani* culture pre-grown in medium containing penicillin G (5,000 IU/ml). Flask 1 (———) algal cell density = $9.1 \times 10^4$, bacterial cell density = $2.6 \times 10^4$, cells/ml; flask 2 (—×—×—) algal cell density = $8.8 \times 10^4$, bacterial cell density = $2.8 \times 10^4$, cells/ml

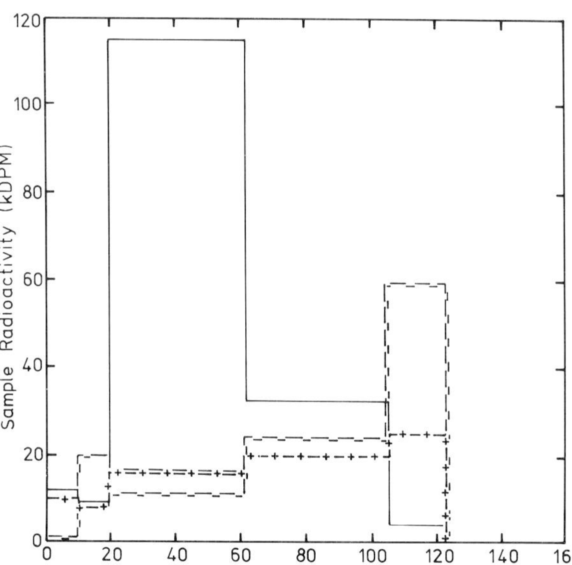


Fig. 5. DOC and POC production and DOC incorporation as a function of particle size. The particulate matter of a seawater sample was fractionated into discrete size classes, suspended in particle-free seawater and light-incubated for a 30-min period. Tracer was introduced as $DI^{14}C$ ($3.3 \times 10^9$ dpm/flask) to one set of samples and the subsequent radioactivity incorporated into DOC (———) and into POC (—+—+—) was plotted against particle diameter. A replicate set of samples was incubated with label added in the form of $DO^{14}C$ ($7.5 \times 10^5$ dpm/flask) and the radioactivity incorporated into POC was measured (– – – –)

ration. The same seawater sample was employed to obtain discrete particle size ranges for incubation with $DI^{14}C$ and $DO^{14}C$, as well as with D-glucose (UL-$^{14}C$) and for incubations with $DI^{14}C$ in the dark (Fig. 6).

In every experiment which employed untreated water samples or algal cultures which had not been rendered axenic, DOC was produced at precisely the same rate at which it was incorporated into POC. Two alternative hypotheses could explain the precision balancing of these two rates. Either the phytoplankton and the microheterotrophs are interactively coupled by a feedback mechanism which ensures homeo-

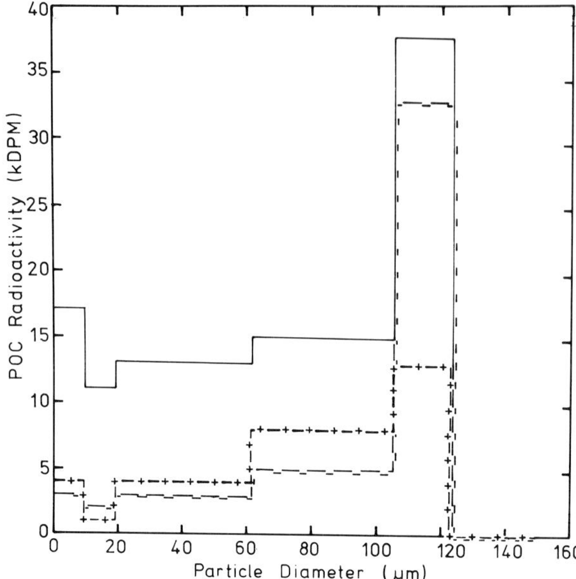


Fig. 6. Radioactivity in POC arising from the incorporation as (UL)-$^{14}$C D-glucose, DO$^{14}$C, or the dark incorporation of DI$^{14}$C. An identical experiment to that described in the legend of Figure 5, but wherein tracer was introduced as DI$^{14}$C, 3.3 × 10$^{9}$ dpm/flask, and the sample incubated in the dark (————), or introduced as (UL) $^{14}$C D-glucose, 1.67 × 10$^{6}$ dpm/flask (—–—–—) or as DO$^{14}$C, 7.5 × 10$^{5}$ dpm/flask (—+—+—)

stasis, or, alternatively, the apparent regulation is an artifact arising in consequence of the DOC-producing species also being the species responsible for the incorporation of DOC. However, if the particles found in a water sample are physically separated into distinct size classes, the larger ones, containing most of the bacterial biomass in these waters (12), only incorporate DOC while smaller size classes produce DOC at rates far exceeding the apparent DOC production of the unfractionated water sample. If the heterotrophic population of the algal cultures was unbalanced prior to an experiment, by pregrowth in medium containing antibiotics, elevated DOC production rates were observed and a fundamental change in the tracer kinetics took place.

We suggest the existence of an ecological unit, composed of at least two interacting species.

## D. References


1. Berman, T.: Release of dissolved organic matter by photosynthesizing algae in Lake Kinneret, Israel. Freshwater Biology, 6, 13-18 (1976).
2. Choi, C.I.: Primary production and release of dissolved organic carbon from phytoplankton in the Western Atlantic Ocean. Deep Sea Res. 19, 731-735 (1972).
3. Fogg, G.E., Nalewajko, C. & Watt, W.D.: Extracellular products of phytoplankton photosynthesis. Proc. R. Soc. B. 162, 517-534 (1965).
4. Hellebust, J.A.: Extracellular products. In: W.D.P. Stewart ed. Botan. Monogr. 10, 838-863 (1974).
5. Nalewajko, C. & Schindler, D.W.: Primary production, extracellular release, and heterotrophy in two lakes in the ELA, northwestern Ontario. J. Fish. Res. Board Can. 33, 219-226 (1976).
6. Pomeroy, L.R.: The oceans foodweb, a changing paradigm. Bioscience. 24, 499-504 (1974).

7. Smith, D.F.: Quantitative analysis of the functional relationships existing between ecosystem components. I. Analysis of the linear intercomponent mass transfers. Oecologia 16, 97-106 (1974).

8. Smith, D.F. & Wiebe, W.J.: Constant release of photosynthate from marine phytoplankton. Appl. Environ. Microbiol. 32, 75-79, (1976).

9. Steeman-Nielsen, E.: The use of radioactive carbon ($^{14}C$) for measuring organic production in the sea. J. Cons. perm. int. Explor. Mer. 18, 117-140 (1952).

10. Thomas, J.P.: Release of dissolved organic matter from natural populations of marine phytoplankton. Mar. Biol. 11, 311-323 (1971).

11. Watt, W.D.: Release of dissolved organic material from the cells of phytoplankton populations. Proc. R. Soc. B. 164, 521-551 (1966).

12. Wiebe, W.J. & Pomeroy, L.R.: Microorganisms and their association with aggregates and detritus in the sea: a microscopic study. Mem. Inst. Ital. IDROBIOL. 29, Suppl. 325-352 (1972).

13. Wiebe, W.J. & Smith, D.F.: $^{14}C$-labelling of the compounds excreted by phytoplankton for employment as a realistic tracer in secondary productivity measurements. Microbial Ecology (In Press).

14. Williams, P.J. Le B. & Yentsch, C.S.: An examination of photosynthetic production, excretion of photosynthetic products, and heterotrophic utilization of dissolved organic compounds with reference to results from a coastal subtropical sea. Mar. Biol. 35, 31-40 (1976).

# Ecological Aspects of Luciferases from Marine Bacteria

J.M. FITZGERALD and J. LEE

## A. Introduction

Though the biochemistry of bacterial luminescence has been extensively studied, many of these investigations have been undertaken on only one or two strains, frequently of the same species. The purpose of our work was to examine numerous properties of luciferases from many representative strains of the four commonly encountered marine species, Photobacterium phosphoreum, P. leiognathi, Vibrio fischeri and Lucibacterium harveyi (6). About twenty characteristics were examined in each of twenty strains. The luminescent and kinetic properties generally show variation between species, and there appears to be a correlation between many of these properties and the habitat of the bacterium producing the enzyme. Luciferases from some strains might be useful in studies of general enzyme kinetics, chemical mechanisms and interbiotic relationships.

## B. Materials and Methods

All L. harveyi isolates were from surface waters and the majority of other isolates from the light organs of fish, collected as described previously (1). Luciferase was extracted (3) after bacteria were grown in broth (4). Measurements of light emitted in vivo and in vitro were made under a variety of conditions, and details will be published elsewhere (2).

## C. Results and Discussion

When the in vitro luminescence reaction is initiated by addition of reduced flavin solution to an air-saturated solution of luciferase and aldehyde, the light intensity (I) rises to an initial steady-state maximum ($1_o$) proportional to enzyme activity and then decays. Specific activity of the luciferase varied with the carbon-chain length of the aldehyde, and with temperature. Tetradecanal gave maximum $1_o$ in all cases. Luciferases from L. harveyi strains differed from other species in giving a greater response with decanal than with dodecanal. Decay curves were constructed by plotting fall in light intensity with time and decay rates under various conditions were calculated from those portions of the curves showing exponential decay. Decay kinetics varied with species, temperature, and aldehyde carbon-chain length. Complexity of decay varied greatly, some species exhibiting single exponential decay over a number of decades followed by a second exponential decay while others showed first-order decay followed by nonexponential decay. Half-lives of a long-lived intermediate (previously described (3) for one strain of L. harveyi) were ascertained by withholding aldehyde from the reaction mixture for measured periods of time and then recording total light output after addition of aldehyde. Half-lives varied with species. Properties of the luciferases differed. Luciferases of P. leiognathi and V. fischeri were intermediate between those of P. phosphoreum on the one hand and those of L. harveyi on the other. Luciferases from strains of P. phosphoreum (all from light organs of deep-sea fish) were readily quenched by butylated hydroxytoluene (BHT), had low temperature coefficients of $I_0$ (and hence presumably low apparent energies of activation), were the least thermostable ($k=0.03s^{-1}$), had an intermediate with a short half-life (15s at $0^o$), and in vivo showed a spectral emission maximum which was blue-shifted compared with enzymes from other species. These properties are consistent with the environmental conditions experienced by fish at depths where temperature may be expected to be an important environmental factor. It is known (5) that fish at such depths (400-800 meters) have rhodopsins adapted to the blue end of the spectrum. Luciferases from L. harveyi on the other hand were negligibly quenched by BHT, had highest temperature coefficients of $I_0$, were most thermostable, and had long-lived intermediates whose half-lives at 0° ranged from 12 to 16 minutes, and over one hour at high enzyme concentration. Between these two groups were the luciferases of P. leiognathi, found as a symbiont in light organs of shallow-water leiognathid fish (6), and of all strains of V. fischeri from light organs of shallow-water fish (1). The luciferases of P. leiognathi however were more markedly quenched by BHT than those of V. fischeri. It is possible that these and other properties may be useful in preliminary classification of luminous bacteria.

To test this hypothesis, specimens of the luminous fish Siphamia cephalotes were caught in Sydney Harbour, and luciferase extracted directly from the light organs. On the basis of thermal stability, half-life of long-lived intermediate, and response to C-10 and C-12 aldehydes, it seemed unlikely that the luminous symbionts in this fish were either P. phosphoreum or L. harveyi. Subsequent taxonomic tests on bacteria isolated from the light organs of Siphamia cephalotes indicated that these bacteria were Photobacterium leiognathi.

The results indicate that some luciferase properties may be useful complementary identification aids. Judiciously selected strains may be of use in studying general enzyme kinetics and the chemistry of bacterial bioluminescence. Some correlation between environmental conditions and luciferase properties appears evident.

## D. References

1. FitzGerald, J.M.: Classification of luminous bacteria from the light organ of the Australian pinecone fish, Cleidopus gloriamaris. Arch. Microbiol. 112, 153-156 (1977).
2. FitzGerald, J.M., & Lee, J.: Manuscript in preparation.
3. Hastings, J.W., & Gibson, Q.H.: Intermediates in the bioluminescent oxidation of reduced flavin mononucleotide. J. Biol. Chem. 238, 2537-2554 (1963).
4. Lee, J. & Murphy, C.L.: Bacterial bioluminescence: absorption and fluorescence characteristics and composition of reaction products of reduced flavin mononucleotide with luciferase and oxygen. Biochem. Biophys. Res. Commun. 53, 157-163 (1973).
5. Nicol, J.A.C.: Some aspects of photoreception and vision in fishes. In: Advances in Marine Biology, Vol.1. Russell, F.S. (ed.) Academic Press, pp. 171-208 (1963).
6. Reichelt, J.L. & Baumann, P.: Taxonomy of the marine, luminous bacteria. Arch. Microbiol. 94, 283-330 (1973).

# Freshwater Microbial Ecology

# Population Shifts in Heterotrophic Bacteria in a Tributary of the Saint John River as Measured by Taxometrics

M.A. HOLDER-FRANKLIN, M. FRANKLIN, P. CASHION, C. CORMIER, and L. WUEST

## A. Introduction

Seasonal and diurnal population shifts in the aerobic, heterotrophic bacteria of the Meduxnekeag River were studied by two taxometric methods, numerical taxonomy (5) and factorial analysis (7). The influence of several chemical and physical parameters on the bacterial populations was also studied.

The homologies established by numerical taxonomy provided a method of grouping the predominant species which signified the bacteria having an apparent ecological advantage at that time. Each strain has a unique position in the matrix or factor space which is projected by the results of 230 tests. This numerical characterization eliminated the necessity for taxonomic purity, but most of the strains were identified as well by classical procedures.

Seventy reference strains, predominantly of American Type Culture Collection origin and all gram negative aerobic and facultatively anaerobic rods were selected on the basis of previous studies and tested with the river isolates. A summary of the seasonal shifts in predominant bacterial strains and one example of diurnal shifts are presented.

## B. Methods

Samples were obtained at the Environment Canada, Naquadat automatic computerized monitor at six hourly intervals during twenty-four hour periods in March, July, September and October. Temperature, dissolved oxygen ($DO_2$), pH, and specific conductivity were monitored hourly.

The 16 samples yielded 1600 isolates which were compared using 230 tests for phenotypic homology by the Jaccard coefficient of Sneath (5) and clustered by the unweighted pair group method with arithmetic averaging (UPGMA). The centroid strains of each cluster were identified and the % G + C of the DNA determined by a rapid method developed for this purpose (2). The testing procedures were facilitated by use of a unique multipoint inoculator, also developed in our laboratory (4). Tests for *Pseudomonas*, *Aeromonas*, *Xanthomonas*, *Vibrio*, *Flavobacterium*, *Enterobacteriacae*, *Cytophaga*, and *Sphaerotilus* were employed, as listed in Bergey's manual (1). Centroid Factor Analysis was carried out on the intertest correlation matrix determined by a group of 46 reference strains, using the four-fold point coefficient as recommended by Sundman (8). Resultant factors were hand rotated as discussed by Comrey (3) using simple structure and positive manifold criteria. Bacterial strains were projected graphically in a three dimensional factor space using the SYMVU program of the Laboratory for Computer Graphics and Spatial Analyses, Harvard University, Cambridge, Massachusetts.

## C. Results

Table 1 contains a summary of the genera which predominated. The summer and early autumn samples were notable in that facultative anaerobes and those which

Table 1. Seasonal Changes in Selected Parameters and Predominant Bacterial Genera

| Month | Temp. °C | Specific Conductivity μ Siemens $cm^{-1}$ | pH | Dissolved Oxygen mg $l^{-1}$ | Nutrients mg $l^{-1}$ C | N | P | Bacterial Genera Cluster Analysis % Total | Diurnal Variation in Bacteria: Factorial Analysis | Cluster Analysis |
|---|---|---|---|---|---|---|---|---|---|---|
| July | 23-25 | 146-153 | 6.8-8.2 | 6-9 | 10.0 | 0.4 | 0.03 | *Flavobacterium* (4.6), *Pseudomonas*[a] (3.9), *Erwinia* (3.4), *Cytophaga* (1.6), *Flexibacter* (0.8), *Aeromonas* (0.5), *Azotobacter* (0.5), *Xanthomonas* (0.5) | Low | Variation in genera and species moderate |
| Sept. | 10-12 | 238-247 | 8.6-8.5 | 10-11 | 6.0 | 0.17 | 0.015 | *Aeromonas* (38), *Flavobacterium* (11.9), *Pseudomonas* (8.9), *Erwinia* (1.8), *Vibrio* (0.6) | Moderate | Species Variation Observed |
| Oct. | 10-12 | 212-216 | 8.7-9.3 | 13-16 | 6.0 | 0.17 | 0.015 | *Pseudomonas* (25.9), *Flavobacterium* (18.6), *Xanthomonas* (4.6), *Aeromonas* (3.1) | Marked | Shown in Figure 1 |
| March | 0-0.1 | 155-161 | 8.1 | 14.0 | N.D. | 0.1 | 0.015 | *Pseudomonas* (8.6) *Flavobacterium* (2.5) | Low | Low |

N.D. = No data

a = non-fluorescent

have a fermentative metabolism predominated. March samples yielded a stable group of fluorescent pseudomonas, and certain weakly fermentative and non-fermentative types. Diurnal variation in the bacterial species, as measured by numerical taxonomy and factorial analysis was very marked in October but not in March and July. The environmental conditions were essentially unchanged during the twenty-four hour period for these two months with the exception of intense algal activity in July, as reflected by changes in $CO_2$, $HCO_3^-$, pH and nutrient levels and a high level of $DO_2$. Algal activity while low in September and October was non-existent in March.

In March, the high dissolved oxygen concentration and low temperature favored Pseudomonas and a group of Flavobacterium. The heavy ice cover in March provided the stable control base needed to show that the river, under conditions which did not vary, maintained a stable population. The temperature and nutrient levels in September were similar to those observed in October, however, the high conductivity (and higher salinity) lowered the $O_2$ levels in September and favored the fermenters.

The diurnal variation of bacteria in October was very marked. In October, the predominance of P. caryophylli was significant and due to a slight mid-day increase in temperature and lowering of $O_2$. Very few chemical and physical parameters changed however. The $CO_2$ levels at midnight were very low. The disappearance of fluorescent Pseudomonas was observed at midnight in July, September and October, and in March the numbers were reduced but a small cluster of biotype C was observed at 12:00 a.m.

Table 2. DNA Analysis (% G + C) of River Isolates

| Isolates | Chemical Method (TC)[d] | Expected [a] |
|---|---|---|
| P. fluorescens A | 62.6 | 60-61 |
| P. fluorescens B | 68.7 | 61-62 |
| P. fluorescens C | 66.0 | 63-64 |
| P. fluorescens F | 59.8 | 59-60 |
| P. stutzeri | 60.6 | 60-66 |
| P. caryophylli | 67.7 | 65.3 |
| P. lemoignei | 56.2 | 69 |
| P. marginata | 66.2 | 68.5 |
| P. cichorii | 60.3 | 59.0 |
| P. putida | 65.0 | 64-66[c] |
| P. alcaligenes | 71.3 | 66.3 |
| A. hydrophila anaerogenes | 62.2 | 57-63 |
| F. capsulatum | 62.1 | 63 |
| F. devorans | 72.0 | 69 |
| F. rigense | 69.3 | 69 |
| F. lutescens | 65.2 | 65 |
| X. campestris | 64.0 | 68.2[b] |

a = buoyant density except where indicated

b = Tm

c = chemical analysis

d = triple column method

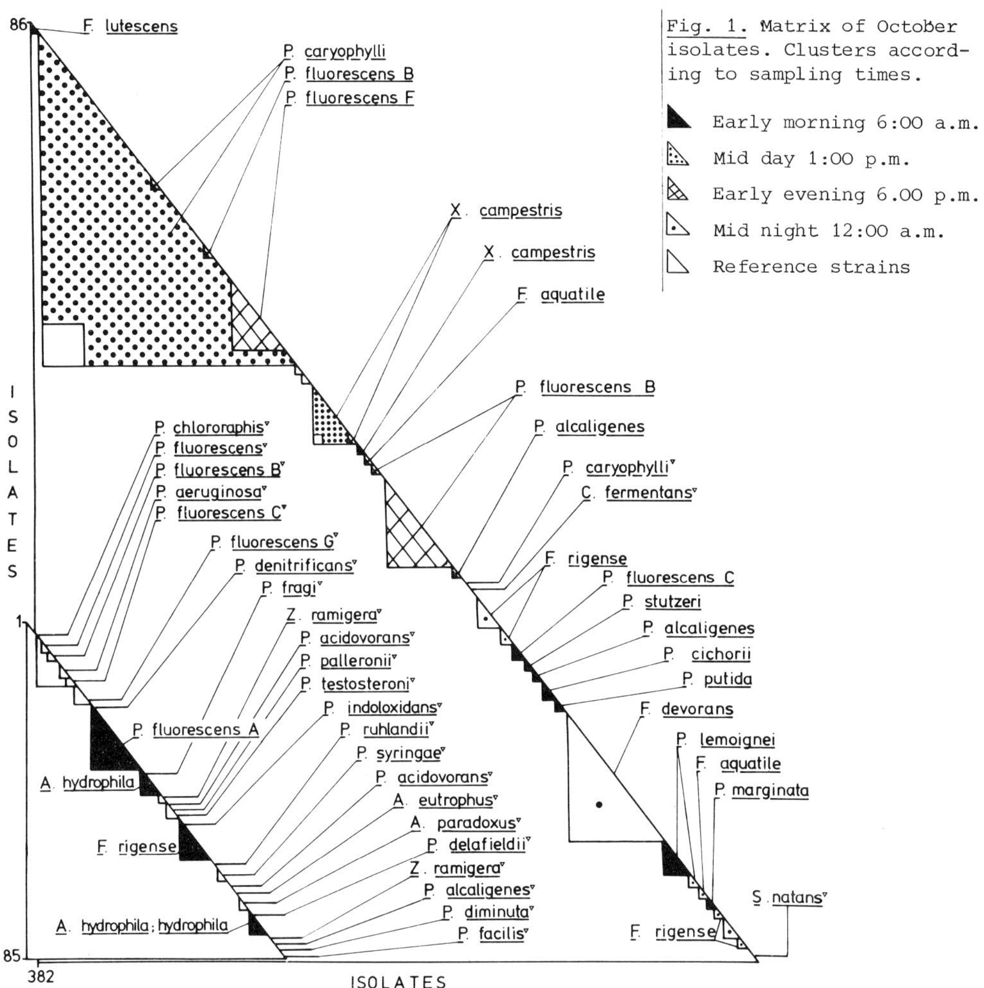


Fig. 1. Matrix of October isolates. Clusters according to sampling times.

- Early morning 6:00 a.m.
- Mid day 1:00 p.m.
- Early evening 6.00 p.m.
- Mid night 12:00 a.m.
- Reference strains

The October data has been selected for more detailed examination. Figure 1 is a diagramatic presentation of a matrix formed by UPGMA clustering of the October coefficients at 75% similarity, presented in sequence (isolates 1 - 382) directly from the computer output without rearrangement. Beginning with strain 1 - the reference strains form the first cluster which also shows the sub-clusters of closely related Pseudomonas reference strains at the 90% similarity level, indicating that the data base was adequate to discriminate these species at the expected level of homology. P. fluorescens A (Bergey biotype I) clustered within the closely related reference strains. A distinctive clustering pattern was observed for each sampling time. The large cluster, P. caryophylli, which appeared in the mid-day sample also contained some very closely related P. fluorescens which were also isolated from the late afternoon sample. The DNA analysis of the centroid strains within the caryophylli and fluorescens cluster (Table 2) indicated that a variety of species and % G + C may be observed within the 75% similarity cluster. The first Xanthomonas campestris cluster contained organisms of the same species from early afternoon and late afternoon in the same cluster. It was a consistent

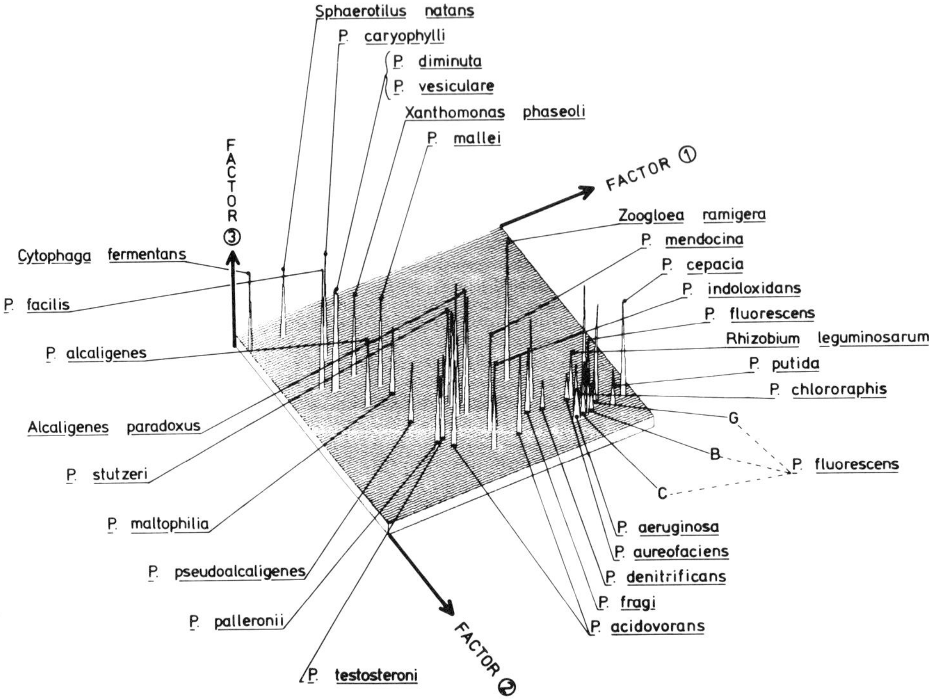


Fig. 2. Factor diagram of reference strains

finding that the majority of the strains within each cluster were from one sampling time and that the clusters were mixed only with strains of the next sampling period. The predominant population i.e. those which clustered or had achieved some supremacy, were changing as the day progressed.

The unidentified cluster between the two designated F. rigense at the bottom of the matrix was recognized as a Flavobacterium but could not be speciated until the DNA analysis was completed and we have since named that cluster F. capsulatum. Several clusters have been identified as the same species but their similarities on the tests used in this analysis have clearly separated them in the matrix, which is an indication of their inherent differences.

The factorial method (Figure 3) described the entire population of isolates in relation to each other on the factor reference frame (Figure 2).

The test results of the reference strains have been utilized to create the factor reference frame for the projection of the coordinates in three dimensions. Each strain is represented by a spire. The diurnal shift is clearly demonstrated.

The rotated factor loadings used for the interpretation were all at the 0.5 or greater level. Factor 1 has been interpreted as nutritional versatility. Many unusual substrates have high loadings on this factor. Factor 2 relates to oxidative pathways of metabolism - utilization of substrates which are intermediates in the tricarboxylic acid cycle and many which combine with acetyl-CoA have high loadings on this factor. Factor 3 was clearly fermentative versus oxidative metabolism. The fermentative characteristic, gas from glucose, was a strong positive on this factor. The oxidase test showed the highest negative loading. Some of the fluor-

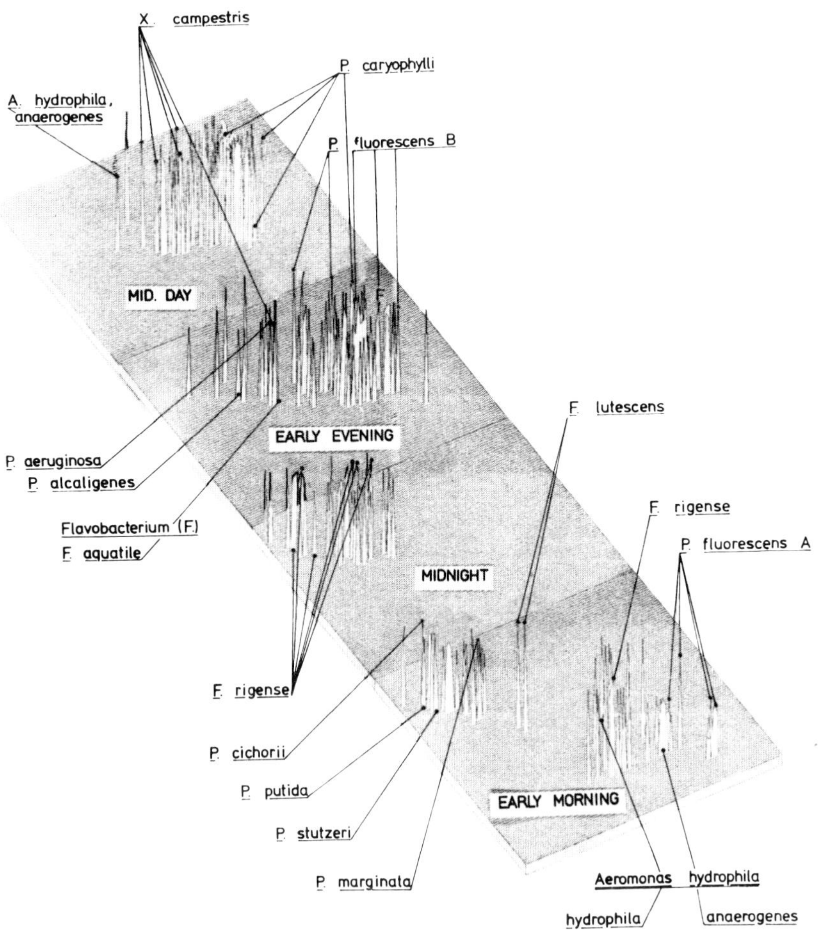


Fig. 3. Factor diagrams of river isolates in reference factor space

escent pseudomonads demonstrated low loadings on this factor as can be seen from the spire height. The highest loading on this factor was 0.5.

The cluster analysis pinpointed the predominant species which were useful high level indicators of bacterial shifts. However, the factorial analysis has greater possibilities for examining the changes in the total population, as isolated, and can be related to external influences with further analysis.

The techniques of numerical taxonomy and factorial ordination have demonstrated a seasonal and diurnal shift in a bacterial population and have shown that bacteria are controlled in low nutrient fresh water rivers primarily by temperature, nutritional versatility, dissolved oxygen and conductivity and less importantly by pH and algal activity.

## D. Acknowledgements

Supported by a grant from the Inland Waters Directorate, Department of Fisheries and Environment, Canada.

## E. References

1. Buchanan, R.E. & Gibbons, N.E.: Bergey's Manual of Determinative Bacteriology, 8th Ed. The Williams and Wilkins Co., Baltimore (1974).
2. Cashion, P., Holder-Franklin, M.A., McCully, J., & Franklin, M.: A rapid method for the base ratio determination of bacterial DNA. Analytical Biochemistry 81 (1977).
3. Comrey, A.L.: A first course in factor analysis. Academic Press Inc. London Ltd. (1973).
4. Kaneko, T., Holder-Franklin, M.A., & Franklin, M.: Multiple syringe inoculator for agar plates. Applied and Environmental Microbiology, 33, (4), 982-985 (1977).
5. Sneath, P.H.A. & Sokal, R.R.: Numerical Taxonomy. W.H. Freeman & Co., San Francisco (1973).
6. Stanier, R.Y., Doudoroff, M., & Palleroni, N.J.: The aerobic pseudomonads a taxonomic study. J. Gen. Microbiol. 43 (2), 159-271 (1966).
7. Sundman, V.: Four bacterial soil poulations characterized and compared by a factor analytical method. Can. J. Microbiol. 16, 455-464 (1970).
8. Sundman, V.: Description and comparison of microbial populations in ecological studies with the aid of factor analysis. Bull. Ecol. Res. Comm. (Stockholm) 17, 135-141 (1973).

# The Kinetics of a Microbial Population

M.C. DEPAUW-GILLET and J. REMACLE

## A. Introduction

Since Monod's studies, much research has focused on the kinetics of microbial populations not only from an intrinsic but also from an applied point of view (4, 11, 16).

The knowledge of growth parameters is very important in man-made biological systems such as purification plants and for elaborating biodegradation models in natural or polluted aquatic ecosystems.

Experimental results have shown that many parameters must be considered because they control microbial development (3, 5, 8, 10, 11, 12, 14, 15), even when predator problems are neglected.

It can be asked how all these parameters are related to the environmental conditions and how they control microbial growth. This study is an attempt to show the relationship between environmental variables such as pH and temperature and the control of some microbial growth characteristics.

## B. Materials and Methods

An Achromobacter strain was isolated from the Belgian river, Sambre, by the dilution method. The cultures were grown in a chemostat (500 ml) fed by the following medium (w/v): yeast extract 0.005 g, $NH_4NO_3$ 0.5 g, Winogradsky's salt base 50 ml, distilled water up to 1000 ml. Sterile glucose 0.5 g $l^{-1}$ was added aseptically after sterilization of the medium. By adjusting the speed of the magnetic-stirrer and the flow rate of air, the dissolved oxygen level was controlled at 4-5 ppm.

The dilution rates were chosen between 0.02 $h^{-1}$ and 0.83 $h^{-1}$. The steady state of the culture was checked after 5 to 6 residence times by estimating the turbidity of the culture. The continuous cultures were replicated three or four times for each dilution rate. The following growth characteristics were examined: bacterial concentration and production, maintenance energy, viability, death rate.

## C. Results and Discussion

Firstly, the bacterial concentration was examined in relation to the dilution rate at 20, 25 and 30$^{o}$C (Table 1). We find the same evolution of cell concentrations in all conditions; a fast increase at the low dilution rates followed by a plateau and the beginning of the decrease of cell concentration at the highest dilution rates.

Thus the highest cell concentrations did not occur at the highest dilution rates as expected by the theory of continuous culture. This fact has been reported earlier (1, 7, 8), and could occur when a limiting factor other than substrate appeared at the high dilution rates (16). For example, the cells could produce growth regulating substance which could be concentrated only in batch culture and in continuous culture at high residence time (7). This assumption is corroborated by our earlier observations with nitrifying bacteria (9, 13).

Table 1. Steady state concentration and production of _Achromobacter_ sp. $\left(\frac{\text{Conc. or Prod}}{\text{Maximum Conc. or Prod.}}\right)$

| D ($h^{-1}$) | 20°C pH | | | | 25°C pH | | | | 30°C pH | |
|---|---|---|---|---|---|---|---|---|---|---|
| | 6.5 | | 7 | | 6.5 | | 7 | | 7 | |
| | CONC. | PROD. | CONC. | PROD. | CONC. | PROD. | CONC. | PROD. | CONC. | PROD. |
| 0.02-0.03 | 0.10 | 0.04 | | | 0.14 | 0.025 | 0.62 | 0.13 | 0.12 | 0.018 |
| 0.03-0.04 | 0.18 | 0.03 | 0.66 | 0.21 | 0.14 | 0.041 | 0.83 | 0.26 | 0.17 | 0.013 |
| 0.04-0.05 | 0.33 | 0.13 | 0.69 | 0.28 | 0.95 | 0.043 | | | 0.16 | 0.014 |
| 0.06-0.08 | 1 | 0.58 | 1 | 0.43 | 0.98 | 0.63 | 1 | 0.17 | | |
| 0.10-0.13 | | | 0.63 | 0.30 | 1 | 1 | 0.88 | 1 | 0.44 | 0.14 |
| 0.13-0.17 | 0.69 | 0.82 | 0.70 | 0.49 | 0.45 | 0.88 | 0.54 | 0.70 | 0.55 | 0.21 |
| 0.20-0.24 | 0.55 | 1 | 0.62 | 1 | 0.32 | 0.64 | 0.49 | 0.86 | | |
| 0.32 | | | | | | | | | 0.89 | 0.64 |
| 0.45 | | | | | | | | | 1 | 1 |
| 0.52 | | | | | | | | | 0.35 | 0.43 |
| 0.64 | | | | | | | | | 0.13 | 0.20 |
| 0.76 | | | | | | | | | 0.08 | 0.15 |
| 0.83 | | | | | | | | | 0.05 | 0.09 |

As shown by the bacterial concentration and the optimal dilution rates, the _Achromobacter_ strain was very sensitive to the temperature and pH changes. At 20°C the bacterial concentration was very low: ten to twenty times lower than at 25°C. At 25°C, the bacterial concentration was twice as high when it was cultivated at pH 7.0 rather than at pH 6.5.

At steady state, the highest cell biomass, 81.2 mg $l^{-1}$ was recorded at 30°C and 0.45 $h^{-1}$. When the cultures were incubated at 20°C and 25°C, the maximal concentrations, 2.3 and 45.5 mg $l^{-1}$, were reached at 0.067 $h^{-1}$ and 0.116 $h^{-1}$ respectively. A previous observation (17) showed that the optimal rate at which the culture could be operated was dependent on the temperature. It must be also noted that our observed concentrations are always low by comparison with other data (1, 5, 6, 7, 14). The mean weight of the bacterial cell was determined at 30°C, all the dilutions being considered (30 determinations) and averages 1.3-1.6 $10^{-10}$ mg d.w. $cell^{-1}$. The cell production vs the dilution rate showed the same pattern as described for the bacterial biomass except at 20°C (Table 1). In this case, the bacterial production was always very low and averaged 0.29 mg $l^{-1}h^{-1}$ against 4.4 and 35.5 respectively at 25°C and 30°C. The drop in the bacterial production at the highest dilutions is corroborated by other data (1, 5).

The maintenance energy was calculated only at 30°C because the substrate consumption was too low at the other temperatures. Pirt's equation (12) was applied. At 30°C, pH 7, the maintenance energy equaled 0.11 $h^{-1}$.

By comparing this data with other results it appears that the value of maintenance energy is rather high. Values close to ours have been published (12) 0.09 $h^{-1}$ and 0.08 $h^{-1}$ (17). The maintenance energy depends on the temperature: the higher temperature, the higher maintenance energy (17), so it can be supposed that the maintenance energy of the Achromobacter strain could be lower at 20°C and 25°C.

The viability of the culture was followed in relation to the dilution rates by modifying: pH and T° (Table 2). The viability showed the same scheme except at 30°C. At 20°C and 25°C, the viability increased sharply until ca 0.9. At this level, the increase became very slow.

Conversely, at 30°C, the viability went up more regularly. At 30°C, 90% of viability was reached only at 0.6 $h^{-1}$ whereas at 20°C and 25°C, it was recorded at ca 0.06 $h^{-1}$.

Table 2. The viability of Achromobacter sp. (%)

| D ($h^{-1}$) | 20°C pH | | 25°C pH | | 30°C pH |
|---|---|---|---|---|---|
| | 6.5 | 7.0 | 6.5 | 7.0 | 7.0 |
| 0.02-0.03 | 0.80 | | 0.44 | 0.75 | 0.08 |
| 0.03-0.04 | 0.77 | 0.75 | 0.62 | 0.86 | 0.09 |
| 0.04-0.05 | 0.89 | 0.77 | 0.86 | | 0.25 |
| 0.06-0.08 | 0.95 | 0.80 | 0.90 | 0.91 | |
| 0.10-0.13 | 0.97 | 0.84 | 0.91 | 0.91 | 0.33 |
| 0.13-0.17 | 0.96 | 0.88 | 0.93 | 0.91 | 0.36 |
| 0.20-0.24 | 0.98 | 0.93 | 0.94 | 0.92 | |
| 0.32 | | | | | 0.48 |
| 0.45 | | | | | 0.56 |
| 0.52 | | | | | 0.64 |
| 0.64 | | | | | 0.90 |
| 0.76 | | | | | 0.95 |
| 0.83 | | | | | 0.98 |

Two models are chosen in order to calculate the death rate in continuous cultures. The first model (14) is written as follows:

$$\frac{1}{V} = 1 + \frac{\gamma}{D} \quad (1)$$

where V: viability $\frac{X_v}{X_v + X_n}$ : dimensionless;

$X_v$: viable cell concentration: cell $l^{-1}$; $X_n$: non viable cells concentration: cell $l^{-1}$; $\gamma$ : cell death rate, $h^{-1}$; D : dilution rate, $h^{-1}$.

The second (5) is a little more complicated because the authors introduce the specific decay rate constant for non viable cells i.e. for non-dividing cells. Therefore, the model is written:

$$\frac{V}{1-V} = \frac{D}{\gamma} + \frac{b_n}{\gamma} \quad (2)$$

$b_n$ : specific decay rate constant of non-viable cells : $h^{-1}$.

Both models are written in a linear form.

Using the first model (14) we observed that the equation described well the culture state at 20° and 25°C. The linearity of the relation was confirmed (Table 3). Some points however were slightly out of the straight line at T 25°C pH 6.5. Besides, when $\frac{1}{D}$ equals 0, the intercept on the ordinate equals ca 1 (1.01 to 1.06) as expected by the model.

Our data also fits the second model (5). The curve is linear mainly at the temperatures 20° and 25°C and $b_n$ is positive (Table 3). Some cells were apparently still alive but had lost the ability to divide. It can be concluded that this model is an improved form of the first. By comparison with the data observed at 20°C and 25°C, some remarks must be made about the evaluation of $\gamma$ and $b_n$ at 30°C. The models cannot describe all the sets of data which do not lie obviously on a single straight line (Fig. 1). This situation is discussed by Grady and Roper (5) who noted that the physiology of the cells is affected by the dilution rate and, consequently, the values of the constants $\gamma$, $b_n$ could be modified. Therefore neither model can forecast the phenomena which occur in the lowest or the highest dilution rates where the physiological processes could be modified. Table 3 shows that the death rates were highest when the growth conditions became more favorable, and it was generally true for both temperature and pH. Both models gave the same conclusions. Concerning $b_n$: decay rate for non-viable cells, the values were always low except at T° 25°C pH 7.0 (Table 3).

Table 3. Death rates of Achromobacter sp cultures

| | | Sinclair and Topiwala model | | Grady and Roper model | | |
|---|---|---|---|---|---|---|
| T° | pH | $\gamma$ | R | $\gamma$ | $b_n$ | R |
| 20°C | 6.5 | 0.003 | 0.83 | 0.005 | 0.00014 | 0.92 |
| | 7 | 0.013 | 0.95 | 0.016 | 0.012 | 0.99 |
| 25°C | 6.5 | 0.006 | 0.97 | 0.016 | 0.025 | 0.92 |
| | 7 | 0.008 | 0.95 | 0.025 | 0.11 | 0.83 |
| 30°C | 7 | 0.24 | 0.91 | 0.32 | 0.0032 | 0.94 |

$\gamma$ = death rate : $h^{-1}$
$b_n$ = specific decay rate constant of non-viable cells : $h^{-1}$
R = correlation coefficient

## D. Conclusion

The aquatic strain of Achromobacter was characterized by a low productivity in comparison with the productivity recorded in other aquatic systems. It is also a stenothermic organism. Indeed, it appears to be sensitive to the culture conditions. Slight changes in the temperature or in pH modify the growth of the strain sometimes to a great extent. The Achromobacter strain was also sensitive to the dilution rate. The bacterial culture developed optimally up to a dilution rate of 0.45 $h^{-1}$ at 30°C but only up to 0.069 $h^{-1}$ at 25°C. Besides, at the high dilution rates, all the bacteria were alive but their growth seemed to be restricted. The most significant discrepancy between viability and growth was observed at 30°C where the maximum steady state concentration and production occurred at 0.45 $h^{-1}$ whereas the highest concentration of viable cells was recorded at 0.83 $h^{-1}$. The ecological implication of this fact is obvious because it emphasises the need to know something of the dynamic behaviour of the microorganisms not only the number of viable cells.

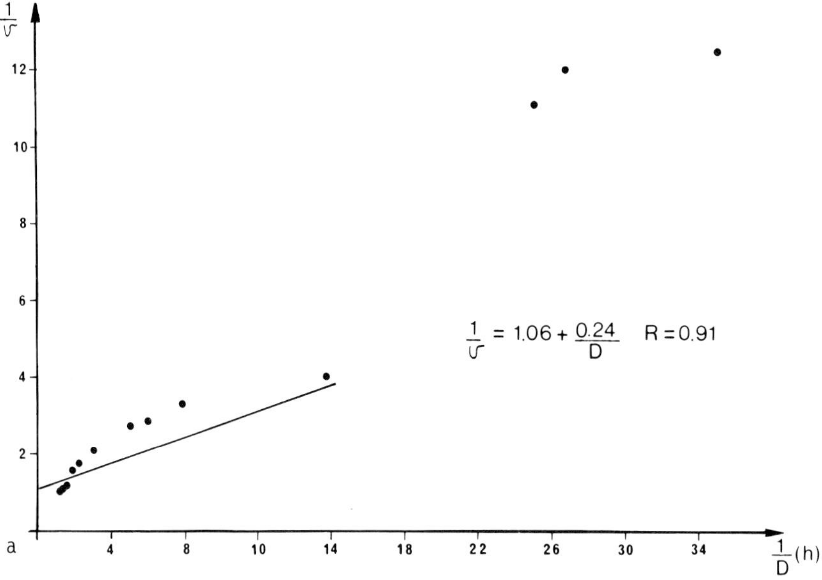


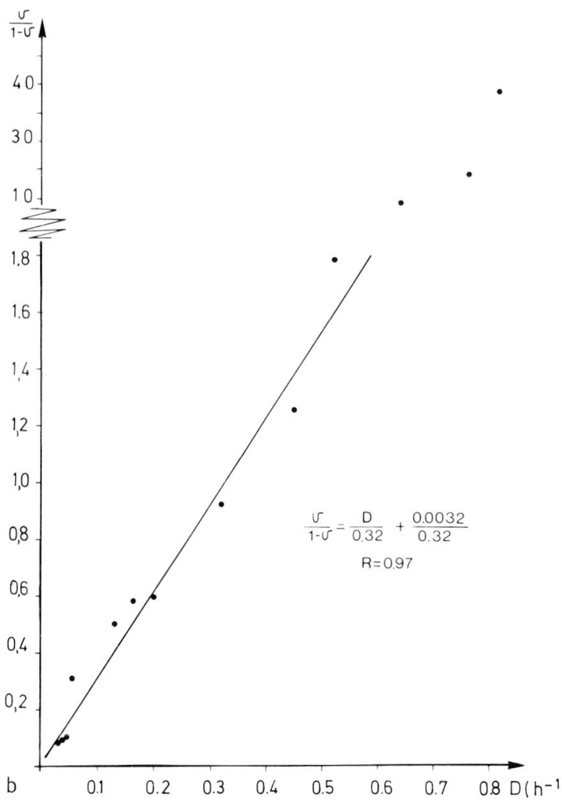


Fig. 1. (a) The Sinclair and Topiwala model ($T^{o}$ $30^{o}C$ pH 7.0); (b) the Grady and Roper model ($T^{o}$ $30^{o}C$ pH 7.0)

## E. References

1. Chiu, S.Y., Erickson, L.E., Fan, L.T. & Kao, I.C.: Kinetic model identification in mixed populations using continuous culture data. Biotechn. Bioeng. 14, 207-231 (1972).
2. Eckenfelder, W.W.: Principles of biological treatment. In "Procedes de traitements biologiques et physico-chimiques des eaux usees" ed. Universite de Liege, Belgique, 1-42 (1977).
3. Edeline, F., & Lambert, G.: A simple simulation method for river self purification studies. Wat. Res. 8, 297-306 (1974).
4. Gaudy, A.F., & Gaudy, E.T.: Biological concepts for design and operation of the activated sludge process. Environmental Protection Agency. Stillwater, Oklahoma, 153pp. (1971).
5. Grady, C.P.L., & Williams, D.R.: Effects of influent substrate concentration on the kinetics of natural microbial populations in continuous culture. Wat. Res. 9, 171-180 (1974).
6. Grady, C.P.L., & Roper, R.E.: A model for the Bio-oxidation process which incorporates the viability concept. Wat. Res. 8, 471-483 (1974).
7. Harrison, D.E.F., & Loveless, J.E.: The effect of growth conditions on respiratory activity and growth efficiency in facultative anaerobes grown in chemostat. Journ. Gen. Microbiol. 68, 35-43 (1971).
8. Herbert, D., Elsworth, R., & Telling, R.C.: The continuous culture of bacteria: theoretical and experimental study. Journ. Gen. Microbiol. 14, 601-622 (1956).
9. de Leval, J., & Remacle, J.: The influence of environmental factors on nitrification. In "Environmental Biogeochemistry" ed. J.D. Nriagu, Ann. Arbor Science Publ. 259-269 (1976).
10. McGrew, S., & Mallette, M.F.: Energy of maintenance in Escherichia coli. J. Bacteriol. 83, 844-850 (1962).
11. Monod, J.: Recherches sur la croissance des cultures bacteriennes. Hermann et Cie Paris. France. 210p. (1949).
12. Pirt, S.J.: The maintenance energy of bacteria in growing culture. Proc. Roy. Soc. London, B, 163, 224-231 (1965).
13. Remacle, J. & de Leval, J.: Approaches to the nitrification studies in river. In "Nitrification and Reducation of Nitrogen Oxides". ASM Conference, ed. W.J. Payne and E.L. Schmidt, 1-10 (1977).
14. Sinclair, C.G., & Topiwala, H.H.: Model for continuous culture which considers the viability concept. Biotechn. Bioeng. 12, 1069-1079 (1970).
15. Sykes, R.M.: Microbial product formation and variable yield. Journ. Wat. Poll. Contr. Fed. 48, 2046-2054 (1976).
16. Tempest, D.W., Hunter, J.R. & Sykes, J.: Magnesium limited growth of Aerobacter aerogenes in a chemostat. Journ. Gen. Microb. 39, 355-366 (1965).
17. Topiwala, H.H. & Sinclair, C.G.: Temperature relationship in continuous culture. Biotechn. Bioeng. 13, 795-813 (1971).

# Interactions of Algae and Heterotrophic Bacteria in an Oligotrophic Stream

G. A. McFETERS, S. A. STUART, and S. B. OLSON

## A. Introduction

Symbiotic communities composed of algae and bacteria have been described in certain restrictive ecosystems, and some authors (1, 2, 5, 6, 8, 9, 19, 24, 26) suggest that in natural aquatic systems heterotrophic bacteria may derive organic nutrients from this source. No such association however, has been reported involving pathogenic bacteria or those used as indicators of sanitary significance.

High numbers of coliform bacteria were found in the water of a pristine alpine stream in Grand Teton National Park, Wyoming, USA. Experiments were begun to determine if the bacteria were associated with a natural, periodic algal community and if extracellular products from the alga would support the growth of indicator bacteria and pathogens.

## B. Materials and Methods

Except for the methods described below, the procedures are reported elsewhere (16). Sample sites were along the outlet stream for Surprise Lake, 2916 m above sea level (MSL). This alpine stream dropped 234 m to Garnet Creek along 700 m of a south facing slope, and was below the timberline with no formal trails evident.

Samples of algae and associated bacteria were prepared for electron microscopy by fixation in 3% (v/v) glutaraldehyde (in 0.1 M $PO_4$, pH 7.3) and treated with 2% osmium tetroxide followed by dehydration through a series of acetone solutions. Infiltration was accomplished by propylene oxide followed by Spurr's solution with shaking for 60 minutes. Thin sections were stained with a saturated solution of uranyl acetate in absolute methyl alcohol followed by a rehydration scheme, stained in Reynold's lead citrate solution, rinsed in 0.02 N NaOH and finally in distilled water.

## C. Results and Discussion

Low numbers of sanitary indicator bacteria have been reported in most streams within Grand Teton National Park, Wyoming, USA (19), except for a small stream that served as the outlet for Surprise Lake that contained greater than 200 to 500 coliform bacteria per 100 ml in midsummer each year. These bacteria originated along the stream rather than from the water leaving Surprise Lake (16).

A thin golden-green benthic algal community containing members of the genera *Chlorella*, *Gleocapsa* and *Stigonema* was observed coincident with the high numbers of water-borne heterotrophic bacteria in July and August. In addition, high numbers of coliform bacteria were found to be physically associated with the algae from the stream surfaces (16). It was suggested that the benthic algal community and the associated bacteria could therefore serve as a possible source of the high numbers of indicator bacteria in the stream each summer.

Experiments were designed to examine the premise that extracellular organic products from the algae could support the growth of various heterotrophic bacteria.

The supernatant of an axenic batch culture of a Chlorella sp. from the natural benthic community (16) was filter-sterilized and used as the sole carbon source for the growth of numerous heterotrophic bacteria, including indicator organisms and pathogens. An incubation temperature of 13C was used, being the temperature observed within the stream in midsummer. The results (Table 1) demonstrate that the bacteria tested (including indicator and pathogenic bacteria) were capable of significant levels of growth using the extracellular products from the alga. Although numerous statements have been made that algae may provide an ecologically important source of carbon compounds for the growth of heterotrophic microorganisms (1, 2, 5, 6, 8, 9, 14, 24, 26) little evidence in support has been published.

Studies recently published, however, show that radioactive ($^{14}C$) algal supernatant was taken up by Escherichia coli (E. coli) and a Klebsiella isolate (16), demonstrating that the bacteria incorporated the algal organic products during active growth.

To determine if a close association existed between algae and bacteria in the benthic community of the stream under study, samples were examined by transmission electron microscopy. Figures 1 and 2 show bacteria closely associated with the algae and in Figure 2 the interalgal space appears to be composed of an extracellular slime matrix surrounding some of the bacteria.

Under natural conditions the interaction between algae and bacteria within aquatic mat communities is probably highly complex. Wright and Hobbie (25) suggested that the algae require 1 to 3 orders of magnitude greater concentration of organic compounds to metabolize heterotrophically than do bacteria, and symbiotic bacteria closely associated with the algae function as a "sink" for such solutes. As a consequence, the bacteria may reduce the level of dissolved organic compounds in natural waters (1, 14, 17, 18, 21, 25). There is some evidence that these bacteria may be under nutrient limitation (17) and that glycolate is an important bacterial nutrient in algal excretions (10, 17, 20, 24).

In the present work the environmental factors responsible for the establishment of this algal, bacterial community in pristine water is unknown. Algal colonization in the Surprise Lake outlet stream occurred when the water temperature rose above 10C in midsummer and disappeared when the temperature dropped. It should be pointed out that the Surprise Lake outlet stream is on a south facing slope, thus receiving more intense solar radiation than other streams, a factor probably contributing to algal growth. In support of this, laboratory experiments have demonstrated the critical role of illumination in determining the quantity of organic products excreted by Chlorella (3, 7, 23). Temperature and illumination

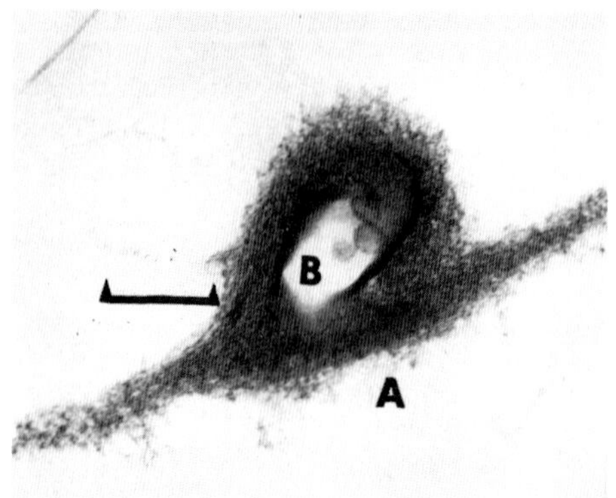


Fig. 1. Electron micrograph of a thin section of a bacterial cell B closely associated with the envelope of an algal cell A. Bar represents 0.2 µm

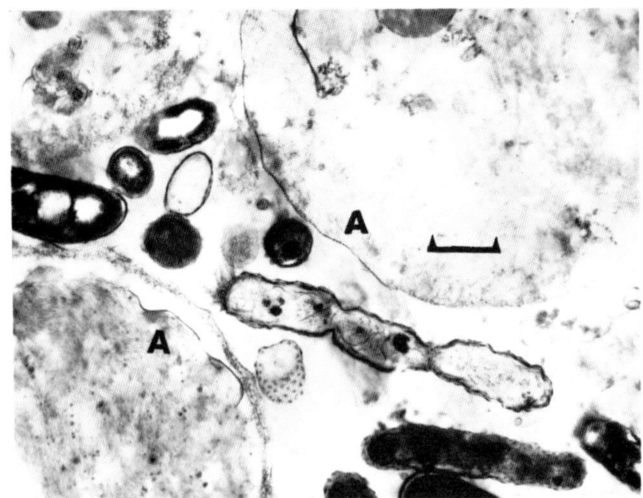


Fig. 2. Thin section through mixed culture of bacteria and algae A. Bar represents 0.5 μm

then are probably among key factors affecting the establishment of the mat community under study.

The evidence presented here and elsewhere (11, 12, 13, 15, 16) supports the idea that heterotrophic bacterial growth in a pristine aquatic environment is possible by the juxtaposition of bacteria with organic matter that has adsorbed upon sediment surfaces or that surrounds algae. The stream bed-water interface may serve as a microzone in which organic compounds from algal and other sources are localized making possible the growth of associated heterotrophic bacteria. The heterotrophic bacteria may be closely associated with algae within a slime matrix, largely produced by the photosynthetic cells. The release of bacteria into the overlying flowing water could be a consequence of active reproduction and the forces exerted by flowing water, resulting in substantial numbers of heterotrophic bacteria being released into the water.

The algal-bacterial association with consequent release of bacteria into the water leads to two further causes for concern, (a) the presence of the bacteria in the flowing water might represent false positive indications of fecal contamination using current water quality criteria; (b) observations of the growth of salmonellae in batch cultures of Chlorella at low temperatures (22). These findings support the contention (4) that pristine waters may contain pathogenic and indicator bacteria in the absence of fecal contamination.

## D. Acknowledgements

We thank David Stuart, John Schillinger and Susan Turbak for consultation and assistance; Susan Zaske for the transmission electron micrographs; Peter Hayden and Frank Betz of the National Park Service for their help.

This work was supported by the National Park Service (Contracts No. CS-12004-B025, CX-6000-3-0087, and 929-0-P20067), the United States Environmental Protection Agency (Training Fellowship No. U-910413-01), and the New York Zoological Society. The use of facilities at the Jackson Hole Biological Research Station is also acknowledged.

## E. References

1. Bauld, J. & Brock, T.D.: Algal excretion and bacterial assimilation in hot spring algal mats. J. Phycol. 10: 101-106 (1974).
2. Bell, W.H., Long, J.M., & Mitchell, R.: Selective stimulation of marine bacteria in algal extracellular products. Limnol. Oceanogr. 19: 833-839 (1974).
3. Belly, R.T., Tansey, M.R., & Brock, T.D.: Algal excretion of $^{14}C$-labeled compounds and microbial interactions in C. caldarium mats. J. Phycol. 9: 123-237 (1973).
4. Fair, J.F., & Morrison, S.M.: Recovery of bacterial pathogens from high quality surface waters. Water Resour. Res. 3: 799-803 (1967).
5. Fogg, G.E.: Extracellular products. p. 475-489 In: Ralewin (ed.) Physiology and Biochemistry of Algae. Academic Press, New York (1962).
6. Fogg, G.E.: The extracellular products of algae in fresh water. Arch. Hydrobiol. Beih. Ergebn. Limnol. 5: 1-25 (1971).
7. Fogg, G.E.: Algal Cultures and Phytoplankton Ecology. Second ed. U. of Wis. Press. 175p. (1975).
8. Fogg, G.E., Nalewajko, C., & Watt, W.D.: Extracellular products of phytoplankton photosynthesis. Proc. Roy. Soc. B. 162: 517-534 (1965).
9. Hellebust, J.A.: Excretion of some organic compounds by marine phytoplankton. Limnol. Oceanogr. 10: 192-206 (1965).
10. Hellebust, J.A.: Extracellular products. p. 838-863. W.P.D. Stewart (ed.) Algal Physiology and Biochemistry. Oxford: Blackwell Scientific Publications (1974).
11. Hendricks, C.W.: Enteric bacterial metabolism of stream sediment eluates. Can. J. Microbiol. 17: 551-556 (1971).
12. Hendricks, C.W., & Morrison, S.M.: Multiplication and growth of selected entheric bacteria in clear mountain stream water. Water Res. 1: 567-576 (1974).
13. Mack, W.N.: Investigations into the occurrence of coliform organisms from pristine streams. Anal. Method. Info. Center, 25, 9474; EPA 670/4-70-002b. (1974).
14. Maksimova, V.I. & Pimenova, M.N.: Influence of concomitant microflora on accumulation of organic compounds in medium during non-sterile culturing of Chlorella. Mikrobiologiya 38: 509-513 (1969).
15. Marshall, K.C.: Interfaces in Microbial Ecology, Cambridge and London Harvard Univ. Press pp. 27-52 (1977).
16. McFeters, G.A., Stuart, S.A., & Olson, S.B.: Growth of heterotrophic bacteria and algal extracellular products in oligotrophic waters. Appl. & Environ. Microbiol. (in press) (1977).
17. Nalewajko, C., & Lean, D.R.S.: Growth and excretion in planktonic algae and bacteria. J. Phycol. 8: 361-366 (1972).
18. Parsons, R.T., & Strickland, J.D.H.: On the production of particulate organic carbon by heterotrophic processes in seawater. Deep-sea Res. 8: 211-222 (1962).
19. Stuart, S.A., McFeters, G.A., Schillinger, J.E., & Stuart, D.G.: Aquatic indicator bacteria in the high alpine zone. Appl. & Env. Microbiol. 31: 163-167 (1976).
20. Tolberg, N.E.: Photorespiration by algae. p. 474-504 In: W.P.D. Stewart (ed.). Algal Physiology and Biochemistry. Oxford: Blackwell Scientific Publications (1974).

21. Vela, G.R. & Guerra, C.N.: On the nature of mixed cultures of C. pyrenoidosa. J. Gen. Microbiol. 42: 123-131 (1966).

22. Ward, C.H., Moyer, J.E., & Vela, G.R.: Studies on bacteria associated with C. pryenoidosa TX 71105 in mass culture. Dev. Ind. Microbiol. 6: 213-222 (1964).

23. Watson, W.D., & Fogg, G.E.: The kinetics of extracellular release of glycolate. J. Experimental Bot. 17: 117-134 (1966).

24. Wright, R.T.: Glycolic acid uptake by planktonic bacteria. Organic matter and natural waters. (ed.) D.W. Hood. Inst. Mar. Sci. University of Alaska, pub. no. 1: 521-536 (1970).

25. Wright, R.T. & Hobbie, J.E.: Use of glucose and acetate by bacteria and algae in aquatic ecosystems. Ecology 47: 447-464 (1966).

26. Wright, R.T., & Shah, N.M.: The trophic role of glycolic acid in coastal seawater. I. Heterotrophic metabolism in seawater and bacterial cultures. Mar. Biol. 33: 175-183 (1975).

# Biological Regulation of Bloom-Causing Blue-Green Algae

E.L. MARTIN, J.E. LEACH, and K.J. KUO

## A. Introduction

Many of the flood control lakes that have been constructed in the eastern Nebraska region of the Great Plains are rapidly undergoing eutrophication, which is directly related to the fact that they receive large amounts of nitrates and phosphates from the runoff of nearby feedlots and fertilized fields. The combination of nutrients with warm temperatures and long day-length of summer and early fall results in substantial blooms of primarily strains of *Microcystis*, *Anabaena* and *Aphanizomenon*. The same types of blue-greens have been implicated as major causes of blooms in many regions of the world (4, 10, 12, 16).

In recent years there have been reports of the isolation of many naturally occurring agents which lyse blue-green algae; some of these (viruses 5, 8, 9, 13; bacteria 3, 4, 7, 14, 15, 17) will be discussed in relation to the work presented in this paper. Specifically, the aims of the investigation reported here were (a) to further characterize cyanophage SM-2, (b) to describe the isolation and partial examination of two algal lytic bacterial agents, and (c) to discuss briefly the significance of these agents with respect to the regulation of blue-green algal blooms.

## B. Materials and Methods

I. *Isolation of the algal lysing agents*. The procedure for the isolation of the cyanophage (SM-2) has been previously described (5). The bacterial agents were isolated by collecting and concentrating water samples from several of the flood control lakes. An Amicon XM 300 Diaflo Ultrafiltration Membrane was used to concentrate each of the water samples 50X, with 0.5 ml portions of the concentrate being transferred on to algal overlays (5) of ASM medium (6) or Modified Hughes medium (1) with 2X $K_2HPO_4$ (MMH). After incubation with 250 fc illumination at 25$^0$C, the lytic bacteria were isolated from plaques which developed on the algal overlays. The isolated colonies were streaked on nutrient agar several times to ensure pure cultures which were tested again on algal overlays to reconfirm their action.

II. *Microscopy*. Similar procedures to those of Mackenzie and Haselkorn (8) were used for electron microscopy of SM-2 and *Synechococcus elongatus*, with grids being viewed on a Philips 201 electron microscope. Lytic bacteria were examined by phase microscopy using a Leitz Ortholux Fluorescence microscope.

III. *One-step and intracellular growth experiments*. *S. elongatus* IU 563 cells were shaken at 200 RPM for 5-7 days in MMH under 250 fc of illumination at 25$^0$C. The cells were centrifuged, washed and resuspended to a density of 2 x $10^8$ cells $ml^{-1}$ in MMH supplemented with 100 $\mu g$ $ml^{-1}$ gelatin. SM-2 was added to the cells at a multiplicity of infection of 10:1 and allowed to adsorb for 1.5 h. After adsorption, the blue-green algal cells and adsorbed virus were pelleted by low speed centrifugation, washed twice and resuspended in 150 ml of growth medium with 100 μg $ml^{-1}$ gelatin. These were incubated as described above. Every 2 h, an 0.1 ml sample was withdrawn, diluted and plated for the one-step growth curve. At the same time a 2.0 ml sample was treated with 0.1 ml chloroform to lyse the cells, diluted and plated for the intracellular growth curve.

IV. Determination of the host range of the lytic bacteria. Using the overlay method (5) to prepare algal lawns, streaked with the two types of bacterial agents, the algal host range was determined. Using the cross-streaking method, the two algal lytic bacteria were further checked to see if they had any effect on test bacteria from our departmental culture collection: *Bacillus subtilis*, *Escherichia coli*, *Neisseria catarrhalis*, *Proteus vulgaris*, *Pseudomonas fluorescens*, *Salmonella typhimurium*, *Staphylococcus aureus* and *Streptococcus faecalis*.

## C. Results and Discussion

Cyanophage SM-2 has been previously characterized (5) as having a host range of *Synechococcus elongatus* IU 563 and *Microcystis aeruginosa* NRC-1, and a structure consisting of a polyhedral head and a long flexible tail. Similar viruses have been isolated from at least three of the flood control lakes.

The one-step growth curve and intracellular growth curve of SM-2 with *S. elongatus* are presented in Fig. 1. The latent period lasts for 15 to 16 h with the first 4 to 5 h being the eclipse period. The rise period ends at about 35 h, 20 h after the latent period, and the average burst size is 200-300 plaque-forming units $cell^{-1}$. The general pattern of the SM-2 cycle is slower than that observed for AS-1 (13) and LPP-1G (9), but somewhat faster than that observed for SM-1 (8).

An uninfected *S. elongatus* cell and a cell after 22 h of infection are shown in Fig. 2 and Fig. 3 respectively. In both of these cells, intact photosynthetic lamellae are observed. The SM-2 virions have accumulated in the nucleoplasm primarily at either end of the cell close to the photosynthetic lamellae. The proliferation of SM-1 (8) and AS-1 (11) occurs in the nucleoplasm and appears to be somewhat similar to SM-2; whereas the proliferation of LPP-1G (9) results in

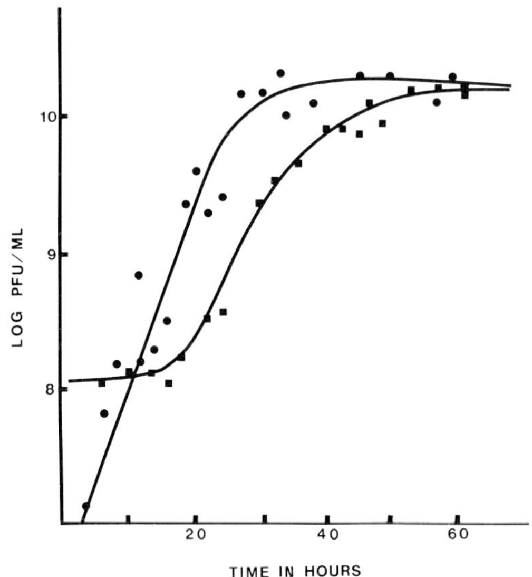


Fig. 1. One-step growth (■) and intracellular growth (●) curves of cyanophage SM-2 and *S. elongatus*

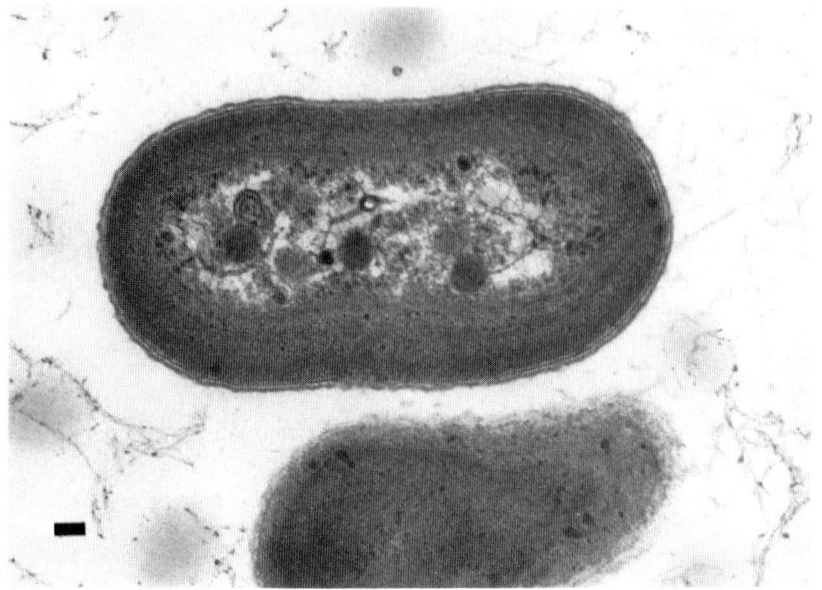

Fig. 2. Uninfected S. elongatus cell. X 40,000. Bar represents 0.1μ

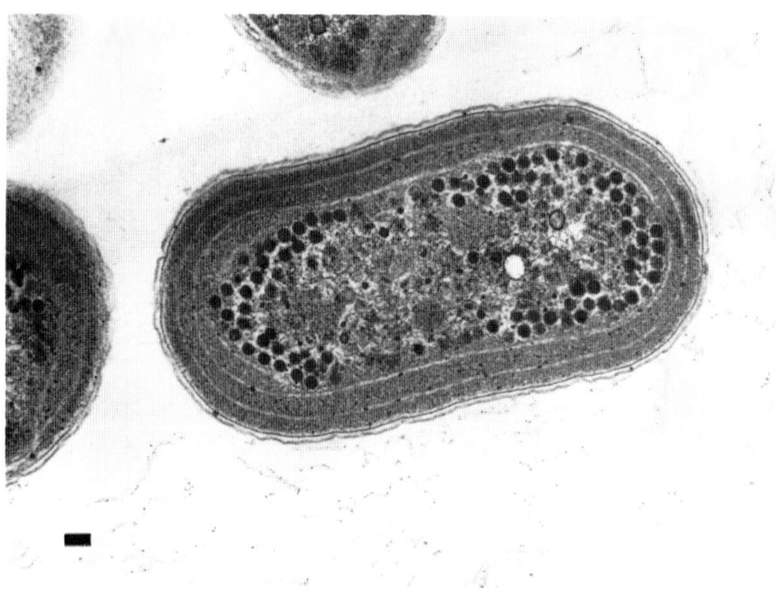

Fig. 3. S. elongatus cell 22 h after infection with SM-2. X 38,100. Bar represents 0.1μ

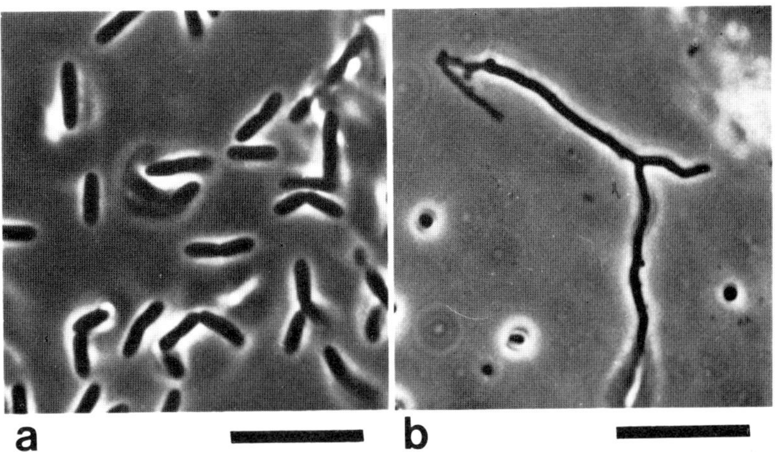


Fig. 4. (a) Alcaligenes isolate, (b) Streptomyces isolate. Bar respresents 10μ

an invagination of the photosynthetic lamellae and viral production in the virogenic stroma.

Using the isolation methods previously described, two main types of blue-green algal lytic bacteria were isolated from several of the flood control lakes. These bacteria were identified using standard methods (2) and found to be an Alcaligenes strain (Fig. 4A) and an Actinomycetales strain (Fig. 4B). While cyanophage SM-2 was very specific in its host range, it can be seen (Table 1) that the two lytic bacteria had increased algal host-ranges, particularly the Actinomycetales isolate. Additionally, the two algal lytic bacteria did not have an inhibitory effect on any of the eight test bacteria, nor in turn were they affected by the test bacteria. Other experiments revealed that the Alcaligenes isolate required host contact for lysis to occur (4, 15, 17); whereas the Actinomycetales isolate was able to lyse many of its hosts with some type of extracellular lytic factor (3, 7, 14). While it is well documented that there are many microbial pathogens for blue-green algae, it has also been proposed that certain bacteria may be mutualistic to the same blue-green algae (10).

Daft and co-workers (4) have found that algal lytic agents (particularly bacteria) and bloom-forming algae co-exist in neutral and alkaline waters that are eutrophic. They further postulated that this correlation is primarily due to the fact that both require similar environmental conditions and not just to the ability of these bacteria to lyse blue-green algae. Work is underway in our laboratory to investigate the role of our algal lytic agents in the eutrophic flood control lakes of eastern Nebraska and to determine if they have potential as algal regulatory agents.

## D. Acknowledgements

Partial support for this project was received from the Office of Water Research and Technology under the P.L. 88-379 Program and from the Water Resources Center, University of Nebraska, Lincoln.

Table 1. Host range of the lytic bacteria[a]

| | Alcaligenes isolate | Actinomycetales isolate |
|---|---|---|
| Blue-Green Algae: | | |
| Anabaena NU 41 | +[b] | + |
| Anabaena NU 380 | + | + |
| Anabaena flos-aquae NU 101 | + | + |
| Anacystis nidulans IU 625 | + | + |
| Lyngbya sp. IU 73 | - | - |
| Microcystis aeruginosa NRC-1 | - | + |
| Plectonema notatum IU 482 | - | + |
| Synechococcus cedrorum IU 1191 | - | + |
| Synechococcus elongatus IU 563 | - | + |
| Green Algae: | | |
| Ankistrodesmus falcatus IU 101 | - | + |
| Chlorella pyrenoidosa IU 252 | - | - |

[a] Abbrevations: IU = Indiana University Culture Collection; NU 41, 101, and 380 = Nebraska algal isolates that have been classified as Anabaena strains.

[b] + = Lysis of algal overlay within 7 days.

## E. References


1. Allen, M.M.: Simple conditions for the growth of blue-green algae on plates. J. Phycol. 4, 1-4 (1968).
2. Buchanan, R.E., Gibbons, N.E.: Bergey's Manual of Determinative Bacteriology, 8th ed., Baltimore: Williams and Wilkins, (1974)
3. Burnham, J.C., Stetak, T., Locher, G.: Estracellular lysis of the bluegreen algal Phormidium luridum by Bdellovibrio bacteriovorus. J. Phycol. 12, 306-313 (1976).
4. Daft, M.J., McCord, S.B., Stewart, W.D.P.: Ecological studies on algal-lysing bacteria in fresh waters. Freshwat. Biol. 5, 577-596 (1975).
5. Fox, J.A., Booth, S.J., Martin, E.L.: Cyanophage SM-2: a new blue-green algal virus. Virology 73, 557-560 (1976).
6. Gorham, P.R., McLachlan, J.S., Hammer, V.T., Kim, W.K.: Isolation and culture of toxic strains of Anabaena flos-aquae (Lyngb.) de Breb. Verh. int. Verein. theor. angew. Limnol. 15, 796-804 (1964).
7. Granhall, V., Berg, B.: Antimicrobial effects of Cellvibrio on blue-green algae. Arch. Mikrobiol. 84, 234-242 (1972).
8. Mackenzie, J.J., Haselkorn, R.: An electron microscope study of infection by the blue-green algal virus SM-1. Virology 49, 505-516 (1972).
9. Padan, E., Shilo, M.: Cyanophages-viruses attacking blue-green algae. Bacteriol. Rev. 37, 343-370 (1973).
10. Paerl, H.W.: Specific associations of the bluegreen algae Anabaena and Aphanizomenon with bacteria in freshwater blooms. J. Phycol. 12, 431-435 (1976).

11. Pearson, N.J., Small, E.A., Allen, M.M.: Electron microscopic study of the infection of Anacystis nidulans by the cyanophage AS-1. Virology 65, 469-479 (1975).

12. Prescott, G.W.: Algae of the Western Great Lakes Area. Dubuque, Iowa: Wm. C. Brown,(1951)

13. Safferman, R.S., Diener, T.O. Desjardins, P.R., Morris, M.E.: Isolation and characterization of AS-1, a phycovirus infecting the blue-green algae, Anacystis nidulans and Synechococcus cedrorum. Virology 47, 105-113 (1972).

14. Safferman, R.S., Morris, M.E.: Evaluation of natural products for algicidal properties. Appl. Microbiol. 10, 289-292 (1962).

15. Shilo, M.: Lysis of blue-green algae by Myxobacter. J. Bacteriol. 104, 453-461 (1970).

16. Singh, R.N.: Limnological relations of Indian inland waters with special reference to waterblooms. Verh. int. Verein. theor. angew. Limnol. 12, 831-836 (1955).

17. Wu, B., Hamdy, M.K., Howe, H.B.: Anti-microbial activity of a myxobacterium against blue-green algae. Bact. Proc. p. 48 (1968).

# Optimization of $N_2$ Fixation in $O_2$-Rich Waters

H.W. PAERL and P.E. KELLAR

## A. Introduction

Several blue-green algae (bacteria) are able to reduce atmospheric $N_2$ to $NH_3$ while photosynthetically evolving $O_2$. It is known that high ambient $O_2$ levels can inhibit $N_2$ fixation (11). Thick-walled cells called heterocysts, which lack $O_2$ evolving photosystem II, protect nitrogenase from $O_2$ inhibition, allowing photosynthesis and $N_2$ fixation to operate simultaneously. Even so, heterocystous blue-greens such as *Anabaena* and *Aphanizomenon* exhibit nitrogenase inhibition when supplied with oxygen at supersaturated levels; such $O_2$ concentrations are often produced during daylight in freshwater algal blooms.

The process of nitrogen-fixation is heavily dependent on light to meet energy demands. Accordingly, maximal nitrogen fixation rates often occur during peak photosynthetic periods (6, 7). A variety of mechanisms are probably operating during this time to allow maximum efficiency of light utilization for both processes, since structural protection by the heterocyst is not completely effective in shielding nitrogenase from rapid buildups of photosynthetically produced $O_2$.

It has been shown in other nitrogen-fixing bacteria that physiological means of protecting nitrogenase exist. High respiration rates appear to be instrumental in allowing *Azotobacter* to fix $N_2$ in aerobic environments (2). In the symbiotic system of soybeans $O_2$ intrusion is prevented by bacterial respiration (1).

In this paper combined results from microscopic observation and physiological experiments on *Anabaena* reveal that this genus has functional associations with heterotrophic bacteria and possesses endogenous light-mediated means of overcoming nitrogenase inactivation during $O_2$ supersaturation. In lakes, both mechanisms are likely to supplement the physical protection afforded by heterocysts, allowing optimal light utilization to occur in $CO_2$ and $N_2$ fixing processes.

## B. Methods and Materials

Samples of lakewater dominated by *A. spiroides* and *A. circinalis* as well as axenic and non-axenic batch cultures of *A. oscillarioides* and *A. flos-aquae* were examined. All cultures were grown on ASM (-N) medium at $18^{o}C$, 2000 lux (cool white illumination), on 12 h : 12 h light-dark cycle. A range of $O_2$ concentrations was introduced by gas phase displacement in 25 ml sealed serum bottles containing 15 ml of cell suspension. Methods and equipment are described elsewhere (Paerl and Kellar, in preparation). The gas phase of control samples was displaced with air, $pO_2$ = 0.2 atm; the gas phases of $O_2$ enriched samples were displaced with a mixture of air and various $O_2$ levels ($pO_2$ = 0.4 atm, $pO_2$ = 0.6 atm. $pO_2$ = 1.0 atm), the air component maintaining non-limiting $CO_2$ concentrations. Both light and dark incubations were carried out in a growth cabinet at lakewater temperatures. Following $O_2$ enrichment nitrogenase activity was monitored at time intervals by acetylene reduction assays (3, 10). The assays were confirmed as representing $N_2$ fixation by $^{15}N_2$ and Kjeldahl nitrogen analyses on parallel samples after growing on nitrogen-free media.

Both light (phase contrast) and scanning electron microscope (SEM) (9) observations were made on freshly sampled lakewater and cultures.

In order to detect reduced conditions in Anabaena and associated bacterial cells, nitro-blue tetrazolium (NBT, $E'_0$ = +50 mV) salts were added at a concentration of 0.01% w/v to samples, and incubated for 15 minutes. Within this time period, NBT is reduced to blue formazan crystals which can be seen deposited in specific cells. Regions of low $O_2$ tension can thus be identified.

## C. Results and Discussion

Immediate inhibition of nitrogenase activity relative to controls ($pO_2$ = 0.2 atm) occurred in all samples enriched with $O_2$. Similar responses were recorded elsewhere (11). During 6-8 h continual oxygenation periods, however, we noticed marked recovery of nitrogenase activity even though elevated dissolved $O_2$ concentrations persisted during this time. In the presence of other bacteria, Anabaena was able to recover in light and dark, with recovery in light being more rapid. Recovery was most rapid when large numbers of bacteria were attached to Anabaena heterocysts.

In most cases bacteria showed preferential attachment to the heterocysts, concentrating in the region of the heterocyst-vegetative cell junction (Fig. 1). Previous experiments (Paerl and Kellar, in preparation) have shown that these bacteria are heterotrophic and show a positive chemotactic response to various amino acids and sugars, believed to be representative algal excretion products (4, 5, 13). Such findings suggest that these bacteria are attracted to and mineralize algal excretion products.

Bacterial association with Anabaena heterocysts has been frequently observed, and is particularly extensive during blooms when surface scums develop (8). These scums often give rise to radical changes in $O_2$ concentration, ranging from daylight supersaturation to night-time deficits. The observation that formazan crystals develop near heterocysts colonized by bacteria indicate that the bacteria provide reduced microzones in the polar regions. Bacteria appear to play an important role in buffering Anabaena against inhibitory buildups of $O_2$ during optimal light periods.

In earlier studies we observed that Anabaena nitrogenase activity was able to recover from $O_2$ inhibition at maximum rates when heterotrophic bacteria were present. Recovery took longer when bacteria were either absent or present in very small numbers. Further experiments reported here, indicate that the period of recovery starts approximately 2-4 hrs after introduction of elevated $O_2$ levels, and the magnitude of recovery appears to be regulated by the abundance of the associated bacteria (Fig. 2). In general, under a variety of conditions bacterial associations lead to optimal $N_2$-fixation rates in Anabaena. The stimulus provided by bacteria is not due to bacterial $N_2$-fixation, as lake and culture waters containing bacteria without Anabaena (separated by sonication and filtration) failed to show significant acetylene reduction rates (Table 1).

At least two species of Anabaena (oscillarioides and flos aquae) grown in axenic cultures also showed an endogenous means of nitrogenase recovery following $O_2$ enrichment. The mechanism appears to be light-mediated. Illuminated samples showed a marked recovery, usually within 2-4 hrs after oxygenation, while dark samples showed persistent inhibition (Fig. 3). Illuminated $O_2$-enriched samples often surpassed controls in nitrogenase activity after 2-4 hrs. Parallel Kjeldahl analyses performed on such occasions substantiated acetylene reduction results, namely that $O_2$

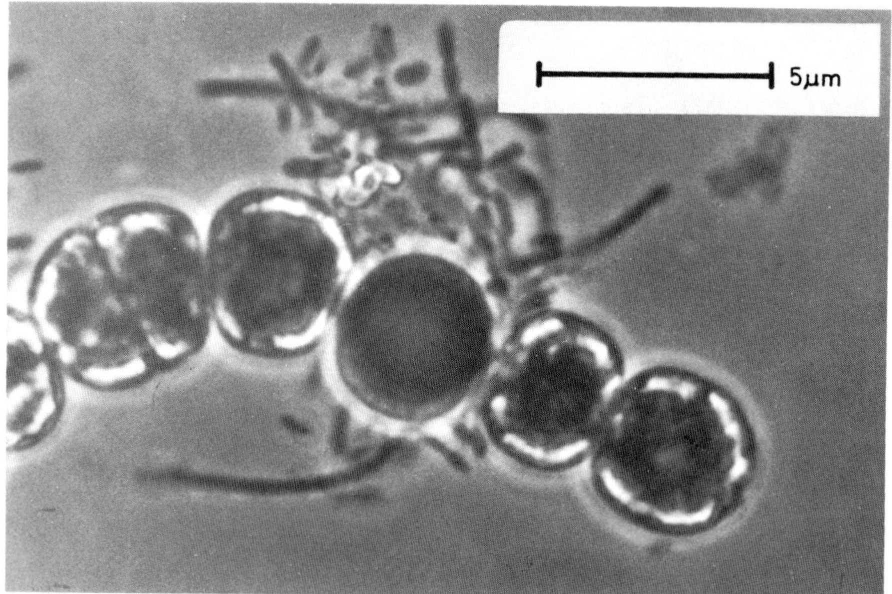


A

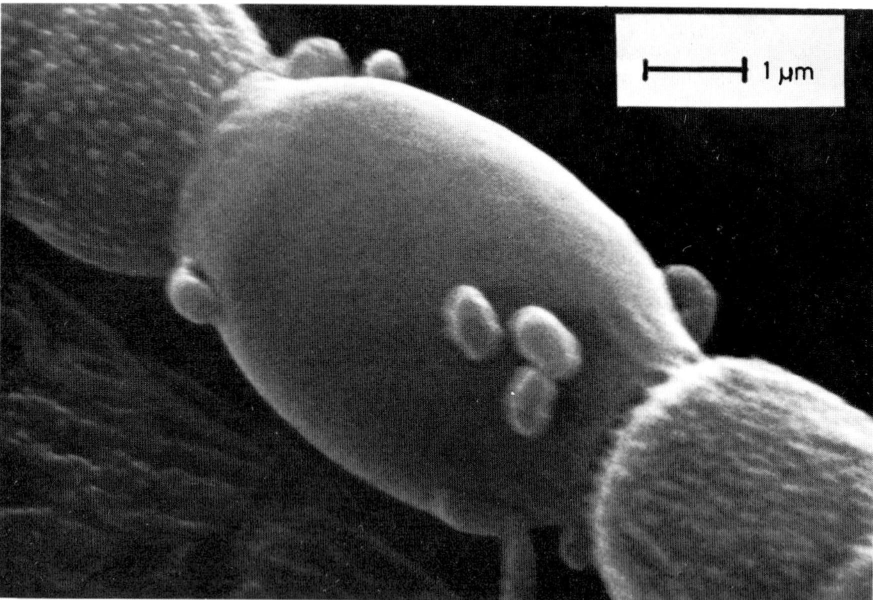


B

Fig. 1. (A) Bacterial accumulation around the heterocyst of Anabaena spiroides, freshly sampled from Lake Ngahewa, New Zealand and viewed by phase contrast optics. The heterocyst is the opaque cell lacking gas vacuoles. (B) Scanning electron micrograph (SEM) of Anabaena flos aquae heterocyst supporting bacteria. Note the distinction in surface morphology between the heterocyst and adjoining vegetative cells. (C) SEM view of Aphanizomenon flos aquae heterocyst supporting large numbers of bacteria. This sample was taken from Clear Lake, California, during an active $N_2$ fixing bloom. (D) High magnification (SEM) view of the polar region joining the heterocyst with vegetative cell. A significantly ($p<0.01$) greater number of bacteria accumulate in polar regions as opposed to the midsection of heterocysts ▶

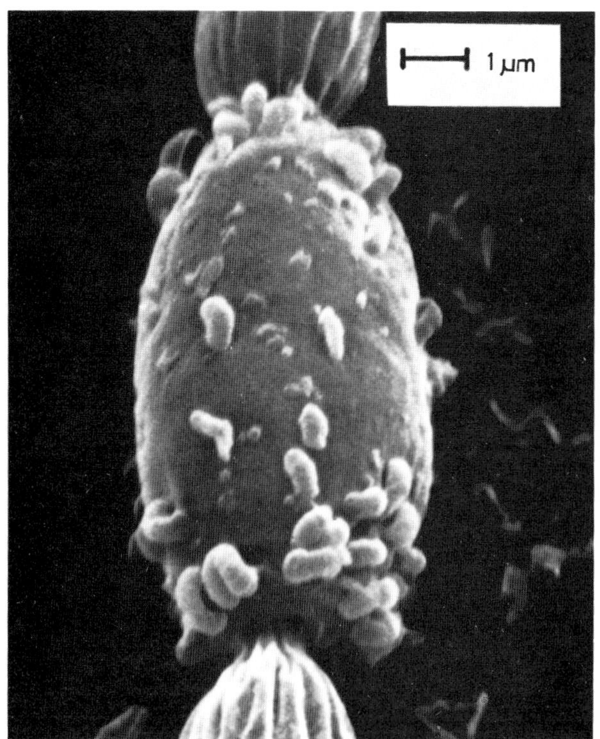


C

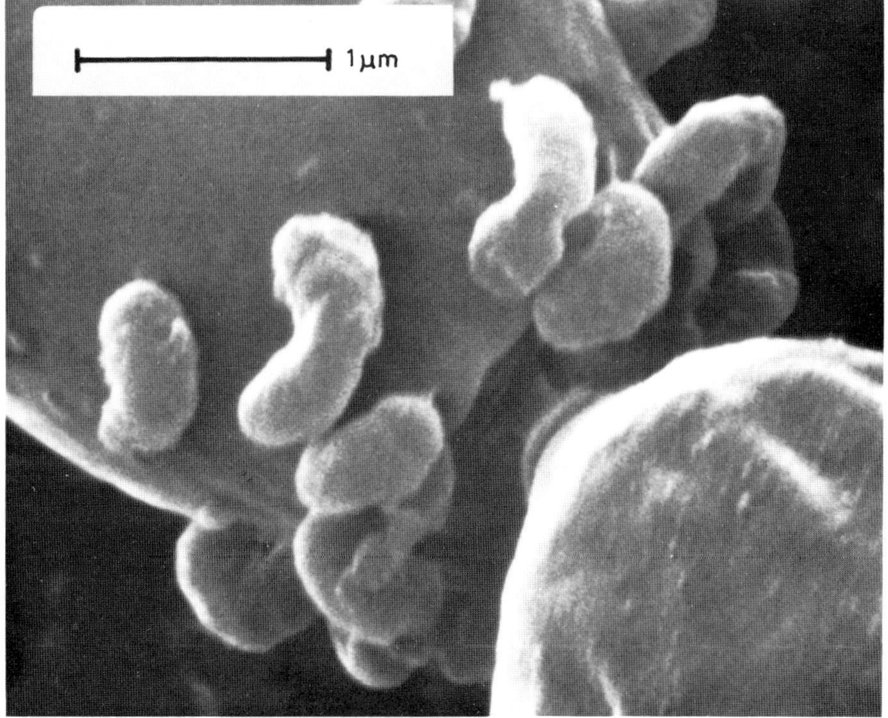


D

enriched samples fixed more $N_2$ than controls after recovery (Table 2). Throughout these experiments $O_2$-enriched samples maintained their elevated $O_2$ levels, when compared to controls.

Why $O_2$-enriched samples assumed higher $N_2$ fixation rates than controls following recovery remains unclear. Parallel $^{14}CO_2$ assimilation studies revealed a profound inhibition of photosynthetic carbon reduction, even after nitrogenase recovery (Fig. 4). These results appear to illustrate the Warburg effect (14), or photorespiration (12), where carbon fixation is inhibited by elevated oxygen levels. At this time light-generated reductant is channeled to alternate electron

Table 1. Nitrogenase activity in (a) *Anabaena oscillarioides* and *A. flos aquae* having associated bacteria, (b) free of bacteria (bacteria removed by sonification and filtration), and (c) associated bacteria alone. These results indicate that the nitrogenase activity is restricted to *Anabaena* and that bacteria stimulate algal $N_2$-fixation without contributing directly to $N_2$-fixation rates themselves

| Culture conditions | n moles $C_2H_4 ml^{-1}, hr^{-1}$ |
|---|---|
| *Anabaena oscillarioides* + bacteria | 1.21 |
| *Anabaena oscillarioides* - bacteria | 0.85 |
| bacteria alone | <0.01 |
| *Anabaena flos aquae* + bacteria | 0.67 |
| *Anabaena flos aquae* - bacteria | 0.58 |
| bacteria alone | <0.01 |

Table 2. Triplicate Kjeldahl analyses demonstrating the ability of *Anabaena oscillarioides* to increase fixed N content (above controls) in cultures receiving elevated $O_2$ levels. Cultures were incubated for 36 hrs under light at which time significant ($p < 0.01$) differences in fixed N could be detected by these analyses

| Sample | µg N.100 $ml^{-1}$ |
|---|---|
| Initial culture | 127.1 |
| $pO_2$ = 0.2 atm | 224.7 |
| " " | 226.1 |
| " " | 215.6 |
| $pO_2$ = 0.4 atm | 283.5 |
| " " | 265.2 |
| " " | 275.8 |

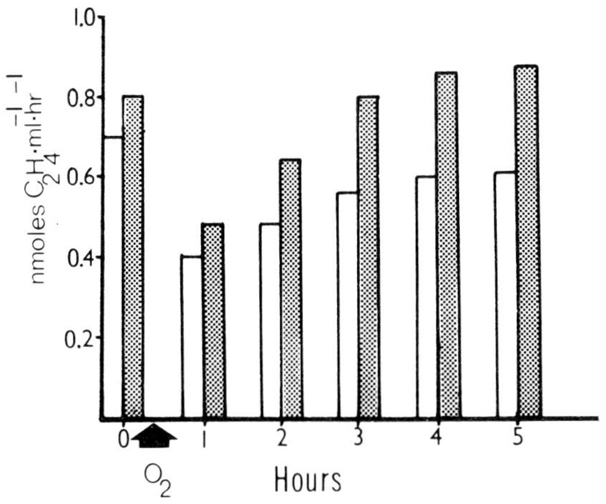


Fig. 2. Recovery of nitrogenase activity in _Anabaena oscillarioides_ following the introduction of elevated oxygen levels ($pO_2$ = 0.4 atm). The arrow indicates the time of oxygenation. In this experiment the rate of recovery of _Anabaena_ in the presence of bacteria was compared to axenic cultures. Bacterial cultures normally showed more rapid recovery than axenic cultures, often exceeding original rates of nitrogenase activity after 4 h

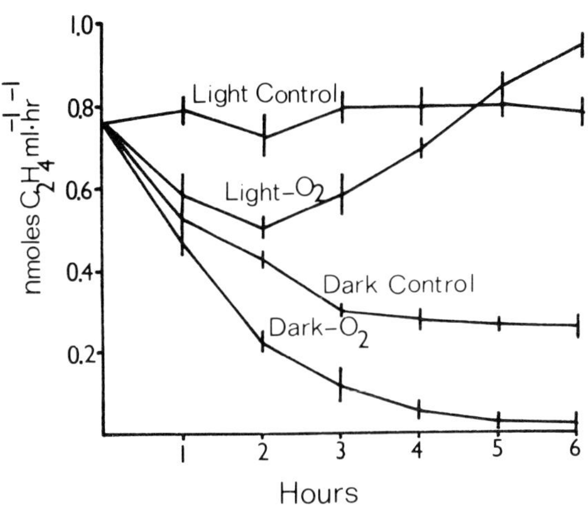


Fig. 3. Light-mediated nitrogenase activity recovery in axenic _Anabaena flos aquae_ cultures supplied with elevated oxygen levels ($pO_2$ = 0.4 atm). Dark recovery appears absent in both this species and _A. oscillarioides_ (not shown in this figure)

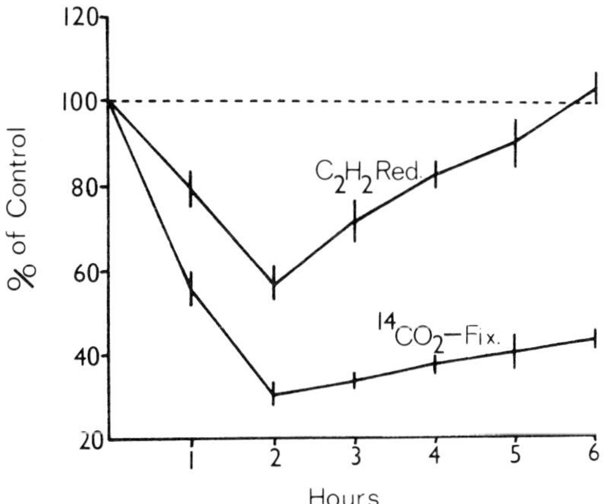


Fig. 4. Comparative inhibition and recovery of nitrogenase activity ($C_2H_2$ reduction) and photosynthetic $^{14}CO_2$ fixation in _A. oscillarioides_ following the introduction of elevated $O_2$ levels ($pO_2$ = 0.4 atm). Treatments were assessed relative to controls ($pO_2$ = 0.2 atm) throughout this experiment. The dotted line represents control relative to treatment values

acceptors such as $O_2$ (12). We suggest that during such periods, reductant may be shunted to nitrogenase as well, maintaining optimal $N_2$ fixation rates.

Observations of NBT deposition on $O_2$-enriched Anabaena filaments undergoing nitrogenase recovery indicate that reduced conditions become more prevalent in photosynthetic cells following oxygenation. Approximately 2 hrs after $O_2$-enrichment, vegetative cells began showing NBT formazan deposition. At roughly the same time nitrogenase activity showed a marked recovery, becoming higher than controls within 4 hrs. The pattern of NBT deposition was distinct from controls, which only showed terminal cells and cells bordering heterocysts as having NBT formazan deposition. It is not clear if these reduced conditions indicate a transfer of potential reductant from vegetative cells to the heterocysts; alternatively such conditions may be reduced enough to stimulate previously inactive nitrogenase in vegetative cells, allowing nitrogen fixation to take place there as well.

In conclusion, Anabaena possesses at least two ways, in addition to heterocyst formation, of protecting the highly reduced nitrogenase enzyme complex from the end product of its own photosynthetic system, $O_2$. Additional protective and recovery mechanisms may also exist. At present it appears that both the generation of reductant and close association with heterotrophic bacteria are important in allowing Anabaena to exploit oxygenated surface waters rich in radiant energy, needed to power both $CO_2$ and $N_2$ reduction. Decreases in algal $CO_2$ reduction during $O_2$ supersaturated daylight hours can stimulate photorespiration (12), the functional significance of which remains obscure. However, if radiant energy can be diverted from $CO_2$ reduction to $N_2$ reduction during such events, as appears to be the case here, Anabaena is making optimal use of this readily available source of energy. We therefore emphasise that the formation of surface scums does not necessarily signify the onset of scenescence. On the contrary, algae present in such scums may be generating photoreductant used in $N_2$ fixation processes, while competitively shading other photosynthetic $N_2$-fixing algae.

## D. Acknowledgements

We thank Mrs J. Simmiss for assistance with the manuscript and Drs A.B. Viner and E. White for helpful discussions. The scanning electron microscopy work was carried out at the Forest Research Institute, Rotorua, New Zealand and the Facility for Advanced Instrumentation, University of California, Davis Campus.

## E. References

1. Bergersen, F.J.: The effects of partial pressure of oxygen upon respiration and nitrogen fixation by soybean root nodules. J. Gen. Microbiol. 29, 113-125 (1962).
2. Dilworth, M.J.: Dinitrogen fixation. Ann. Rev. Plant Physiol. 25, 81-114 (1974).
3. Flett, R.J., Hamilton, R.D. & Campbell, N.E.R.: Aquatic acetylene-reduction techniques: solutions to several problems. Can. J. of Microbiol. 22, 43-51 (1976).
4. Fogg, G.E.: The production of extracellular nitrogenous substances by a blue-green alga. Proc. R. Soc. London Ser. B. 139, 372-397 (1952).
5. Fogg, G.E. & Pattnaik, J.: The release of extracellular nitrogenous products by Westiellopsis prolifica. Phykos 5, 58-67 (1966).
6. Ganf, G.G. & Horne, A.J.: Diurnal stratification, photosynthesis, and nitrogen-fixation in a shallow equatorial lake (Lake George, Uganda) Freshwat. Biol. 5, 13-39 (1975).

7. Granhall, U. & Lundgren, A.: Nitrogen fixation in Lake Erken. Limnol. Oceanogr. 16, 711-719 (1971).

8. Paerl, H.W.: Specific associations of the bluegreen algae Anabaena and Aphanizomenon with bacteria in freshwater blooms. J. Phycol. 12, 431-435 (1976).

9. Paerl, H.W. & Shimp, S.L.: Preparation of filtered plankton and detritus for study with scanning electron microscopy. Limnol. Oceanogr. 18, 802-805 (1973).

10. Stewart, W.D.P., Fitzgerald, G.P., & Burris, R.H.: In situ studies on $N_2$-fixation using the acetylene reduction technique. Proc. Nat. Acad. Sci. 58, 2071-2078, (1967).

11. Stewart, W.D.P. & Pearson, H.W.: Effects of aerobic and anaerobic conditions on growth and metabolism of bluegreen algae. Proc. Roy. Soc. London Ser. B. 175, 293-311 (1970).

12. Tolbert, N.E.: Photorespiration. In: Algal Physiology and Biochemistry. Botanical Monographs V.10. Stewart, W.D.P. (ed.) London: Blackwell Scientific Publications, Ltd., pp. 474-504 (1974).

13. Walsby, A.E.: The extracellular products of Anabaena cylindrica Lemm. I. Isolation of macromolecular pigment-peptide complex and other components. Br. Phycol. J. 9, 371-381 (1974).

14. Warburg, O. Über die geschwindigklet der photochemischen kohlensaurezersetzung in lebenden zellen. Biochem. Z. 103, 188-217 (1920).

# Taxonomy and Ecology of Algae in Fishponds and Fishpens of Laguna de Bay, Philippines

M.R. MARTINEZ

## A. Introduction

The use of algae as natural feed for fishes has been extensively practised in the Philippines (3, 8, 9). Several studies have been conducted on algal composition of fishponds as they relate to the productivity of milkfish (Chanos chanos Forsk.) (1, 4, 6-9). No extensive algal taxonomy and ecology however, has been done on fishponds and fishpens.

Fishponds are small bodies of water made by excavating a piece of land, while fishpens are innovative, net-enclosed fish-rearing cages set in a bigger body of water, e.g., a lake. The phytoplankton serve as the natural feed, but artificial food such as rice bran, chopped vegetables and shrimp are often added.

## B. Materials and Methods

The fishponds and fishpens are situated on hard, muddy ground in Los Banos, Laguna, Luzon, Philippines, between $14^{o}$ 12.9' N latitude and $121^{o}$ 10' E longitude, about 60 kilometers southeast of Manila.

Samples of surface water were collected in 1 l plastic bags bi-weekly from January to December, 1974. Surface water temperature was recorded at the collecting sites, while pH and conductivity measurements were made in the laboratory with a Corning pH meter and a Wheatstone Conductivity Bridge Model no. 31 with a pipette conductivity cell, respectively. Algal counts were carried out within 4 h of sampling and population densities estimated in a haemacytometer as outlined (2). The recorded monthly algal counts are averages of 12 samplings from two fishponds and two fishpens.

## C. Results and Discussion

There were 156 species of algae identified in the fishponds and fishpens which included: 27 Euglenophyta, 24 Cyanophyta, 51 Chlorophyta, 3 Pyrrophyta, and 51 Chrysophyta. These organisms will be described in later papers.

The temperature, conductivity and pH readings of the study areas exhibited spatial and periodic fluctuations. The conductivity readings recorded in microohms $cm^{-1}$ (EC x $10^{6}$ at $25^{o}C$) reflects salinity (5) and was the most changeable factor with values varying as much as 500 units; pH and temperature were less variable. Temperature and conductivity readings were closely related; low conductivity measurements were obtained from September to December (526-899, EC x $10^{6}$ at $25^{o}C$) when the temperatures were relatively high (26.5 - $30.9^{o}C$). Thus, solar radiation probably affected conductivity or salinity by increasing evaporation.

Values for the physico-chemical parameters in the fishponds were higher than those in the fishpens except for conductivity (Figs. 1 and 2). The conductivity values (EC x $10^{6}$ at $25^{o}C$) in the fishpens ranged from 542.55 to 1,210.0 while the fishponds had readings ranging from 505.48 to 715.50. Occasional higher conduct-

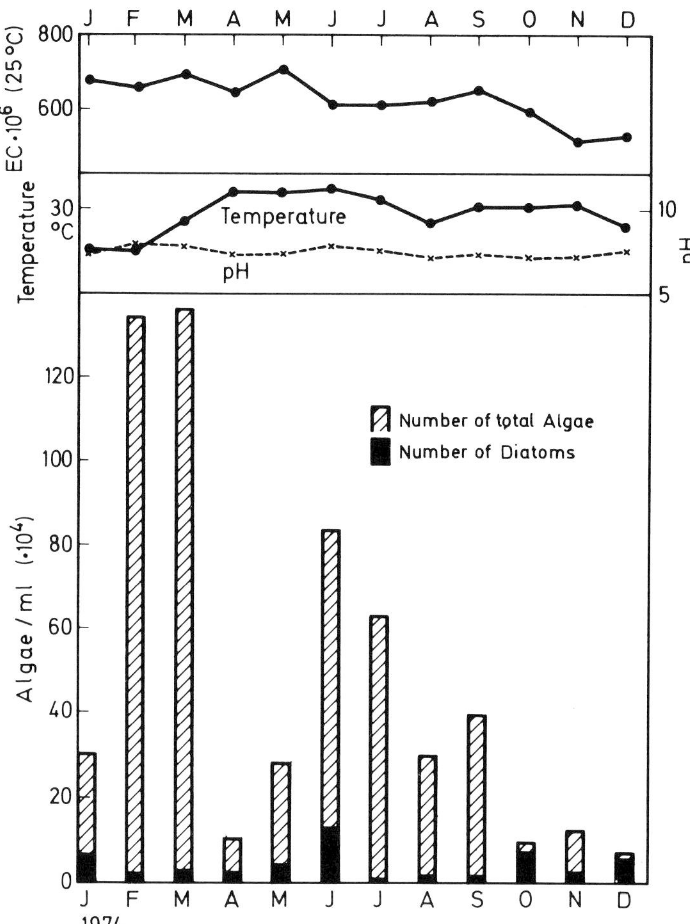


Fig. 1. Fluctuation of algal population with pH, temperature and conductivity readings in the surface waters of the fishponds, Laguna de Bay

ivity readings noted in the fishponds may be due to the disturbance of the soft sediments of the ground by weeding or the presence of supplementary feed; high conductivity measurements in the fishpens may be the result of mixing with water of Laguna de Bay and the intrusion of nutrients from the surrounding sewage disposal systems and some agricultural activities.

Temperature values for both rearing places ranged from 25-32.6$^{o}$C. The average pH recorded were 7.47 and 7.30 for the fishponds and fishpens, respectively.

The fishponds had higher species diversity, but lower population densities than the fishpens. Of the 156 species identified, 153 were recorded in fishponds. There were two peaks of algal counts, one in February to March and the other in June through September (Fig. 1). The highest population was recorded in March (1.39 x $10^6$ cells $ml^{-1}$) and ninety percent was attributed to two species of Spirogyra, while forty percent of the total counts in June (0.83 x $10^6$ cells $ml^{-1}$) was due to Dictyosphaerium pulchellum Wood. As the number of total algae decreased however, from October to December the diatoms increased relatively in density. The common diatoms observed were of the pennate forms, e.g., Navicula, Fragilaria, Nitzschia, Pinnularia and Synedra spp.

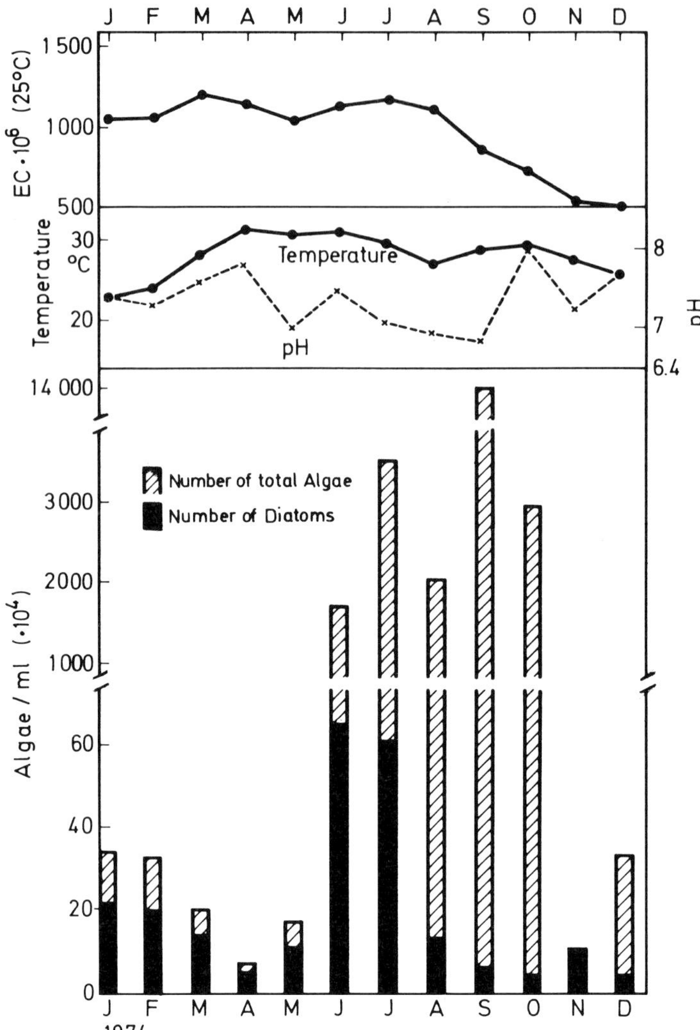


Fig. 2. Fluctuation of algal population with pH, temperature and conductivity readings in the surface waters of the fishpens, Laguna de Bay

Simple and multiple regression analysis showed that algal distribution, taking all independent variables individually and together, was affected more by pH except percent occurrence of diatoms which varied more with conductivity readings. There was a direct linear relationship between total algae and pH by up to 50%, while change in the relative abundance of diatoms was an inverse function of conductivity readings.

Ninety-eight algal species recorded in the fishpens were mostly diatoms belonging to the order Centrales, e.g., Cyclotella, Coscinodiscus and Melosira species. The number of total algae in the fishpens began to increase in June, reached its peak in September and declined in November (Fig.2). Increase in population coincided with increase in density of Microcystis aeruginosa Kuetz. emend. Elenkin (140 x $10^6$ cells $ml^{-1}$ in September), and a decline in the percentage of diatoms from 66.55% in May to 0.40% in September. There was a greater number of individuals within the species or division observed in the fishpens than in the fishponds which may be due to greater stability of the physico-chemical parameters in the former.

Among the parameters observed in the fishpens, only pH affected the change of the number of total algae with time by about 49%. There was an inverse relationship between the two factors at pH 6.5-7.75 but as the pH went up the number of algae showed direct relationship with pH.

There seems to be a greater effect of pH on the distribution of algae in waters of both the fishponds and fishpens. Probably the most important effect of pH is on the form carbon was made available to the organism.

## D. References


1. Esguerra, R.S.: Enumeration of algae in Philippine bangus fishponds and in the digestive tract of the fish with notes on conditions favorable for their growth. Phil. J. Fisheries. 1, 171-192 (1957).
2. Martinez, M.R., Chakroff, R.P., Pantastico, J.B.: Notes on direct phytoplankton counting technique using the haemacytometer. Phil. Agric. 59, 43-45 (1975).
3. Megia, T.G.: Plankton as an aquatic food resources. Agric. Industrial Life. 1, 5-6 (1952).
4. Rabanal, H.R. & Montalban, H.: The growing of algae or "lumot" in bangus fishponds. Fisheries Gazette. 5, 3-21 (1961).
5. Richards, L.A.: (ed.) Diagnosis and Improvement of Saline and Alkali Soils, USDA, Washington, D.C. Agricultural Handbook No.60, 1954, pp. 80-91.
6. Ronquillo, I.A. & de Jesus, A.: Notes on growing of lab-lab in bangus nursery ponds. Phil. J. Fisheries. 5, 99-102 (1957).
7. Rosel, R.S. & Arguellas, A.S.: Soil types and growth of algae in bangus fishponds. Phil. J. Science. 61, 1-8 (1936).
8. Vicencio, Z.T.: Studies on algal food habits of milkfish, Chanos-chanos Forsk. Unpubl. M.S. Thesis, U.P. Diliman, Quezon City (1964).
9. Villadolid, D.V. & Villaluz, D.K.: A preliminary study on bangus cultivation and its relation to algal culture in the Philippines. Dept. Agric. Nat. Res. Popular Bulletin No.20 (1950).

# Biodegradation in Aquatic Environments

# The Formation of Oil Emulsifying Agents in Hydrocarbonclastic Bacteria

G.D. FLOODGATE

## A. Introduction

The rate of degradation of oil spilt in the marine environment depends upon a large number of factors, including the state of dispersion of the oil (4). The degree and kind of emulsion formed depends upon the conditions prevailing when the oil is spilt. Among these is the presence of emulsifying substances capable of reducing the interfacial tension, so that small stable droplets are formed. Natural emulsifiers from biological sources including micro-organisms have been known for a long time (6), and are present in the surface film of the sea (5). This paper investigates the properties of some natural emulsifiers from bacteria.

## B. Methods and Materials

I. Organisms. The original mixed culture was derived from an enrichment growing on crude Kuwait oil in sand taken from a beach at Newborough, Anglesey, Wales. Two hydrocarbonoclastic strains were isolated, and maintained on 0.3% peptone sea water agar. Both were gram negative. One, a motile rod, formed a mucoid colony (1-2 mm) and was labelled BOB. The other, a nonmotile rod, formed a smaller slightly yellow colony and was labelled SOB.

II. Apparatus and chemicals. Glassware was cleaned by soaking in 3N hydrochloric acid, followed by distilled water and then deionised water. For gas liquid chromatography (GLC) a Perkin-Elmer F11 dual column chromatograph was used as described (3). All chemicals used were analytical grade.

III. Sea water. One batch of aged sea water was used throughout the series of experiments. Fortified sea water was made from aged sea water by adding $10^{-2}$M potassium nitrate and 7 x $10^{-4}$M disodium hydrogen phosphate; these concentrations having been found (1) to give the maximum rate of degradation of crude oil. The pH was adjusted to 7.2 - 7.6.

IV. Measurement of emulsification. The method of assessing the degree of dispersion was based on the "swirling beaker test" (2).

## C. Results

I. The presence of surfactants in aged sea water. It might be expected that the ageing of sea water would destroy any surfactants present. When, however, 0.2 ml of Forties crude oil was shaken overnight in 100 ml of aged sea water at 25$^{o}$C at 150 revolutions min$^{-1}$ on a rotary shaker, the oil spread as a surface film. In the control flask, containing 100 ml of deionised water and 0.2 ml of oil, a number of tar balls were formed.

When aged sea water was compared with freshly collected sea water from the Menai Straits, using the same amount of Forties oil, it was found that both flasks had surface films of oil after shaking for 24 h. Similar results were obtained with Ekofisk and Kuwait crudes.

II. Emulsification of several crude oils by mixed and pure cultures.
1. Mixed cultures. In order to examine the emulsifying effect of the mixed culture on a number of crudes, a series of Erlenmyer flasks containing 100 ml of fortified sea water sterilised by filtration were prepared. Half the flasks were inoculated with 0.1 ml of the mixed culture, the rest remained sterile. Either Kuwait, Ekofisk, Brega or Louisiana crude oil (0.2 ml) was added to each of two flasks. All flasks were incubated on a rotary shaker at 150 revolutions $min^{-1}$ at $25^{o}C$. After 20 h, all the inoculated flasks showed a fair degree of emulsification, which by 48 h was complete. The sterile controls were unchanged except for the Kuwait oil which had formed tar balls.

The experiment was repeated using a weighed quantity of Brega oil which was recovered after 3 d using dichloromethane, dried with anhydrous sodium sulphate and weighed. The measured oil was examined by GLC. It was concluded that although the alkanes had diminished in concentration, at this stage about 90% of the oil was still present. It would appear that the bacteria had grown at the expense of the alkanes and had formed surfactants which had emulsified the rest of the oil.
2. Pure cultures. Similar experiments with pure cultures of BOB and SOB showed that both organisms rapidly emulsified Forties and Brega crude oils with a similar pattern of results to the mixed cultures.

III. Formation of emulsifying agents on substrates other than oil. To demonstrate that the emulsifying agents were formed on substrates other than crude oil, a series of media were prepared containing 0.5% of various substrates in fortified sea water. In the case of the amino acids and casamino acids a concentration of 0.1% was used. All the media were sterilised by filtration and inoculated with a 24 h culture of BOB or SOB. They were then incubated on a rotary shaker at 150 revolutions $min^{-1}$ at $25^{o}C$ for a period of 7-9 d. The exception was hexadecane; in this case the flasks were incubated for about one month. The bacteria were removed by centrifugation and the spent media clarified by passage through a 0.45μ Millipore filter. To 100 ml of the filtrate, 0.2 ml of Forties crude oil was added and the presence of emulsifying agents detected by the swirling beaker test. The degree of emulsification was judged against a flask without filtrate.

In a number of cases, the bacteria which had been deposited by centrifugation were resuspended in 100 ml of sterile aged sea water, 0.2 ml of Forties crude oil added and examined for emulsifying ability against an aged sea water control. The results are given in Table 1. It will be seen that BOB was generally more effective than SOB in producing emulsification and there did not appear to be any relationship between the chemical structure of the substrate and the formation of emulsifiers. It can also be seen that the cells themselves were able sometimes to emulsify the oil to some extent.

When the emulsion from either the spent medium or the cells was examined microscopically, it was seen to be composed mainly of droplets with an average diameter of 10 μ. An incomplete emulsification however, was sometimes observed. Some of the oil formed globules which were approximately 1.0 mm in diameter. These appeared to contain water droplets suggesting that the globules consisted of a water in oil emulsion and were small tar balls.

IV. The effect of the emulsifier on the rate of oil degradation. In order to find out the effect of the presence in the sea water of preformed emulsifying agents, the organisms were first grown on a substrate known to promote production of substantial amounts of emulsifier e.g. gluconate or serine. The bacteria were removed by centrifugation and the spent medium resterilised by passage through a 0.22 μ millipore filter. Oil was then added to equal volumes of the spent and freshly prepared media. After inoculation with an equal amount of bacterial suspension the flasks were placed on the rotary shaker. At intervals one flask of spent medium and one of fresh medium were centrifuged to remove the bacteria, the oil separated by 10 ml of dichloromethane, dried with anhydrous sodium sulphate and examined in the gas chromatograph. It was noted that after one hour the oil in the spent medium was emulsified. After one day microscopic examination of the culture showed many more bacteria in the spent medium than in the fresh. After two days or more, the spent medium was obviously turbid while the fresh medium was not or only faintly so. GLC showed that the n-alkanes in the spent medium were utilised much more rapidly than in the fresh medium.

Table 1. Presence of emulsifier in spent medium and cells after growth on various substrates in fortified sea water

| | Organism | | | |
|---|---|---|---|---|
| | BOB | | SOB | |
| Substrate | cells | spent medium | cells | spent medium |
| Glucose | - | - | - | + |
| Fructose | N.T. | - | N.T. | - |
| Lactose | - | - | - | - |
| Glycerol | N.T. | + | N.T. | - |
| Mannitol | - | - | - | - |
| Starch | + | + | - | - |
| Acetate | + | + | n.a. | + |
| Tartrate | N.T. | + | N.T. | - |
| Gluconate | + | +++ | + | ++ |
| Hexadecane | N.T. | ++ | N.T. | - |
| Glycine | N.T. | - | N.T. | - |
| Alanine | - | - | + | - |
| Serine | - | +++ | - | - |
| Casamino acids | + | + | n.a. | + |

+++ No oil on the sides of the flasks; medium light brown and opaque.
++ Little oil on the sides of the flasks; a few mm. size droplets on the surface; most oil as minute droplets in light brown, opaque medium.
+ Some oil on the sides of the flasks; many mm size oil droplets on the surface; some cloudiness in the main body of the liquid.
- No different to the control.
N.T. not tested.
n.a. not available.

V. The chemical nature of the emulsifier. It is known that fatty acids are produced at an early stage in the oxidation of n-alkanes (4) and these are known to have surface active properties. In spite of many attempts, however, fatty acids were not detected. There was some indication that the emulsifier was polysaccharide in nature since much of the activity was precipitated by acetone, and was anthrone positive.

## D. References


1. Atlas, R.M., Bartha, R.: Degradation and mineralisation of petroleum in sea water; limitation by nitrogen and phosphorus. Biotech. Bioeng. 14, 309-318 (1972).
2. Benyon, L.R.: Oil spill dispersants. Workshop on oil spill clean up. Institute of petroleum. London. pp. 2-6 (1970).
3. Davis, S.J., Gibbs, C.F.: The effect of weathering on crude oil residue exposed at sea. Water Res. 9, 275-285 (1975).
4. Floodgate, G.D.: Biodegradation of hydrocarbons in the sea. In: Water Pollution Microbiology. Mitchell, R. (ed.). London-New York, Wiley & Sons, pp. 153-171 (1972).
5. Liss, P.S.: Chemistry of the Sea. Surface Microlayer. In: Chemical Oceanography. 2nd edition. Riley, J.P., Skirrow, G., (ed.) London-New York. Academic Press. pp. 193-243 (1975).
6. Zajic, J.E., Westlake, D.: Bioemulsifiers. C.R.C. Critical Reviews in Microbiology, 5, 39-66 (1976).

# An Assessment of the Biodegradation of Petroleum in the Arctic

R.M. ATLAS

## A. Introduction

Development of Arctic petroleum reserves is necessary for meeting world energy needs. However, extraction and transport of these petroleum resources may result in contamination of various Arctic ecosystems with petroleum hydrocarbons. There have already been numerous crude and refined petroleum spillages in northern ecosystems. The fate of such spilled oil will depend in large part on the characteristics of the ecosystem contaminated and on local climatological conditions.

Arctic ecosystems have several features that are unique to polar regions. Temperatures are very low; there are alternate seasons of no daylight and complete sunlight: aquatic ecosystems are covered with ice many months of the year: terrestrial ecosystems are underlain by permafrost.

This paper is based on studies on the fate of oil in several Arctic ecosystems including near and offshore marine environments, estuaries, freshwater lakes, and various types of soil.

## B. Materials and Methods

We have used a general approach for assessing petroleum biodegradation in the Arctic by exposing samples from representative Arctic ecosystems to petroleum hydrocarbons, enumerating the microbial populations, including hydrocarbon utilizers, measuring the formation of metabolic products and following composition changes in the exposed petroleum.

Exposure to petroleum hydrocarbons has been in vitro or in situ. Prudhoe crude oil has been used in our work. For in vitro experiments, samples collected from various ecosystems were incubated at $5^{o}C$. For in situ exposure, petroleum was spilled within boundaries. In soil, the boundary was plastic sheeting (5), in estuaries and lakes, plexiglas cylinders (2), and in nearshore marine environments, plastic lined flow-through chambers (4). In some cases, nitrogen and phosphorus fertilizers were added to assess nutrient limitations (3, 4).

Estuarine studies were conducted in Prudhoe Bay, site of a large Arctic oilfield, and Elson Lagoon near Pt. Barrow, Alaska. Marine studies were conducted in eastern Chukchi and western Beaufort Seas and included nearshore and offshore sites. Freshwater studies in several lakes near Pt. Barrow, Alaska and studies on a variety of soils near Pt. Barrow, typical of coastal tundra, including high center polygon, low center polygon and meadow tundra soils, were carried out.

Microbial populations were enumerated from these ecosystems before and after exposure to petroleum hydrocarbons. Estimation of hydrocarbon utilizers was made using oil agar media with 1% Prudhoe crude oil (1). Heterotrophic microorganisms were enumerated using marine agar 2216 for marine and estuarine ecosystems, and trypticase soy agar for soil and freshwater ecosystems. Incubation was at $5^{o}C$ for up to 1 month. Ratios of oil utilizers to heterotrophs were calculated (6).

Hydrocarbon biodegradation potentials were determined for marine and estuarine samples (Roubal and Atlas, manuscript in preparation). Ten ml samples were incubated with 50 μl of Prudhoe crude oil spiked with $^{14}C$ hexadecane (sp. act. 0.4 μCi $ml^{-1}$ oil) in 60 ml serum vials for 42 days at 5°C. The $^{14}CO_2$ produced was collected and quantitated by liquid scintillation counting.

Petroleum exposed in situ was recovered by solvent extraction with diethyl ether in separating funnels for aqueous samples and a Soxhlet apparatus for soil samples. Recovered oil was quantitated by gravimetric analysis. Changes in composition of the oil were measured by gas chromatography for specific components and by column chromatography and mass spectrometry for structural classes of hydrocarbons (4).

## C. Results and Discussion

Higher numbers of naturally occurring hydrocarbon utilizing microorganisms were found in soil and sediment than water and ice in early summer (Table 1). A slightly lower percentage of the heterotrophs were hydrocarbon utilizers in offshore marine environments than in estuarine ecosystems. A higher percentage of heterotrophs were hydrocarbon utilizers in terrestrial ecosystems. Arctic plants contain relatively high concentrations of lipids and hydrocarbon compounds which contribute to their frost hardiness. These compounds become substrates for soil microorganisms. When Prudhoe crude oil was introduced during summer, both numbers of hydrocarbon utilizers and heterotrophs rose in all ecosystems examined.

Table 1. Comparison of occurrence of hydrocarbon utilizing microorganisms in several Arctic ecosystems

| ECOSYSTEM | Location | Sample Type | Estimated no of hydrocarbon utilizers[1] | Estimated % of heterotrophs[2] |
|---|---|---|---|---|
| Estuary | Prudhoe Bay | water | $1x10^{2}$ | 0.4 |
| | Elson Lagoon | water | $5x10^{0}$ | 0.05 |
| | Elson Lagoon | ice | $1x10^{-1}$ | 0.01 |
| | Elson Lagoon | sediment | $3x10^{3}$ | 0.1 |
| Marine | Offshore: | | | |
| | Pt. Barrow | water | $2x10^{0}$ | 0.01 |
| | Pt. Barrow | sediment | $2x10^{3}$ | 0.01 |
| Freshwater | Lake: | | | |
| | Pt. Barrow | water | $1x10^{1}$ | 0.5 |
| | Pt. Barrow | sediment | $1x10^{4}$ | 1.1 |
| Soil | Barrow | polygon top | $1x10^{4}$ | 3.0 |
| | Barrow | polygon trough | $3x10^{4}$ | 10.0 |
| | Barrow | meadow | $2x10^{5}$ | 10.0 |

[1] Enumerated at 4°C (no. $ml^{-1}$ or $g^{-1}$).

[2] Estimated percent of viable heterotrophs capable of hydrocarbon utilization enumerated at 4°C.

Table 2 shows the hydrocarbon biodegradation potentials based on $^{14}C$ hexadecane for estuarine and marine samples at two times of year. The biodegradation potentials were dramatically lower in September than April; during September they were almost nil and probably reflect a nutrient limitation. Available fixed forms of nitrogen are largely used up by phytoplankton during the Arctic summer and are very low in late summer. The results indicate that the fate of oil liberated in a spill in the Arctic will be dependent on the time of year, and estimates of rates of degradation may not be extrapolated from one month to the next.

Table 2. Hydrocarbon biodegradation potential in various aquatic Arctic ecosystems ($^{14}CO_2$ from $^{14}C$ hexadecane spiked Prudhoe crude oil)

| ECOSYSTEM | Location | Sample Type | Hydrocarbon Biodegradation Potential April | Hydrocarbon Biodegradation Potential September |
|---|---|---|---|---|
| Estuary | Prudhoe Bay | ice | 2000 | - |
| Estuary | Prudhoe Bay | water | 5200 | 60 |
| Estuary | Prudhoe Bay | sediment | 5300 | 170 |
| Estuary | Elson Lagoon | ice | 2200 | - |
| Estuary | Elson Lagoon | water | 3000 | 20 |
| Estuary | Elson Lagoon | sediment | 5100 | 130 |
| Marine | Offshore: | | | |
| Marine | Pt. Barrow | ice | 1700 | - |
| Marine | Pt. Barrow | water | 2200 | 60 |
| Marine | Pt. Barrow | sediment | 2100 | 80 |

Recovery of oil residues following in situ exposure in various Arctic ecosystems also indicates that short duration measures of degradation rates must be used with great caution in predicting the long term fate of petroleum hydrocarbons (Table 3). For example, during 1 month's exposure in a nearshore marine ecosystem, 22% of experimentally spilled oil was lost, an estimated removal rate of 264 g $m^{-2}$ $month^{-1}$. During the following 11 months, however only an additional 2% of the oil was lost, an estimated removal rate of only 288 g $m^{-2}$ $year^{-1}$. During the first month, most of the evaporative losses occur, causing an initially high loss rate. During many of the following months, low temperatures precluded significant degradation in surface waters. Addition of nitrogen and phosphorus fertilizers resulted in increased rates of oil disappearance indicating oil biodegradation is nutrient limited in this ecosystem. Degradation was further limited by available oil surface area.

Rates of oil loss were lower on ice than on open water at the same time of year. Even abiotic losses were low on, or under ice. In terrestrial ecosystems, soil type is important in rate of loss. In well drained polygon top soils, greater rates of loss were found during 1 year than in poorly drained polygon troughs. In meadow soils, 29% of oil spilled at 12 $lm^{-2}$ could be recovered in the upper 8 cm after 5 years. The loss rates include abiotic evaporation, vertical movement below 8 cm, and biodegradative losses.

Table 3. Oil losses (abiotic and biotic) from *in situ* exposure of Prudhoe crude oil in various Arctic ecosystems

| ECOSYSTEM | Location | Sample Type | Oil % | Loss $g\ m^{-2}$ | Exposure Time |
|---|---|---|---|---|---|
| Estuary | Prudhoe Bay | water | 60 | 108 | 2 months |
| Estuary | Elson Lagoon | water | 35 | 63 | 1 month |
| Estuary | Elson Lagoon | ice | 9 | 16 | 1 month |
| Marine | Nearshore: | | | | |
| Marine | Barrow | water | 22 | 264 | 1 month |
| Marine | Barrow | water | 24 | 288 | 1 year |
| Freshwater | Lake: | | | | |
| Freshwater | Pt. Barrow | water | 25 | 45 | 1 month |
| Soil | Barrow | polygon top | 54 | 5832 | 1 year |
| Soil | Barrow | polygon trough | 15 | 1620 | 1 year |
| Soil | Barrow | meadow | 71 | 7668 | 5 years |

Gas chromatographic analyses of the recovered oils generally showed initial rapid abiotic removal of the light hydrocarbons, followed by slow biodegradative losses. Both mass spectral and column chromatographic analyses of the residual oil generally showed exposed oil to have the same relative hydrocarbon class composition regardless of time of exposure. This implies that rates of degradation are proportional to the percent class composition in the oil.

In summary, microorganisms capable of oil biodegradation are present in all the ecosystems examined. The relative importance of hydrocarbon utilizers in the indigenous population varies greatly between ecosystems. When exposed to crude oil, the numbers of hydrocarbon utilizers increased and hydrocarbon utilizers became a dominant group in the heterotrophic population. Rates of degradative loss, however, varied greatly between ecosystems and between different times of year. Estimates of degradation based on short term exposures greatly overestimate rates of removal of petroleum hydrocarbons. Almost no degradative losses occur during winter in surface waters and soils; oil biodegradation is restricted to a few months of the year in these ecosystems. Even paraffins persist in Arctic ecosystems for many years. Rates of removal are certainly dependent on local climatological conditions, especially temperature, on available nutrients in the ecosystem, and on compositional and physical characteristics of the contaminating oil. Petroleum hydrocarbons may persist for decades following contamination of aquatic and terrestrial Arctic ecosystems.

## D. Acknowledgment

Portions of this work have been supported by the United States Office of Naval Research, Energy and Resources Development Administration, National Oceanographic and Atmospheric Administration and Bureau of Land Management. Contributions to this work have been made by M. Busdosh, A. Horowitz, G. Roubal and A. Sexstone.

## E. References

1. Atlas, R.M., Bartha, R.: Abundance, distribution and biodegradation potential of microorganisms in Raritan Bay. Environ. Poll. 4, 291-300 (1973).
2. Atlas, R.M., Schofield, E.A.: Petroleum biodegradation in the Arctic. In: Impact of the Use of Microorganisms on the Aquatic Environment. Bourquin, A.W., Ahearn, E.G., Meyers, S.P. (eds.). Corvallis, Ore.: Environmental Protection Agency, EPA 660/3-74-001, pp. 183-198 (1975).
3. Atlas, R.M., Busdosh, M.: Microbial degradation of petroleum in the Arctic. In: Proceedings of the 3rd International Biodegradation Symposium. Sharpley, J.M., Kaplan, A.M. (eds.). London: Applied Science Pub., pp. 79-86 (1976).
4. Horowitz, A., Atlas, R.M.: Continuous open flow-through system as a model for oil degradation in the Arctic Ocean. Appl. Env. Microbiol. 33, 647-653 (1977).
5. Sexstone, A., Atlas, R.M.: Mobility and biodegradability of crude oil in Arctic tundra soils. Dev. Ind. Microbiol. 18, 673-684 (1977).
6. Walker, J.D., Colwell, R.R.: Enumeration of petroleum degrading microorganisms. Appl. Environ. Microbiol. 31, 198-207 (1976).

# Anaerobic Breakdown Processes of Organic Matter in Freshwater Sediments

TH.E. CAPPENBERG, E. JONGEJAN, and J. KAPER

## A. Introduction

Anaerobic mineralization can be defined as a biological process in which organic matter is converted to methane and carbon dioxide, in the absence of molecular oxygen. This process can play an important role in the carbon cycle in aquatic environments. A part of this process involves the formation and breakdown of fatty acids.

Further information on the breakdown kinetics of lower fatty acids was needed in our work on anaerobic mineralization, as part of the studies on the carbon cycle in Lake Vechten. Acetate and lactate, for example, are intermediate metabolites in methanogenesis. Their pool sizes do not change appreciably in mud and, therefore, the rate of their dissimilation must equal the rate of their formation (4, 5, 7, 8). The dissimilation rate of an intermediate is the product of the turnover rate constant and its pool size. In this paper some experiments using radio gas chromatography as a tool to obtain accurate data on this process are described.

Data on the ecological relationships between bacterial species involved in the anaerobic mineralization were also obtained from mixed continuous culture studies. The chemostat provides a model system which can be used widely in ecological and physiological studies (6,16,17). The influence of the $H_2$-producing, sulfate-limited Desulfovibrio desulfuricans on the fermentation of limiting amounts of $H_2$ by Methanobacterium formicicum was investigated in mixed continuous cultures. A symbiotic relationship described supplements the existing information on the ecological interactions between bacteria in the breakdown of organic matter in sediments.

## B. Materials and Methods

I. Mud sampling. Undisturbed mud-cores were taken by means of a modified Jenkin surface-mud sampler from the deepest part (12 m) of Lake Vechten which is a slightly eutrophic sandpit (4.7 ha), and can be characterized as a seepage lake.

II. Incubation technique and analyses of $^{14}C$-labelled acetic acid, methane and carbon dioxide. The samples were incubated anaerobically at the sediment temperature (8 - 10$^{o}$C) in 1.5 ml Suba-Seal-stoppered vials placed in a shaking waterbath. The analysis of acetic acid in aqueous solution was performed on a Packard-Becker 417 gas chromatograph fitted with a flame ionization detector (FID) using helium as the carrier gas.

For analysis of $^{14}C$-acetic acid the effluent of the gas chromatographic column was divided by a 1:1 splitter into two streams: one leading to the FID, the other via a 1/6" o.d. stainless steel tube to a Packard 894 gas proportional counter. For further details on the analysis of the labelled compounds see (9).

III. Continuous culture studies. A Desulfovibrio desulfuricans strain and a Methanobacterium formicicum strain, both isolated from mud of Lake Vechten according to the roll-tube techniques, were used. The organisms were identified according to the criteria proposed for sulfate-reducing bacteria (15), and for methane-producing bacteria (20), respectively and were cultured as described (6). Culture experiments were conducted in all-glass, water-jacketed vessels fitted with multi-socked flat-flanged lids and Quick-fit glass components as described earlier (6).

## C. Results and Discussion

I. Turnover and breakdown of acetate and lactate. A mud sample (about 1 g wet weight) was removed from a mud core at a depth of 5 cm and transferred anaerobically to a 1.5 ml vial. At zero time 200 µl of 2-$^{14}$C-acetate (1 µCi ml$^{-1}$) were injected into the vial, the contents rapidly mixed, and incubated at 10$^{o}$C. At appropriate time intervals 25 µl aliquots of the supernatant were removed for radio gas chromatographic analysis.

The $^{10}$log of the cpm (counts per minute) in the acetate fraction when plotted against time gave a straight line with a correlation coefficient r = 0.955. From the regression coefficient a turnover rate constant (k) was calculated as k = 0.244 ± 0.015 hr$^{-1}$. A similar experiment with mud from a depth of 1 cm gave a k = 0.07 hr$^{-1}$. The turnover of an intermediate such as acetate is the product of its turnover rate constant and its concentration. The average acetate pool size during the experiment was 0.3 ± 0.05 mg l$^{-1}$ at 5 cm depth and 0.6 ± 0.05 mg l$^{-1}$ at 1 cm depth, giving for the depths turnovers of 0.07 mg l$^{-1}$ hr$^{-1}$ and 0.04 mg l$^{-1}$ hr$^{-1}$ respectively. The higher turnover of acetate found in the lower layers in the mud of Lake Vechten is in agreement with our earlier observation in that the maximum abundance of acetate-fermenting methanogenic bacteria occurs at a depth of 4-6 cm (4, 5).

The breakdown of acetate was measured in a number of experiments in which mud was incubated at the environmental temperatures and in which $^{14}CH_4$ and $^{14}CO_2$ produced from 1-$^{14}$C-acetate, 2-$^{14}$C-acetate or U-$^{14}$C-acetate were analysed. From the results summarized in Table 1 it can be concluded that of the methane formed from acetate 84% originates from the methyl carbon and the rest from the carboxyl carbon. On the other hand, of the carbon dioxide formed from acetate 78% originates from the carboxyl carbon and the rest from the methyl carbon. The formation of methane from the carboxyl carbon of acetate is probably due to $H_2/CO_2$-fermenting methanogenic bacteria which can use the $CO_2$, produced by acetate-splitting bacteria, as a substrate.

It is not yet known what kind of organisms play a role in the anaerobic conversion of the methyl carbon of acetate into $CO_2$. By radio gas chromatography we found that incubation with $CH_4$ of mud samples or of pure cultures of Desulfovibrio desulfuricans does not result in formation of $CO_2$ (9). Similar results have been reported (18). It is possible that acetate is converted into two molecules of $CO_2$ by the anaerobic sulfur-reducing and acetate-oxidizing Desulfuromonas acetoxidans, recently isolated organism (14). If this occurs in the mud of Lake Vechten the following scheme of the breakdown of acetate may be described:

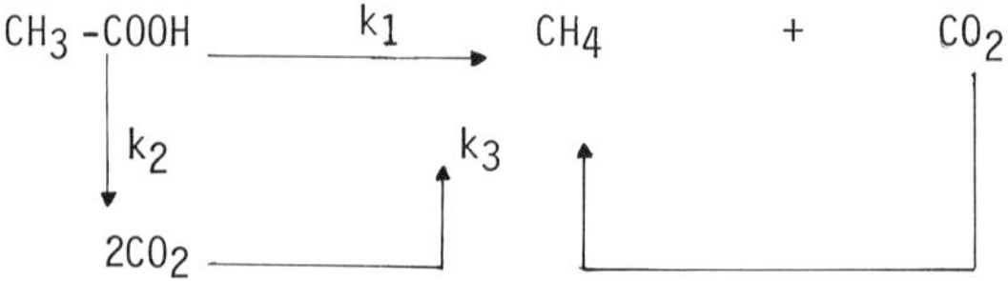


For specifically labelled substrates this leads to the following two relations: with carboxyl carbon-labelled acetate;

$$_1F_{CO_2} = \frac{^{14}CO_2}{^{14}CO_2 + {^{14}CH_4}} = \frac{k_1 + k_2}{k_1 + k_2 - k_3} \cdot \frac{e^{-k_3t} - e^{-(k_1+k_2)t}}{1 - e^{-(k_1+k_2)t}} \quad (1)$$

with methyl carbon-labelled acetate;

$$_1F_{CO_2} = \frac{^{14}CO_2}{^{14}CO_2 + {^{14}CH_4}} = \frac{k_2}{k_1 + k_2 - k_3} \cdot \frac{e^{-k_3t} - e^{-(k_1+K_2)t}}{1 - e^{-(k_1+k_2)t}} \quad (2)$$

From these relations it follows that $F_{CO_2}/_2F_{CO_2} = 1+k_1/k_2$; and from Table 1, that at 5 cm depth $_1F_{CO_2}/_2F_{CO_2} = 0.86/0.24 = 3.6$: thus leading to $k_1/k_2 = 2.6 \pm 0.3$. This together with the turnover rate constant at 5 cm depth ($k_1 + k_2 = 0.244\ hr^{-1}$) indicates a value of $k_1 = 0.068 \pm 0.009\ hr^{-1}$, and of $k_2 = 0.18 \pm 0.02\ hr^{-1}$. A value of $k_3$ can be found by substitution in (1) of the known values for $_1F_{CO_2}$ and $k_1 + k_2$; thus a $k_3 = 0.24\ hr^{-1}$ is obtained. Alternatively, $k_3$ can be determined from an experiment wherein mud is incubated with $^{14}CO_2$. For the mud at 1 cm depth similar results may be obtained with $_1F_{CO_2}/_2F_{CO_2} = 0.67/0.32 = 2.2$ (Table 1), and with the turnover rate constant of $k_1 + k_2 = 0.07\ hr^{-1}$, indicating the following k values: $k_1 = 0.032\ hr^{-1}$; $k_2 = 0.38\ hr^{-1}$; $k_3 = 0.66\ hr^{-1}$.

The label distribution between $^{14}CH_4$ and $^{14}CO_2$ in relation to the depth in the mud was studied. Samples from different depths from one mud column were incubated with 2-$^{14}$C-acetate (0.2 μCi/ml) and the $^{14}CH_4$ and $^{14}CO_2$ formed, were analysed radiogas chromatographically.

From Fig. 1 it is clear that oxidation of the methyl carbon of acetate to $CO_2$ is more important in the upper layers of the mud. Therefore, it seems likely that part of the carbon of acetate is converted to $CO_2$ by anaerobic sulfur-reducing organisms, since by oxidation of FeS free sulfur is available in the upper layers

Table 1. Label distribution between $^{14}CO_2$ and $^{14}CH_4$ from $^{14}$C-acetate in mud from a depth of 5 and 1 cm in the mud core, respectively

| Substrate | $F_{CO_2} = \frac{^{14}CO_2}{^{14}CO_2 + {^{14}CH_4}}$ -5cm | -1cm |
|---|---|---|
| 1-$^{14}$C-acetate | 0.86 ± 0.03 | 0.67 ± 0.03 |
| 2-$^{14}$C-acetate | 0.24 ± 0.01 | 0.31 ± 0.01 |
| U-$^{14}$C-acetate[a)] | 0.57 ± 0.01 | 0.51 ± 0.01 |

a) U-$^{14}$C-acetate is a 1:1 mixture of 1-$^{14}$C- and 2-$^{14}$C-acetate. Therefore the calculated values of $F_{CO_2}$, based on the values from the specifically labelled acetates, are (0.86 + 0.24)/2 = 0.55 ± 0.02 and (0.67 + 0.31)/2 = 0.49 ± 0.02 respectively, which is in excellent agreement with the observed values of 0.57 ± 0.01 for a depth of 5 cm and 0.51 ± 0.01 for a depth of 1 cm.

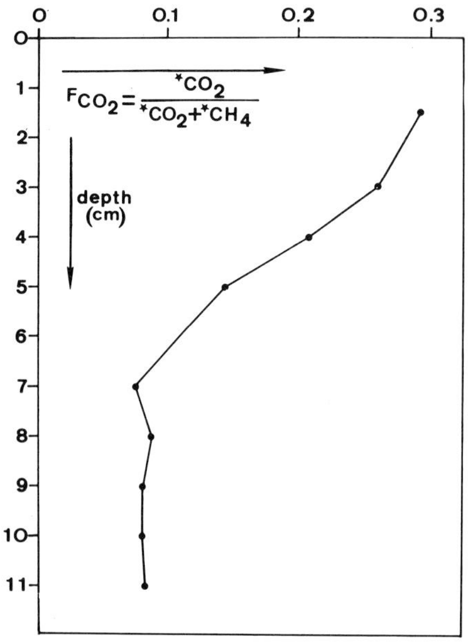


Fig. 1

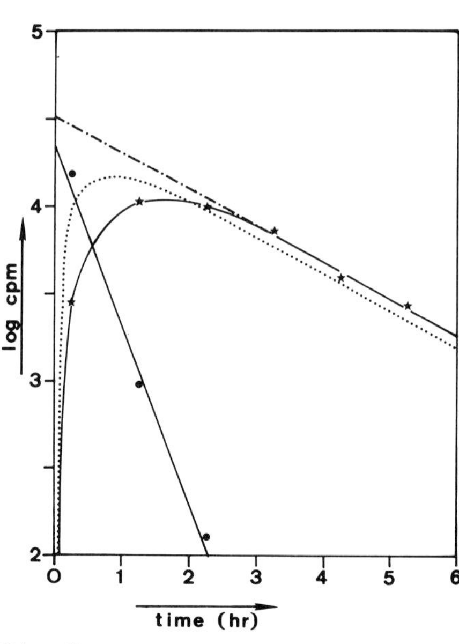


Fig. 2

Fig. 1. Label distribution between $CO_2$ and $CH_4$ ($F_{CO_2} = \frac{*CO_2}{*CO_2 + *CH_4}$) in relation to the depth in the mud, after incubation of mud samples with 2-$^{14}$C-acetate during 1 h at 10 C

Fig. 2. Breakdown of Na-U-$^{14}$C-L-lactate in mud. ●——● log (cpm) in the lactate fraction; *——* log (cpm × 1.5) in the acetate fraction; -·-·-·- tangent to the acetate curve; ······· calculated curve using Eq. (4)

of the mud in Lake Vechten. Moreover addition of sulphate to the mud stimulated this (7, 18). Since the turnover rate constants, pool sizes and breakdown pathways of acetate are depth dependent generalizations regarding the turnover and breakdown of acetate in mud of Lake Vechten are rather difficult. One should be careful in drawing conclusions from the label distribution between $CH_4$ and $CO_2$ from the anaerobic breakdown of complex labelled organic substrates such as glucose and cellulose in respect to intermediates in methanogenesis (e.g. 13, 18, 19).

The turnover of lactate was measured as disappearance of U-$^{14}$C-lactate from mud samples (for methods see 7, 8). In Fig. 2 the log of cpm of the acetate and lactate fractions is plotted against time (the cpm value of acetate is multiplied by 1.5 because only 2/3 of the lactate radioactivity can theoretically be present in the acetate fraction).

Apparently lactate is broken down according to first order kinetics with k=2.37 $hr^{-1}$. The acetate curve shows a rapid formation during the first hour and a slow decrease later on, with a maximum occurring after approx. 2 hr.

If the lactate breakdown can be described according to:

$$\underset{A}{\text{lactate}} \xrightarrow{k_1} \underset{B}{\text{acetate}} \xrightarrow{k_2} CH_4 + CO_2$$

then equations (3) and (4) are valid:

$$A/A_o = e^{-k_1 t} \qquad (3)$$

$$B/A_o = \frac{k_1}{k_1 - k_2} \cdot (e^{-k_2 t} - e^{-k_1 t}) \qquad (4)$$

with $A_o$ as the lactate activity at t=0. When $k_1 > k_2$, eq. (4) after some lapse of time is reduced to eq. (5):

$$B/A_o = \frac{k_1}{k_1 - k_2} \cdot e^{-k_2 t} \qquad (5)$$

This means that after some time the acetate curve in Fig. 2 becomes a straight line with a slope of - 0.0434 $k_2$. As can be seen this is indeed the case. From the tangent to the acetate curve (- - - - - -) we calculate $k_2 = 0.48\ hr^{-1}$. When these values for $k_1$ and $k_2$ are substituted however, into eq. (4) the resulting curve (........) does not fit the observed values. Notably the maximum is too high and occurs too early. Therefore we may conclude that the breakdown of lactate is more complicated than in this simple model. Other products are formed simultaneously, e.g. propionic and butyric acid, so further investigation of this problem is necessary.

The gas chromatography- gas proportional counting system described here provides an accurate tool for the simultaneous analysis of turnover rate and breakdown of lower fatty acids in bottom deposits. A substrate relationship between sulfate-reducing and methane-producing bacteria is evident since labelled methane is obtained from mud incubated with labelled lactate.

II. Continuous culture experiments. Further information on the role of these bacteria in the non-methanogenic and the methanogenic phases in mineralization can be obtained from continuous culture studies. It is well known that part of the methanogenesis in sediments is dependent on non-methanogenic species for their supply of $H_2$. Sulfate reducers are capable of producing $H_2$ from the oxidation of lactate, via pyruvate, when growing on limiting amounts of electron-acceptor sulfate (10). *Methanobacterium formicicum*, which grows only formate or $H_2 + CO_2$ as substrates, can act as a natural sink of the produced $H_2$ and thus maintain a low partial pressure of hydrogen in the ecosystem. This metabolic coupling is manifested as interspecies hydrogen transfer of electrons in the form of molecular hydrogen from $H_2$-producing heterotrophs to a $H_2$-utilizing terminal electron acceptors such as methanogenic bacteria.

Two chemostats, one containing *Desulfovibrio* under sulfate-limitation and the other containing *Methanobacterium* under formate-limitation, were operated over a range of dilution rates ($D < D_c = 0.32\ hr^{-1}$ and $D < D_c = 0.04\ hr^{-1}$, respectively, for *Desulfovibrio* and *Metanobacterium*) to provide a source of bacteria for a third growth-vessel. The behaviour of the two bacterial species together was examined by establishing a number of steady-state mixed population levels, and by measuring the concentrations of bacteria and substrates in all the stages (Table 2).

In the mixed growth-vessel the population of *Methanobacterium* increased, and that of *Desulfovibrio* remained more or less constant. The concentration of $H_2$ decreased, whereas that of methane increased stoichiometrically. This indicates that $H_2$ produced by *Desulfovibrio* is used by *Methanobacterium*, so that a symbiotic relationship does exist between the two species. This is in agreement with a previous observation (2), in which batch culture growth of *D. desulfuricans* without sulfate was inhibited when small amounts of $H_2$ from lactate or ethanol were produced; on the other hand, in the presence of a $H_2$-using methanogenic bacterium good growth

Table 2. Oxidation of lactate by *D. desulfuricans* under sulfate-limitation ($D=0.03\ hr^{-1}$, $s_r=2.5$ mmol $SO_4^{2-}$) without and with *M. formicicum* under formate-limitation ($D=0.03\ hr^{-1}$, $s_r=2$ mmol formate)[a]

| | | cell number per l | $H_2S$ prod. mmol | $H_2$ prod. mmol | $CH_4$ prod. mmol | lactate used mmol | acetate prod. mmol |
|---|---|---|---|---|---|---|---|
| *Desulfovibrio* | alone | $1.7 \times 10^{11}$ | 2.4 | 23.9 | - | 6.7 | 7.1 |
| *Methanobacterium* | alone | $9 \times 10^{9}$ | - | - | 0.5 | - | - |
| *Desulfovibrio* | plus | $1.5 \times 10^{11}$ | 2.3 | - | - | 6.5 | 6.9 |
| *Methanobacterium* | | $7.5 \times 10^{11}$ | - | - | 6.7 | - | - |

*Desulfovibrio* alone

$$3\ \text{lactate} + SO_4^{2-} + H_2O \longrightarrow 3\ \text{acetate} + 3CO_2 + H_2S + 2H_2 + 2H_2O$$

*Methanobacterium* alone

$$4\ \text{formate} \longrightarrow CH_4 + 3CO_2$$

Together

$$3\ \text{lactate} + SO_4^{2-} + 4\ \text{formate} + 2\ H_2 + 1/2\ CO_2 \longrightarrow 3\ \text{acetate} + H_2S + 2CH_4 + 6CO_2 + 2H_2O$$

a) Data obtained in steady-state conditions, which were assumed to have been reached when there had been no significant change in cell number for a minimum of four volume changes.

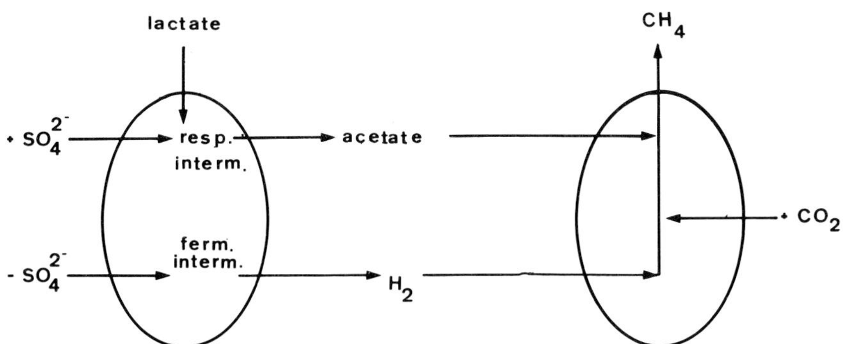


Fig. 3. Schematic representation of the possible relationships between *Desulfovibrio* and *Methanobacterium* in freshwater sediments (resp. interm. = respiration intermediates, ferm. interm. = fermentation intermediates)

occurred obviously due to removal of $H_2$ by methanogenesis. Yields (g bacterial mass $g^{-1}$ substrate) recorded for *Methanobacterium* alone or with *Desulfovibrio* in our studies were not significantly different (2.1 versus 2.4 g $g^{-1}$, respectively), but no corrections were made on maintenance coefficients (m) or on true molar growth yield ($Y_G$) as cultivation of *Methanobacterium* on $H_2$-limitation in the chemostat was impossible, so conclusions can not be drawn at the moment.

Illustrated in Fig. 3 are the possible relationships between the sulfate-reducing and methanogenic species in the freshwater sediments. From the earlier studies employing mixed continuous cultures with a lactate-limited *Desulfovibrio* and an acetate-limited *Methanobacterium* (6) a commensalism between the two species could be described, i.e., the acetate-fermenting *Methanobacterium* could utilize the acetate produced by *Desulfovibrio* which, in turn, was not affected by the presence of the former. As lactate is always present in the mud of Lake Vechten (4, 5), but sulfate is depleted during summer stratification by the activity of sulfate reducers (3, 11), a seasonal variation in the products formed by *Desulfovibrio* can be expected, i.e., in summer more $H_2$ and in winter more acetate. Field observations (4) showed that the maximal abundance of sulfate reducers in the mud is in the 0-2 cm layer, and that of $H_2/CO_2$-fermenting methanogenics is in the 2-3 cm layer. These latter bacteria may thus serve as a hydrogen sink and provide favourable growth conditions for the sulfate reducers. Employing selective inhibitors, and by $^{14}C$-acetate incubation of mud samples and in pure cultures of *Desulfovibrio* role of $H_2$ or acetate as a supplementary source of energy for sulfate reducers in the natural environment could not be established (5, 8, 12). The observation that hydrogenase is located in the periplasmic space of *D. gigas* (1), and may function as a hydrogen-binding protein required for the transfer of low levels of hydrogen between microorganisms lends further support to the concept of interspecies hydrogen transfer. The chemostat studies however are model systems and may not represent the in situ activities of the bacteria. A more natural approach i.e. use of auto-radiographical and anti-body fluorescent techniques is desirable.

## D. References

1. Bell, G.R., Le Gall, J., Peck, H.D.,Jr.: Evidence for the periplasmic location of hydrogenase in Desulfovibrio gigas. J. Bact. 120, 994-997 (1974).
2. Bryant, M.P., Campbell, L.L., Reddy, C.A., Crabill, M.R.: Growth of Desulfovibrio in lactate or ethanol media low in sulfate in association with $H_2$ -utilizing methanogenic bacteria. Appl. Environ. Microbiol. 33, 1162-1169 (1977).
3. Cappenberg, Th.E.: Ecological observations on heterotrophic, methane oxidizing and sulfate reducing bacteria in a pond. Hydrobiologia 40, 471-485 (1972).
4. Cappenberg, Th.E.: Interrelations between sulfate-reducing and methane-producing bacteria in bottom deposits of a fresh-water lake. I. Field observations. Antonie van Leeuwenhoek 40, 285-295 (1974a).
5. Cappenberg, Th.E.: Interrelations between sulfate-reducing and methane-producing bacteria in bottom deposits of a fresh-water lake. II. Inhibition experiments. Antonie van Leeuwenhoek 40, 297-306 (1974b).
6. Cappenberg, Th.E.: A study of mixed continuous cultures of sulfate-reducing and methane-producing bacteria. Microb. Ecol. 2, 60-72 (1975).
7. Cappenberg, Th.E.: Methanogenesis in the bottom deposits of a small stratifying lake. In: Microbial Production and Utilization of Gases ($H_2$,$CH_4$,CO) Symp. Proc. Schlegel, H.G., Pfennig, N., Gottschalk, G. (eds.) E. Goltze-Verlag, Göttingen, pp. 125-134 (1976).
8. Cappenberg, Th.E., Prins., R.A.: Interrelations between sulfate-reducing and methane-producing bacteria in bottom deposits of a fresh-water lake. III. Experiments with $^{14}C$-labeled substrates. Antonie van Leeuwenhoek 40, 457-469 (1974).
9. Cappenberg, Th.E., Jongejan, E.: Microenvironments for sulfate reduction and methane production in freshwater sediments. In: Biogeochemistry of defined microenvironments in aquatic and terrestrial systems. Proc. Third. Int. Symp. Environm. Biogeochem.Krumbein,W.(ed.),in press (1977).
10. Gall, J. le, Postgate, J.R.: The physiology of sulphate reducing bacteria. Adv. Microb. Physiol. 10, 81-133 (1973).
11. Gemerden, H. van.: On the bacterial sulfur cycle of inland waters. Thesis. Univ. Leiden, pp. 110, (1967).
12. Khosrovi, B., Macpherson, R., Miller, J.D.A.: Some observations on growth and hydrogen uptake by Desulfovibrio vulgaris. Arch. Mikrobiol. 80, 324-337 (1971).
13. Nelson, R.D., Zeikus, J.G.: Rapid method for the radioisotopic analysis of gaseous end products of anaerobic metabolism. Appl. Microbiol. 28, 258-261 (1974).
14. Pfennig, N., Biebl, H.: Desulfuromonas acetoxidans gen. nov. and sp. nov., a new anaerobic, sulfur-reducing, acetate-oxidizing bacterium. Arch. Microbiol. 110, 3-12 (1976).
15. Postgate, J., Campbell, L.L.: Classification of Desulfovibrio species, the nonsporulating sulphate-reducing bacteria. Bact. Rev. 30, 732-738 (1966).
16. Tempest, D.W.: The cultivation of micro-organisms. I. Theory of the chemostat. In: Methods in Microbiology, Norris, J.R., Ribbons, D.W. (eds.) Vol. 2, Academic Press, New York, pp. 259-276 (1970).
17. Veldkamp, H.: Mixed culture studies with the chemostat. In: Applications and New Fields; monograph from the 6th Intern. Symp. on Cont. Culture of Microorganisms; Oxford 1975. Dean, A.C.R., Ellwood, D.C., Evans, C.G.T., Melling, J. (eds.) Ellis Horwood, Chichester, pp. 315-328 (1976).

18. Winfrey, M.R., Zeikus, J.G.: Effect of sulfate on carbon and electron flow during microbial methanogenesis in freshwater sediments. Appl. Environ. Microbiol. 33, 275-281 (1977).
19. Winfrey, M.R., Nelson, D.R., Klevickio, S.C., Zeikus, J.G.: Association of hydrogen metabolism with methanogenesis in Lake Mendota sediments. Appl. Environ. Microbiol. 33, 312-318 (1977).
20. Wolfe, R.S.: Microbial formation of methane. Adv. Microbial Physiol. 6, 107-146 (1971).

# Microbial Ecology of Soils

# Effect of Solid Particles on Growth and Metabolic Activity of Microorganisms

Z. FILIP

## A. Introduction

Microorganisms influence or even change their environment, whereas their growth and metabolic activity are strongly limited by environmental conditions. In an effort to obtain an explanation of the complex relationships between microorganisms and the environment, it is necessary to turn from the classical topographical ecology to the biochemical ecology of microorganisms (1). Such an approach involves the study of the inorganic and organic colloids as the site of physico-chemical reactions in soils. The colloid particles of some clays probably played an important role even at the beginning of life on Earth by contributing to protein formation (2). Clays, as microecological factors, may have influenced the first well organized forms of life, the microorganisms. It is not only for historical reasons that we should attempt to understand these relationships better. It is because clays and other solid particles are in close contact with microorganisms in soils and interactions between particles and microorganisms have been established in other environments. The purpose of this paper is to give information about some of our results.

## B. Material and Methods

The methods used have been described in detail elsewhere (4).

## C. Results and Discussion

Examination of a soil suspension under the microscope usually shows microorganisms forming larger or smaller aggregates with mineral particles, mostly clays. Some of the microbial cells are adsorbed at the clay surface, or are aggregated with other cells. If a very fine clay suspension is present, the microbial cells may be coated by a thin mineral layer (7). In our experiments we have attempted to determine whether microbial activity in culture is influenced by the presence of clays, or other solid particles, and whether enrichment of sands or soils with clay minerals exerts some influence on the biological activity of amended substrates.

Montmorillonite, for example, when added to cultures of *Saccharomyces cerevisiae* did not influence cell growth under anaerobic conditions. If oxygen was present, however, the growth of this yeast both in stationary and shake cultures was enhanced by addition of 0.5% clay. The growth increase over the controls was up to 90% for *S. cerevisiae* and up to 235% for *Candida utilis*. The presence of montmorillonite also resulted in an increase in carbon dioxide production in relation to oxygen consumption, causing an increase in the RQ values, perhaps due to a slight increase in glycolytic utilization of the substrate, manifested in an increased formation of intermediate breakdown products which could be used for the synthesis of cellular matter. In well aerated cultures of *Aspergillus niger*, for example, the RQ value was 0.87 for the control and 1.40 for cultures where 20 mg montmorillonite per 2 ml buffered solution were added. During the cultivation of A. niger in shake cultures with glucose and $NaNO_3$ as the only carbon and nitrogen sources, cell matter yield was 0.3 $gl^{-1}$ in control flasks after 3 days, 0.8 $gl^{-1}$ after 6 days and 3.2 $gl^{-1}$ after 10 days. At the same time 2.7 g, 9.7 g and 7.4 g of biomass per 1 liter of medium, i.e. up to ten times more than in controls, was produced in the cultures with 0.5% montmorillonite added. In stationary cultures and cultures shaken for four days and then incubated stationary, the difference was not as high, but in general, the positive effect of clay added was very similar. Over a three week growth period marked differences could be seen in stationary cultures of A. niger with and without montmorillonite added. The mycelium was of white colour in the control flasks and the nutrient solution only slightly yellow, whereas both the mycelium and nutrient solution were dark brown in the fungal cultures where montmorillonite was present.

To study this phenomenon which may have some significance in humification, further experiments were carried out using different soil fungi, e.g. Epicoccum nigrum, Stachybotrys chartarum, Hendersonula toruloidea, Aspergillus sydowi, these have been found to produce humic acid-like-polymers from glucose or other simple aliphatic compounds (5, 6). After 15-20 days' incubation, the stationary cultures of E. nigrum gradually turned red brown and then dark brown as the humic acid-type-polymers formed. The addition of montmorillonite accelerated the color change, and the intensity of the dark color was proportional to the concentration of the clay. With 1% montmorillonite, the extinction value of centrifuged solution after 20 days was eight times higher, and after 30 days up to twelve times higher than the values of the controls. The cultures containing clay also contained much higher amounts of humic polymers which could be recovered by precipitation at pH 2. From the control 120 $mgl^{-1}$ were obtained and 380 mg, 570 mg and 980 mg per liter from E. nigrum stationary cultures containing 0.25%, 0.5%, and 1.0% montmorillonite respectively. The isolated humic acid-type-polymers contained about 54% C and 8% N on an ash free basis and the addition of clay did not affect these percentages. The ash content, however, was different and ranged up to 60% for the 1.0% clay treatment. From infrared analysis the ash present in polymers after they were thoroughly purified was found to be identical with the clay minerals added. This means that resistant organic-mineral complexes were produced in clay amended fungal cultures. X-ray diffraction tests were also carried. They showed an expansion of the montmorillonite interlayer spacing from the original 12 Å to 18 Å, during the first 10 days of cultivation. From the 20th to 30th day the montmorillonite interlayer spacing contracted to 12 Å again. These changes could also be recognized if the clay mineral was in dialysis tubes, so that a sorption of some low molecular weight substances, which could be phenolics, could be supposed.

In separate experiments the sorption of some phenolic acids, usually produced during the first stages of fungal growth, could be measured at different pH values. The sorption of these substances, which can act as growth inhibitors, could support the fungal growth in the first stages of cultivation. Later, when exchanged for ammonium released during cell autolysis, the same substances could support the dark polymer formation because they polymerize autooxidatively in water solutions above pH7.

Increased cell counts were generally noted after addition of clay (bentonite) to sand or soil samples. Usually the highest clay amendment (0.5% = 200 $qha^{-1}$) was most effective. Respiratory and ammonification activity of soil microflora, however, did not differ in the amended and nonamended samples. Immediately after addition of some readily utilizable carbon or nitrogen sources, the microbial activity increased quickly in clay amended samples. The total amount of humic acids was usually not influenced in soil samples enriched with casein (1%) and incubated with different bentonite concentrations. The humic acid fraction however, which is firmly bound to the mineral part of the soils and which is extractable only after acid hydrolysis was sharply increased (up to 100%) by the clay additions. This is of interest for arable soils, where such mineral-bound humic acids help to retain optimal physico-chemical conditions, important for soil fertility.

In some experiments (8) positive effects of clay minerals on the survival rate of soil microorganisms exposed to extremely undesirable conditions were established. In our experiments similar effects were observed in cultures of municipal waste compost microflora amended with glass beads, humic acid preparation or bentonite. During the first phase of cultivation at 30°C, $CO_2$-evolution by compost microflora was between 120-180 $mgd^{-1}$. After the temperature was raised suddenly to 70°C, the $CO_2$-evolution dropped quickly except in cultures where humic acid was added. The respiratory activity was very low during the high temperature period in the controls (10-40 mg $CO_2$ $d^{-1}$). At the same time maximal $CO_2$-release was at 195 $mgd^{-1}$ in cultures with glass beads, 275 $mgd^{-1}$ in cultures with humic acid and 380 $mgd^{-1}$ in cultures with bentonite addition. These results provide good evidence for a favorable effect of solid particles, especially clays, on the activity of municipal waste compost microflora under high temperatures.

The data summed up in this paper show that clay minerals and other solid particles influence cell growth and activity when added to cultures of microorganisms and to soil samples. The role of clay minerals in the ecology of soil microorganisms has been discussed elsewhere (3, 4, 9). Further research should be done to obtain a better understanding of the relationships between microorganisms and their microenvironment. These relationships are theoretically interesting and of practical importance.

## D. References


1. Alexander, M.: Biochemical ecology of microorganisms. Ann. Rev. Microbiol. 25, 361 (1971).
2. Bernal, J.D.: The Physical Basis of Life. London: Routledge & Kegan Paul, (1951).
3. Filip, Z.: Clay minerals as a factor influencing the biochemical activity of soil microorganisms. Folia microbiol. 18, 56 (1973).
4. Filip, Z.: Wechselbeziehungen zwischen Mikroorganismen und Tonmineralen und ihre Auswirkung auf die Bodendynamik. Habil.-Thes., Univ. Giessen, 172 p. (1975).
5. Filip, Z., Semotan, J., Kutilek, M.: Thermal and spectrophotometric analysis of some fungal melanins and soil humic compounds. Geoderma 15, 131 (1976).
6. Haider, K., Martin, J.P., Filip, Z.: Humus biochemistry. In: Soil Biochemistry, Vol. IV. E.A. Paul & A.D. McLaren (eds.). New York: M. Dekker, Inc., p.196 (1975).
7. Marshall, K.C.: Interfaces in Microbial Ecology. Cambridge, Mass. & London, Engl.: Harvard Univ. Press, p.71, (1976).
8. Marshall, K.C.: Clay mineralogy in relation to survival of soil bacteria. Ann. Rev. Phytopathol. 13, 357 (1975).
9. Stotzky, G.: Clay minerals and the microbial ecology. Trans. New York Acad. Sci. 30, 11 (1967).

# Ecological Variability of Soil Microorganisms

E.N. MISHUSTIN

## A. Introduction

The work of our laboratory over many years involving a large group of Soviet investigators shows that the majority of soil microorganisms reproduce optimally in particular zones. Although every soil type has a set of dominant organisms, there are also some microorganism species that are widespread and even those organisms dominant in a particular soil can be isolated from any soil.

It has been of interest to study whether some microorganisms apparently adapt more successfully than others to a specific set of conditions in a soil. We have been particularly interested in the adaptation of microorganisms to temperature and osmotic pressure, and we have considered the morphological, biochemical and physiological properties of these organisms.

## B. Materials and Methods

Optimal temperature for growth was assessed by measuring the diameter of macrocolonies for bacteria, actimomycetes and fungi and the increase in biomass in liquid culture of a number of bacteria.

Cell osmotic pressure was determined by the volumetric technique. Freshly grown bacterial cells were kept in salt solutions of different concentrations and then centrifuged in slightly modified Troumsdorf test tubes. According to change in cell volume the isotonic concentration of NaCl was established allowing the cell osmotic pressure to be determined. For bacteria without capsules the technique was satisfactory.

## C. Results and Discussion

The maximum and optimal growth temperature of non-sporogenous bacterial isolates from a number of zones were assessed. The results in Table 1 indicate that the isolates of ammonifying bacteria of the genera *Pseudomonas*, *Achromobacter* and *Mycobacterium* have a higher optimal temperature range if from the dry steppe and subtropics than those from the colder areas, e.g. the tundra. In contrast, *Azotobacter chroococcum* strains appear to have a similar optimal temperature range wherever isolated. Other isolates discussed later also showed this phenomenon.

Table 2 gives the optimal and maximal temperatures of a series of sporogenous isolates from the same zones and soils. The isolates fall into two groups, *Bacillus agglomeratus*, *B. cereus* and *B. mycoides* have a lower optimum temperature requirement in the colder zones and show a well pronounced adaptive response to climatic conditions. Other isolates, *B. megaterium* and *B. mesentericus* do not.

In our opinion the reasons for this difference in adaptation to temperature is related to the succession of organisms on dead plant organic matter. Plant matter is first colonised by non-sporogenous bacteria and some fungi (mainly *Mucor*). These then die out and *Bacillus* strains appear in large numbers, the pioneers being

Table 1. Reaction of non-sporogenous bacteria to temperature

| Zones | Type of Soil | °C Temperature optimum | | | °C Temperature maximum | | |
|---|---|---|---|---|---|---|---|
| | | 1 | 2 | 3 | 1 | 2 | 3 |
| Tundra | Northern podzol | 24-27 | 23-27 | n.a. | 34-35 | 34-36 | n.a. |
| High mountain zone | primitive soil | 24-27 | 26-28 | n.a. | 34-35 | ab.37 | n.a. |
| Taiga | soddy-podzolic soil | 27-30 | 27-30 | 27-35 | ab.35 | ab.37 | ab.42 |
| Steppe | chernozem | | | 27-35 | | | |
| Dry steppe | brown and grey soils | 35-37 | 34-37 | 30-37 | over 45 | ab.39 | ab.42 |
| Sub-tropics | red soils | 28-35 | 28-34 | 28-35 | over 45 | 42-45 | 42-44 |

n.a. not available

1. Pseudomonas Achromobacter spp.
2. Mycobacterium spp.
3. Azotobacter

Table 2. Reaction of sporogenous bacteria to temperature

| Zone | Type of Soil | °C Temperature optimum | | | | | °C Temperature maximum | | | | |
|---|---|---|---|---|---|---|---|---|---|---|---|
| | | 1 | 2 | 3 | 4 | 5 | 1 | 2 | 3 | 4 | 5 |
| Tundra | northern podzol | 27-34 | 28-34 | 25-26 | 34-40 | 37-43 | ab.38 | ab.40 | ab.33 | ab.47 | ab.50 |
| High mountain zone | primitive soil | 30-34 | 25-27 | 25-27 | 34-40 | 37-43 | ab.42 | 38-40 | 32-33 | ab.47 | ab.50 |
| Taiga | soddy-podzolic soil | 32-37 | 28-32 | 28-32 | 32-40 | 37-43 | ab.43 | ab.42 | ab.40 | ab.47 | ab.50 |
| Steppe | chernozem | n.a. | n.a. | 30-34 | n.a. | n.a. | n.a. | n.a. | ab.40 | n.a. | n.a. |
| Dry steppe | brown and grey soils | ab.40 | 34-40 | 35-38 | 34-40 | 37-43 | ab.45 | 46-47 | ab.44 | ab.47 | ab.50 |
| Subtropics | red soils | ab.40 | 37-40 | 28-42 | 37-43 | 37-43 | ab.47 | ab.45 | 46-47 | ab.47 | ab.50 |

n.a. not available

1. B. agglomeratus
2. B. cereus
3. B. mycoides
4. B. megaterium
5. B. mesentericus

B. agglomeratus, B. cereus, B. mycoides. B. megaterium, B. mesentericus and actinomycetes appear later.

This succession of groups is related to the physiological properties of the bacteria. Non-sporogenous bacteria are not fastidious and reproduce rapidly. Once the decaying plant material has added to it microbial protein and vitamins the sporing bacilli begin to reproduce and they utilise the plant residues through their enzymes. The actinomycetes also begin to increase in numbers provided the pH is not acid.

It is the microorganisms involved in the first phase of organic matter decomposition, the non-sporogenous bacteria and some bacilli (B. agglomeratus, B. cereus, B. mycoides) which show an adaptive response to temperature. This group seeks to utilize the substrate as quickly as possible and therefore has had to adapt to the environment. For other bacilli rapid reproduction is not necessary in the initial stages of organic matter decomposition.

Azotobacter and similar microorganisms occupy a specific biological niche and adaptation to environmental temperature is probably not necessary for them.

The optimum and maximum temperatures of the anaerobic nitrogen fixers, Clostridium pasteurianum and Cl. acetobutylicum were also studied by V.T. Yemtsev in our laboratory. The results indicated that the strains had adapted to their environment but the optimal temperature range of Cl. pasteurianum was consistently lower than Cl. acetobutylicum. The difference in optimal temperature range for strains isolated in cold climates compared with those from warmer areas was small. Cl. acetobutylicum is involved in organic matter degradation late in the process and this may account for its lesser adaptation to temperature.

Some preliminary studies were also carried out on the fungi of the genera Mucor and Penicillium and on some actinomycetes. The isolates were not identified to species level. A temperature adaption was obvious in the fungi but not in the actinomycetes and correlates with the finding that the fungi are amongst the first colonisers of organic matter. The actinomycetes on the other hand are late colonisers.

There appears then to be a temperature adaptation by many microorganisms in the soils studied.

We were also interested in osmotic adaptation because in the European part of the USSR for example the increase in mean annual temperature from north to south is associated with a decrease in precipitation. Soil organisms in the southern zone are therefore growing in drier soils than in the north. It was of interest then to see if organisms in these drier soils had higher cell osmotic pressures. Experiments with B. mycoides strains from four zones suggest that they do (Table 3).

Table 3. Osmotic pressure in cells of B. mycoides cultures of different origin

| Type of soil from which the cultures are isolated | Value of osmotic cell pressure (atm.) |
|---|---|
| Podzols and soddy-podzolic | ab. 2.0 |
| Northern chernozems | 5.4-7.2 |
| Southern chernozems | 7.2-9.0 |
| Chestnut | 14.4-16.2 |

We also found there was variation in morphology of macro-colonies of B. mycoides in different zones. Strains from northern podzols and soddy podzolic soils were predominantly of the folded type; in the chernozems folded forms with strictly oriented strands occurred and in chestnut and grey soils (1) smooth forms.

It was of interest to see if physiological and morphological properties of soil organisms changed as the organisms had adapted to different conditions.

An increase in temperature optimum in bacteria from soil in the southern zone is related to the possession of more active enzymes. Van Hoff's law is illustrated by the reaction rate increasing 2-3 times with a temperature rise of $10^{o}C$. Southern zone isolates in many cases have an optimum $10^{o}C$ above northern zone isolates and show energetic reproduction and produce a larger biomass with an active enzyme system. Under favourable conditions the microbial processes in the southern zone soils are more energetic than those in the north.

The enzyme complex composition also undergoes some adaptation at high temperature. In southern B. mycoides cultures, proteolytic activity is enhanced and catalase activity is inhibited substantially. The oxidising system may also be lost in adaptation to higher temperatures.

Changes to other physiological processes have been noted particularly in fermentation and end-products (2). In southern cultures of Cl. pasteurianum and Cl. acetobutylicum the yield of butyric acid was noticeably reduced in fermentation products and alcohol content increased compared with the same cultures from northern locations. The same trend was present to a lesser extent for Cl. butyricum and Cl. butylicum. The percentage ratio between acids and alcohols formed by Clostridium cultures is given in Table 4 and shows that the ratio diminishes in cultures from north to south.

Table 4. Ratio between acids and alcohols in fermentation products in Clostridium spp. isolated from different zones

| Culture origin | Cl. pasteurianum | Cl. butyricum | Cl. butylicum | Cl. acetobutylicum |
|---|---|---|---|---|
| Tundra and high mountain soils | 19.0 | 20.0 | 5.0 | 4.0 |
| Taiga | 8.0 | 16.5 | 4.0 | 3.8 |
| Dry steppes | 4.5 | 15.0 | 2.5 | 2.5 |
| Subtropics | 3.2 | 14.7 | 2.0 | 2.1 |

Nitrogen fixation also changes in Clostridium with zone. Per unit of carbon utilised southern cultures assimilate less nitrogen than those from soils of a colder climate, a difference which was more marked for Cl. pasteurianum and Cl. butyricum.

A number of cultures of Clostridium were subjected to pyrolysis gas chromatography. Pyrograms of the same species, but of different origins differed slightly suggesting some changes in the strains.

This study is fragmentary so far and calls for additional investigation. It does seem however, that many soil microorganisms show ecological variation by adapting to conditions in particular zones, giving rise to ecological races.

## D. References

1. Mishustin, E.N., Komlavi, D.N., Smirnova, G.A.: Mikrobiologiya No.4, 747-750, (1963).
2. Mishustin, E.N., Yemtsev, V.T.: Pochvennye azotfiksiruyushchiye bakterii roda *Clostridium*. Publ. house "Nauka", (1974).
3. Mishustin, E.N.: Microbial Ecology, 2, 97-118, (1975).

# Aerobic Bacteria in the Horizons of a Beech Forest Soil

T. KAURI

## A. Introduction

An investigation of the bacterial populations of an acid brown earth in a 90-year-old beech forest (*Fagus sylvatica* L.) in Kongalund was undertaken. The beech forest is a characteristic forest type in Scania, South Sweden.

The classification is that given for a nearby site (5) with a modification for the horizon below $A_{00}$. This layer is a mixture of humus and mineral particles and is for lack of better classification (8) for this sort of soils in Sweden designated $A_{01}/A_1$ (B. Nihlgård, pers. comm.). The humus could be called moder in German classification (8).

The present report is part of an investigation during a three year period.

## B. Materials and Methods

Soil cores were taken with a metal bore from two different sites; four subsamples were taken from each horizon at each site. All eight subsamples were pooled to yield one sample from each horizon and were treated aseptically. From thoroughly mixed and sieved (2 mm) samples, 20.0 g portions were removed and suspended in 190 ml Winogradsky's standard salt solution (7). Suspensions from each horizon were treated in a homogenizer (MSE ATO-mix) for three min ($A_{00}$ samples four min) at high speed (12,000 rpm) and dilutions prepared in standard salt solution. For estimating bacterial spores, the 1:10 dilutions were heated at 80°C for 10 min.

Both bacteria and bacterial spores were counted on surface spread plates (five replicate plates) of soil extract agar, modified after Holm & Jensen (2). This modification consisted of adding NaOH during the boiling of the soil suspension (2 g NaOH/1000 g soil and 1000 ml dist. water) which kept the pH at *c.* 5 during the heat treatment. Without NaOH, the pH was lowered to 3.5. The final pH of the medium was adjusted to 6.7. To avoid fungal growth, sterile filtered cycloheximide (final concentration of 70 mg per litre) was added to the medium. The plates were incubated at 20°C for two weeks.

Five to six samplings were carried out per year.

## C. Results and Discussion

The numbers of bacteria decreased in the deeper horizons (Tab. 1). The most conspicuous difference was between horizons $A_{00}$ and $A_{01}/A_1$. $A_{00}$ showed distinctly higher counts of bacteria than the other layers. Compared with plate counts from the litter layer of certain other beech forest soils (2, 3), the bacterial numbers were somewhat higher, although the soil pH values were lower in Kongalund.

The differences between the other layers were smaller. Quite conspicuous seasonal variations were found ($A_{01}/A_1$). Twice a year, bacterial counts showed peaks that occurred approximately at the same time as the leaf-fall in autumn and the leaf-burst along with the outburst of ground vegetation in spring.

The numbers of bacterial spores were relatively high, as reported frequently for acid soils (1). However, calculated in per cent of the corresponding bacterial numbers, they were lower than those found by, for instance, Jagnow (3). Especially for $A_{00}$, the percentage of spores was very low.

Only the bacterial numbers (Table 1) were positively correlated ($p<0.01$) to the moisture content and to the organic matter (loss on ignition) in the soil between different horizons.

Table 1. Mean values of soil bacteria and bacterial spores (cfu - colony forming units) per g dry weight in different horizons of a beech forest. Some soil characteristics are also given

| Brown Forest[a] Soil Horizons | Bacteria cfu/g DW | Spores cfu/g DW | Moisture[b] Content % of WW | Loss on[c] Ignition % of DW | pH[d] |
|---|---|---|---|---|---|
| $A_{00}$ | $2.10^9$ | $2.10^6$ | 69 | 81 | 4.50 |
| $A_{01}/A_1$ | $3.10^7$ | $4.10^6$ | 32 | 15 | 3.90 |
| $A_1$ | $1.10^7$ | $2.10^6$ | 23 | 8 | 3.95 |
| (B) | $5.10^6$ | $8.10^5$ | 18 | 5 | 4.10 |

a) B. Nihlgard (pers. comm.)
b) drying at 105°C for 24 h
c) 850°C for 4 h
d) glass electrode, soil and water (1:10, w/v)

In a forest ecosystem, litter is a major component in the energy flow to the soil. During the summer of 1973, the forest was invaded by a fast-growing population of Dasychira pudibunda larvae (6). In consequence, the whole forest became defoliated several weeks before the normal time of leaf-fall. This apparently was reflected in the bacterial populations. For $A_{01}/A_1$ and (B), the usual peaks in bacterial numbers during autumn were lower than previously, and $A_1$ had an increase in bacterial numbers earlier than during the previous year. A decrease in bacterial counts was found in the top layer.

The variation in bacterial populations found in the present work can hardly be due to sudden occasional weather changes. A beech forest, with its distinct litter layer and a thick canopy during the summer months, has a special microclimate (4) with fluctuations in environmental conditions smaller than more exposed ecosystems.

## D. References

1. Goodfellow, M.: Properties and composition of the bacterial flora of alpine forest soil. J. Soil Sci. 19, 154-167 (1968).
2. Holm, E., Jensen, V.: Aerobic, chemoorganotrophic bacteria of a Danish beech forest. Oikos 23, 248-260 (1972).
3. Jagnow, G.: Seasonal amount of fungal mycelium, numbers of aerobic bacteria and bacterial spores and numbers of saccharolytic anaerobic bacteria and bacterial spores in a beech and spruce forest soil of the Solling. In: IV Colloquium Pedologia. Dijon. I.N.R.A. Ann. Zool. Ecol. Special Issue, 303-313 (1970).

4. Nihlgård, B.: The microclimate in a beech and a spruce forest - a comparative study from Kongalund, Scania, Sweden. Bot. Not. 199, 333-352 (1969).
5. Nihlgård, B.: Precipitation, its chemical composition and effect on soil water in a beech and a spruce forest in south Sweden. Oikos 21, 208-217 (1970).
6. Nilsson, I.: The influence of a leaf-eating insect (*Dasychira pudibunda* L. Lepidoptera) on internal plant nutrient transports and tree growth in a beech forest (*Fagus sylvatica* L.) in southern Sweden. Submitted to Oikos (1977).
7. Pochon, J.: Manuel technique d'analyse microbiologique du sol. Paris (1954).
8. Troedsson, T., Nykvist, N.: Marklära och märkvard. Almqvist & Wiksell. Stockholm (1973).

# Virulent, Temperate and Defective Bacteriophages from Gram-Positive Soil Bacilli

W.J. KELLY and D.C. REANNEY

## A. Introduction

The recent controversy over recombinant DNA studies has focussed attention on the "natural barriers" which prevent or minimise the interchange of genetic material between taxonomically distinct species. In this paper we report some results of a series of experiments designed to explore: (a) the incidence of lysogeny among gram-positive soil bacilli and (b) host ranges of a group of phages for members of the genus *Bacillus*. Because most phages and bacteria were isolated from essentially the same soil sample(s) such data may shed some light on natural networks of polynucleotide exchange.

## B. Materials and Methods

I. *Host organisms*. The source of *Bacillus* strains is listed in Table 1.

II. *Prophage induction*. Growing cultures of organisms were treated (a) with $H_2O_2$ to give a final concentration of 0.44 mM (17), or (b) with mitomycin C to give a final concentration of 0.4 μg/ml (16), or (c) uv. "Lysates" were clarified by low speed centriguation and putative virus material sedimented at 70,000 g for 2 hours.

III. *Host range studies*. Pilot screenings for host range involved the "spotting" of "lysates" on to bacterial lawns. When zones of clearing were observed the "lysate" was serially diluted and spotted on to the relevant host. Gradual "fade-out" was taken to indicate a killer effect; individual plaques suggested productive phage multiplication.

IV. *Antibiotic sensitivity assay*. Standard Biolab discs were placed on bacterial lawns and incubated overnight at 37$^o$C; concentrations were: ampicillin 25 μg, Bacitracin 10 units; chlorotetracycline 30 μg, gentamycin 10 μg, pencillin 5 units, polymyxin 300 units; tetracycline 30 μg per disc. For determination of mercury resistance, $HgCl_2$ was added to a final concentration of 1.25-20 μg/ml. Lead resistance was determined by adding $Pb(NO_3)_2$ to agar at concentrations of 100-2000 μg/ml.

V. *Electron microscopy*. Purified phage preparations were stained with PTA and examined in an Hitachi HSB or a JEM-100B electron microscope.

## C. Results and Discussion

Virulent phages for *Bacillus* isolated in our lab to date include 22 isolated on *B. pumilus* W43, 2 on *B. stearothermophilus*, 5 on *B. circulans* and 3 on *B. licheniformis* (13). The ability of these phages to produce zones of clearing on multiple *Bacillus* spp. is well established (13). That such clearing is due, in most instances tested, to productive phage multiplication is shown in Table 2.

When tested against the soil bacteria listed in Table 1 the data shown in Table 3 was obtained. The only *B. pumilus* strain detected in this common soil flora was susceptible to 28/32 phages. The relative inability of *B. cereus* and *B. metaterium* to sustain the growth of the majority of these phages is consistent with our previously published data. Although these phages are classified as "virulent" many probably originated from lysogenic bacteria. K1 for example, will lysogenise *B. pumilus* W. 43 *in vitro*.

To determine the incidence of lysogeny among our cultures, selected strains were induced as described. Table 4 shows that 10/18 strains, including one strain of *B. stearothermophilus*, contained a defective temperate phage similar if not identical to that described by others (6). In those cases where the defective phage and a "typical" phage have been jointly induced from the same cell the defective particle is normally present in excess.

Table 1. Source of *Bacillus* strains

| | | | |
|---|---|---|---|
| (a) | International Cultures | | |
| | *B. cereus* | NCIB | 7464 |
| | *B. circulans* | NCIB | 8144 |
| | | NCIB | 9374 |
| | *B. coagulans* | NRS | T2007 |
| | *B. licheniformis* | NCIB | 6346 |
| | *B. megaterium* | NCIB | 8291 |
| | *B. pumilus* | NCIB | 2595 |
| | | NCIB | 8738 |
| | | NCIB | 9369 |
| | *B. stearothermophilus* | NCIB | 8919 |
| | | NCIB | 8923 |
| | | NRS | T85 |
| | | NRS | T91 |
| | | NRS | T91-p |
| | | ATCC | 7953 |
| | *B. sphaericus* | NCIB | 8217 |
| | *B. subtilis* | NCIB | 3610 |
| | | ATCC | 6633 |
| (b) | Local isolates | | |
| | *B. cereus* | T9, 7 isolates | |
| | *B. licheniformis* | W37 | |
| | | DW2 | |
| | *B. megaterium* | 9 isolates | |
| | *B. pumilus* | W29 | |
| | | W43 | |
| | | 1 isolate | |
| | *B. stearothermophilus* | TP-1 | |
| | *B. sphaericus* | W32 | |
| | | 8 isolates | |
| | *B. subtilis* | Mil | |
| (c) | A group of 31 isolates which either have not been fully categorised or do not easily fit the taxonomic criteria. | | |

Table 2. Relative efficiency of plating values of phages K13, K18, C3 and D6

| Bacterial Strain | Code | K13.W43 | K13.W32 | C3. W43 | K18.W43 | D6. T91 |
|---|---|---|---|---|---|---|
| B. stearothermophilus | NRS T91 | $4.0 \times 10^{-2}$ | $9.0 \times 10^{-2}$ | $7.0 \times 10^{-2}$ | - | 1.0 |
| | ATCC 7953 | 1.0 | 0.8 | 1.0 | - | - |
| | NRS T85 | 0.2 | $8.0 \times 10^{-2}$ | 0.3 | - | 1.0 |
| | TP-1 | 2.9 | 0.5 | 1.8 | - | 1.0 |
| B. sphaericus | W32 | 1.0 | 1.0 | 1.3 | 0.6 | - |
| | NRS 344 | 0.7 | 1.1 | - | - | - |
| B. subtilis | NRS 744 | 0.2 | $5.0 \times 10^{-2}$ | 1.0 | - | 0.2 |
| | ATCC 6633 | - | - | $1.9 \times 10^{-5}$ | $6.0 \times 10^{-6}$ | - |
| B. licheniformis | W37 | 1.0 | 1.0 | 1.0 | - | 1.0 |
| B. pumilus | W43 | 1.0 | 1.0 | 1.0 | 1.0 | $1.0 \times 10^{-7}$ |

\- = not attempted

From unpublished work by M. Vissers

Table 3. Susceptibility of soil bacilli to virulent phages

| Organisms | | Lytic responses /32 phages |
|---|---|---|
| B. megaterium | 1 | 4 |
| | 2, 3, 8, 9 | 2 |
| | 4, 5, 6, | 1 |
| | 7 | 3 |
| B. sphaericus | 1, 7 | 0 |
| | 2 | 8 |
| | 3, 4, 5, 6 | 1 |
| | 8 | 5 |
| B. cereus | 1, 2, 7 | 1 |
| | 3, 4, 5, 6 | 0 |
| B. pumilus | 1 | 28 |
| B. licheniformis DW 2 | | 20 |
| B. subtilis Mil | | 1 |
| Unclassified gram positive | sp. 1 | 15 |
| | 2 | 12 |
| | 3 | 14 |
| | 4 | 7 |

In a systematic survey 51 Bacillus strains were induced and "lysates" tested against each other strain in turn. The results are given in Table 5. As in a similar study (1) only a small fraction of the attempted lytic responses gave a positive result. As previously reported (13) responses tended to fall into one of two categories (a) zero response or very limited activity or (b) host ranges spanning multiple species. Most wide "host ranges" are evidently due to the killing effect of the defective particle and should be set against the ability of the "virulent" phages to cause productive infection in multiple species (Table 3).

Table 4. Properties of bacteriophages induced from selected *Bacillus* spp.

| Organism | Code | Phage Type A | Phage Type B | Phage Type Defective |
|---|---|---|---|---|
| *B. cereus* | 1 | | √ | |
| | 2 | | ? | |
| | 4 | | √ | |
| | 5 | | √ | |
| | 6 | | √ | |
| *B. coagulans* | NRS T 2007 | √ | | |
| *B. licheniformis* | DW 2 | | | √ |
| *B. pumilus* | NCIB 2595 | | | √ |
| | 8738 | | | √ |
| | 9369 | √ | | |
| | W43 | | | √ |
| | 1 | | √ | √ |
| *B. sphaericus* | NCIB 8217 | | | √ |
| | W32 | | | √ |
| *B. stearothermophilus* | NCIB 8919 | | √ | |
| | NRS T91 | | | √ |
| *B. subtilis* | NCIB 3610 | | | √ |
| | ATCC 6633 | | | √ |

A = Contractile tail phages
B = Non contractile tail phages

The results do not indicate that all temperate phages for *Bacillus* have limited host ranges but do demonstrate the inherently greater difficulty of obtaining temperate viruses for such studies from organisms which carry defective phages. It is worth noting that the phenomenon of pseudolysogeny, which is common in this genus, means that phage genes can be transmitted "paravertically" (10) without the need for chromosomal integration.

These data, in conjunction with previous studies, confirm that phages can promiscuously infect *Bacillus* species. Although some of the gaps spanned do not seem large by the biochemical criteria used in taxonomy, two *Bacillus* "species" infectable by a common phage may be genetically further apart than members of such enteric "genera" as *Escherichia* and *Klebsiella* (14, 15). It is unfortunate that the only ready criterion for assessing host range is the production of zones of clearing which may indicate (a) productive infection (b) killing by a bacteriocin (c) killing by the defective particle (d) induction of a killer activity by a phage. The cell-to-cell transfers most likely to be of adaptive value - i.e. the transduction of plasmids - would not have shown up in the present study.

The widespread distribution of the defective temperate phage is one of the puzzles of *Bacillus* genetics. It has been shown (12) that, at least in *B. subtilis*, these particles package only cellular DNA. The particle would seem admirably adapted to transduction were it not for its apparent inability to inject DNA and its killer effect. The common occurrence of the defective particles, however, suggests that some selective advantage is conferred on bacilli by these genes. Phage carriage, by converting DNA into an extracellular state, preserves genes under stresses e.g. cold or exposure to antibiotics which would kill cells (14).

Table 5. Susceptibility of various strains to induced phages

| Bacterial Strain | No. Strains lysed/51 | % | No. Species lysed/9 | Phage type |
|---|---|---|---|---|
| *B. cereus* NCIB 7464 | 1 | 2 | 1 | ND |
| *B. cereus* 1 | 5 | 10 | 2 | Temperate |
| 2 | 11 | 22 | 4 | Temperate |
| 3 | 2 | 4 | 1 | No particles seen |
| 4 | 0 | 0 | 0 | Temperate |
| 5 | 2 | 4 | 1 | Temperate |
| 6 | 6 | 12 | 3 | Temperate |
| 7 | 4 | 8 | 1 | No particles seen |
| *B. circulans* NCIB 9374 | 3 | 6 | 3 | ND |
| *B. megaterium* 4 | 8 | 16 | 3 | No particles seen |
| *B. licheniformis* DW2 | 33 | 65 | 8 | Defective |
| *B. pumilus* NCIB 2595 | 9 | 18 | 4 | Defective |
| 8738 | 37 | 73 | 6 | Defective |
| 9639 | 0 | 0 | 0 | Temperate |
| *B. pumilus* W 29 | 6 | 12 | 2 | ND |
| W 43 | 6 | 12 | 3 | Defective |
| *B. pumilus* 1 | 14 | 27 | 4 | Temperate + Defective |
| *B. stearothermophilus* NCIB 8157 | 2 | 4 | 1 | ND |
| *B. sphaericus* NCIB 8217 | 4 | 8 | 2 | Defective |
| W 32 | 22 | 43 | 5 | Defective |
| *B. subtilis* NCIB 3610 | 17 | 33 | 7 | Defective |
| ATCC 6633 | 11 | 22 | 6 | Defective |
| Mil | 7 | 14 | 5 | ND |

ND = Not done

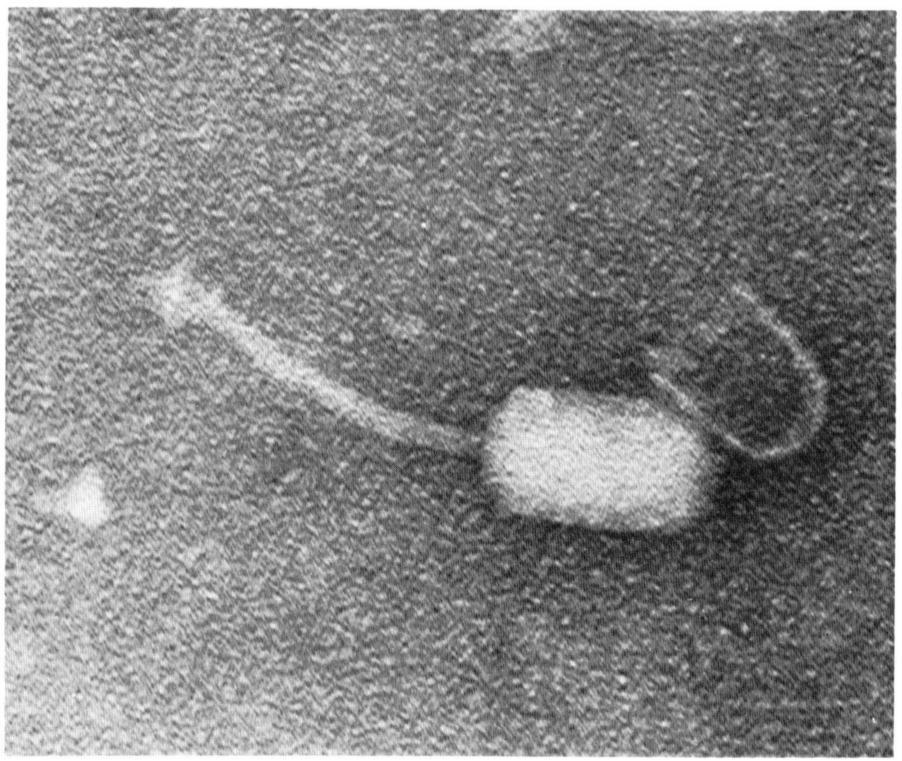

Fig. 1. A temperate phage with B2 morphology induced from B. cereus 1; magnification = 462,000. (Electron microscopy by A. Day)

One unarguable result of widespread induction of the defective temperate is an ability to seed soil with "packaged" cell genes from almost every known Bacillus species (Tables 4, 5). It is not inconceivable that heterologous gene mixings occur by something akin to transformation when new bacilli colonise soil pre-seeded with such encapsulated cell DNA.

Next we intended to survey the ability of our phages to transduce chromosomal genes. Recent data however suggested that transduction of plasmid DNA is more likely to be the prime vehicle of adaptive change in this genus than direct transduction of cellular genes (14). Among staphylococci the transduction of plasmid DNA is much more efficient than that of cell genes and is known to be responsible for the spread of antibiotic resistance (11). Plasmid-like circular DNA has been found in Bacillus (8). To date the only known genetic markers carried by Bacillus plasmids are changes in sporogenicity (7) and production of and resistance to a killer activity (9). We attempted to detect plasmids in our Bacillus cultures by screening for more convenient markers such as resistance to antibiotics and/or heavy metals and were encouraged in this direction by the report that some members of another Gram-positive soil genus, Streptomyces, contain plasmids which govern both production of and resistance to antibiotics (18).

Because Bacillus spp. produce bacitracin and polymyxin the finding of resistance to these antibiotics was not unexpected. Also B. cereus has been reported to produce a $\beta$ lactamase (5) hence its observed relatively high resistance to penicillins was not surprising. The most significant aspect of our data was the finding of Hg resistance in one strain of B. coagulans. In most of the Gram-negative organisms we have screened and in staphylococci comparable levels of Hg resistance are almost always associated with an extra-chromosomal element. We are currently attempting to correlate these putative genetic markers with the presence or absence of plasmid-like circular DNA in our cultures.

The data presented indicate that phages provide an extensive potential for cell to cell passage of polynucleotide in this genus. Recent studies suggest that Bacillus spp. may transmit genes to and receive genes from members of other genera (2, 3, 4). If these in vitro experiments have in vivo parallels it follows that the interchange of extra-chromosomal DNA in nature may be far more extensive than is normally assumed.

## D. References


1. Ackermann, H-W. & Smirnoff, W.A.: "Lysogeny in Bacillus thuringiensis" 3rd Int. Congr. Virol. Abstr. C406, p.266 (1975).
2. Courvalin, P., Weisblum, B. & Davies, J.: "Aminoglycoside-modifying enzyme of an antibiotic-producing bacterium acts as a determinant of antibiotic resistance in E. coli". Proc. Natl. Acad. Sci. (USA) 74, 999-1003 (1977).
3. Domardskii, I.V., Levadnaya, T.B., Sitnikov B.S., Rassadin, A.S. & Denisova, T.S.: Dokl. Akad. Nauk, SSR 226 143 (Russian) (1976).
4. Ehrlich, S.D.: "Replication & expression of plasmids from Staphylococcus aureus in Bacillus subtilis". Proc.Natl.Acad.Sci. (USA), 74, 1680-1682 (1977).
5. Hill, P.: "The production of pencillins in soils & seeds by Penicillum crysogenum & the role of penicillin β-lactamase in the ecology of soil Bacillus". J. Gen. Microbiol. 70, 243-252 (1972).
6. Kelly, W.J.: "Induction studies in the genus Bacillus". B.Sc. Hons. thesis University of Canterbury. (1975).
7. Lovett, P.S.: "Plasmid in Bacillus pumilus & enhanced sporulation of plasmid negative variants". J. Bacteriol. 115, 291-298 (1973).
8. Lovett, P.S., & Bramucci, M.G.: "Plasmid DNA in Bacillus subtilis & Bacillus pumilus". J. Bacteriol. 124, 484-490 (1975).
9. Lovett, P.S., Duvall, E.J. & Keggins, K.M.: "Bacillus pumilus plasmid pPL10: Properties & insertion into B. subtilis 168 by transformation". J. Bacteriol. 127, 817-828. (1976).
10. Nahmias, A.J. & Reanney, D.C.: "The evolution of Viruses". Annu. Rev. Ecol. Systematics 8, 29-49 (1977).
11. Novick, R.P., & Morse, S.I.: "In vivo transmission of drug resistance factors between strains of Staphylococcus aureus". J. Exp. Med. 125, 45-59 (1967).
12. Okamoto, K., Mudd, J.A. & Marmur, J.: "Conversion of Bacillus subtilis DNA to phage DNA following mitomycin C induction". J. Molec. Biol. 34, 429-437 (1968).
13. Reanney, D.C.: "Extrachromosomal elements as possible agents of adaptation and development:. Bacteriol. Rev. 40, 552-590. (1976).
14. Reanney, D.C.: "Genetic engineering as an adaptive strategy". 29th Brookhaven Symposium in Biology, Brookhaven National Laboratory N.Y. in press. (1977).
15. Seki, T., Oshima, T. & Oshima, Y.: "Taxonomic study of Bacillus by DNA hybridisation and interspecific transformation". Int. J. Syst. Bacteriol. 25, 258-270, (1975).

16. Subbaiah, T.V., Goldthwaite, C.D. & Marmur, J.: Nature of bacteriophages induced in Bacillus subtilis, p. 435-436 in V. Bryson & H.J. Vogel eds. "Evolving Genes and Proteins". Academic Press, N.Y. (1965).

17. Stickler, D.J., Tucker, R.G. & Kay, D.: "Bacteriophage-like particles released from Bacillus subtilis after induction with hydrogen peroxide". Virology, 26, 142-145 (1965).

18. Wright, L.F. & Hopwood, D.A.: "Identification of the antibiotic determined by the PSCP1 plasmid of Streptomyces coelicolor A3 (2)". J. Gen. Microbiol. 95, 96-106 (1976).

# Growth and Activity of the Nematode-Trapping Fungus *Arthrobotrys conoides* in Soil

J. EREN and D. PRAMER

## A. Introduction

Although nematode-trapping fungi are taxonomically diverse, they are ecologically a natural group united by their ability to capture and consume animals of microscopic dimension (1, 3, 9). Prey are trapped by specialized hyphal structures which vary with species, but include mucilaginous knobs or networks that act by adhesion, and constricting or nonconstricting rings that capture nematodes by occlusion (5, 8). Despite their remarkable morphological adaptation, nematode-trapping fungi are not obligate predators. They will grow as saprophytes on various laboratory media, but many do not form traps in pure culture. If nematodes are added, however, there is hyphal differentiation and traps are rapidly formed in response to the presence of prey. Under favorable conditions, nematodes are captured and consumed in large numbers.

Initial attempts to exploit the activity of nematode-trapping fungi for the control of disease were made almost 40 years ago by plant pathologists working in Hawaii on root-knot of pineapple (6, 7). In their first series of experiments, pure cultures of six different species of predaceous fungi were added to nematode-infested soil, but no convincing evidence of a reduction in nematode damage to plants was obtained. However, a useful degree of control was achieved when soil was supplemented with organic matter (chopped green pineapple tops) rather than inoculated with fungi. In explanation of this effect it was suggested that incorporation into soil of green plant tissue provided saprophytic nematodes with food. Development of these nematodes promoted multiplication of nematode predators, including the predaceous fungi. Being indiscriminate in their taste, the fungi caught and consumed not only saprophytic nematodes but the root-knot organism as well, with the result that nematode damage to plants was lessened. In the course of these studies no evidence that control was attributable directly to nematode-trapping fungi was provided, but the food chain postulated has on many occasions been erroneously cited as fact.

Variations of these pioneering studies have been performed periodically by others in various parts of the world. Encouraging results frequently have been reported for laboratory and greenhouse studies of plant protection by nematode-trapping fungi, but the results of field tests generally have been inconclusive and discouraging (10).

The degree of success achieved in attempting biological control is largely a function of available knowledge concerning the system under study. We know that predaceous fungi are able to destroy nematodes under certain conditions, but growth and trap formation are prerequisites to nematode capture, and this initial development depends, not on predation, but, rather, on the ability of the fungi to compete as saprophytes in soil. Empirical attempts at biological control using predaceous fungi will undoubtedly continue to be made, but the probability of success will be increased greatly if they are based on a better understanding of how environmental variables influence the abundance, distribution, and activity of nematode-trapping fungi in soil. Toward this end, we at Rutgers developed an immunofluorescent staining technique (4) and used it to measure the effects of soil conditions on growth and predation by the nematode-trapping fungus *Arthrobotrys conoides*.

## B. Materials and Methods

*A. conoides* produces sticky hyphal loops which are initially discrete and later compounded into extensive networks which capture nematodes by adhesion and entanglement. The culture was maintained on cornmeal agar, and inocula were prepared by growing the fungus in liquid medium for 7 days at 28° C. The mycelium was harvested and homogenized for 1.5 min in a Waring blender, producing a suspension of hyphal fragments with an average individual length of 200 μ. The density of suspensions of hyphal fragments was determined microscopically using a hemocytometer, and adjusted to desired levels by dilution.

The nematode used throughout these studies was the free-living species, *Panagrellus redivivus*. It was cultivated on moist oatmeal in Petri plates, or grown axenically in flasks of liquid medium (2). Nematodes were recovered from soil using Baermann funnels or sieves, and enumerated by direct microscopic count.

The experimental soil was Lakewood sand. It had a pH of 4.3; a moisture-holding capacity of 28%; 93%, 4%, 3% and 0.8%, sand, silt, clay and organic matter, respectively; and a population by plate count of $5 \times 10^5$ bacteria and $3 \times 10^4$ fungi $g^{-1}$. Five g quantities of treated and untreated soil samples were placed in separate series of Petri plates (35 x 12 mm) and adjusted to 60% of moisture-holding capacity. Some plates were inoculated with *A. conoides* at a level of $10^4$ hyphal fragments per g. soil and other plates served as uninoculated controls. The total length of *A. conoides* hyphae present per g soil was measured by immunofluorescent staining (4) immediately after inoculation and at intervals of 1, 3 and 6 weeks storage at 28° C in a humid atmosphere. An increase in total length indicated hyphal extension and vegetative growth, whereas a decrease suggested hyphal deterioration and lysis.

## C. Results and Discussion

The fate of hyphal fragments of *A. conoides* in heat-sterilized and unsterilized samples of soil, with and without added glucose (0.1%), as determined by the immunofluorescent staining technique, is illustrated by Fig. 1.

Hyphal fragments added to unsterilized soil did not persist. However, if glucose was provided, the initial rate of decline was reduced, but, after 6 weeks, survival did not differ significantly in soil with or without a sugar supplement. Hyphal fragments added to sterilized soil increased in length during the first week of incubation and then decreased. There was very rapid development of *A. conoides* in sterilized soil which received glucose. More than an 8-fold increase in linear growth was recorded in 3 weeks, but the beneficial effect was transient, and after 6 weeks, the abundance of *A. conoides* filaments in supplemented, sterilized soil was approaching that supplied initially. There was little or no development of the fungus in soil with or without added glucose unless the indigenous population was first eliminated by sterilization.

To obtain direct evidence in support of a conclusion that *A. conoides* does poorly if it must compete with other microorganisms for sugar in unsterilized soil, 15 species of fungi and 12 species of bacteria were randomly isolated from Lakewood sand. Mixtures of these organisms in combination with *A. conoides* were returned to sterilized soil. The soil fungi and bacteria were added at levels of $2 \times 10^4$

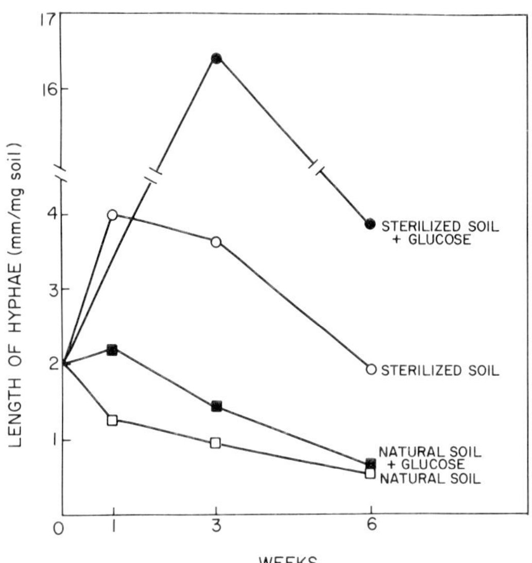


Fig. 1. Influence of sterilization and added glucose on A. conoides in soil. The soil was heat-sterilized and supplemented with 0.1% glucose

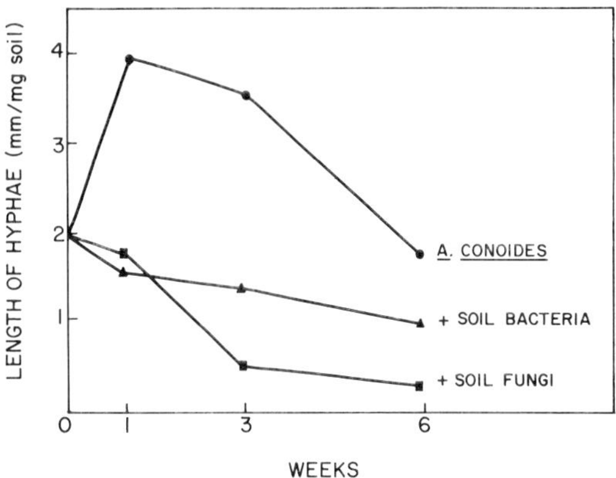


Fig. 2. Influence of added bacteria and fungi on development of A. conoides in sterilized soil. The bacteria and fungi were first isolated from the experimental soil and then returned to sterilized soil at levels of $2 \times 10^5$ and $2 \times 10^4$, respectively

and 2 x $10^5$ cells g-1 soil, respectively, and their influence on the development of A. conoides was measured by immunofluorescent staining.

The results illustrated in Fig. 2 confirmed the previous observation that there is limited linear growth of A. conoides in unsupplemented sterilized soil. The addition of populations of either soil bacteria or soil fungi had an unfavorable effect. The mixture of fungal species was more antagonistic than the bacterial mixture, possibly because Lakewood sand has an acid reaction which would favor fungi and depress the activity of bacteria.

Fig. 3 illustrates the results of a study designed to measure the competitive saprophytic ability of A. conoides in soil that was not sterilized but was supplemented with organic matter more complex than glucose and more typical of agricultural practice. Although the same total amount of organic matter (0.5%) was added in each case, the nitrogen content of the soil supplements differed greatly (oat straw, 0.4%N; alfalfa meal, 3% N; and dried blood, 14.5% N).

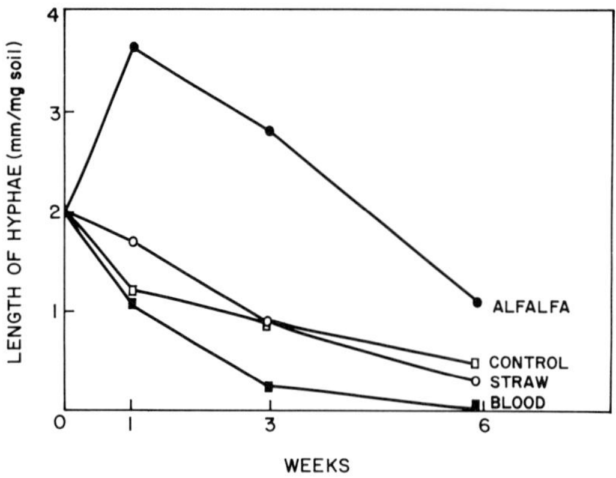


Fig. 3. Influence of organic matter on development of A. conoides in unsterilized soil. The soil was supplemented at a level of 0.5% organic matter in each case

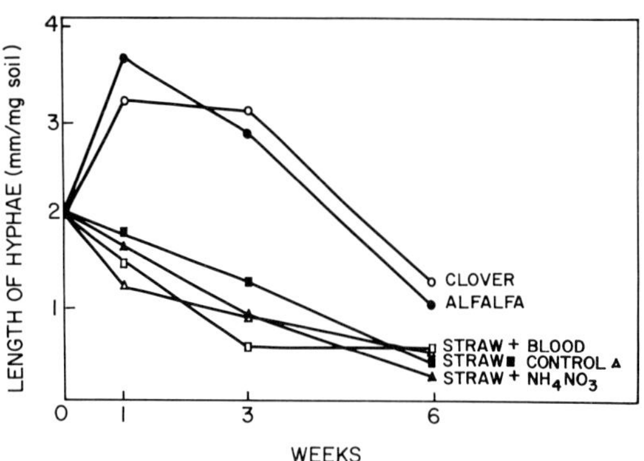


Fig. 4. Influence of organic matter on development of A. conoides in unsterilized soil. Alfalfa, clover and straw were added at a level of 0.5%. The straw was augmented with dried blood (0.09%) or $NH_4NO_3$ (0.03%) to establish in the straw treated soils a C/N ratio equivalent to that of the leguminous plant tissue

Alfalfa meal supported rapid growth of A. conoides in unsterilized soil. Hyphal extension was at a maximum after 1 week, but it was not sustained at that level and suffered a sharp subsequent decline. Straw had no significant effect on growth or survival of A. conoides in unsterilized soil, and dried blood had an adverse influence. The reaction of soil treated with alfalfa and dried blood increased to pH 6.4 and pH 7.7, respectively. In the latter case, ammonia produced by deamination of blood protein may have reached toxic levels and caused lysis of fungal filaments.

The beneficial effect of alfalfa may be explained by a C/N ratio which is more favorable for fungal growth than the C/N ratios of straw and blood. Alternatively, A. conoides may be biochemically better equipped than other soil microorganisms to metabolize alfalfa, and have little or no advantage in competing for straw or blood. To determine which of these two possibilities was operative, a comparison was made of the development of A. conoides in unsterilized soil samples supplemented with alfalfa or clover, or with straw plus inorganic ($NH_4NO_3$) or organic (dried blood) nitrogen, to establish in the straw treated soils a C/N ratio equivalent to that of the leguminous plant tissue.

As illustrated in Fig. 4, alfalfa and clover both supported rapid vegetative growth, but the effect did not persist, and a decrease in survival of A. conoides was noted after 3 and 6 weeks. Soil samples supplemented with straw alone or straw in combination with inorganic or organic nitrogen did not support growth of A.

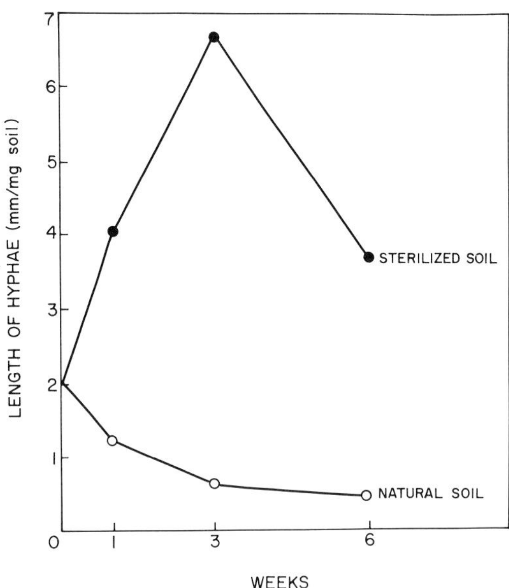


Fig. 5. Influence of sterilization and added nematodes on A. conoides in soil. P. redivivus was used at a level of 100 worms $g^{-1}$ in sterilized and unsterilized soil samples inoculated with A. conoides

conoides. The fate of the fungus in these soils was similar to that in soil which received no supplement.

Since different results were obtained for different supplements formulated to have equivalent C/N ratios, the beneficial effects were not due solely to the elemental composition of leguminous plant tissue. Nevertheless, alfalfa and clover may have acted indirectly rather than directly. They may not be suitable substrates for A. conoides, but able instead to support growth and multiplication of saprophytic soil nematodes. Increased numbers of nematodes would be expected to favor nematode-trapping fungi, and to provide A. conoides with a competitive advantage for development in soil. This line of reasoning is analogous to that employed many years ago by Hawaiian plant pathologists to explain the control of root-knot of pineapple achieved by treatment of nematode infested soil with green plant tissue (6, 7). It suggests a linear sequence of links comprising a food chain in which plant tissue is consumed by nematodes, and the nematodes are then consumed by nematode-trapping fungi. If, in fact, this were the case, nematodes alone should support growth of their predators, and in the presence of prey the need to supplement soil with plant tissue should be obviated. However, when this possibility was tested directly by adding P. redivivus (100 worms g-1) to samples of sterilized and unsterilized soil inoculated with A. conoides, the fungus was able to take advantage of the presence of prey in sterilized soil only. There was no favorable effect of added nematodes on A. conoides in unsterilized soil (Fig. 5).

It is not clear why A. conoides was constrained in unsterilized soil and able to act predaceously in sterilized soil. Possibly, it was due to destruction by heat sterilization of soil microorganisms that interfered with the process of predation, or there may have been a temperature activated release to a deficient soil solution of nutrients required for the growth of A. conoides, or both. It is very clear, however, that no direct or simple relationship exists between the activity of nematode-trapping fungi and the mere presence of prey in natural soil.

In the course of these studies, A. conoides had displayed an ability to grow saprophytically on leguminous plant tissue but not predaceously on nematodes in unsterilized soil. It remained necessary, however, to combine these treatments and examine the fate of nematodes in unsterilized soil which was not only inoculated with A. conoides, but supplemented with alfalfa meal as well. Table 1 lists the results of a series of tests in which the effects of various individual and combined treatments on predation of P. redivivus by A. conoides in unsterilized soil were measured by the enumeration of surviving nematodes.

Table 1. Influence of treatment on numbers of nematodes in soil

| Treatment | Sample | Numbers of nematodes/125 g soil* 1 week | 2 weeks | 4 weeks |
|---|---|---|---|---|
| 1. Nematodes | 1 | 2,700 | 1,900 | 1,000 |
| | 2 | 2,300 | 1,450 | 1,150 |
| 2. A. conoides | 1 | 23 | 28 | 24 |
| | 2 | 15 | 16 | 35 |
| 3. Nematodes and | 1 | 1,40C | 1,500 | 450 |
| A. conoides | 2 | 2,200 | 1,300 | 1,000 |
| 4. Alfalfa | 1 | 19 | 60 | 500 |
| | 2 | 30 | 48 | 850 |
| 5. Alfalfa and | 1 | 9,500 | 6,300 | 12,000 |
| nematodes | 2 | 7,100 | 7,700 | 10,800 |
| 6. Alfalfa, | 1 | 1,080 | 650 | 1,600 |
| nematodes, and A. conoides | 2 | 1,200 | 530 | 1,250 |

* The experimental soil contained an average of 25 indigenous nematodes and it was inoculated with 8,150 P. redivivus, to yield a total population of 8,175 nematodes/125 g at time zero.

Soil supplemented with nematodes only (treatment 1) did not support their growth and the population of worms decreased from an initial level of 8,150 to approximately 1,000 after 4 weeks. A. conoides had little or no significant influence on the number of nematodes indigenous to the soil (treatment 2), but when the fungus and P. redivivus both (treatment 3) were added to soil, A. conoides appeared to have caused a slight decrease in numbers of nematodes.

Analyses of the alfalfa-treated soil samples provided the most significant results. The indigenous nematode population of soils that received alfalfa increased more than 25-fold in 4 weeks (treatment 4). P. redivivus multiplied in alfalfa-treated soil (treatment 5) but the magnitude of its response to alfalfa was not as great as that of the indigenous nematode population (treatment 4). There was a marked decrease in the number of nematodes in soil samples that received alfalfa and hyphal fragments (treatment 6) of A. conoides. The effect was evident after 1 week and most pronounced after 2 weeks. However, analyses after 4 weeks revealed an increase in the number of nematodes: the effect had been transient and the worm population had more than doubled. This failure of A. conoides and added organic matter to produce a prolonged beneficial effect and hold nematode populations at reduced levels for more than 2 weeks is consistent with the short-lived effects of organic matter previously described and discussed.

The sequence of experiments described here produced an accumulation of results which support the position that under some conditions at least, predaceous fungi can control nematode populations in unsterilized soil. In the case of A. conoides, inoculation with the fungus was itself inadequate. Most attempts at soil inoculation have little chance of success, particularly if the inoculant was isolated in the first place from soil. Moreover, for predaceous fungi to capture and kill nematodes, they must first grow and produce traps. It is not their presence but their growth that must be initially encouraged. Certain types of organic matter appear able to provide that encouragement. Glucose does not, unless the soil is first sterilized, but leguminous plant tissue is selectively colonized by A. conoides and used by the fungus to develop saprophytically in soil which is not sterilized. Nematodes added to soils supplemented with leguminous plant tissues were captured and consumed by A. conoides. The beneficial effect was transient, however, and it could not be explained satisfactorily by a simple linear food chain. Added organic matter was used by prey as well as by predator, and an understanding of its mode of action must be sought in the complex system of interacting variables which relate the growth of nematode-trapping fungi to the destruction of their prey. Nevertheless, the immunofluorescent staining technique was the key to unlocking a storehouse of information concerning the ecology of A. conoides, and there is cause for optimism. Much remains to be learned, however, and information gained in the laboratory ultimately will have to be field tested to determine if predaceous fungi can indeed protect plants from nematode diseases of economic significance.

## D. References

1. Cooke, R.C., Godfrey, B.E.S.: A key to the nematode-destroying fungi. Brit. mycol. Soc. Trans. 47, 61-74 (1964).
2. Cryan, W.S., Hansen, E., Martin, M., Sayre, F.W., Yarwood, E.A.: Axenic cultivation of the dioecious nematode Panagrellus redivivus. Nematologica 9, 313-319 (1963).
3. Duddington, C.L.: The Friendly Fungi. London: Faber and Faber, 1957, p. 188.
4. Eren, J., Pramer, D.: Application of immunofluorescent staining to studies of the ecology of soil microorganisms. Soil Sci. 100, 39-45 (1966).
5. Estey, R.H., Tzean, S.S.: Scanning electron microscopy of fungal-nematode trapping devices. Tran. Br. mycol. Soc. 66, 519-522 (1976).
6. Linford, M.B.: Stimulated activity of natural enemies of nematodes. Science 85, 123-124 (1937).
7. Linford, M.B., Yap, F., Oliveira, J.M.: Reduction of soil populations of the root-knot nematode during decomposition of organic matter. Soil Sci. 45, 127-141 (1938).
8. Nordbring-Hertz, B.: Scanning electron microscopy of the nematode-trapping organs in Arthrobotrys oligospora. Physiol. Plant. 26, 279-284 (1972).
9. Pramer, D.: Nematode-trapping fungi. Science 144, 382-388 (1964).
10. Pramer, D.: Predaceous fungi and the biological control of nematode pests. In: Proceedings 13th Annual Meeting, Agricultural Research Institute. National Academy of Sciences, Washington, D.C., pp. 47-53 (1964).

# The Influence of Profile Depth on Vertical Distribution of Soil Fungi in a Humid Tropical Environment

A.B. OKUK and E.U. OKPALA

## A. Introduction

Soil organic matter content, moisture, aeration, temperature, pH and the other edaphic factors have been studied in relation to distribution of fungi in soil. Decrease in fungal numbers with depth has been reported (5, 7, 12, 13). Some work correlates decrease in organic matter with depth, with decrease in fungi with depth (1, 3). Other work suggests that distribution of fungi may be independent of organic matter distribution. For example, fungal numbers decreased in peat soils with depth although the organic matter content of the soil was consistent throughout the profile (8). In tropical soils it was found that presence of organic matter in sub-surface soil layers did not result in higher fungal numbers (2).

The work reported here is an attempt to correlate organic matter distribution in a tropical soil in Nigeria with fungal numbers. It is the first of a series of studies on a number of soils.

## B. Materials and Methods

I. Soil sampling. A pit (90 cm x 120 cm x 180 cm) was dug at the edge of a site, burnt annually to limit poisonous vipers and to improve regrowth of the grass *Landettia* sp. used for mat production.

Samples were collected with a sterile auger at 15 cm intervals from the bottom of the pit up by pushing the auger 8 cm horizontally into the profile. The samples from the four sides of the pit were placed in polythene bags and the four samples from each depth pooled (6) and sieved ( 2mm mesh). Drying was not necessary as collection was made between October 1972 and February 1973 when soils around Nsukka are caked.

Samples were processed within 4 h of collection. Moisture content was estimated following drying at 105$^{o}$C.

II. Dilution pour plates. Methods used were similar to those described (9, 10). Soils were diluted to $10^{-4}$ and glucose nitrate agar (GNA) was the plating medium. Streptomycin (4 $\mu$g ml$^{-1}$) was added to each plate after pouring and before setting. Fast growing colonies were aseptically removed within the first four days after counting, and new colonies subsequently counted. Counts were made finally on the 8th day except for samples from the deeper horizons which were counted on the 10th day.

III. Soil organic matter content and soil pH. The method for estimating oxidisable organic matter was as described (4). pH was determined by placing sieved samples in grooves in a slab. Each was moistened with distilled water, excess water poured off and Universal Indicator added. The colour change was compared with a chart.

## C. Results

The greatest number of fungi were found between 15-60 cm (Table 1) with numbers between 15-30 cm, 30-45 cm, and 45-60 cm all showing significant differences. There were no significant differences in numbers in samples taken between 60-136 cm. A significant drop occurred at 136 cm.

Examination of the effect of pH and organic matter on distribution of fungi showed that pH remained relatively constant through the profile (Fig. 1).

Table 1. Number of fungal colonies $g^{-1}$ over dry soil at various depths of the soil profile

| Depth of Profile cm | | Fungal colonies 1000/g dry soil |
|---|---|---|
| 0 - 15 | a. | 18.50 |
| 15 - 30 | b. | 29.50 ab*** |
| 30 - 45 | c. | 35.25 bc*** |
| 45 - 60 | d. | 28.00 cd*** |
| 60 - 75 | e. | 26.26 |
| 75 - 90 | f. | 23.00 |
| 90 - 105 | g. | 20.00 |
| 105 - 120 | h. | 17.25 |
| 120 - 135 | i. | 15.00 |
| 135 - 150 | j. | 10.75 ij* |

Figures are means of 4 replicates
L.S.D., Po 5 = 3.28*; Po 1 = 4.43**; Po 01 = 5.90***
a, b - i, j, treatments compared for significant differences.

Table 2. Fungi dominant at various depths of the soil profile

| Profile depths, cm. | Identified Fungi |
|---|---|
| 0 - 30 cm | Curvularia tetramera, Rhizopus nigricans, Cladosporium sp. Fusarium sp. Trichoderma viride, Zygorhynchus sp. Penicillium sp. Alternaria sp. |
| 30 - 60 cm | Mortierella sp. Chaetomium globossum, Trichoderma viride, Trichoderma sp. (white) Penicillium sp. |
| 60 - 90 cm | Mortierella sp. Trichoderma sp. (white) Penicillium sp. |
| 90 - 120 cm | Pullularia pullulans, Penicillium sp. |
| 120 - 150 cm | Trichoderma sp. (white) Penicillium sp. (slow developers). |

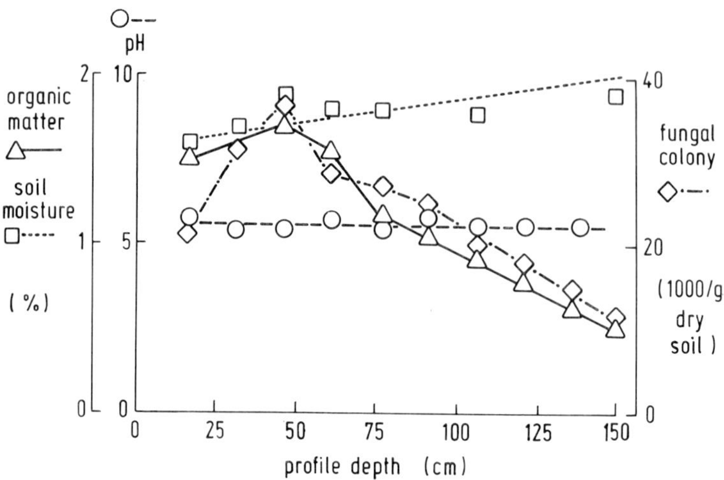


Fig. 1. Effect of soil organic matter, soil moisture and pH on fungal distribution in a soil in the Nsukka area

Organic matter content was highest at about 40 cm depth (Fig. 1) at which depth the greatest number of fungi were found. There appears to be a positive correlation between fungal numbers and organic matter content. The fungal types dominant at each level are shown in Table 2.

No explanation can be offered as to why fungi from the lower horizons grew slowly.

An attempt was made to correlate fungal distribution with soil moisture. Soil moisture was lowest at the top of the profile and increased with depth (Fig. 1), but as it was the dry season, moisture content was low: 1.6% - 2.0%. The moisture content did not appear to affect distribution.

## D. Discussion

Organic matter distribution appears to be the main factor affecting distribution of fungi in the soil at the time of sampling this particular soil.

The very slight variations in water content do not appear to affect fungal numbers. The presence of fungi at 150 cm may be due to the gravelly nature of the soil (57.6% coarse sand).

One of the interesting findings was the slow growth of fungi from lower depths. It has been suggested (11) that slow growing fungi may occur in the upper horizons, but are suppressed by faster growing strains. The reason for their slow growth remains to be investigated.

## E. References


1. Brown, J. C. Fungal mycelium in dune soils estimated by a modified impression slide technique. Transactions of the British Mycological Society 41, 81-88 (1958).
2. Eicker, A. Vertical distribution of fungi in Zululand soils. Transactions of the British Mycological Society 55, 45-57 (1970).
3. Gray, P.H.H. & Taylor, C.B. A Microbiological study of podsoil profile 11. Laurentian soils. Canadian J. of Research 13, 251-255 (1935).
4. Jackson, M. L. Soil Chemical Analysis. Constable & Co. Ltd., London. (1958).

5. Sewell, G.W.F. Studies of fungi in a Calluna heathland soil 1. Vertical distribution in soil and on root surfaces. Transactions of the British Mycological Society 42, 343-353 (1959).

6. Schmithener, A.F & Williams, L.S. Methods for analysis of soil-borne pathogens and associated fungi. Botany & Plant Pathology Mimio. series No. 29 (1958).

7. Stenton, H. The soil fungi of Wicken fen. Transactions of the British Mycological Society 36, 304-314 (1953).

8. Timonim, M.I. The micro-organisms in profiles of certain virgin soils in Manitoba. Can. J. of Research 13, 32-146 (1935).

9. Timonim, M. I. The interaction of higher plants and soil micro-organisms 1. Microbial population of rhizosphere of seedlings of certain cultivated plants. Can. J. of Research, 18, 307-317 (1940).

10. Waksman, S.A. & Fred. A tentative outline of the plant method of determining the number of micro-organisms in the soil Soil Science 41, 27-28 (1922).

11. Warcup, J.H. The ecology of soil fungi. Transactions of the British Mycological Society, 34, 376-399 (1951).

12. Warcup, J.H. Studies on the occurrence and activities of fungi in a wheat field soil. Transactions of the British Mycological Society 40, 237-262 (1957).

13. Yong, C. & Stenton, H. A study of the Phycomycetes in the soils of Hong Kong. Transactions of the British Mycological Society, 47, 127-139 (1964).

# Short-Term Measurement of Denitrification Rates in Soils Using $^{13}N$ and Acetylene Inhibition Methods

J.M. TIEDJE, M.K. FIRESTONE, M.S. SMITH, M.R. BETLACH, and R.B. FIRESTONE

## A. Introduction

Denitrification has long been recognized as a basic component of the nitrogen cycle: it results in the loss of combined nitrogen, a limiting nutrient for primary productivity. Though the importance of this loss to agricultural and natural ecosystems is well known, denitrification remains the least understood transformation in the nitrogen cycle. In addition there has been recent concern about the denitrification intermediates, $N_2O$ and $NO_2^-$. The former catalyses destruction of the ozone layer through its photochemical product NO and the latter reacts with secondary amines to form nitrosamines which are potent carcinogens. There has been much recent interest also in use of denitrification to rid waste products of excessive nitrogen. Despite the general significance of this process, research progress has been limited because of the absence of direct, sensitive, and precise methodology to assay denitrification in nature.

We have used two new and complementary methods which overcome this previous methodological limitation. One method employs the radioactive isotope of nitrogen, $^{13}N$, because of the greatly improved sensitivity it provides. The specific activity is $1.04 \times 10^{12}$ mCi/g-atom which provides a minimum detectable amount of $3 \times 10^{-14}$ µmoles. The other uses acetylene to inhibit $N_2O$ reductase thereby causing stoichiometric accumulation from nitrate of nitrous oxide, which can be easily measured by gas chromatography (1, 4). In this paper we report on the evaluation of and information obtained from the use of these two methods to measure rates and products of denitrification in fresh soils.

## B. Materials and Methods

I. $^{13}N$. $^{13}N$ has a half-life of 10 minutes, emits positrons (1.19 MeV) which, upon annihilation, formed two 0.511 MeV gamma rays. We generate $^{13}NO_3^-$ at the MSU Sector Focused Cyclotron by using 0.7 ml of water as the target, employing the $^{16}O$ (p, $\alpha$)$^{13}N$ reaction. We used a 0.7 to 4 µA, 11.5 to 14.5 MeV proton beam bombarding for 10 minutes. A pneumatic rabbit is used to transport the target in and out of the beam. The nitrogen composition of the product is >80% nitrate with lesser quantities of nitrite and ammonium present. The latter two contaminants can be removed by oxidation and volatilization respectively.

Soils were incubated in either sealed flasks or flasks in which soil slurries were constantly flushed with helium. The latter incubation system and its subsequent gas collection and detection system are shown in Figure 1. The $^{13}NO_3^-$ (∿1 mCi, 4.4 pg) with or without $^{14}NO_3^-$ carrier was added to the soil slurry which was mixed by a magnetic stirrer and continuously stripped of product gases with helium. The flushed gases passed first through a trap which consists of tubing immersed in liquid nitrogen, and subsequently through a second trap which consists of tubing packed with Molecular Sieve and immersed in liquid nitrogen (2). The detectors are NaI crystals attached to photomultiplier tubes. The data are continuously

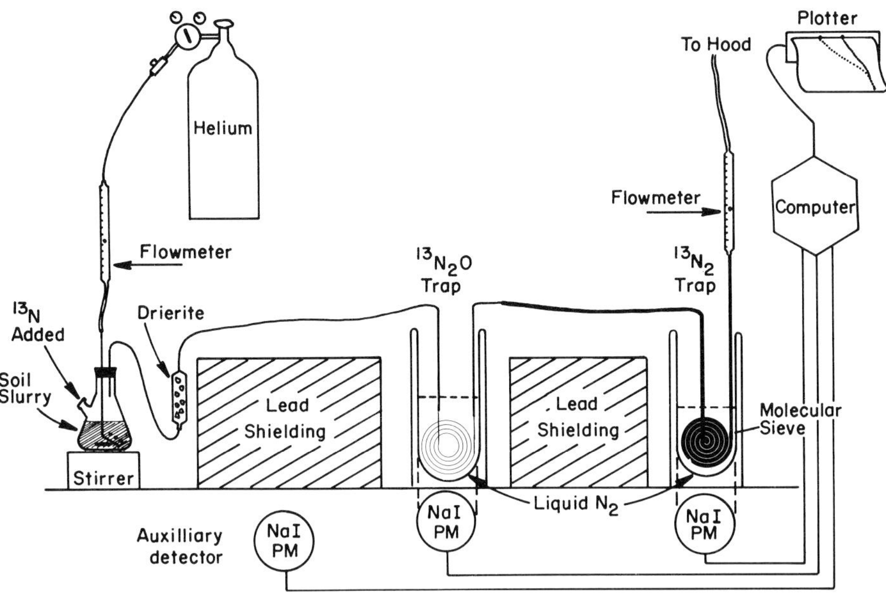


Fig. 1. Flushing and differential trapping system used to study $^{13}N_2O$ and $^{13}N_2$ produced by denitrification of $^{13}NO_3^-$ and $^{13}NO_2^-$ in soils

collected and fed to a computer where the data are corrected for decay and background.

Sealed flasks containing either soil slurries or field moist soil were incubated with $^{13}NO_3^-$. The headspace gases were monitored by gas chromatography which was connected to a proportional counter to determine the radioactivity. The proportional counter was also connected to the computer which made the decay and background corrections.

II. Acetylene inhibition method. Soil slurries or field moist soils were incubated in sealed Erlenmeyer flasks. The flasks were twice evacuated and flushed with helium or argon to achieve anaerobic conditions. The slurries were incubated on a rotary shaker. Headspace gas samples were periodically removed by syringe and analyzed by gas chromatography. A Carle gas chromatograph equipped with micro-thermistor detector and added operational amplifier was used to quantitate $N_2O$ in concentrations >100 ppm. A Perkin-Elmer gas chromatograph with a $^{63}Ni$ electron capture detector was used for measuring $N_2O$ concentrations from ambient to 100 ppm (3). $N_2O$ headspace concentrations were corrected for $N_2O$ dissolved in the aqueous phase.

## C. Results

Rates of $^{13}N_2O$ and $^{13}N_2$ production determined by the flushing system are shown in Figure 2. At 31 min, 20 ppm $^{14}NO_2^-$-N was added. The rate of labeled gas produced slowed, reflecting the lowered specific activity of this substrate pool. This evidence indicates that denitrifiers were responsible for the labeled gas production. The very early (<10 min) lag and subsequent linear rate have been observed in all samples studied.

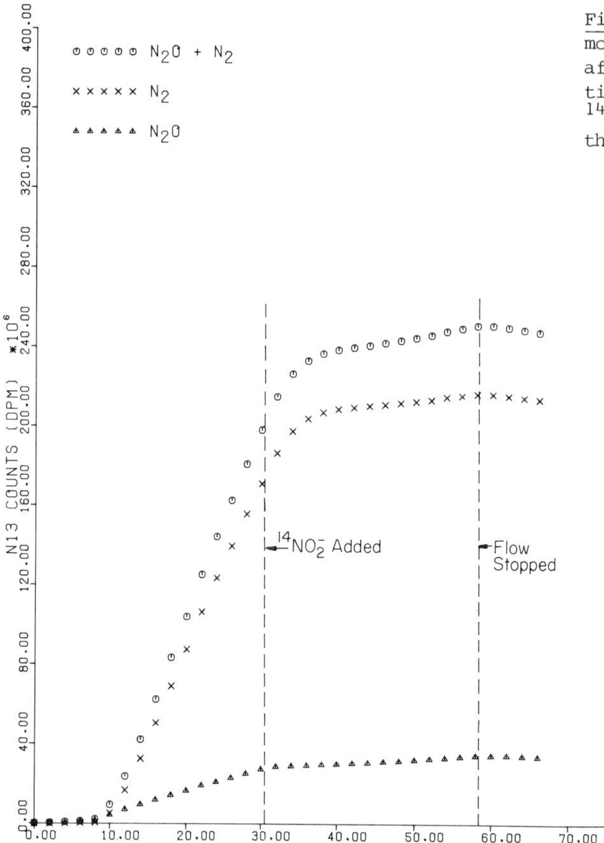


Fig. 2. $^{13}N$ gases produced, as monitored by the flushing system, after onset of anaerobic conditions at -5 min. At 31 min 20 ppm $^{14}NO_2^-$-N was added and at 58 min the flushing gas flow was stopped

The pattern of denitrification rate as measured by the acetylene inhibition method is shown in Figure 3. A linear phase is observed for the first 100 min in most soils. Thereafter the rate accelerates, usually achieving a second linear phase after 300 to 400 min. The effect of the protein synthesis inhibitor, chloramphenicol, on the change in rates for a loam (Brookston) soil and a sand which has been enriched for denitrifying activity is also shown in Figure 3. Clearly, chloramphenicol inhibits the acceleration in rate especially in the sand. The difference noted between the soils is probably due to the binding and thus inactivation of the chloramphenicol by the reactive soil constituents which would be expected to be more prevalent in a loam than a sand. It is of note that in the sand the first linear phase is extended for a much longer period.

If soil is incubated anaerobically for longer periods than discussed so far, an additional accelerated rate is sometimes observed. This usually occurs after 10 hours. On occasion an increase in denitrifier population during this period has been shown. Whether or not this third phase occurs seems to depend on carbon availability. Added carbon has a far more pronounced effect during this phase than for either of the previous two.

The products of denitrification vary with phase. The major product in Phase I, is $N_2$, as is shown in Figure 2, while the major product in Phase II is commonly $N_2O$. These findings have been confirmed by both the $^{13}N$ and acetylene inhibition methods ($N_2O$ production compared in paired flasks, one with acetylene and the other without).

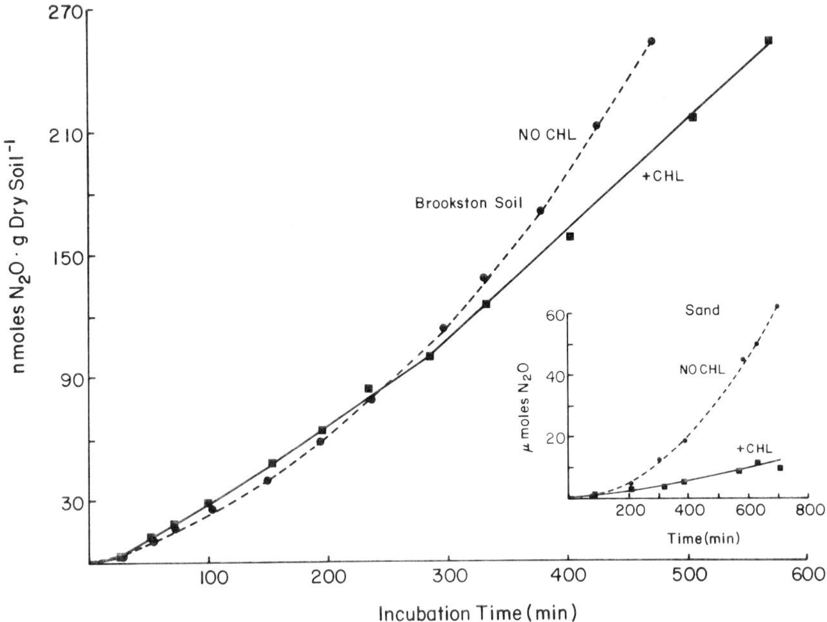


Fig. 3. Rate of denitrification in a Brookston loam soil as measured by the acetylene inhibition method in the presence and absence of chloramphenicol. The inset shows results of the same experiment done with a sand which had been enriched by denitrifier activity

Denitrification rates obtained by the two methods have been compared. Examples of these comparisons for $^{13}N$ and acetylene inhibition methods are, respectively: Brookston loam soil, 0.51 vs, 0.49 nmoles gas · g soil$^{-1}$ min$^{-1}$ ; muck soil, 2.15 vs 1.88, Spinks loamy sand, 0.10 vs 0.20. The source, handling, temperature, and nitrate additions were as nearly identical as possible. The only difference was that the $^{13}N$ method required a flushing system while the acetylene inhibition method required a sealed flask. The rates obtained did correlate with the organic matter content of these three soils.

By use of the $^{13}N$ assay the degree of inhibition of $N_2O$ reductase by acetylene have been verified. Interestingly, with low nitrate concentrations (<1 ppm) more acetylene was required for adequate inhibition. Generally 0.1 to 0.15 atm acetylene gave >90% inhibition at low nitrate concentrations while 0.01 atm was adequate at high nitrate concentrations.

## D. Discussion

There appears to be a consistent pattern of change in denitrifier activity following the onset of anaerobiosis which has recognizable and explicable features and is summarized in Table 1. We do not mean to imply that these time intervals apply to all soils nor that all soils necessarily exhibit all distinct phases (for example III may be absent or II may be a continuum with III).

The equilibrium phase is seen only with the $^{13}N$ method and is thought to be due to time lags associated with removal of oxygen, mixing of added label, and transport of the product gas to the traps. Phase I is considered to reflect the activity of the native denitrifying enzyme concentration before de novo synthesis in response to the new assay environment. This explanation is supported by the clearly linear rate seen by both methods for this period and the lack of effect by chloramphenicol on this rate. The subsequent exponential pattern is suggested to be due to active synthesis of new denitrifying enzymes since this is the first period in which chloramphenical shows an effect. Eventually a linear rate is noted which is probably after the maximum concentration of denitrifying enzymes has been synthesized by the indigenous population and before cell division has become significant. This linear phase is described as Phase II. The higher rate of this phase indicates that the native denitrifying biota in agricultural soils are partially repressed in dentrifying enzyme synthesis.

Subsequent to Phase II, there may be an increase in denitrifier population density. The greatly increased rate after this longer period of incubation and the strong effect of added carbon on the rate are consistent with this explanation.

The elucidation of these phases suggests that as moisture and thus aeration states change in the soil microenvironment, the denitrifying population will show a temporal response similar to the one described here. Thus in nature denitrification could occur as spurts of activity following rainfall, with the quantity of nitrate lost being controlled by the mass of denitrifying enzymes in oxygen limited sites.

This pattern also implies that incubation studies of longer than 10 hours will entail an enrichment for denitrifiers which is not likely to occur in nature because of the lack of carbon, nitrate (by this time), and length of anaerobic period required. Thus the rates obtained from such studies, usually called potentials, are probably unrealistically high and do not resemble any condition that occurs in nature.

The change in the predominant products from $N_2$ to $N_2O$ was unexpected. This observation is important to estimates of the impact of denitrification on the ozone layer. This drastic change in $N_2O/N_2$ ratios suggests that it is unlikely that a constant ratio will be of use to describe $N_2O$ losses from a variety of soils under a variety of conditions.

We have found that these two new methods for short-term rate estimates of denitrification can be used for measurement of denitrification rates in both Phase I and

Table 1. Generalized description of the phases of denitrification in soil after onset of anaerobic conditions

| Phase | Time[a] | Explanation | Shape of curve |
|---|---|---|---|
| Equilibrium | 0 to 10-15 min | Adjustment to anaerobiosis and assay conditions | |
| I | 10-15 min to 100 min | Native enzyme activity | Linear |
| Transition | 1.5 to 4 hr | Active derepression of enzyme synthesis | Exponential |
| II | 4 to ∿10 hr | Full derepression of denitrifying enzymes by indigenous community | Linear |
| III | >10 hr | Growth | Logarithmic |

[a] Generalized time intervals; individual soils may vary from these values.

II. The rates obtained by both methods for the three soils under similar conditions were sufficiently similar, given the heterogeneity of soil, to be encouraging. A point of caution in using the acetylene inhibition method is the important effect of nitrate concentration on degree of inhibition. Commonly in nature nitrate concentrations are low and thus high concentrations of acetylene, at least 0.1 atm, would probably be required to achieve the desired inhibition.

## E. Acknowledgements

This work was supported in part by the U.S. Department of Agriculture Regional Research Project NE-39 and by an Unrestricted Grant Award from Eli Lilly Research Laboratories. Published as Journal Article No.8229 of the Michigan Agricultural Experiment Station.

## F. References

1. Balderston, W.C., Sherr, B., Payne, W.S.: Blockage by acetylene of nitrous oxide reduction in *Pseudomonas perfectomarinus*. Appl. Environ. Microbiol. 31, 504-508 (1976).
2. Gersberg, R., Krohn, K., Peck, N., Goldman, C.R.: Denitrification studies with $^{13}N$ labelled nitrate. Science 192, 1229-1231 (1976).
3. Rasmussen, R.A., Krasnec, J., Pierotti, D.: $N_2O$ analysis in the atmosphere via electron capture gas chromatography. Geophys. Res. Letters 3, 615-618 (1976).
4. Yoshinari, T., Knowles, R.: Acetylene inhibition of nitrous oxide reduction by denitrifying bacteria. Biochem. Biophys. Res. Comm. 69, 705-710 (1976).

# Nitrogen Fixation and Mineralisation in Kauri *(Agathis australis)* Forest in New Zealand

W.B. SILVESTER

## A. Introduction

Although forest ecosystems apparently conserve nutrients by efficient cycling, losses of nutrients occur continuously at a slow rate and may be replaced by weathering and rainfall input. In New Zealand, Agathis forest remnants show enormous accumulation of nitrogen to 16tN $ha^{-1}$ , with half or more in the living and dead organic matter (8). With a paucity of legumes in New Zealand and non-legume symbiotic diazotrophs confined to pioneer locations (6), significant asymbiotic nitrogen fixation was shown to be associated with rhizosphere of mycorrhizal short roots of many native conifers (7). This activity was shown to be non-specific and to extend out into the decaying leaf litter in which the feeder roots penetrate. Despite the very large accumulation of total nitrogen in this forest there is evidence of nitrogen deficiency (3) and the mineralisation of organic nitrogen appears to be an important limitation to plant growth.

## B. Materials and Methods

Nitrogenase activity was measured by acetylene reduction assay using 10% acetylene in air (7). Litter fall was measured at four sites over two years using $1m^2$ traps emptied monthly. Total nitrogen was determined by kjeldahl digestion using copper/selenium catalyst and distillation of ammonia. Ammonia was extracted from leaf litter by shaking for 24 hours in 1:10 ratio of litter to 1 molar KCl and distilling the ammonia liberated by addition of MgO.

$^{15}N_2$ uptake was conducted using a gas mixture of $Ar/O_2/N_2$:60/20/20 with 72 and 40 atoms percent excess $^{15}N_2$. The material was twice extracted in 70% ethanol by grinding in a pestle and mortar and the alcoholic extract kjeldahled. $^{15}N$ enrichment was assayed in an GEC/AE1 MS10 mass spectrometer, with all enrichments measured against standard unenriched material.

## C. Results

I. Structure of litter profile. Mor humus under kauri may accumulate to a depth of 2m adjacent to large trees with an average depth of 46cm at the Trounson site. The litter is clearly defined into litter L, fermenting F and humus H layers and the fine feeding roots of kauri are confined to the bottom F and top H layers. The distribution of this material and its nitrogen content are given in Table 1. This organic layer overlying the organic soil has an initial very high C/N ratio which with increasing age is reduced from 82 to 36 on average at this site, there is a very large amount of immobilised nitrogen especially in the H layer.

II. Nitrogenase activity. 1. *In situ* activity. A typical profile of nitrogenase activity is given in Fig. 1 which shows the very close relationship with pH and the specific location within the F layer. The nitrogenase activity shown of 10 n moles $C_2H_4g^{-1}h^{-1}$ is a good average for this material. However forest litter

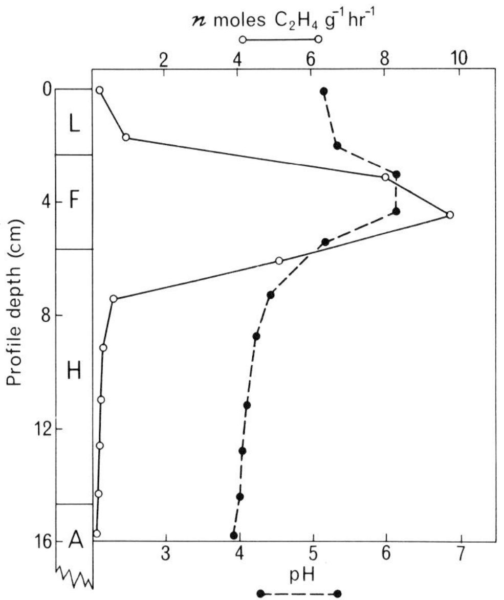


Fig. 1. Nitrogenase activity and pH profiles of litter under kauri forest at Trounson Park

under kauri is heterogeneous with contributions from other species and from wood and reproductive organs. This is demonstrated in Table 2 where various components of the fermenting layer were identified and assayed separately.

2. $^{15}N_2$ uptake. $^{15}N_2$ was used along with parallel acetylene reduction assays to determine the relationship between the two processes and showed that there was a consistent ratio of 3.26 (Table 3) for decaying kauri leaves.

3. Assessment of annual nitrogen input. The mean rate of acetylene reduction in all material has been estimated as 4.30 n moles $C_2H_4g^{-1}h^{-1}$ and for all F material as 3.39 n moles $C_2H_4g^{-1}h^{-1}$. Using the conversion factor to nitrogen fixation given above and the estimate of total L of 15t $ha^{-1}$ and F of 40t $ha^{-1}$ then nitrogen input by fixation (assuming activity for one half of the year) is 2.42 kg N $ha^{-1}a^{-1}$ for L and 5.10 kg N $ha^{-1}a^{-1}$ for F a total of 7.52 kg N $ha^{-1}a^{-1}$. A variety of other measurements confirm this value and extensive work in other forest sites (1) indicate that natural forest systems have an annual nitrogen fixation in decomposing litter of 5-20kg N $ha^{-1}a^{-1}$.

Table 1. Forest floor litter biomass and nitrogen content under kauri at Trounson Park. Based on mean depth of 46cm and 30% of H as roots

| Layer | Density | Depth cm | Volume $m^3ha^{-1}$ | Mass t $ha^{-1}$ | N Content % | N Content t $ha^{-1}$ | C/N |
|---|---|---|---|---|---|---|---|
| L | 0.05 | 3 | 300 | 15 | 0.55 | 0.08 | 82 |
| F | 0.10 | 4 | 400 | 40 | 0.80 | 0.32 | 56 |
| H | 0.18 | 39 | 2730 | 491 | 1.25 | 6.14 | 36 |
| | | 46 | 3430 | 546 | | 6.54 | |

C/N based on C = 45% (kauri values range 43-49%)

4. Substrate addition. A variety of nutrients were added to leaves by allowing leaves to stand on filter paper soaked in a nutrient solution. The results (Table 4) show that while glucose stimulated higher activity the main stimulation was by addition of phosphate. Further experiments have shown that lime addition enhances both respiratory activity of decaying litter and nitrogenase activity.

III. Nitrogen mineralisation. As the feeding roots of kauri are located almost exclusively within the litter layers of the forest floor mineral nitrogen content in the layers is of great importance. Early work on extraction showed that no nitrate nitrogen was present and that ammonium nitrogen was not readily extracted. Subsequently it was found that to displace the ammonium nitrogen, 24 hours of constand agitation in 1 molar KCl was required. Ammonium levels show considerable variation during the year with levels varying from 0-80 $\mu gNg^{-1}$. Ammonium levels showing a peak in the top H layer are increased by addition of lime and show a close inverse correlation with C/N ratio (Tables 5 and 6).

Table 2. Nitrogenase activity, total nitrogen content and proportion of each component in forest litter at Trounson Park 28th November 1974. Number of samples in parenthesis

| Material | Acetylene reduction n Moles $C_2H_4g^{-1}h^{-1}$ | | N Content % | Proportion % |
|---|---|---|---|---|
| Kauri leaf | (4) | 7.2 | 0.73 | 32 |
| Taraire leaf | (4) | 2.6 | 1.20 | 23 |
| Kauri male cones | (1) | 6.6 | 0.86 | 12 |
| Kauri female cones | (3) | 1.5 | 0.24 | 8 |
| Kauri bark | (4) | 0.2 | 0.26 | 7 |
| Kauri twig | (2) | 1.2 | 0.28 | 18 |

Table 3. $^{15}N_2$ uptake by decaying kauri leaves. Parallel $^{15}N_2$ uptake and acetylene reduction was performed at 24°C. Four replicates used in all cases, means and confidence limits presented ($p = 0.05$)

| | Time of Exposure to $^{15}N_2$ h | Atoms % excess $^{15}N$ | $ngN\ g^{-1}h^{-1}$ | $ngC_2H_4g^{-1}h^{-1}$ |
|---|---|---|---|---|
| 1. | 20 | 0.294 ± 0.019 | 73.8 | |
| | 40 | 0.514 ± 0.005 | 82.8 | |
| | 60 | 0.659 ± 0.047 | 83.5 | |
| | 80 | 0.948 ± 0.014 | 86.0 | |
| | 100 | 0.873 ± 0.057 | 85.8 | 280.0 |
| | | Ratio $\frac{C_2H_4}{N_2}$ | 3.29 | |
| 2. | 12 | | 84.5 | 273.9 |
| | | Ratio $\frac{C_2H_4}{N_2}$ | 3.24 | |

Table 4. Effect of addition of various components of nutrient solution upon acetylene reduction by decaying kauri leaves. Results are expressed as percentage of untreated control

| Treatment | Acetylene reduction percentage of control |
|---|---|
| Glucose | 190 |
| Cations | 70 |
| Phosphate | 310 |
| Ammonium | 32 |
| Total Nutrient | 25 |

Table 5. Effect of addition of lime to $NH_4$ N levels in decaying kauri litter

| Material | Lime addition $g\ m^{-2}$ | pH | $NH_4$-N $\mu g\ g^{-1}$ |
|---|---|---|---|
| F | 0 | 5.4 | 4.5 |
| | 160 | 6.7 | 18 |
| H top | 0 | 4.5 | 21 |
| | 160 | 6.20 | 43 |
| H bottom | 0 | 4.05 | 8 |
| | 160 | 4.30 | 24 |

Table 6. $NH_4$-N levels in decaying leaves of known age and C/N level. Leaves from trees at two sites Sharp's Bush & Thompson's bush were aged on the trees, removed and incubated in a moist chamber for 30 days and final $NH_4$-N levels measured

| Site | Leaf age years | Total N % | C/N | $NH_4$-N $\mu g\ g^{-1}$ |
|---|---|---|---|---|
| Sharps | 0-1 | 0.69 | 68 | 16 |
| | 2-3 | 0.58 | 81 | 5 |
| | 4-5 | 0.43 | 109 | 3 |
| Thompsons | 0-1 | 0.98 | 48 | 42 |
| | 2-3 | 0.85 | 55 | 10 |
| | 4-5 | 0.77 | 61 | 5.5 |

## D. Discussion

The total organic build up and nitrogen accumulation on the floor of kauri forest is greatly in excess of the maxima quoted in a review (5) of terrestrial nutrient cycling. This excessive accumulation of nitrogen is partially explained by the nitrogen fixation input and by the extremely slow rate of decomposition and mineralisation and absence of nitrification such that solution losses are minimal.

The present work on kauri shows that the process of nitrogen fixation is quantitatively important especially over the life span of kauri forest which may be 500 years or more, during which time nitrogen input could be 5t $ha^{-1}$.

The top litter layers in forest represent a very large carbohydrate source, which is both highly aerobic and often acidic. In kauri, the significant rise in pH during early decomposition and the high C/N ratio make this material an ideal site for nitrogen fixing aerobes. It is likely that this phenomenon is widespread in forest ecosystems. It is significant that the rate of nitrogenase is so dramatically increased by phosphorus especially as phosphorus has been identified along with nitrogen as one of the elements critical in kauri nutrition (3).

The primary nutritional requirement for *Agathis* growth both in Australia (4) and New Zealand (3) is nitrogen and the present work shows that total nitrogen values in many forests are very large. Mineralisation is therefore of much greater immediate significance to growth of trees than is total nitrogen and nitrogen fixation. Feeding roots of kauri quickly become located in the decomposing organic layer after tree establishment, and in most forests are confined to the F/H. The C/N ratio of this material varies from 30 - 60 and nitrification is normally absent (significant nitrate has been detected once). Ammonia levels may be high in summer, but fall rapidly in autumn and are often undetectable in winter. The extraction techniques required to remove all $NH_4$-N from this material indicate that the $NH_4^+$ is firmly bound to the organic exchange complex. Large amounts of KCl are required to displace the $NH_4^+$ and quantitative extraction is achieved only by shaking for 24 hours or vacuum infiltration of KCl. Despite the high C/N ratio, significant mineralisation does occur in the root zone, and these results stress the fact that components of the litter possess widely varying C/N ratio and thus microsites in the root zone will show net mineralisation while others show immobilisation.

The situation of the plant roots and the discrete particles of the organic material mean that although there is a net production of mineral nitrogen, this may not be readily available to plant roots which are unable to penetrate the exchange sites. The soil solution levels of $NH_4$-N are extremely low and thus the situation of tree roots in forest means that uptake of $NH_4$-N will be strongly dependent on mycorrhizal penetration of organic material, a situation analogous to the phosphorus uptake phenomena associated with roots in mineral soil. Roots of kauri and all New Zealand conifers are strongly mycorrhizal and it is likely that their significance is as great to nitrogen nutrition as has been shown for phosphorus uptake (2).

## E. Acknowledgements

This work was funded by Research grants from University Grants Committee and N.Z. Forest Service. The assistance of Theresa Orchard, W. Toner, Margaret Spooner and R. Serra in gathering and processing material is gratefully acknowledged.

## F. References

1. Bucha, J.: Nitrogen fixation in decaying forest litter. Unpublished M.Sc. thesis, University of Auckland (1975).
2. Morrison, T.M. & English, D.A.: The significance of mycorrhizal nodules of Agathis australs. New Phytol. 66, 245-250 (1967).
3. Peterson, P.J.: Variation in the mineral content of kauri (Agathis australis) leaves with respect to leaf age, leaf position and tree age. New Zealand J. of Science, 4, 669-678 (1961).
4. Richards, B.N., & Bevege, D.I.: Nutrient requirements of kauri pine in a lateritic podzolic soil in Southern Queensland. Australian Forestry 32, 55-62 (1968).
5. Rodin, L.E. & Brazillvich, H.I.: Production and Mineral Cycling in terrestrial vegetation. Edinburgh & London: Oliver and Boyd, (1967).
6. Silvester, W.B.: Dinitrogen fixation by plant associations excluding legumes. In: A Treatise on Dinitrogen Fixation; Section IV Agronomy & Ecology. Hardy R.W.F. & Gibson, A.H. (ed.) New York & London, Wiley pp. 141-190 (1977).
7. Silvester, W.B., & Bennett, K.J.: Acetylene reduction by roots and associated soil of New Zealand conifers. Soil Biol. Biochem. 5, 171-179 (1973).
8. Silvester, W.B., & Orchard, Theresa: Final Report on Nitrogen Cycling in Kauri Forest. Internal Report, Forest Research Institute, New Zealand (1977).

# A Naturally Occurring Nitrogen-Fixing Rhizosphere Association in the Manawatu

H.C.M. WHITEHEAD, D. ELLIOTT, K.L. GILES, S. ROGAL, and J.J. PATEL

The recently described nitrogen-fixing rhizosphere association between *Spirillum* and *Zea mays* (1) has caused much interest. While this association appears to be one of tropical soils where high temperatures prevail, its description triggered investigation as to whether any such naturally occurring associations existed in temperate New Zealand soils. Because of the established root exudate pattern of C4 grasses it was thought probable that any such association would be on the roots of C4 species. The micro-aerophilic nature of most nitrogen-fixing organisms suggested that water-logged soil would be an appropriate habitat for such an organism, and because of the relatively low air temperatures in New Zealand a dark exposed soil might be expected to be the warmest and hence the most appropriate habitat for such associations. In accordance with these three main assumptions several lagoon sites and field drain sites, kept constantly wet by poor drainage, were sampled for associations yielding positive acetylene reduction activity.

Plants were brought back to the laboratory and immediately assayed for acetylene reduction activity. The roots were well washed in running water and the whole plants placed in 500 ml glass containers. Acetylene reduction activity was assayed with a 10% acetylene atmosphere using a Perkin-Elmer gas chromatograph with a 37 cm Poropak T column. The acetylene to ethylene ratio was used in calculations of acetylene reduction activity. The plants were kept in light at $20^{o}C$ during all assays. Four of the twelve samples taken were found to have positive acetylene reduction activity. In order to obviate any trivial explanation for this activity, root samples were taken from the plants, rewashed and again assayed. This was in order to exclude any aerial blue-green algae which might have been on the leaves of the plants. After considerable washing and retesting it was found that only one of the twelve samples had a stable and repeatable acetylene reduction activity and was associated with the roots of the C3 grass *Agrostis stolonifera*. A rate of 40 nM ethylene $g^{-1}h^{-1}$ wet weight of roots was detected.

Roots of *Agrostis stolonifera* were surface sterilised with 1% hypochlorite solution for 10 minutes and placed on a semi-solid Dobereiner medium in order to isolate any included bacteria(1). Colonies of these bacteria were streaked on solid agar and four bacterial strains isolated. *In vitro* acetylene reduction assays using these four strains indicated that two of them were active in the acetylene reduction assay. By creating various combinations of the four strains it was possible to alter the *in vitro* acetylene reduction activity one hundred fold (Fig.1).

The isolates were identified or tentatively labelled as follows: Isolate a, grew fast on N-deficient Burk's medium, the colonies being 3-4 mm in diameter within 24-36 h; colonies were flat, and almost transparent; the cells were Gram -ve slightly curved with polar flagellum: it did not liquefy gelatine or hydrolize starch, but almost all other carbon sources were utilised: the isolate was tentatively identified as a *Pseudomonas*. Isolates Nos b and d were very similar: they both produced 1-2 mm size colonies which were mucoid and raised above the agar surface on N-deficient Burk's medium: the colonies did not spread or grow further: cells were Gram -ve to Gram variable, initially appearing branched and typically pleomorphic, breaking into small oval to coccoidal cells within 5-7 days; these isolates were tentatively identified as Arthrobacter strains.

Isolate c; cells large predominantly blunt, rod-shaped with lipoid bodies: the cells appear in pairs but had no flagella, and almost all cells turned into circular cysts within 4-6 days: on Burk's medium the colonies grew big with copious amounts of extracellular slime: the slime turned a cinnamon colour but the pigment did not fluoresce under uv: most sugars were utilised as carbon sources but the organism did not liquify gelatine or hydrolyse starch. The isolate was identified as Azotobacter beijerinckii.

Attempts have been made to re-establish the rhizosphere association with sterile rooted cuttings of Agrostis stolonifera in enclosed glass containers in sterile potting mix. Results have been rather variable, indicating the importance of establishing a strong and uniform root system to allow re-establishment of acetylene reduction activity. The pH of the medium was also of importance and it was found that pumice:peat (50/50) giving a final pH of about 6.8-7 was very much more effective than reassociation in vermiculite (pH 7.8-8.2). Further tests on the effect of pH still remain to be done. In those cases where an active association was formed the combination of bacteria giving optimum acetylene reduction activity was similar to that in vitro. The presence of the pseudomonad and the non-fixing Arthrobacter (isolate b) was associated with the increased acetylene reduction activity of both Azotobacter beijerinikii (isolate c) and the Arthrobacter strain (isolate d) (Fig.2).

| Species | nM ethylene/hr |
|---|---|
| 1,4,3 | 4·07 |
| 3 | 5·50 |
| 2,4,3 | 32·6 |
| 1,3 | 46·0 |
| 4,3 | 86·0 |
| 2,3 | 649 |
| 1,2,3 | 887 |

**Carbohydrate source: glucose**

Fig. 1

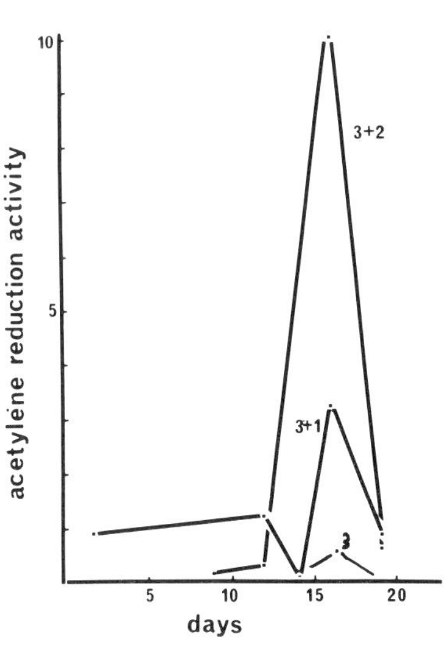


Fig. 2

Fig. 1. In vitro fixation rates of mixtures of the isolated root bacteria

Fig. 2. Stimulation of (3)'s fixation in rhizosphere by the Pseudomonad and the non-fixing Arthrobacter

## A. References

1. Day, J.M., Neves, M.C.P., & Dobereiner, J.: Nitrogenase activity on the roots of tropical forage grasses. Soil Biol. Biochem 7, 107-112 (1975).
2. Cooper, R.: Bacterial fertilizers in the Soviet Union. Soils and Fertilizers 12, 327-333 (1959).

# The Utility of $^{15}N$ in Soil Ecology Research

A.G. NORMAN

Most soil microbiological investigations are not ecological in the full sense of the word, but are somewhat more limited in purpose.

They are often directed to a microbial domain occupied by roots of living plants, grossly disturbed from time to time by plowing, planting, cultivating or harvesting a crop. Much less attention has been given to the microbiology of undisturbed soil ecosystems, such as grassland or forest. Moreover the investigators, efforts are often concentrated on the net activities of the soil microflora rather than interactions between microbial groups. One must recognize too the great complexity of the soil environment, and its structural discontinuity, in that it is comprised of a myriad of micro-habitats, greatly affected by moisture changes which determine the lebensraum for many types of organisms and the concentration of solutes surrounding them. All this has tended to focus attention on the short-term microbiological events that may influence plant growth, primarily nutrient cycling and on the long-range events that influence the quality of the soil environment, primarily its organic matter or humus status. Nitrogen transformations are paramount in both of these. The textbook diagram of the nitrogen cycle is something akin to a road map in that it shows that one can get from one place to another, but it tells nothing of the traffic flow or the constraints upon it. That is why, when the stable nitrogen isotope 15N became available, there were high hopes that it would be possible to quantify the processes and to better understand the flux from plant-unavailable to plant-available pools.

Although a great deal has been learned in these stable isotope studies, in some cases employing the carbon isotope $^{13}C$ concurrently, it is probably a fair assessment to say that the confident early expectations have not been fully realized. This is not for lack of effort, for, in spite of the cost of the procedures, a large number of researchers over the last 30 years have carried out experiments in the laboratory, greenhouse and field using $^{15}N$ as a tracer to clarify some part of the nitrogen cycle. The subject has been well reviewed from time to time, and most recently, comprehensively, by Hauck and Bremner (6). They properly call attention to some aspects that limit the interpretation of $^{15}N$ tracer data, not the least of which is that the early assumption that the extent of discrimination between $^{14}N$ and $^{15}N$ would be negligible. For some purposes it is, particularly if the tracer compound added differs substantially in $^{15}N$ content from atmospheric (0.366 atmos %). Small, but significant, enrichment in the $^{15}N$ content of the organic fraction of soils is well authenticated. In the other direction synthetic nitrogen fertilizers may have lower than normal $^{15}N$ content (6). Controversies start when such differences are ignored. However, exchange phenomena that complicate some work with radioisotopes do not have to be considered.

In the past 10 years more than 300 papers reporting stable nitrogen isotope studies in soil have appeared. Incubation studies of rather short duration seem to have given place to experiments of longer duration in which a plant or crop has been present. The purpose of many of the latter has been to provide information as to the effectiveness, in terms of growth or yield, of different forms, rates or times of application of nitrogen fertilizer. Relatively few bear closely on microbial ecology. Those directed towards the elucidation and quantification of nitrogen transformations have given results that can be disappointingly difficult to interpret. The crux of the situation is the microbiological lability of nitrogen in its various forms. The labeled and unlabeled traffic on the two-way pathways becomes mixed so that there is biological interchange of the labeled nitrogen additions with the unlabeled soil nitrogen. A further complication arises when plants are present because rhizosphere events may be quite different from those brought about by the microflora at large. Inescapably, however, there cannot be dissimilation without concurrent assimilation, decomposition without concurrent synthesis, mineralization without concurrent immobilization. It is the traffic flow in the opposite direction that progressively tends to blur the sharp detail that the $^{15}N$ data would otherwise provide. In mineralization studies, therefore, the information gained is a net figure. Additional nitrogen is retained in an organic form in the active biomass, which in turn may later be decomposed. The fate of the biomass nitrogen, released or reutilized, is a part of the soil organic matter story slowly becoming understood.

One approach has been to add $^{15}N$ labeled ammonium or nitrate plus an available carbon source such as glucose, cellulose or acetate, sometimes labeled with $^{14}C$. Subsequent mineralization of the nitrogen initially incorporated into microbial tissue can be followed as can the amount of carbon retained (4, 8, 9, 13, 16). The nitrogen turnover rate, initially high, declines with time. It is a fair comment that experiments of this type have mostly been of relatively short duration - weeks rather than months or years.

Many investigators have followed the decomposition of plant materials in soil, either unenriched plant materials with the addition of a tagged nitrogen source, or more informatively, $^{14}C$ and $^{15}N$ labeled plant materials, previously grown in a controlled environment chamber. Mostly these have been incubation studies with progress followed by $CO_2$ evolution and periodic sampling of the inorganic and organic nitrogen. With double-labeled plant material, it is initially possible to distinguish between products from the indigenous humic material and the incorporated plant residues, but with time this separation becomes blurred. Moreover in this type of incubation study nitrogen losses often occur, usually ascribed to dentrification, but better described as N unaccounted for. This can include ammonia fixed by clay minerals and the sum of experimental errors. Some relatively long time experiments of this type have been reported. For example, in one 5-year study double labeled barley tops and roots were incorporated in soil in aerated columns, periodically leached to recover mineral nitrogen. After 5 years, only 12 and 15 percent of the carbon in the tops and roots, respectively, remained but 30 and 57% of the nitrogen was retained (2). In a different sequence, tagged mineral nitrogen supplied to an oat crop, was followed through eleven successive crops, each returned to the soil at maturity (10). Availability of the incorporated nitrogen steadily declined for each cropping period.

The microbial ecology of the flooded rice paddy is patently complex. Numerous papers have appeared describing experiments in which $^{15}N$ has been used but the intent in many of the studies was determination of the response to and recovery of forms of nitrogen fertilizers applied at various rates and times. The utilization of inorganic nitrogen by rice is higher in flooded soil than in upland rice, probably due to the persistence of available N in the ammonium form in flooded soil (3). Unequivocal evidence of significant nitrogen fixation in tropical rice soils has been obtained using both $^{15}N$ and acetylene reduction procedures (11, 15, 18).

Fixation has been shown to be stimulated by the presence of crop residues, not to be dependent on light, and to be higher in planted than unplanted soil (11, 15).

Nitrogen transformations in forested soils or pastures have received far less attention than in cropland. In a coniferous stand much of the nitrogen is retained in the soil humus and litter layers. Nutrient cycling is a tightly closed system. If $^{15}N$ fertilizer is applied only a small fraction appears in the current year needles with quite slow later redistribution within the tree (14).

Some unique ecological situations have been clarifed with the aid of $^{15}N$. Nitrogen fixation by bacteria in the phyllosphere of Douglas fir has been shown to occur and to be appreciable (7). Scattered reports of nitrogen fixation by bacteria in the phyllosphere of other species occur in the literature but recently confirming evidence has been obtained on leaves of Guatemala grass both by acetylene reduction and by exposing whole plants to $^{15}N$ in situ, leading to an estimate of 260 - 400 g $ha^{-1}yr^{-1}$(1).

Nitrogen fixation in other unusual environments has similarly been established. In Antartica, for example, fixation was proved to be associated with *Nostoc commune* and certain lichens which contain *Nostoc* as a phycobiont (5). Rapid fixation has also been shown to occur in desert algal crusts (12) or cold desert soils (17).

These last several observations on nitrogen fixation have been included to indicate the value of the nitrogen isotope to clarify microbial activities in special situations and are divergent from my main theme which is to assess the utility of $^{15}N$ procedures in identifying, following and quantifying the nitrogen traffic on the major biological pathways in soil. My conclusion, in part for reasons already stated, is that $^{15}N$ procedures have been more successful in providing answers to a diversity of agronomic questions than to some of the elegant but diffuse issues underlying microbial ecology. It is possible to provide good information on the efficiency of use of fertilizer nitrogen, subsequent recovery in later crops, losses in drainage water, etc. and to construct in effect a nitrogen fertilizer balance sheet. These procedures have not proved effective in arriving at a detailed description of the nitrogen budget in a soil or soil-plant ecosystem. Models have been attempted but they require acceptance of assumptions or assertions that are most difficult to quantify. $^{15}N$ procedures have been successfully used in studies involving nitrogen fixation, both symbiotic and nonsymbiotic, though in the latter case the quantitative aspects in soil remain obscure. The biogeochemistry of nitrogen is still not well understood. Only a small portion of the total nitrogen in an ecosystem is plant-available - what there is is maintained by and is a product of the living microbial biomass.

## A. References

1. Bessems, E.P.H.: Nitrogen Fixation in the phyllosphere of Gramineae. Agric. Res. Rep. (Wageningen) 786, 1-68 (1973).

2. Broadbent, F.E. & Nakashima, T.: Mineralization of carbon and nitrogen in soil amended with carbon-14- and nitrogen-15- labeled plant material. Soil Sci. Soc. Amer. Proc. 38 (2), 313-315 (1974).

3. Broadbent, F.E. & Reyes, O.C.: Uptake of soil and fertilizer nitrogen by rice in some Philippine soils. Soil Sci. 112 (3), 200-205 (1971).

4. Chichester, F.W., Legg, J.O., & Stanford, G.: Relative mineralization rates of indigenous and recently incorporated $^{15}N$-labeled nitrogen. Soil Sci. 120, (6), 455-460 (1975).

5. Fogg, G.E., & Stewart, W.D.P.: *In situ* determinations of biological nitrogen fixation in Antartica. Brit. Antarctic Surv. Bull. 15, 39-46 (1968).

6. Hauck, R.D., & Bremner, J.M.: Use of tracers for soil and fertilizer nitrogen research. Adv. Agron. 28, 219-266 (1976).
7. Jones, K.: Nitrogen fixation in the phyllosphere of the Douglas fir. Ann. Bot. (London) 34 (134), 239-244 (1970).
8. Kai, H., Admad, Z., & Harada, T.: Factors affecting immobilization and release of nitrogen in soil and chemical characteristics of the nitrogen newly immobilized. Soil Sci. Plant Nutr. 19 (4), 275-286 (1973).
9. Ladd, J.N., & Paul, E.A.: Changes in enzyme activity and distribution of acid-soluble, amino acid-nitrogen in soil during nitrogen immobilization and mineralization. Soil Biol. Biochem. 5 (6), 825-840 (1973).
10. Legg, J.O., Chichester, F.W., Stanford, G., & DeMar, W.H.: Incorporation of nitrogen-15-tagged mineral nitrogen into stable forms of organic nitrogen. Soil Sci. Soc. Amer. Proc. 35 (2),273-276 (1971).
11. MacRae, I.C., & Castro, T.F.: Nitrogen fixation in some tropical rice soils. Soil Sci. 103 (4), 277-278 (1967).
12. Mayland, H.F., & McIntosh, T.H.: Distribution of nitrogen fixed in desert algal-crusts. Soil Sci. Soc. Amer. Proc. 30 (5), 606-609 (1966).
13. McGill, W.B., Shields, J.A., & Paul, E.A.: Relation between carbon and nitrogen turnover in soil organic fractions of microbial origin. Soil Biol. Biochem. 7 (1), 57-63 (1975).
14. Nommik, H.: The uptake and translocation of fertilizer $^{15}N$ in young trees of Scots pine and Norway spruce. Stud. Forest Suecica 35, 3-18 (1966).
15. Rao, V.R., Kalininskaya, T.A., & Miller, Y.M.: The activity of non-symbiotic nitrogen fixation in soils or rice fields studied with $^{15}N_2$. (Russ) Mikrobiologiya 42 (4), 729-734 (1973).
16. Shields, J.A., Paul, E.A., Lowe, W.E., & Parkinson, D.: Turnover of microbial tissue in soil under field conditions. Soil Biol. Biochem. 5 (6), 753-764 (1973).
17. Skujins, J., & Eberhardt, P.J.: In situ $^{15}N$ studies of the nitrogen cycle in cold desert soils. Agron. Abs. p.94 (1973).
18. Yoshida, T., & Aneajas, R.R.: Nitrogen fixing activity in upland and flooded rice fields. Soil Sci. Soc. Amer. Proc. 37 (1), 42-46 (1973).

## FOOTNOTE TO REFERENCES

The citations above do not adequately reflect the world-wide contributions to this topic in the past ten years. They were selected only as supporting statements made in the paper.

# Effect of Climate on Oxidation and Reduction of Sulfur Compounds in Soils

R.J. SWABY and R. FEDEL

## A. Introduction

From our present knowledge it is impossible to predict accurately from soil properties or cropping history the proportion of added sulfur compounds oxidising or reducing in Australian soils.

## B. Materials and Methods

Soils and methods were similar to those published by Vitolins and Swaby (1), Swaby and Vitolins (2) and Swaby and Fedel (3).

## C. Results and Discussion

Table 1 gives the percentage of soils in all and in each climatic zone showing high to nil oxidation of sulfur.

Table 1. Influence of Climate on Sulfur Oxidizing Activity of Soils results expressed as percentages

| Activity | All Climates | Tropical Wet | Temperate Moist | Temperate Dry | Subtemperate Wet | Subtemperate Moist | Arid Hot |
|---|---|---|---|---|---|---|---|
| High | 18 | 17 | 24 | 3 | 59 | 44 | 0 |
| Moderate | 27 | 33 | 35 | 16 | 30 | 37 | 24 |
| Low | 27 | 25 | 27 | 31 | 7 | 19 | 45 |
| Nil | 27 | 25 | 14 | 50 | 4 | 0 | 31 |
| No. of Soils | 360 | 60 | 96 | 121 | 27 | 27 | 29 |

Of 360 Australian soils examined only 18% of them oxidised sulfur at a high level of activity, while 27% exhibited moderate, low or no oxidation. When the soils were classified according to climate, sulfur oxidation was high in 59% of Subtemperate-Wet soils ranging down to 0% of Arid-Hot. Thus the longer the moist season the greater the incidence of high levels of sulfur oxidation and usually the longer the dry season the greater the occurrence of nil oxidation. Table 2 gives the percentage of soils in the various climates containing from abundant to no *Thiobacillus thiooxidans*.

In 375 soils, *Th. thiooxidans* was abundant in only 12%, common in 30%, rare in 24% and absent in 34%. They were abundant in 44% of the Subtemperate-Wet soils grading down to 0% of the Arid-Hot soils, showing that persistence of non-sporing *Th. thiooxidans* was related to the wetness of the climate and that death of their cells by dehydration was proportional to the length of the dry season.

Table 2. Influence of Climate on Incidence of Thiobacillus Thiooxidans in Soils
Results expressed as percentages

| Incidence | All Climates | Tropical Wet | Temperate Moist | Temperate Dry | Subtemperate Wet | Subtemperate Moist | Arid Hot |
|---|---|---|---|---|---|---|---|
| Abundant | 12 | 3 | 15 | 8 | 44 | 22 | 0 |
| Common | 30 | 53 | 39 | 13 | 38 | 33 | 17 |
| Rare | 24 | 15 | 32 | 21 | 11 | 33 | 28 |
| None | 34 | 28 | 14 | 58 | 7 | 12 | 55 |
| No. of Soils | 375 | 60 | 102 | 130 | 27 | 27 | 29 |

Table 3. Influence of Climate on Sulfur Reducing Activity of Soils under Aerobic Conditions

Results expressed as percentages

| Substrate | Activity | All Climates | Temperate Dry | Temperate Moist | Tropical Wet |
|---|---|---|---|---|---|
| Sulfur | High | 86 | 68 | 89 | 97 |
| | Moderate | 4 | 7 | 4 | 0 |
| | Low | 2 | 7 | 0 | 3 |
| | Nil | 8 | 18 | 7 | 0 |
| | No. of Soils | 86 | 28 | 28 | 30 |
| Sulphate | High | 0 | 0 | 0 | |
| | Moderate | 0 | 0 | 0 | |
| | Low | 0 | 0 | 0 | |
| | Nil | 100 | 100 | 100 | |
| | No. of Soils | 56 | 28 | 28 | |
| Cystine | High | 61 | 36 | 46 | 100 |
| | Moderate | 4 | 7 | 4 | 0 |
| | Low | 1 | 4 | 0 | 0 |
| | Nil | 34 | 53 | 50 | 0 |
| | No. of Soils | 86 | 28 | 28 | 30 |

The percentage of soils in the different climatic zones which evolved varying levels of $H_2S$ from three sulfur substrates incubated aerobically are presented in Table 3.

When sulfur was the substrate added to 86 soils, 86% evolved $H_2S$ with high activity, showing that aerobic sulfur reducers occurred widely, particularly in the Tropical-Wet region. Loss of $H_2S$ by aerobes from added sulfate was nil from all 56 soils. With cystine, $H_2S$ activity was high from 61% of them and nil from 34%. Highly active losses of $H_2S$ occurred from 100% of Tropical-Wet soils at one extreme and from 36% of Temperate-Dry at the other end, while no $H_2S$ was evolved from approximately half the soils of Temperate areas. As with sulfur, likewise with cystine, more soils from the hot, moist climates lost $H_2S$ more actively than from the cooler, dry areas. Table 4 provides similar data to Table 3, but for soils incubated under waterlogged conditions.

Table 4. Influence of Climate on Sulfur Reducing Activity of Soils under Anaerobic Conditions

Results expressed as percentages

| Substrate | Activity | All Climates | Temperate Dry | Temperate Moist | Tropical Wet |
|---|---|---|---|---|---|
| Sulfur | High | 24 | 14 | 35 | |
| | Moderate | 8 | 4 | 11 | |
| | Low | 13 | 14 | 11 | |
| | Nil | 55 | 68 | 43 | |
| | No. of Soils | 56 | 28 | 28 | |
| Sulfate | High | 34 | 4 | 11 | 83 |
| | Moderate | 0 | 0 | 0 | 0 |
| | Low | 0 | 0 | 0 | 0 |
| | Nil | 66 | 96 | 89 | 17 |
| | No. of Soils | 86 | 28 | 28 | 30 |
| Cystine | High | 45 | 35 | 54 | |
| | Moderate | 9 | 11 | 7 | |
| | Low | 5 | 4 | 7 | |
| | Nil | 41 | 50 | 32 | |
| | No. of Soils | 56 | 28 | 28 | |

Only 24% of 56 anaerobic soils evolved $H_2S$ from sulfur very actively, thus differing from the results for porous aerated soils, although blackening of the soils with FeS was observed frequently. More Temperate-Moist soils lost $H_2S$ than Temperate-Dry ones. The percentage of soils losing no $H_2S$ was greater under anaerobic than aerobic conditions and likewise greater from drier than moister climatic areas. High volatilisation of $H_2S$ from sulfate occurred in 34% of 86 waterlogged soils, but no losses from 66% of them. High production of $H_2S$ was recorded more often from the wetter climates than the drier ones, while nil production occurred more frequently in the drier climates than the wetter ones. Even at its worst, the loss of $H_2S$ never exceeded 0.06% of sulfur added or 0.09% of cystine-S, which is negligible. Table 5 shows the percentage of soils in the various climates in which sporing and non-sporing mesophiles and thermophiles were responsible for loss of $H_2S$ from three substrates in media incubated aerobically.

Spores of aerobic mesophiles evolved $H_2S$ when all 86 soils were inoculated into sulfur medium, so climate had no effect on the distribution of these ubiquitous organisms. Spores of aerobic thermophiles caused this in only 33% of all soils, the incidence being related to the wetness of the climate. Both sporers were found more often in pasteurised than in unheated soils, suggesting that vegetative cells may have antagonised sulfide producers. Many bacterial genera were isolated, including some species of Desulfovibrio but fewer of Desulfotomaculum. Sulfate-reducers were found in only a small percentage of soils, particularly from wetter climates. Sometimes pasteurisation enhanced detection of spores. With cystine, mesophiles were also detected more often than thermophiles, and their cells more often than their spores for all climates. Again many genera were implicated in evolving $H_2S$ from cystine on primary isolation, but soon lost this activity on further subculturing. Table 6 presents similar data to Table 5, but derived from anaerobic media.

Table 5. Influence of Climate on Sulfur Reducing Organisms under Aerobic Conditions

| Substrate | Climate | Spor.- Meso. | Spor.- Thermo. | Non-Sp. & Sp.-Meso. | Non-Sp. & Sp.-Thermo. | No. of Soils |
|---|---|---|---|---|---|---|
| Sulfur | All Climates | 100 | 33 | 93 | 20 | 86 |
| | Temp.-Dry | 100 | 14 | 92 | 32 | 28 |
| | Temp.-Moist | 100 | 18 | 92 | 28 | 28 |
| | Tropic-Wet | 100 | 64 | 96 | 0 | 30 |
| Sulphate | All Climates | 12 | 1 | 9 | 0 | 86 |
| | Temp.-Dry | 7 | 0 | 11 | 0 | 28 |
| | Temp.-Moist | 4 | 0 | 11 | 0 | 28 |
| | Tropic-Wet | 23 | 3 | 7 | 0 | 30 |
| Cystine | All Climates | 13 | 1 | 60 | 16 | 86 |
| | Temp.-Dry | 4 | 0 | 60 | 14 | 28 |
| | Temp.-Moist | 7 | 4 | 60 | 7 | 28 |
| | Tropic-Wet | 27 | 0 | 60 | 23 | 30 |

Table 6. Influence of Climate on Sulfur Reducing Organisms under Anaerobic Conditions

| Substrate | Climate | Spor.- Meso. | Spor.- Thermo. | Non-Sp. & Sp.-Meso. | Non-Sp. & Sp.-Thermo | No. of Soils |
|---|---|---|---|---|---|---|
| Sulfur | All Climates | 97 | 19 | 100 | 2 | 86 |
| | Temp.-Dry | 96 | 7 | 100 | 4 | 28 |
| | Temp.-Moist | 92 | 11 | 100 | 4 | 28 |
| | Tropic-Wet | 100 | 37 | 100 | 0 | 30 |
| Sulphate | All Climates | 13 | 3 | 13 | 0 | 86 |
| | Temp.-Dry | 11 | 0 | 7 | 0 | 28 |
| | Temp.-Moist | 4 | 4 | 4 | 0 | 28 |
| | Tropic-Wet | 26 | 6 | 26 | 0 | 30 |
| Cystine | All Climates | 34 | 3 | 69 | 4 | 86 |
| | Temp.-Dry | 21 | 7 | 54 | 11 | 28 |
| | Temp.-Moist | 29 | 0 | 75 | 0 | 28 |
| | Tropic-Wet | 50 | 0 | 77 | 0 | 30 |

In almost all 86 soils, anaerobic, sporing mesophiles were responsible for $H_2S$ from sulfur more often than anaerobic, sporing thermophiles and climate had no marked effect, indicating the universal distribution of these organisms. Data for sulfate reduction under anaerobic conditions paralleled those previously obtained aerobically, showing that dissimilatory microorganisms were uncommon, except in the Tropical-Wet zones, where Desulfovibrio occurred more often than Desulfotomaculum. In cystine medium, anaerobic mesophiles were detected more frequently than anaerobic thermophiles, as found for aerobic groups; but the incidence of mesophiles alone was higher than occurred with aerobes. Tropical-Wet soils contained the spores of anaerobic mesophiles more often than Temperate soils. Temperate-Dry soils rarely contained anaerobic, thermophilic sulfide producers, while soils from other climates contained none. As with the aerobes, many anaerobic genera produced $H_2S$ from cystine on primary isolation only.

## D. References


1. Vitolins, M.I., Swaby, R.J.: Activity of sulfur-oxidising microorganisms in some Australian soils. Aust. J. Soil Res. 7, 171-183 (1969).
2. Swaby, R.J., Vitolins, M.I.: Sulfur oxidation in Australian soils. Trans. 9th Int. Congr. Soil Sci., Adelaide 4, 673-681 (1968).
3. Swaby, R.J., Fedel, R.: Microbial production of sulfate and sulfide in some Australian soils. Soil Biol. Biochem 5, 773-781 (1973).

# Physiology of P Release from Fungi-Implications for the P Cycle

D.J.W. BURNS and R.E. BEEVER

## A. Introduction

Although microorganisms play an essential role in P cycling within soils details of their involvement is not known because of the complexity of the soil system. Organic P can be present free in the soil, and in living organisms (the microorganism pool): these two pools need to be distinguished (1). Study is also needed of the effects of environmental stresses on the loss of P from the microorganism pool to the soil. We have tested the effects of some stresses on P loss in a model system. Cells of Neurospora crassa were uniformly labeled with $^{32}P$ and exposed to various stresses in liquid culture under closely controlled conditions. Because fungi exist in soil both in resting and actively growing forms, we have examined the behaviour of ungerminated conidia and of young actively growing mycelium (germlings).

## B. Materials and Methods

The N. crassa strain, the composition of the P-free base medium (BM) and most of the experimental methods have been described (2). Cells were uniformly labeled with $^{32}P$ by supplementing the appropriate growth media with $^{32}P_i$. Sufficient radioactivity was added such that $10^6$ cells (the standard number $ml^{-1}$) contained c 200,000 c.p.m. The dry weight of $10^6$ conidia was c 26 μg and of $10^6$ germlings c 39 μg. Both contained c 1.6% P (d.w.). The conidial suspension was allowed to stand at room temperature for 1 h before use; germlings were grown for 2.5 h in medium BM+50 μM $P_i$. By 2.5 h the $P_i$ concentration had dropped to 44 μM and hence BM+44 μM $P_i$ was chosen as our standard experimental medium.

## C. Results and Discussion

I. Physiology of P loss. When $^{32}P$-labeled cells were resuspended in growth medium there was only a very slow release of radioactivity into the medium (Fig. 1). Adding chemicals such as saponin and tannic acid (broadly representing two classes of higher plant compounds), or making the medium anaerobic by bubbling nitrogen through it, increased the rate of loss from germlings but not from conidia. Freezing or drying of both conidia and germlings for a period prior to resuspension caused an initial loss of 10-30% of the total P followed by a continuing slower loss. Extensive P loss was not necessarily associated with cell death. For example, germlings treated with saponin or tannic acid lost similar amounts of P but differed markedly in viability (Table 1).

The chemical nature of the material lost during the treatments was investigated in two ways. Firstly, use of a selective $P_i$ precipitation method showed that in treatments causing appreciable P loss, from 30 to 70% of the material lost was not $P_i$ (Table 1). In a second more detailed study, samples of the material lost from germlings subjected either to freezing or to drying were separated by 2-dimensional

Table 1. Viability of, and P lost from, N. crassa cells subject to various treatments

| Treatment[a] | CONIDIA | | | | GERMLINGS | | | |
|---|---|---|---|---|---|---|---|---|
| | % viability[b] | % P lost by 60 min[c] | % $P_i$ of P lost[d] | Expt No. | % viability[b] | % P lost by 60 min[c] | % $P_i$ of P lost[d] | Expt No. |
| | | | | | 100 | 0.9 | 64 | 4 |
| Control | 100 | 0.7 | 96 | 1 | - | 0.8 | 58 | 5 |
| | 100 | 0.5 | 73 | 2 | 100 | 0.7 | 60 | 6 |
| Tannic | 82 | 0.5 | 90 | 1 | 3* | 16 | 69 | 4 |
| | - | 0.6 | 59 | 2 | - | 22 | 66 | 5 |
| Saponin | 87 | 0.9 | 64 | 1 | 83 | 12 | - | 4 |
| | - | 0.7 | 39 | 2 | - | 13 | 31 | 5 |
| Nitrogen | 85 | 0.6 | 84 | 2 | 97 | 2.3 | 88 | 6 |
| | | | | | - | 1.8 | 84 | 5 |
| | 30* | 33 | 49 | 1 | 2* | 31 | 45 | 4 |
| Drying | - | 14 | 44 | 2 | - | 42 | 42 | 5 |
| | 91 | 15 | 37 | 3 | - | 37 | 43 | 7 |
| | 77* | 22 | 63 | 1 | 75 | 11 | 42 | 4 |
| Freezing | - | 1.5 | 42 | 2 | - | 31 | 35 | 5 |
| | 114 | 4.0 | 33 | 3 | - | 22 | 44 | 7 |

[a] Treatments as described in Fig. 1.

[b] At 60 min a sample was diluted in water and appropriate volumes plated on sucrose-sorbose medium. The number of colonies present after 3 days incubation is expressed as a % of that present in the control treatment. Treatments that differ significantly from the control are marked with an asterisk (Dunnett's multiple comparison test, 1% level).

[c] Expressed as a % of the total P present in the cells at the beginning of the treatment. The control treatments for expts 3 and 7 are not listed as P loss was not measured.

[d] The $P_i$ in a subsample of the 60 min filtrate was precipitated (6) and the loss of counts from the supernatant expressed as a % of the total.

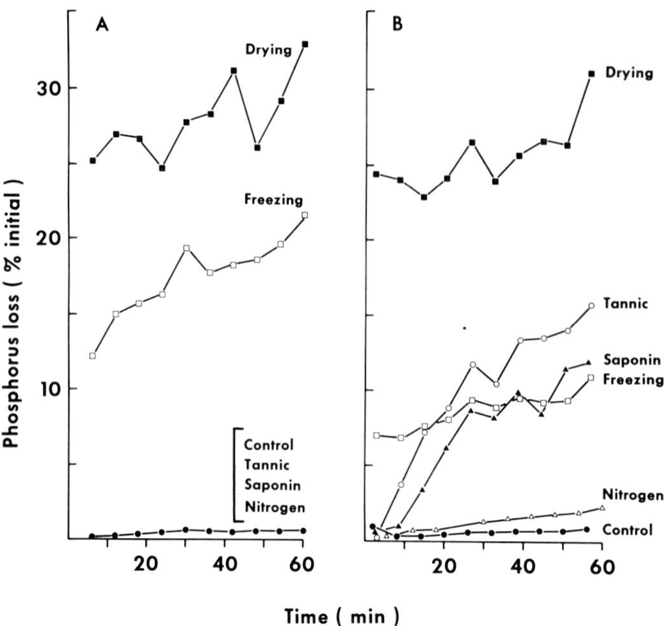


Fig. 1 A and B. Loss of $^{32}P$ compounds from N. crassa cells subject to various treatments. (A) Conidia; (B) germlings. Treatments were as follows: control, cells were resuspended at zero time in non-radioactive BM+44 μM $P_i$ at 30°C; tannic, cells were resuspended in medium plus tannic acid (B.D.H., 1 g $l^{-1}$); saponin, cells were resuspended in medium plus saponin (B.D.H., white, 1 g $l^{-1}$); nitrogen, nitrogen gas was bubbled through the medium for 15 min before resuspending the cells and bubbling was continued through the experiment; drying, a sample of cells was centrifuged, the supernatant removed, and the pellet dried under an air stream for 10-15 min before the cells were resuspended in medium; freezing, another sample of cells was pelleted and the pellet placed at -20°C for 10-15 min before resuspension. At the times indicated, samples of the suspension were removed, filtered through glass fiber filters, and the radioactivity in the filtrate measured

TLC. The two treatments gave a very similar separation pattern; Fig. 2 shows an autoradiograph of the material from cells stressed by drying. The pattern of P compounds present closely resembles that of the acid-soluble ester + $P_i$ fraction of rapidly growing germlings. (This soluble ester fraction comprises about 24% and $P_i$ about 6%, of the total P of germlings.) These observations suggest that the P released following freezing or drying comes directly from the soluble P fraction of the germlings. However hydrolysis of other P metabolites must play a role on those occasions when an amount of P greater than the total present in the ester pool is released.

II. Relevance to the P cycle in the soil. Fractionation of the organic P in soils by standard methods seldom reveals the presence of acid-labile esters. Using sensitive radioactive methods hot water can be shown to extract the soluble ester pool of soil microorganisms (5). Furthermore when soils are subjected to γ-irradiation or sonication a similar group of P esters is released into the soil solution and can be extracted by cold water (5). Our results show that the environmental influences occurring in the soil are likely to cause loss of esters from microorganisms into this 'soil soluble ester pool'. Because such loss does not

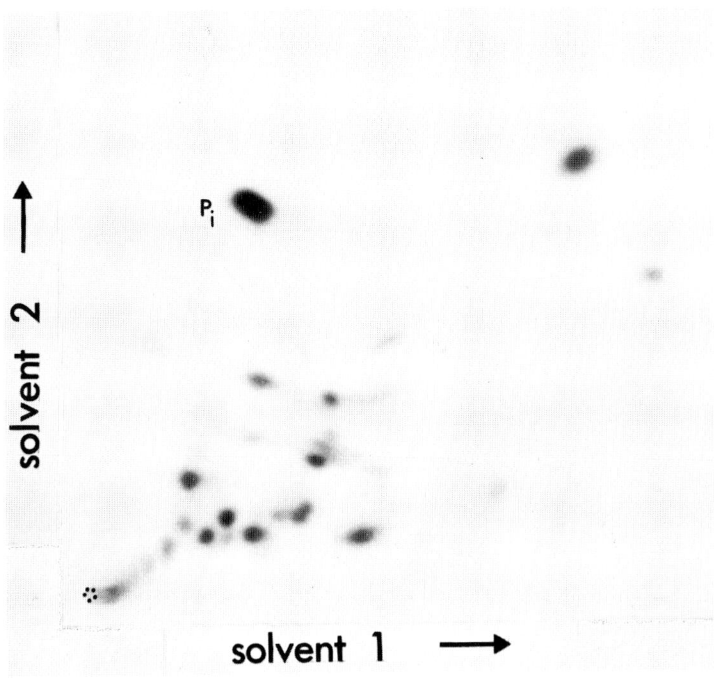


Fig. 2. Autoradiograph of a TLC separation of the P compounds lost on rewetting dried N. crassa germlings. Germlings were dried as described in Figure 1, resuspended in water for 5 min, and the cells filtered off. The filtrate was freeze-dried, 8% trichloroacetic acid added, the solution left at $4^{o}C$ for 4 h to inactivate any phosphatases and then the acid was removed by ether extraction. The aqueous fraction was dried on a rotary film evaporator and the residue was dissolved in water. Appropriate aliquots were fractionated by TLC using the P ester separation method (3) except that cellulose thin layers were used rather than paper

necessarily result in loss of viability, microorganisms may even function as soluble ester 'generating systems'. One can estimate the magnitude of the soil soluble ester pool as follows: the microorganism P pool in soils amounts to about 10 μg P (g dry soil)$^{-1}$. If the microorganism ester fraction constitutes 10% of this and if on average 1% of the esters are subject to loss at any time due to environmental stress, then the soil soluble ester pool could amount to 10 ng (g dry soil)$^{-1}$. In some situations such as the thawing of frozen soil, or the rewetting of dried soil, the pool could be much larger. In addition to knowing the magnitude of the soil soluble ester pool evaluation of the role of these esters in the P cycle depends on knowledge of the rates of ester turnover. Consideration must be given to the movement of ester P through the soil as there is evidence (5) that esters bind less readily to soil than does $P_i$. Recently there has been an increasing interest in the possible role of sulfate esters in the sulfur cycle of soils (4). Will we see a similar interest in P esters in the future?

We thank Philippa Clark for excellent technical assistance and Alison Marshall for statistical analysis. R.J. Redgwell guided us in the difficult art of separating P esters.

## D. References

1. Beever, R.E, & Burns, D.J.W.: Microorganisms and the P cycle: some physiological considerations. In: Reviews in Rural Science III. Blair, G.J. (ed.) Armidale: U.N.E., in press (1977).
2. Burns, D.J.W. & Beever, R.E.: Kinetic characterisation of the two phosphate uptake systems in the fungus Neurospora crassa. J. Bacteriol. 132, 511-519 (1977).
3. Bieleski, R.L.: Levels of phosphate esters in Spirodela. Plant Physiol. (Lancaster) 43, 1297-1308 (1968).
4. Fitzgerald, J.W.: Sulfate ester formation and hydrolysis: a potentially important yet often ignored aspect of the sulfur cycle of aerobic soils. Bacteriol. Rev. 40, 698-721 (1976).
5. Martin, J.K.: Organic phosphate compounds in water extracts of soils. Soil Sci. 109, 362-375 (1970).

# Microbial Ecology and Decomposition in Arctic Tundra and Subarctic Taiga Ecosystems

P.W. FLANAGAN

## A. Introduction

Northern forest sites have low decomposition rates (K) while those further south and in tropical areas have higher K values (3, 14, 16). It has been suggested that in arctic tundra ecosystems litter is decomposed at very slow rates resulting in organic accumulation and in a depauperate N, P, and K available nutrient pool (8). Some workers suggest that decomposition processes are slowed by the adverse climatic conditions of northern climes more than are primary production processes (2, 13, 15, 16, 20); which fits the observation that there is more accumulation of dead organic matter in some northern ecosystems compared with southern systems. The concept has consequently developed that northern ecosystems in general are limited functionally by climatically-controlled immobilization of microbial activities, organic remains and essential nutrients.

There is no clear indication however from field data and decomposition studies on tundra areas that northern ecosystems are accumulating organic matter. The present assessment of annual decay rates at Pt. Barrow, Alaska is very close to the measure of total net annual primary production (9). Whether or not decomposition is quicker in warmer than in colder climates is not in question. The question is whether the balance between decomposition and primary productivity changes with latitude, and if so, are decay rates lagging production rates in the colder climates? If primary producer activity decreases with decreasing temperature are secondary production and decomposition processes amongst microorganisms proportionally more limited by low temperatures? This situation would imply that microorganisms are less cold tolerant and/or cold adaptable than are producer cryptograms and phanerogams.

The present study was undertaken seven years ago to examine if microorganisms in northern ecosystems function at low temperatures at rates as high as those of temperate organisms at their respective habitat temperatures. It was of particular interest to see if psychrophilism was prevalent in microorganisms of northern ecosystems and if northern ecosystem microorganisms could recycle enough organic matter each year to balance annual organic matter production.

## B. Study Sites

The tundra sites at Eagle Summit and Pt. Barrow have been described (1, 5) as has the spruce forest site (10). A fifth site in a stand of paper birch *Betula papyrifera* included in this study was on the campus of the University of Alaska, Fairbanks and, is underlain by fine silty loess soil on top of birch creek schist of precambrian age. *Alnus* sp. intermingle with the dominant trees and *Calamagrostis* sp. prevails in open areas while the entire forest floor is covered in summer by a carpet of *Equisetam arvense*. Patches of *Hylocomium splendens* occur amongst the *Equisetum* especially close to stem bases.

## C. Methods

Fungal and bacterial biomass was estimated using described (12) modified (10, 19) methods. Fungi were isolated (17) on potato dextrose agar and malted soil extract agar. Fungi from temperate ecosystems (peat bogs) were provided by Pamela Latter, Institute of Terrestrial Ecology, Merlewood, Cumbria, UK., and by Dr. Paul Dowding, Trinity College, Dublin and isolations were made (17) in Western Ireland.

Decay rates were measured by inserting indigenous plant remains and cellulose strips in litter bags into the upper 10 cm of organic soil. Bags were collected for reweighing each year for 3 years.

$NH_4^+$, $NO_2^-$ and $NO_3^-$ were estimated using publicised methods (4) and available P after ammonium flouride extract and by the molybdenum blue method.

The effect of temperature on growth of selected isolates was determined by measuring increase in weight of mycelium on following incubation at 5 degree intervals from -5°C to 55°C at specific times. Yield of mycelium (grams/fungus/gram substrate) and fungal respiration at different temperatures was measured in mineral vitamin solution with added glucose in Gilson respirometers (8, 10).

Only those fungi able to grow optimally at or below 20°C were considered truly psychrophilic (11). Those able to metabolise at or below 0°C and with growth optima in the 25°C-35°C range were designated cold tolerant mesophiles. Organisms unable to grow below 2°C with optimal metabolic activity in the 30°C-40°C range were considered true mesophiles. Ten representative organisms of each of the above categories, grown from single spores or single 3-5 mm lengths of hyphal tip obtained from liquid potato dextrose culture, were placed in mineral salt-vitamin solution for measurement of their growth, respiration and yield (7, 8, 10). Growth, yield coefficients, and respiration of some major fungal species from latitudinally separated research regions were also determined at a variety of temperatures (7, 8, 10). Respiration rates in root-free organic remains from the upper 10 cm of organic soil were monitored at a variety of temperatures (8, 10) to establish the optimum temperatures for microbial activity by site and for comparison with measure of decay in litter bags.

## D. Results and Discussion

Bacterial biomass (direct counts) never constituted more than ten percent of the total microbial biomass in any of the study areas. Fungi were the predominant decomposers in all sites. On the average, live hyphae accounted for 70%, 65%, 85% and 50% of all hyphae at Eagle Summit, Pt. Barrow, the birch forest and spruce forest respectively.

No significant differences in metres (m) of live mycelium $g^{-1}$ organic matter were detected between tundra and birch sites, but spruce forest had significantly fewer m. of live hyphae (Table 1).

No significant difference was observed between sites in the rate of cellulose decay. The levels of annual decay were lowest in the spruce forest and equally higher in the birch and tundra sites (Table 1). Organic matter and litter of the latter sites had similar and higher readily extractable fractions (in hot water and 80% ethanol) than the spruce litter materials. The tundra sites were very much richer in soluble N and P than the two forest sites despite the relatively higher decay rate in the birch forest. Barrow tundra had 2 and 4 times more organic matter than the birch and spruce sites respectively (Table 2).

Of 38 and 36 fungal isolates 22% and 10% were psychrophilic at Pt. Barrow and Eagle Summit respectively, and 55% and 10% of fungi in these sites were cold

Table 1. Annual Decomposition and Productivity of Fungi and Plants

| Site | Decomposition* $g\,m^{-2}\,y^{-1}$ | Decomposition* % wt loss cellulose | $\bar{x}$ m. live mycelium per g organ- matter | calculated microbial productivity $g\,m^{-2}\,y^{-1}$ | measured primary productivity $g\,m^{-2}\,y^{-1}$ | annual microbial yield per g primary productivity | Indicated rate of organic accumulation $g\,m^{-2}\,y^{-1}$ |
|---|---|---|---|---|---|---|---|
| Eagle Summit | 235 | 3.4±5 | 570 | 133 | 203 [a] | .57 | -5 |
| Pt. Barrow | 310 | 3.3±4 | 650 | 141 | 275 [b] | .56 | -35 |
| Birch forest | 350 | 5.2±1.3 | 580 | 178 | 375 [c] | .51 | +25 |
| Spruce forest | 198 | 3.4±1.0 | 300 | 117 | 180 [d] | .65 | -18 |

* Litter bags inserted in litter layer and at 5-10 cm depth

(a) (1)

(b) (5)

(c) Litter tray data

(d) 1977 estimate, taiga biome colleagues

Table 2. Site organic matter and nutrient content

| Site | Kg $m^{-2}$ organic matter to 10 cm | % of organic matter readily extractable* | mg $m^{-2}$ soluble-extractable N | mg $m^{-2}$ soluble-extractable P |
|---|---|---|---|---|
| Eagle Summit | 10.5 | 14.5 | 118.5 [a] | 7.4 [a] |
| Pt. Barrow | 18.0 | 13.5 | 3600.0 [b] | 22.0 [b] |
| Birch forest | 4.3 | 14.8 | .422 | .190 |
| Spruce forest | 8.6 | 2.5 | .156 | .095 |

* Warm water followed by 80% Ethanol

(a) (1)

(b) (6)

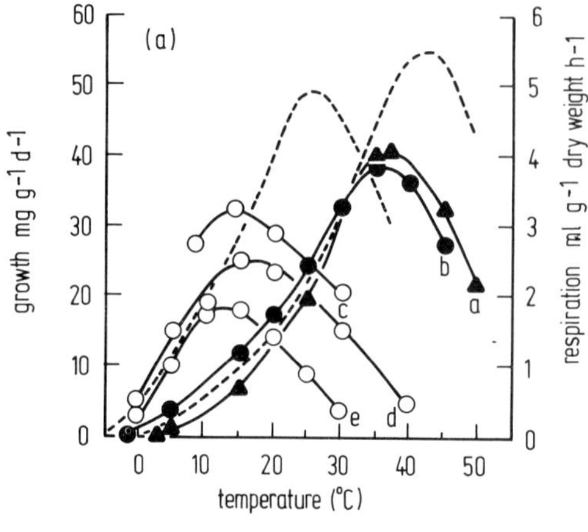


Fig. 1. (a) Fungal growth (solid lines) and respiration (fractured lines) vs. temperature. a Mesophilic fungi (average of 10 different species typified by (Rhizopus stolonifera). b Same for cold tolerant mesophiles typified by Cladosporium herbarum. c-e Psychrophilic fungi, Cylindrocarpon magnusianum, Mycelium sterilium and Chrysoporium pannorum respectively. (b) Annual amount of decay and respiration vs. temperature in tundra (▲ ▒ ), taiga birch (o □ ), and spruce (● ■ ) forests. (c) Fungal yield vs. temperature. Points represent the average reading for 10 representative organisms in the categories shown on the figure

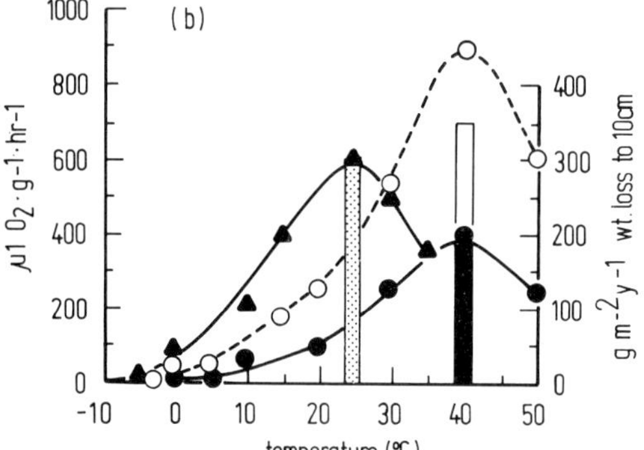


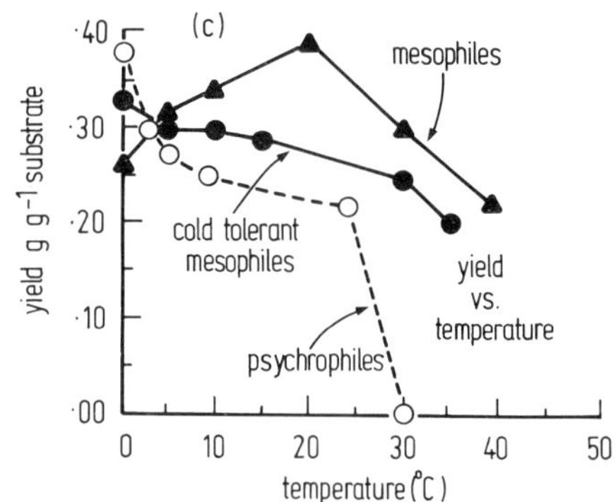


tolerant mesophiles. The remainder were mesophilic. Spruce and birch forest mycoflora were composed of 40% mesophilic and 60% cold tolerant mesophilic fungi. Only one psychrophilic form was isolated in these taiga sites.

Decomposer microorganisms in the organic remains of tundra sites respired optimally at 25°C while those of taiga forest floors had optimum respiratory activity above 40°C. (Fig. 1 (b)). Optimum temperature for growth of major species components from these sites was 15°C and 35°C respectively. The low temperature thresholds for respiration and growth of major tundra psychrophilic fungal components was -6°C and 20°C while for taiga cold tolerant fungi they were 0°C and +3°C. (Fig. 1 (a)). Yield of fungal tissue (g fungus/g substrate) was higher at low temperatures than at optimum growth and respiration temperatures for all organisms tested (10 in each of the categories psychrophilic, cold tolerant mesophilic and mesophilic) (Fig. 2). Optimum temperatures for growth respiration and yield differed for each individual organism and markedly between the temperature related categories: psychrophiles, mesophiles and cold tolerant mesophiles (Fig. 2 (a)). Fungi from temperate European sites were all mesophilic. No cold tolerant or psychrophilic forms occurred amongst the temperate site species screened.

Calculated microbial productivity varied significantly between sites. The primary productivity: microbial productively ratio however was similar between sites. The spruce forest fungal productivity was lowest but in relation to primary productivity was highest (Table 1).

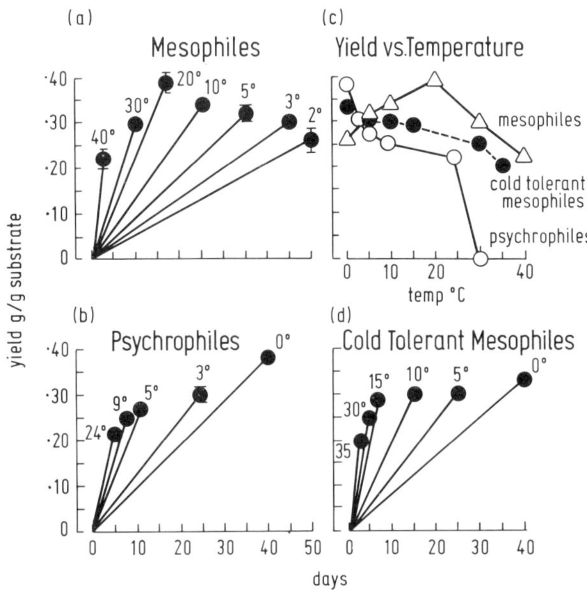


Fig. 2 a-d. Fungal yield (g fungal tissue / g available substrate) at various temperatures and times. Solid points on (a, b and d) represent the average reading for 10 representative organisms in the respective categories. I represents the standard error of the means for the 10 isolates. Where I is not shown standard error is at limit of •. In (c) a collage, all data for all isolates in each category are represented by individual curves

Annual amount of decay was highest in the birch forest but the level of decay was not much above those measured in tundra sites where consistently the annual level of decay was as high or higher than the measured annual primary productivity at the sites (Table 1).

Except for the birch site data, there is no indication of accumulation of organic matter in any of the ecosystems. Accumulated organic matter is highest in the tundra sites, lowest in the birch forest and of intermediate level in the spruce forest.

The discrepancy between the measures of annual net primary productivity and decay of organic matter is small and due, no doubt, to the margin of error implicit in the techniques used to measure and calculate productivity and decomposition. Still there is reasonable argument that microbial decay processes and productivity in these northern ecosystems are quite capable of balancing the input to the systems from primary productivity. It would appear, then, that the ecosystems studied are in, or very close to, a steady state.

The lower fungal biomass in organic remains in the spruce forest floor is reflected in the lower respiration rate in these substrates and in their lower annual decay rates. Because of the similarity in temperature responses of mycoflora from spruce and birch forest floors, similarity of location and the low levels of available N and P in both, it may be interpreted that lower decay rates in the spruce forest floor are not related primarily to low temperature attenuation of decomposers or nutrient deficiency but are more likely related to the low quality of spruce forest litter where the readily available carbon compounds are very low (Table 2) and the content of lignocellulose, waxy compounds, and moss remains is high.

Psychrophilic and low temperature tolerant fungi in the tundra are clearly responsible for the rates of decay in tundra sites where the mean annual temperature is below - 10°C (5).

The contour of the growth and respiration vs. temperature curves in tundra psychrophilic fungal species shows them functioning optimally at 15°C and 25°C respectively and their rate of activity is markedly higher than that of cold tolerant and temperate mesophiles in the same temperature range. The combination of cold tolerant and psychrophilic fungi in tundra assures efficient decay at temperatures where mesophilic fungi would barely metabolize and allows advantage to be taken of short periods when temperatures are relatively high.

Because exponential responses to temperature change show that the rate of positive response to temperature increases at higher temperatures, the linear response of psychrophiles to temperature increases at low temperature suggests special adaptations to low temperatures. At low temperatures, psychrophilic fungi attain higher maximal yields at faster rates than cold tolerant and temperate mesophiles. Psychrophilic forms, however do not have as efficient yield from substrates above 5°C as cold tolerant and temperate mesophiles. It appears that psychrophiles are adapted to maximise retention of energy at very low temperatures; an adaptation with obvious survival value.

The marked difference between optimal temperature for respiration, and that for growth is in keeping with the variation in yield at different temperatures (Fig. 2). Apparently, in northern ecosystems, as temperatures decline more of the substrate is converted to biomass and as temperatures increase less is used in biomass produdction irrespective of whether an organism is psychrophilic or a cold tolerant mesophile. The mesophilic fungi attain maximum yield potential at 20°C and decline in efficiency as temperatures increase or decrease. These observations are of importance in calculation of annual microbial productivity within and between ecosystems.

Annual microbial productivity should be calculated based on some average biomass estimate, substrate availability and integration of known temperature/yield relationships across the site annual temperature transient.

Optimum temperatures for mesophilic activity (35°C-45°C) rarely occur in temperate field sites, in fact temperatures above 25°C are rare whereas in northern ecosystems field temperatures especially in the upper soil surfaces are frequently in the 15°C-20°C range. In this temperature range psychrophilic fungi and cold tolerant mesophiles function at higher rates than mesophilic fungi and the latter do not attain an equal rate of activity until temperature is in the 25°C-30°C range.

The question of whether northern latitude microorganisms metabolize (grow and/or respire) as fast at low temperatures as temperate ecosystem mesophilic microorganisms at higher temperatures can be answered. Mesophiles respire, grow and yield more at higher temperatures than psychrophiles at their preferred lower temperatures but neither at low or high temperature do they exceed the metabolic activities of cold tolerant mesophiles from northern climes.

The presence of the latter group of fungi together with psychrophiles assumes overall that microbial metabolism and decay processes proceed during the biokinetic season in northern latitudes at rates equal to those occurring in temperate ecosystems functioning at higher field temperatures. The biokinetic season in temperate ecosystems, however, is much longer than that in northern latitudes so that decay processes in the former continue throughout the year while in the latter they cease abruptly when soil temperatures reach -5°C. Nonetheless, microbial activity during the short season decreases organic matter sufficiently to balance the income from primary productivity.

These decay rates, mycofloral activities and the surprisingly high levels of available N and P in tundra soils contradict the view (18) that it is cold and nutrient deficiency that causes organic accumulation in tundra ecosystems. It also appears that northern forest microorganisms recycle as much organic matter annually as is produced. Low decay rates of organic matter on spruce forest floors in this study are attributed to their low chemical quality. Accumulated organic matter in northern ecosystems may not be associated with imbalances between decay and production rates but reflect the degree and/or duration of imbalances that occurred during the period preceding the attainment of steady state.

## E. Acknowledgements

Without the data and time freely given by my colleagues in the U.S. Tundra Biome and taiga forest studies this paper would not have been possible. Special thanks are due to Drs. S.F. MacLean Jr., Robert White and F. Stuart Chapin III for critical discussion of the study approach, and field data. Research was financed partially by National Science Foundation Grants BMS 75-13998, and 9V29342 both to the University of Alaska to support the taiga forest research program and the U.S. Tundra Biome program respectively.

## F. References

1. Anderson, J.H.: Plants, Soils, Phytocenology and Primary Productivity of the Eagle Summit Tundra Biome Site. U.S. Tundra Biome Data Report 74-42 pp. 211, (1974).
2. Bell, M.K.: Decomposition of herbaceous litter. In: Biology of Plant Litter Decomposition. Vol. 1 C.H. Dickinson and G.J.F. Pugh (ed.). Academic Press, New York. pp. 37-67 (1974).
3. Bray, J.R. and E. Gorham.: Litter production in forests of the world. In: Advances in Ecological Research, G.B. Gregg, (ed.). p. 101-152, (1964).
4. Bremner, J.D.: In: Methods of Soil Analysis. Vol. 2. C.A. Block (ed.), American Soc. Agronomy publication, Agronomy 9. (1965).
5. Bunnell, F.L., MacLean, S.F. Jr. & Brown, J.: Barrow, Alaska U.S.A. In: Structure and Function of Tundra Ecosystems. T. Rosswall and O.W. Heal (eds.) Ecol. Bull. (Stockholm) 20: 73-124 (1975).
6. Chapin, E.F. III, Miller, P., and Coyne, P.: Carbon and nutrient budgets, quantities and turnover rates. Chap. 19. In: An Arctic Ecosystem: The Coastal tundra of Northern Alaska, Brown et al., (eds.). Dowden Hutchinson and Ross. Stradsburgh, Pa. (1977). (In Press).
7. Flanagan, P.W.: Physiological groups of decomposer fungi. In: The Fungal Community. D.T. Wicklow & G.C. Carroll, (eds.). Marcel Dekker Inc., New York. (1977). (In Press).
8. Flanagan, P.W. & Bunnell, F.L.: Decomposition models based on climatic variables, substrate variables, microbial respiration and production. In: The Role of Terrestrial and Aquatic Organisms in Decomposition Processes. J.M. Anderson & A. Macfadyen (eds.). Blackwell Sci. Publ. p.437-457 (1976).
9. Flanagan, P.W. & Bunnell, F.L.: Microbial activities and decomposition. In: Arctic Ecosystem: The Coastal Tundra of Northern Alaska. J. Brown et al., (eds.). Dowden, Hutchinson and Ross, Inc. Stradsburgh, Pa. (1977) (In Press).
10. Flanagan, P.W. & Van Cleve, K.: Microbial biomass, respiration and nutrient cycling in a black spruce taiga ecosystem. In: Soil Organisms as Components of Ecosystems. U. Lohm, (ed.). Ecol. Bull. (Stockholm). 23: (1977) (In Press).
11. Griffin, D.M.: Ecology of Soil Fungi. Syracuse Univ. Press. 193pp. (1972).

12. Hanssen, J.F., Thingstad, T.F., & Goksoyr, J.: Evaluation of hyphal lengths and fungal biomass in soil by a membrane filter technique. Oikos 25, 102-107 (1974).

13. Heal, O.W. & French, D.D.: Decomposition of organic matter in tundra. In: Soil Organisms and Decomposition in Tundra. A.J. Holding et al., (eds.). IBP Tundra Biome Steering Committee. Stockholm, pp. 279-310 (1964).

14. Jenny, H.W., Gessel, P., & Bingham, F.T.: Comparative study of decomposition rates of organic matter in temperate and tropical regions. Soil Sci. 68, 419-432 (1949).

15. MacLean, S.F. Jr.: Primary production decomposition and the activity of soil invertebrates in tundra ecosystems: A hypothesis. In: Soil Organisms and Decomposition in tundra. A.J. Holding et al., (eds.). IBP Tundra Biome Steering Committee. Stockholm, p. 197-206 (1974).

16. Olson, J.S.: Energy storage and the balance of producers and decomposers in ecological systems. Ecology 44 (2), 233-331 (1963).

17. Parkinson, D. & Williams, S.T.: A method for isolating fungi from soil micro-habitats. Plant and Soil 13 (4), 347-355. (1961).

18. Rodin, L.E. & Baszilevich, N.J.: Production and Mineral Cycling in Terrestrial Vegetation. Oliver and Boyde, Edinburgh. pp. 288 (1967).

19. Trolldenier, G.: The use of fluorescence microscopy for counting soil micro-organisms. Bull. Ecol. Res. Comm. (Stockholm) 17, 53-59. (1973).

20. Waksman, S.A. & Gerretsen, F.C.: Influence of Temperature and Moisture upon the Nature and Extent of Decomposition of Plant Residues by Microorganisms. Ecology 12 (1), 33-60 (1931).

# Microbial Ecology of Animals

# "New" Viral Zoonoses: Past, Present and Future

C.E.G. SMITH

During the past 20 years or more, there have been, at intervals, severe outbreaks of viral disease apparently "new" to medicine. To some extent their recognition can be accounted for by improved general and medical communications through WHO and other international channels, but in almost every case where an explanation has been advanced, these outbreaks have been attributed to man's own activities - often interference with previously little frequented areas because of population pressure and/or agricultural developments. With the world population continuing to increase markedly despite all efforts at control and with an estimated 460 million people without sufficient to eat for good health (3) these pressures are unlikely to diminish during the remainder of this century, therefore it seems reasonable to expect further outbreaks of severe viral diseases, some of them "new".

As examples, I shall consider Kyasanur Forest disease, O'nyong nyong fever, Bolivian haemorrhagic fever, Marburg and Marburg-like diseases.

Kyasanur Forest disease, a severely prostrating febrile illness with gastrointestinal and haemorrhagic complications, bradycardia and hypotension, followed in some cases by encephalitis, was first recognized in Mysore in 1956, in an outbreak of about 500 cases with 50 deaths. It was associated with a forest epizootic in monkeys, at first thought to be yellow fever, and was caused by a togavirus of the tick-borne Russian spring-summer encephalitis complex which was transmitted by *Haemaphysalis* ticks(12). Subsequent analysis (1) strongly suggests that the following sequence of events caused this outbreak. (a) The infection was maintained enzootically in the forest and probably beyond by forest floor ticks and mammals or birds. (b) An increased human population, which had more than doubled in 10 years, and shortage of arable land led to extensive cattle grazing in the forest. (c) The presence of the cattle led to a number of floral and faunal changes including a large increase in *Haemaphysalis* ticks. This was because the adult stage feeds on large mammals, and the population of the host of the adult stage is the most important factor in controlling the tick population. (d) The *Haemaphysalis* population became infected and transmitted the virus to monkeys, which amplified it, and to man in the forest.

O'nyong nyong fever broke out in north west Uganda in 1959 and spread south and east to involve over 1 million cases (9). The disease, which resembled dengue with severe arthralgia, was caused by a virus which was transmitted by *Anopheles funestus* and *Anopheles gambiae* (11). Ross River virus has caused outbreaks of similar disease in Australia (2). Both of these alphaviruses are closely related to chikungunya virus which caused a series of very large epidemics in Indian cities in 1963-4 (8). It was transmitted by *Aedes aegypti* and its virulence for man may well be increased by repeated man - *A. aegypti* transmission. Chikungunya seems to be widely distributed in Africa and Asia but the cause of the emergence of the variant O'nyong nyong virus remains unexplained.

Bolivian haemorrhagic fever is caused by Machupo, an arenavirus, related to that of Argentinian haemorrhagic fever, a similar severe febrile disease with haemorrhagic manifestations, hypotension, bradycardia, oliguria and central nervous system involvement. It was first recognized as a single outbreak in a village of about 2,500 people in which there were 470 cases and 142 deaths (7). The arenaviruses cause prolonged viruria in rodents which is the source of infection for man. The outbreak was associated with an intense population peak in the grassland rodent, Calomys callosuswhich invaded the town and its houses in large numbers. Rodent control brought the outbreak to an end (6).

The first outbreaks of Marburg disease (10) occurred in 1967 in Germany and Yugoslavia. Laboratory workers who had had contact with blood or tissues of Cercopithecus aethiops monkeys imported from Uganda developed a severe haemorrhagic and febrile disease with intense headache, nausea, diarrhoea, a skin rash, mental confusion and sometimes coma. The infection spread to doctors, nurses and post-mortem attendants in the hospitals and, of a total of 31 cases, seven died. One case, six weeks into convalescence, appears to have transmitted the infection to his wife by the venereal route. In 1975, three cases occurred in South Africa. The index case had been infected somewhere in Africa and transmitted the infection to his travelling companion and to a nursing sister. Only the original case died but the nursing sister had a persistent infection of the eye (4). This outbreak was caused by a virus identical with Marburg virus which is a structurally unique large rod-shaped RNA virus. The two cases of persistent infection with Marburg virus (one leading to transmission) indicate that some "new" infections may be capable of transmission outside the area of the initial outbreak.

Nothing more was heard of Marburg infections until, in 1976, large outbreaks of similar disease broke out in the Sudan and Zaire, both caused by an antigenically different but structurally similar virus (Ebola virus). The source and mode of transmission to the index case or cases remains obscure.

In the Sudan, an outbreak with an overall case mortality approaching 50% started in late June in Nzara. The evacuation of one patient to the hospital at Maridi led to an outbreak of 229 cases mainly in the hospital staff and their families. Four cases were evacuated to Juba and one hospital case resulted. The Zaire outbreak (of similar severity) appears to have started at the end of September at a Mission in Yambuku, 90 km north of Bumba. It spread to some 43 villages within a radius of 50 km. Two cases (one a nurse) were infected from a patient evacuated to Kinshasa. In England, one laboratory worker in a high-security laboratory became infected by pricking his finger through a protective glove, and was seriously ill. Strict barrier nursing (using protective clothing) and quarantine were highly effective in bringing both outbreaks to an end. It became clear that transmission from person to person was relatively uncommon unless there was close physical contact with blood or body fluids containing high virus concentrations. Injections given with the same syringe and needle to numbers of persons considerably amplified both epidemics.

The first questions to ask are whether these diseases were indeed new - it seems unlikely: if not, why were they not recognized sooner; and why were they recognized when they were? The 1956 outbreak of Kyasanur Forest disease was recognized mainly because attention was drawn to it by a high monkey mortality, but it was investigated and identified primarily because of the proximity of the Virus Research Institute, Poona, and because the team there at the time were interested in such outbreaks. Similarly, O'nyong nyong fever, except perhaps for the scale of the epidemic, might well have been dismissed as yet another outbreak of dengue-like disease without serious consequences, had not the Virus Research Institute, Entebbe, been ready and able to investigate it. In the case of Bolivian haemorrhagic fever it is fairly clear that the outbreak which was brought to notice (and investigated by the Middle American Research Unit) because of its severity, had been preceded by sporadic cases in forest workers and the like. These had not been investigated or diagnosed because of the low availability of modern medical care and lack of laboratory facilities for diagnosis. The first occurrences of Marburg disease received the fullest investigation from the start because they occurred in laboratory workers in highly developed countries. The Ebola outbreaks were brought to notice because of their great severity, because they were spreading rapidly among hospital staff, because the whole health services of the epidemic areas appeared to be in jeopardy, and because there appeared to be risk (widely reported in the local and world Press) of much wider spread. It is clear that sporadic cases could have escaped medical attention or have been dismissed (in the absence of available diagnostic resources) as yellow fever, typhus, typhoid or some other severe febrile disease.

We can conclude that unless "new diseases" which occur in relatively remote areas cause large epidemics, or affect hospital staffs, or occur in the "parish" of a virus research institute, they are unlikely to be investigated or their cause discovered. There are probably a number of candidate causes for "new" diseases in the International Catalogue of Arboviruses (5) (e.g. Crimean haemorrhagic fever - Congo virus in Africa, where sporadic cases were recognized only because they occurred in the Entebbe "parish") and every year more are discovered.

How can we improve the early recognition of outbreaks of severe viral infections and thus their control? With increasing world travel, both in quantity and speed, these outbreaks are important not only to the countries where they originate, but also to faraway countries to which cases may be introduced. Such imported cases are primarily a hazard to hospital and laboratory workers but, should one of these "new" diseases be readily transmissible from man to man by the respiratory route, the risk of much wider spread would be real. We need only remember that pandemic influenza strains may originate from wild animal or bird sources, and the havoc and cost of pandemics.

It is probable that the emergence of a "new" zoonosis is usually due to a change in the ecology of its maintenance cycle or to changes in the ecology of neighbouring areas. New contacts with man, or changes in virulence, may be induced by increases in the population of maintenance hosts or related species and/or by establishment of the infection in a new maintenance host. The larger the scale of man-made environmental changes and the more they involve areas little frequented by people other than long-term indigenous forest dwellers, the greater must be the probability of emergence of a zoonosis ("old" or "new"). The intrusion of agriculture, particularly the cultivation of food crops attractive to rodents, into previously underdeveloped areas obviously increases the hazard of rodent-maintained infections; extensive food storage inadequately rodent-proofed has a similar effect. Irrigation or other water developments, including those that reduce the salinity of surface waters, increase the hazard of mosquito-borne infections; and the introduction of large domestic mammals (especially cattle) into new territory may enhance the risk of tick-borne infections. More subtle changes probably account for those outbreaks with less obvious explanations.

Clearly the first priority for early recognition of potentially dangerous outbreaks must be to educate the health and administrative personnel of largely undeveloped areas (particularly in the tropics) of the need for some form of surveillance and reporting of acute febrile disease in all new agricultural ventures involving intrusion into undeveloped territory - particularly cases occurring in hospital staffs. This need not be elaborate or involve expensively trained staff. Outbreaks (or even cases) of febrile illness can be reported by policemen, foremen, teachers, or villagers given simple but clear instructions. There is of course no point in reporting unless there is a more highly trained person to respond and investigate and most important, to feed back interpretation of the data to those who report, so that they co-operate in the knowledge of what is done with their reports.

What is next needed is limited investigational facilities and a small but well-trained team which can be called on at short notice at national, state, or provincial level depending on the size of the country. These facilities need not be elaborate or expensive but should have a close relationship with, and when necessary a first call on, any available microbiological laboratory resources. The team should keep reporting under review, assess and advise on communicable disease problems as they become apparent and, when it considers that an incident requires on the spot investigation, it should be able to go to it with minimum delay carrying all necessary equipment with it. It should be equipped, well trained and disciplined to collect specimens safely from cases or corpses of dangerous infectious diseases, trained to make an epidemiological assessment and able to institute emergency control measures with such local support as is available.

Outside expert assistance is often needed in the control and investigation of such outbreaks. The nature of the help required will vary with circumstances but if the infection is a highly dangerous one, only a well-equipped and well-trained team should be sent under the control of an able, experienced and tactful leader. It must be self-sufficient in terms of immediate medical care for its members, equipment (including protective clothing, containers, field sterilizers, etc.), materials, and if necessary, camping equipment, electricity generators, fuel, etc. Adequate supplies must be carried to enable the team to effectively equip the local hospital and health authorities to control the outbreak. This implies formidable logistic problems, the most important of which are communications, transport and the dissemination of information.

Because of its neutrality, WHO can do much to help arranging, co-ordinating and supporting such teams in the field with assistance from various laboratory and financial resources. However, there is an urgent need for WHO to have outline plans and check-lists of requirements for mounting these exercises, to have emergency funds which are available at very short notice so as to eliminate delays, and for more well-trained staff to assist with logistic, communication, financial and diplomatic problems.

## A. References


1. Boshell, M.J.: Kyasanur Forest disease: ecological considerations. Amer. J. trop. Med. Hyg., 18, 67-80 (1969).
2. Doherty, R.L., Barrett, E.J., Gorman, B.M., & Whitehead, R.H.: Epidemic polyarthritis in eastern Australia 1959-1970. Med. J. Aust., 58, 5-8 (1971).
3. F.A.O. The state of food and agriculture 1974. Rome Food and Agriculture Organization. (1975).
4. Gear, J.S.S., Casel, G.A., Gear, A.J., Trappler, B., Clausen, L., Meyers, A.M., Kew, M.C., Bothwell, T.H., Sher, R., Miller, G.B., Schneider, J., Koornhof, H.J., Gomperts, E.D., Isaacason, M. & Gear, J.H.S.: Outbreak of Marburg virus disease in Johannesburg. Brit. Med. J., 4, 489-493. (1975).
5. International Catalogue of Arboviruses Ed. Berge, T.O. U.S. Department of Health, Education and Welfare Publication No. (CDC) 75-8301 (1975).
6. Johnson, K.M.: Epidemiology of Machupo virus infection. Amer. J. trop. Med. Hyg., 14, 816-818 (1965).
7. Mackenzie, R.B., Beye, H.K., Valverde, C.L. & Garron, H.: Epidemic hemorrhagic fever in Bolivia. Amer. J. trop. Med. Hyg., 13, 620-625 (1964).

8. Sharma, H.M., Shanmugham, C.A.K., Iyer, S.P., Ramachandra Rao, A. & Kuppuswami, S.A.: Report on a random survey conducted to assess the prevalence of a 'dengue-like' illness in Madras City - 1964. Ind. J. med. Res, 53, 720-728 (1965).

9. Shore, H.: O'Nyong nyong fever: an epidemic virus disease. Trans. R. Soc. trop. Med. Hyg., 55, 361-373 (1961).

10. Smith, C.E.G.: Lessons from Marburg disease. Scient. Basis Med. A. Rev., pp. 58-80 (1971).

11. Williams, M.C., Woodall, J.P. & Gillett, J.D.: O'Nyong nyong fever: an epidemic virus disease. Trans. R. Soc. trop. Med. Hyg., 59, 186-197 (1965).

12. Work, T.H.: Russian spring-summer virus in India: Kyasanur Forest disease. Progr. med. Virol., 1, 248-279 (1958).

# Ecology of Mosquito-Borne Viruses in India and South-East Asia

K.M. PAVRI

## A. Introduction

This presentation is confined to the mosquito-borne arboviruses of public health importance in India and South East Asia. These are Japanese encephalitis (JE) virus, West Nile (WN) virus, and the four serotypes of dengue (DEN) virus, which are all flaviviruses (group B) and chikungunya (CHIK) virus which is an alphavirus (group A). JE and WN viruses together with St. Louis encephalitis (SLE) virus and Murray Valley encephalitis (MVE) virus form a complex of closely related agents. The severe infections with JE, SLE and MVE are characterized by encephalitic symptoms which are rather infrequent in cases of WN virus infections except for the earlier reports from Israel (15). JE, SLE and MVE viruses would possibly have been considered as different serotypes of the same virus had it not been for their unique geographic distribution, each confined to a single continent. The distribution of WN virus extends to three continents. The four dengue virus serotypes are similarly considered to form another complex, which exhibits a serological overlap with other group B viruses including the JE-WN complex. Both DEN and CHIK viruses cause somewhat similar disease including the more severe manifestations of haemorrhagic fever (HF).

## B. Epidemiological Features

Table 1 presents some important epidemiological features of these viruses. DEN appears to share more common epidemiological features with CHIK virus rather than with members of its own group, in that they share the same mosquito vector(s) and seem to involve man directly rather than tangentially as is the case with JE-WN virus zoonoses. In Malaysia interesting studies demonstrate a sylvatic cycle involving monkeys and a canopy mosquito Aedes niveus, (Rudnick, personal communication) but the problem remains to relate this zoonotic, perhaps primeval, cycle of dengue with the large-scale urban outbreaks which continue to occur in India and South East Asia. Although the foci of prevalence are the same a most intriguing feature of CHIK epidemiology, distinguishing it from DEN, has been noted in India, Burma and Sri Lanka. After causing large-scale epidemics CHIK gradually seems to disappear from the area and is not encountered for 7-8 years or even 2-3 decades (2, 7, 16). The amazing part of this difference is that whereas DEN viruses maintain high endemicity despite the presence of circulating group B antibodies (suspected to be cross-protective) CHIK is unable to do so even in the absence of specific or cross-reacting antibodies in susceptible populations. One possibility is that CHIK virus infects all the susceptible hosts and is reintroduced in the area by migrant populations infected at other places only when ecological factors are suitable. Another explanation might be its continuous maintenance in an occult cycle. It may not be too far-fetched to consider such a cycle being maintained through transovarial transmission in some mosquito which mainly bites nonsusceptible hosts. Transovarial transmission of La Cross virus in mosquitoes has been documented (17).

Table 1. Some important epidemiological features of mosquito-borne viruses in India and South East Asia

| Virus | Field isolation or susceptibility experiments | Arthropods involved | Mechanism(s) of maintenance and spread to humans | Areas of prevalence | Types of prevalence |
|---|---|---|---|---|---|
| JE | Domestic animals (pigs) birds, bats: man | Culicine Anopheline | Zoonotic cycle: Man tangential host | Mostly rural: semiurban (?) | Endemic with occasional outbreaks/ epidemics |
| WN | Domestic and wild animals, birds, bats: man | Culicine Anopheline (& Ticks ?) | Zoonotic cycle: Man tangential host | Urban and rural | Highly endemic; sporadic cases |
| DEN | Man, subhuman primates | Culicine: A. aegypti A. albopictus | Man-mosq-man (Sylvatic cycle with canopy mosq.) | Mostly urban or semiurban | Highly endemic with occasional epidemics |
| CHIK | Man, subhuman primates * | Culicine A. aegypti A. albopictus | Man-mosq-man (?) (Occult cycle ?) | Mostly urban or semiurban | Explosive epidemics (gradually diminishing endemicity ?) |

* The virus has been isolated from bats and a bird in Africa

## C. Pitfalls in Interpretation of Serological Data

During epidemics it is not possible to provide laboratory proof of etiology in all cases. The use of a battery of viral antigens to provide unambiguous serological evidence for a particular virus is also impracticable. Therefore interpretation of serological data, particularly in areas where several arboviruses coexist, becomes complicated.

CHIK virus belongs to the Semliki-Mayaro-Getah complex.

Haemagglutination (HI) antibodies reacting to CHIK virus have been reported in domestic animals in some countries in SE Asia. In India, such positive sera from horses were later found to give maximum reaction with Getah virus (11) Hotta et al. (4) questioned the significance of CHIK HI antibodies detected in animals from Indonesia and also their specificity.

Neutralizing antibodies against JE virus were reported in some sera of dengue haemorrhagic fever (DHF) cases collected in 1965-66 during DFH epidemics (5), but it was only in 1969 that a large outbreak of JE was recognized in Chiangmai Valley in Thailand (3). There are reports of significant elevations in HI antibody titres either for JE virus or DEN-2 virus in paired sera from suspected HF patients in Surabaja, Indonesia (6). The isolation of JE virus from the blood of a child suffering from shock and haemorrhagic lesions is also reported (9). The ecological features of JE virus are different from those of DEN and CHIK viruses in tropical Asia so that overlap of such infections might be less frequent. A much greater confusion arises with WN virus which maintains endemicity both in urban and rural areas. During recent years it has been associated only with sporadic cases and has been associated only with sporadic cases and has been recognized in India by serendipity during ecological studies on other viruses. During JE/DEN studies it was associated with a case of encephalitis at Vellore (1) and with dengue-like fever at Jaupur (8). WN virus was also isolated from a febrile case during epidemlogical studies on Kyasanur Forest disease (10). Theiler and Downs presented retrospective serological evidence of WN virus being mainly responsible for dengue-like illness in Beirut in 1903. They also mentioned the reports from Israel in the early 1950's when this virus was associated with severe dengue-like illness and sometimes encephalitis in older people (15).

## D. Interaction of Viruses with Environmental Factors

I would like to emphasise the need to consider the spatiotemporal relationship with variability in behaviour of viruses. Has WN virus mellowed with age, or by

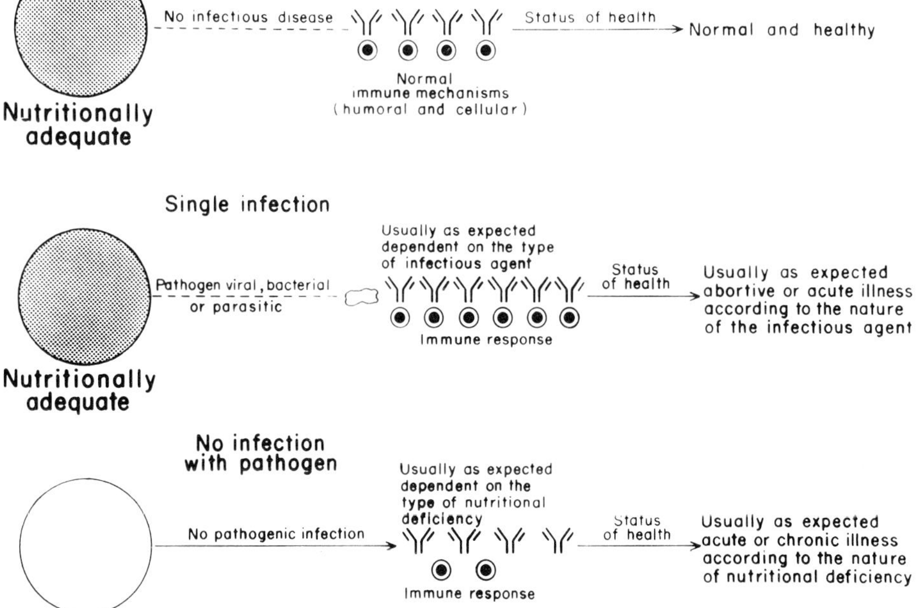


Fig. 1. Possible disease patterns resulting from interaction between nutritional and immune status of hosts and the types of infectious agents harboured by them

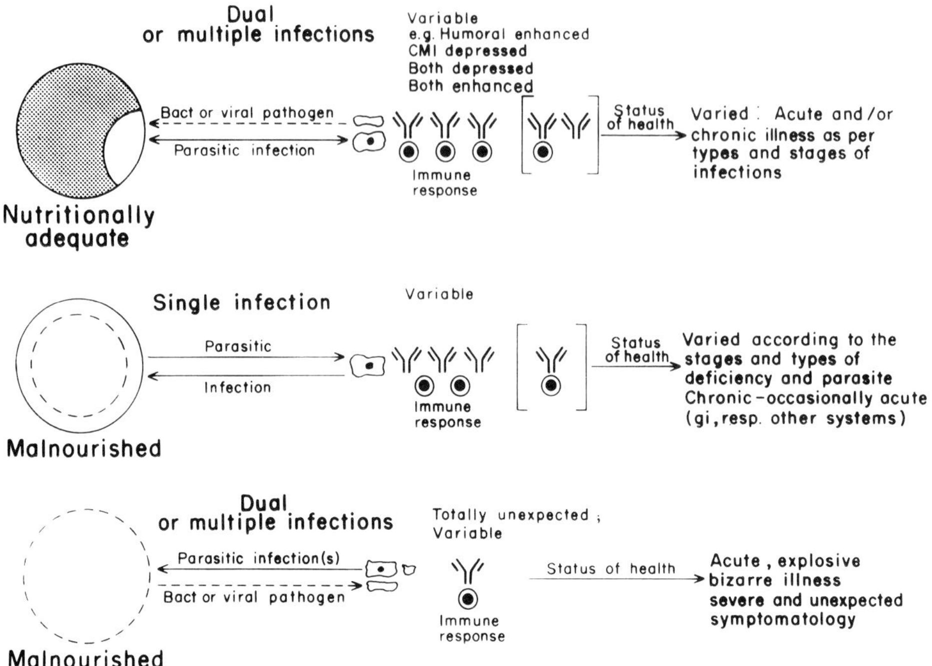


Fig. 2 (legend see Fig. 1)

contrast, have viruses like DEN, CHIK and Crimean haemorrhagic fever-Congo become more virulent in certain areas and in clusters? This could perhaps be explained through genetic factors either of the virus or the host, or alternatively, through interaction with some extraneous environmental factors. I have selected the alternative involving parasites and viruses for DHF and shock syndrome (12). A retrospective study on sera from Bangkok received by us in 1957 and 1958 showed statistically significant concentrations of IgE (associated with parasitic infections and not with viral ones) in the 1958 sera from DHF patients as compared to the 1957 survey sera from a similar age group (14). The concept of 'multiple etiology' has been extended further (13). In developing countries like India, and particularly among the lower socio-economic group, the external environment is poor in nutritional standards, but rich in infectious agents, a situation which is bound to have an effect on the internal ecology of the host. Possible disease patterns resulting from interaction between nutritional and immune status of hosts and the types of infectious agents harboured by them are depicted in Figures 1, 2 and 3. Such interactions might be infrequent in the developed countries, but there too, the effects of stress, industrial pollutants and drugs on pathogenesis of viral infections might manifest themselves in unusual ways.

I close with a beautiful and apt thought on interdependence expressed by John Donne:

> "No man is an island entire of itself. Every man is a piece of the continent - a part of the main ... any man's death diminishes me ... and, therefore, never send to know for whom the bell tolls, it tolls for thee".

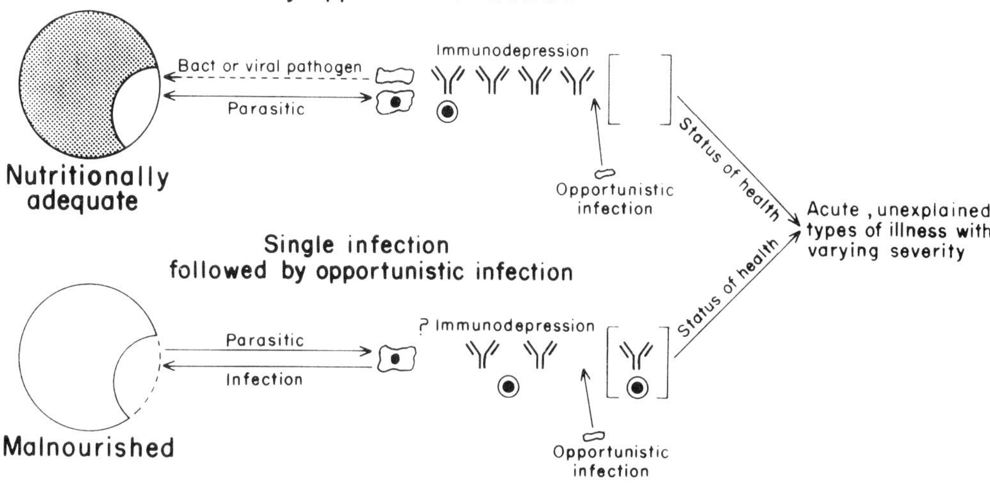


Fig. 3 (legend see Fig. 1)

## E. Acknowledgement

I am thankful to Doctors N. P. Gupta, K. Banerjee, F. M. Rodrigues, S.N. Ghosh for helpful suggestions.

## F. References

1. Carey, D. E. Rodrigues, F. M., Myers, R. M. and Webb, J.K.G.: Arthropod-borne viral infections in children in Vellore, South India, with particular reference to dengue and West Nile viruses. Indian Paediatrics 5, 285-296 (1968).
2. Editorial: Epidemic of fever with severe joint pains: Chikungunya virus. Maharashtra Med J. XX, 219-220 (1973).
3. Grossman, Richard A., Gould, Douglas J., Smith, Thomas J., Johnsen, Dennis O. and Pantuwatana, Suntaree: Study of Japanese encephalitis virus in Chiangmai Valley, Thailand. Amer. J. Epidem. 98, 111-120 (1973).
4. Hotta, Susumu, Aoki, Hideo, Samoto, Susumu and Yusui, Taxuko: Virologic-epidemiological studies on Indonesia. III. HI antibodies against selected arboviruses (Group A and B) in human and animal sera collected in Surabaja, East Java in 1968. Kobe J. Med. Sci. 16, 235-250 (1970).
5. Igarashi, Akira, Fukai, Konosuke, Ahandrik, Sompop and Tuchinda, Prakorb: Antibody against Japanese encephalitis virus in sera of dengue haemorrhagic fever patients in Thailand. Biken J. 11, 41-49 (1968).
6. Iswara, Atasiati, Soeharto and Biroum: Anti-arbovirus HI antibodies in sera from suspected haemorrhagic fever patients found in Surabaja, Indonesia in 1968-69. Kobe J. Med. Sci. 16, 211-214 (1970).
7. Khai Ming, C., Soe Thein, U. Thaung, Tin U. Khin Nwe and Diwan, A. R.: Serologic survey for certain childhood viral infections in Rangoon in 1971. J. Trop. Med. Hyg. 77, 260-266 (1974).

8. Mathew, Thankam, Suri, N. K., Bhola, S. R., Suri, J. C., Arora, R. R. and Lahiri, S. K.: Serological investigations of an epidemic of fever by group B arboviruses in Jaipur (1973). Indian J. Med. Res. 64, 1136-1142 (1976)

9. Nguyen-Thi Kin-Thoa, Lam-Thi Cam-Thu, Ngo-Thi-Vien and Nguyen-Xuan-Quang: Isolation of a strain of Japanese B encephalitis from the blood of a young patient suffering from cardiovascular collapse. Bull. Soc. Path. Exot. 67, 341-346 (1974).

10. Paul, Sharda Devi: Isolation of a West Nile virus from a human case of febrile illness. Indian J. Med. Res. 58, 1177-1179 (1970).

11. Pavri, Khorshed M.: Virological and serological studies on cases of febrile illness with haemorrhagic manifestations in Calcutta. Bull. Wld. Hlth. Org. 35, 60 (1960).

12. Pavri, Khorshed M.: Double aetiology involving parasites and viruses proposed for dengue haemorrhagic fever and shock syndrome. Indian J. Med. Res. 64. 713-729 (1969).

13. Pavri, Khorshed, M.: Oration delivered as a recipient of Basanti Devi Amir Chand Prize 1976. 15 October, 1976.

14. Pavri, Khorshed, Sheikh, B. H., Ghosh, S. N. and Chodankar, V. P. : Immunoglobulin E (IgE) in sera of patients of Dengue Haemorrhagic Fever (DHF) Indian J. Med. Res. (October 1977).

15. Theiler, Max and Downs, W. G.: The arthropod-borne viruses of vertebrates. Yale University Press, New Haven and London, 1973.

16. Vasenjak-Hirjan, J., Hermon, Y. and Vitarana, T.: Arbovirus infections in Ceylon. Bull. Wld. Hlth. Org. 41, 243-249.

17. Watts, D. M., Pantuwantana, S., De Foliart, G. R., Yuill, T. M. and Thompson, W. H.: Transovarial transmission of La Crosse virus (California Encephalitis Group) in the mosquito, *Aedes triseriatus*. Science 182, 1140-1141 (1973).

# The Ecology of Chikungunya Virus in Southern Africa

B. M. McINTOSH

Throughout the hot, wooded regions of Africa chikungunya virus is the most frequent cause of infection in man by Alphaviruses. Furthermore, there are factors present in its ecology which suggest that chikungunya virus also possesses some potential as a cause of even greater future human morbidity. In its ecology this virus shows certain notable similarities to yellow fever virus in that both viruses frequently infect wild primates while canopy-frequenting mosquito species with a high feeding preference for these vertebrates act as vectors. Significant ecological differences are, however, evident since it seems probable that enzootic foci of yellow fever virus are confined to moist forest while those of chikungunya also exist in comparatively dry, open, woodland savanna. Consequently, chikungunya virus enjoys a more extensive distribution in Africa, extending into eastern as well as southern Africa, where moist forest is limited and yellow fever absent.

Southern Africa may be defined as that part of the continent south of the northern boundaries of southwest Africa, Rhodesia and the Zambezi River where it flows through central Mocambique. Within this region chikungunya virus is restricted to the tropical and subtropical zones. The known southern limits of the distribution of chikungunya virus in Southern Africa correlate with the 20$^{o}$C isotherm and it seems that in some unknown way temperature plays a vital role in its distribution. Recognized human epidemics have occurred in rural areas within the wooded savanna, and have tended to occur at infrequent intervals. Since its original isolation in 1956, during an epidemic in the lowlands of eastern Transvaal Province in South Africa, outbreaks have appeared in 1962 in Rhodesia and in 1975, 1976 and 1977 in the same area in South Africa as the original outbreak. The last 3 years were exceptionally wet and this has presumably led to this recent increase in human infection.

Viral activity was however detected during an interepidemic period when evidence was found of recent infection in a wild monkey population in South Africa in 1964 and it is likely that the virus persists in southern Africa despite the sporadic nature of recognized human infection (1). This is also suggested from antibody surveys in Mocambique. In the southern half of Mocambique immune human beings were found at all 13 localities sampled in 1957. In more recent surveys at Mopeia on the Zambezi River in 1973, immune rates of from 50-80% were found among children although there were no records of recent infection in the area and it is evident that some human infection in this remote region goes unrecorded.

There is no information to indicate how the virus survives in nature. The only suggestive evidence is of infection in bats in Africa. The virus has been isolated twice from the salivary glands of _Scotophilus_ spp. in Senegal, West Africa, and 2 species of South African bats of the genera _Tadarida_ and _Pipistrellus_ circulated virus at significantly high levels after inoculation of virus.

Although there is no evidence to suggest that wild primates are involved in viral maintenance there is now ample evidence to show that they are important natural hosts in as much as they undoubtedly actively participate in viral transmission cycles. In West Africa the virus has been isolated from the vervet monkey

Fig. 1. Distribution of chikungunya virus in southern Africa as determined by virus isolations from man (×), and antibody in man, monkeys and baboons (•). Shown also are the years and localities of recorded human outbreaks

*Cercopithecus aethiops*, the bushbaby, *Galago senegalensis*, and the baboon, *Papio papio* (2). In southern Africa antibody has frequently been found in wild populations of vervets and baboons. Inoculation of virus into these 2 species also showed that they became viremic at levels which readily infect mosquitoes and the virus has indeed frequently been transmitted experimentally between vervet monkeys by mosquitoes. The repeated isolation of the virus from canopy-frequenting mosquito species, feeding readily on wild primates, such as *Aedes africanus* in east and west Africa and the *Aedes furcifer/taylori* group in west and southern Africa has also implicated wild primates.

The small size of their populations in southern Africa would seem however to preclude wild primates as maintenance hosts and the available evidence suggests that their importance in the human disease is as a viral amplifying host, their infection leading to massive feed-back of virus to vector populations with the subsequent involvement of rural human populations. This possibility first became apparent in southern Africa after the Rhodesian epidemic in 1962 when antibody was found in sera from all 13 vervets and 4 baboons sampled from the epidemic area after the epidemic (3). A similar situation was found to exist after the 1976 epidemic in South Africa (4). On that occasion all 92 baboons tested on 2 farms where human infection had been intense were found to be immune (Table 1).

In southern Africa members of the *Aedes furcifer/taylori* group have been implicated as vectors in both wild primate and human infection (4). Suspicions in this respect first arose because of their prevalence and feeding habits as observed in field studies in the epidemic area after the Rhodesian outbreak although virus was not isolated from these or other mosquito species during the epidemic. These suspicions were strengthened when experimental transmission tests with wild-caught females, which probably consisted largely of *Ae. furcifer*, showed them to be vectors, which was confirmed later in tests with a laboratory colony of *Ae. furcifer*. These later tests showed that from a group of *Ae. furcifer* ingesting 5.8 - 6.5 dex of virus from a viremic vervet, 2 out of 8 mosquitoes transmitted virus to vervets. More direct evidence concerning these species was obtained during the 1976 epidemic in South Africa when 16 isolations of the virus were made from 538 females collected on the 2 farms previously referred to (Table 1), where human and baboon infection had been very high.

Table 1. Immune rates to chikungunya virus in man and baboons after an epidemic in the wooded savanna, South Africa, 1976

| Farm | Immune Rates (%) Man Juvenile | Man All ages | Baboon All ages |
|---|---|---|---|
| Lillie | 100 (9)[a] | 100 (19) | 100 (21) |
| Hope | 56 (9) | 87 (46) | 100 (71) |
| | 78 (18) | 94 (65) | 100 (92) |

(a) - figure in parenthesis is total number tested.

Studies on the feeding habits of the Ae. furcifer/taylori group in South Africa have shown them to feed readily on baboons and vervets and it is likely that these animals are the preferred hosts (Table 2) (4). Feeding occurred mainly in the canopy but also to a significant extent on the ground. They also fed readily on man on the ground and alighting rates on man as high as 120 mosquitoes per hour have been recorded. Their distribution and activity are however decidedly sylvatic, and only rural human populations seem likely to be bitten to a significant degree and human infection in southern Africa has in fact been restricted to such populations. These species breed in tree-holes, they occur widely in the wooded savanna where their distribution correlates with the known distribution of the virus, and it is apparent from their feeding habits that they could readily transfer virus from wild primates to man. Present evidence therefore identifies Ae. furcifer, and possibly Ae. taylori also, as the important epidemic vectors in southern Africa.

In regard to potential vectors it is worth commenting on Aedes aegypti, a proven vector elsewhere in Africa and also in Asia. While this is one of the commonest species found breeding in tree-holes in the wooded savanna there is no evidence from southern Africa to implicate this species as an important vector of chikungunya virus. However, it is at present also difficult to entirely exclude it, as adults from sylvatic populations have not been collected in adequate numbers and have not therefore been sufficiently sampled for virus either during epidemics or interepid-

Table 2. Numbers of females of the Aedes furcifer/taylori group attracted to man or baboons at 2 horizontal levels

| Bait<br>Level | Man[a]<br>ground | Baboon[b]<br>ground | Baboon[b]<br>canopy 10 m |
|---|---|---|---|
| Number mosquitoes | 1891 | 38 | 410 |
| Collecting time (hours) | 134 | 8 | 8 |
| Number mosquitoes per trap/hour | 14 | 5 | 51 |

(a) Collected by hand in test-tubes.
(b) Collected in suction traps.

emic periods. Until they have been adequately studied, the habits and possible vectorship role of the sylvatic savanna population of this species must remain obscure. As mentioned earlier it seems likely that chikungunya virus has a great potential as a cause of future human morbidity in southern Africa. This is believed to be so because increasing urbanization in the future could lead to vast populations of the domestic form of Ae. aegypti which in turn could produce urban epidemics on a scale similar to those already experienced in Asia.

Observations in South Africa have indicated that certain microhabitats within the wooded savanna favour transmission of chikungunya virus through their influence on the local prevalence of wild primates and members of the Ae. furcifer/taylori group. The exceptionally large trees of the riverine gallery is one such habitat. These trees are not only utilized by vervets and baboons as dormitories but they also provide larval habitats on a generous scale. Another microhabitat is the large granite outcrop, rising some 50-150 metres above the surrounding plain which is frequently used by baboons as a safe overnight refuge. Furthermore with the increasing pressures of civilization building-up to their disadvantage, it now appears that baboons are tending to concentrate more and more at these outcrops. These granite hills are prolifically supplied with caves and holes which our studies have shown provide a humid resting site for mosquitos including members of the Ae. furcifer/taylori group, thereby probably enhancing their longevity and increasing the chances of successful virus transmission.

## A. References

1. McIntosh, B.M.: Antibody against chikungunya virus in wild primates in southern Africa. S.A.J. Med. Sc. 35, 65-74 (1970).
2. Robin, Y.: Institut Pasteur, Dakar. Personal communication.
3. McIntosh, B.M., Paterson, H.E., McGillivray, G., De Sousa, J.: Further studies on the chikungunya outbreak in Southern Rhodesia in 1962. I. Mosquitoes, wild primates and birds in relation to the epidemic. Ann. Trop. Med. Parasit. 58, 45-51 (1964).
4. McIntosh, B.M., Jupp, P.G., Dos Santos, I.: A rural epidemic of chikungunya in South Africa with involvement of Aedes (Diceromyia) furcifer (Edwards) and baboons Papio ursinus (Kerr). S. Afr. J. Sci. 73, 267-269 (1977).

# Considerations on Arctic Arboviruses

E. KURSTAK

## A. Introduction

Arboviruses in arctic and subarctic regions have received little attention until now. The recent development of northern regions in several countries creates the need for urgent research on such viruses and the infections they cause.

The problem of arboviruses of arctic regions was raised during a Symposium on arboviruses in Helsinki in 1975. The following year - 1976 - an International Committee on Arctic Arboviruses (ICAA) was created with membership of Canada, Finland, U.S.A., U.S.S.R. and the World Health Organization (W.H.O.). University of Montreal (Canada) houses the general secretariat of this committee. In order to promote research on these viruses, the ICAA organized the Second International Symposium on Arctic Arboviruses in Mont Gabriel (Canada) in May 26-28th, 1977. The theme of this symposium was the comparison between arboviruses from arctic and tropical regions (1).

The objectives of the ICAA are the co-ordination of investigations on the ecology, epidemiology, molecular biology, diagnosis and control of mosquito-borne and tick-borne viruses which may induce encephalitis or other febrile illnesses affecting man and other animals in circumpolar zones. The ICAA will take the necessary steps to establish a bank of information on these viruses and every three years an international symposium will be organized under its auspices. The third symposium will be held in the U.S.A. in 1980.

## B. Isolation of Arboviruses in Northern Regions

In the Canadian arctic and subarctic regions (north of $60^{o}$N) California encephalitis virus (CEV), snowshoe hare subtype, was first isolated from *Aedes canadensis* mosquitoes (2). Subsequently CEV has been repeatedly isolated (2, 3) from *Aedes communis* and 3 other *Aedes* in the north boreal forest zones ($66^{o}$N, $138^{o}$W), and the arctic tidewater region of Inuvik ($69^{o}$N, $135^{o}$W). Two of these species are *Aedes hexodontus* and *Aedes punctor*. It appears that small mammals, particularly snowshoe hares, *Lepus americanus*, arctic ground squirrels, *Citellus undulatus* and red squirrels, *Tamiasciurus hudsonicus*, are reservoirs of CEV. The natural cycle of infection by this bunyavirus during successive summers involving *Aedes* mosquitoes as vectors and small mammals as reservoirs occurs from the boreal forests through the open woodlands to the tundra regions. The situation is similar in Alaska (U.S.A.) where the snowshoe hare subtype of CEV was repeatedly isolated (3).

Since 1969, in the northern zones of the European and of the Far East parts of the U.S.S.R., 241 strains of Tyuleniy virus (Togaviridae, Flavivirus), 3 Uukuniemi group viruses (Bunyaviridae) were isolated from *Ixodes putus** and *Ixodes signatus* ticks, the main parasites of seabirds (4). It appears that *I. putus* ticks are the major vectors of arboviruses. As many as 236 virus strains were isolated

* *I. putus* is the name used by U.S.S.R. workers for the tick known elsewhere as *I. uriae*.

from *I. putus* (Zaliv Terpeniya virus-85, Tyuleniy virus-53, Okhotskiy virus-47, Sakhalin virus-43, and Paramushir virus-8).

Tick-borne Uukuniemi and Humhinge viruses and mosquito-borne Inkoo (CE group) virus are prevalent in Scandinavia (South of 62°N).

## C. Current Status on Arboviruses of Polar Regions

I. *Mosquito vectors*. The experimental injection of snowshoe hare subtype of CEV to *Culiseta inpornata* mosquitoes provokes infection. Viral replication is detected after 27 days of incubation at 0°C or 6 days at 13°C. Demonstration of viral antigens and virions in the salivary glands of the infected mosquitoes was possible by direct immunofluorescence. In field conditions the infection rates for infected species of *A. canadensis* and *A. communis* are 1:503 and 1:2564, respectively. Neutralizing antibodies to the snowshoe hare subtype of CEV are detectable in 14% of mammals in the northern Yukon Territory of Canada. One of the means of overwintering of CEV in the Canadian Arctic is by transovarial transfer of the infection as suggested earlier. Experimental transmission of arboviruses has also been demonstrated by intrathoracic injection and feeding of arctic *A. communis* mosquitoes. Intracytoplasmic virions are demonstrated by immunoelectron microscopy. These data demonstrate their potential as natural vectors in polar regions (2, 3).

II. *Tick vectors*

1. *Togaviridae - Flavivirus*. Tyuleniy virus, one of the important Togaviruses isolated from *Ixodes putus* ticks principally on Tyuleniy and Commodore Islands (U.S.S.R.), is pathogenic for suckling-mice after intracerebral injection and can also replicate in chick embryos and BHK-21 cells. These viruses can experimentally infect *Aedes aegypti* and *Culex molestus* mosquitoes and several species of birds. In man, Tyuleniy virus can cause a febrile illness after visit to colonies of birds. The infection rate of adults of *I. putus* varies from 1:35 to 1:600. Serological data demonstrated that the natural reservoirs of Tyuleniy virus are birds, mainly Brünnich's guillemots, common murres, pelagic and red-faced cormorants, and fulmars. Detection of antibodies against this virus in fur seals, cows and man suggests their involvement in the dissemination of Tyuleniy virus (4).

2. *Bunyaviridae - Uukuniemi and Sakhalin-groups*. Two viruses of these groups were isolated at high latitudes in the U.S.S.R., namely Zaliv Terperniya virus (Uukuniemi group) and Sakhalin virus (Sakhalin group), both from *I. putus* ticks. Another Bunyavirus (ungrouped) isolated is Paramushir virus which infects *I. putus* and *I. signatus* ticks. The infection rates of ticks range from 1:50 to 1:10,000 (4).

Zaliv Terpeniya virus can infect suckling mice by intracerebral inoculation and chick and duck embryo cell cultures. Under natural conditions rodents can be infected by the ingestion of ticks and dead bodies of infected birds.

Sakhalin virus as Zaliv Terpeniya virus infects suckling mice and can replicate in primary duck and human embryo cell cultures, as well as in BHK-21, HeLa and Vero cell lines. Viruses related to Sakhalin virus were isolated in Gull Island (Alaska) and on the Newfoundland coast (Canada). The presence of antibodies against viruses related to Sakhalin virus in several species of birds in the Far East part of the U.S.S.R. suggests their important role in the dissemination of the virus.

Paramushir virus, isolated on Tyuleniy and Commodore Islands, present several properties similar to Bunyaviruses. Among laboratory animals only suckling mice are sensitive to the virus by intracerebral inoculation. Replication of the virus with cytopathic effect (CPE) occurs in L-cells, HeLa and pig embryo kidney cell cultures. Without CPE, replication can be detected in chick, duck, human embryo fibroblast cultures and in BHK-21 cells (4, 5, 6).

## D. Conclusions

The role and importance of arctic and subarctic arboviruses in human and veterinary medicine is not sufficiently known. It is recommended by the International Committee on Arctic Arboviruses to generate new research on all the aspects of arboviruses and their infections in circumpolar regions which hitherto were left insufficiently investigated. This new knowledge will contribute directly to the upgrading of public health measures in polar and subpolar regions, thereby improving the work output and general well-being of human residents both settled and transient in these areas, through prevention of arthropod-borne virus infections.

## E. Acknowledgements

Work sponsored by the Ministry of Indian and Northern Affairs of Canada.

## F. References

1. Kurstak, E.: Arctic and Tropical Arboviruses. New York: Academic Press, in press (1978).
2. McLean, D.M., Crawford, M.A., Ladyman, S.R., Peers, R.R. & Purvin-Good, K.W.: California encephalitis and Powassan virus activity in British Columbia, 1969. Amer. J. Epidemiol. 92, 266-272 (1970).
3. McLean, D.M.: Arbovirus vectors in the Canadian Arctic. In: Arctic and Tropical Arboviruses. Kurstak, E. (ed.). New York: Academic Press, in press (1978).
4. Lvov, D.K.: Arboviruses of high latitudes in the U.S.S.R. In: Arctic and Tropical Arboviruses. Kurstak, E. (ed.). New York: Academic Press, in press (1978).
5. Lvov, D.K., Timopheera, A.A., Gromashevsky, V.L., Gostinshchikova, G.V., Veselovskaya, O.V., Chervonsky, V.I., Fomina, K.B., Gromov, A.I., Progrebenko, A.G., & Zhermer, V.Y.: Zaliv Terpeniya virus, a new Uukunieni group arbovirus isolated from *Ixodes* (*Ceratixodes*) *putus* (Pick. - Camb. 1878), on Tyuleniy island (Sakhalin regions) and Commodore islands (Kamchatsk region). Arch. Ges. Virusforsch. 41, 165-169 (1973).
6. Ritter, D.G. & Feltz, E.: On the natural occurence of California encephalitis virus and other arboviruses in Alaska. Canad. J. Microbiol. 20, 1359-1366 (1974).

# The Ecology of Influenza Viruses

R.G. WEBSTER, W.J. BEAN, JR., and V.S. HINSHAW

## A. Introduction

The ecology and natural history of influenza viruses in animals (including man) is poorly understood. The origin of human pandemic strains and the significance of the multitudes of influenza A viruses in animals - particularly in avian species - remains to be elucidated (1). Increasing evidence suggests that the new pandemic strains of human influenza virus originate from the influenza viruses present in lower mammals and birds - either by direct transmission or after genetic recombination between mammalian or avian strains and the influenza A virus prevalent in man at that time (3, 11).

This paper will deal with the ecology of influenza viruses in pigs and ducks and their role as reservoirs of influenza viruses for other species - including man.

## B. The Role of Pigs in the Ecology of Influenza Viruses

The only definitive evidence for the role of animal influenza viruses in human disease comes from the recent studies on swine influenza viruses. Within recent years, Hsw1 N1 viruses have been implicated in human disease on the basis of serological studies and virus isolations (6, 7, 9). During 1976, Hsw1 N1 viruses were isolated from military recruits at Ft. Dix, New Jersey (4).

I. Studies on swine influenza viruses in U.S.A. Studies on the prevalence of swine influenza viruses /Hsw1 N1/ in pigs in USA in 1976-77 showed that these viruses were isolated from pigs during each month of the year. Studies on two farms in Wisconsin during the peak of virus activity in swine in November and December, 1976, resulted in the isolation of Hsw1 N1 viruses from both humans and pigs which were concurrently suffering from respiratory disease with typical clinical symptoms of influenza. (B.C. Easterday - personal communication) Similar studies on 14 farms in other states where Nsw1 N1 virus activity in pigs was demonstrated, revealed no evidence of interspecies transmission of these /Hsw1 N1/ viruses to humans or other animals or birds on the farm (2).

II. Characterization of recent swine influenza virus isolates. The antigenic and genetic relatedness of representative Hsw1 N1 viruses from pigs and people were compared with A/NJ/8/76 /Hsw1 N1/ - a human isolate from recruits at Fort Dix (Table 1). The recent swine and human isolates from one farm in Wisconsin /A/Sw/Wis/56/76 and A/Wis/263/76/ were antigenically indistinguishable from the virus isolated from recruits at Fort Dix /A/NJ/8/76/. The earlier swine isolates, A/Sw/Iowa/15/30 and A/Sw/Wis/1/67, were antigenically distinguishable from the recent swine and human isolates suggesting that antigenic drift has occurred in the Hsw1 N1 influenza viruses since 1930. Antigenic analysis of 364 swine influenza virus isolates showed that the majority (80%) of the viruses were similar to A/NJ/8/76 but the remainder resembled classical swine viruses.

Analysis of the RNAs (2) from Hswl N1 viruses (Figure 1) showed that A/Wis/263/76 and A/Sw/Wis/56/76, which were isolated from a human and a pig, on the same farm, possessed identical patterns but differed from the recent swine isolate, A/Sw/Tn/1/75, and the human isolate, A/NJ/8/76. Although the Hswl N1 viruses from 1967-76 revealed slight differences in the migration of bands 4 and 6, the most

Table 1. Hemagglutination-inhibition reactions of human and swine (Hswl Nl) influenza viruses

| Antigen | Source | Antisera* | | |
|---|---|---|---|---|
| | | A/Sw/Tn/1/75 | A/Sw/Tn/21/76 | A/NJ/8/76 (H)/ A/eq/Pr/1/56 (N) |
| A/Sw/Iowa/15/30 | Pig | 20 | 20 | 20 |
| A/Sw/Wis/1/67 | Pig | 40 | 40 | 40 |
| A/Sw/Wis/56/76 | Pig[a] | 320 | 320 | 160 |
| A/Wis/263/76 | Human[a] | 320 | 320 | 320 |
| A/SW/Wis/11/76 | Pig[b] | 80 | 80 | 40 |
| A/Wis/301/76 | Human[b] | 640 | 1280 | 640 |
| A/NJ/8/76 | Human | 320 | 320 | 320 |

*Hl titer = reciprocal of serum dilution causing inhibition of four hemagglutinating units.

Postinfection ferret sera were used for these studies.

[a,b] Indicates that viruses were from pigs and man on the same farm.

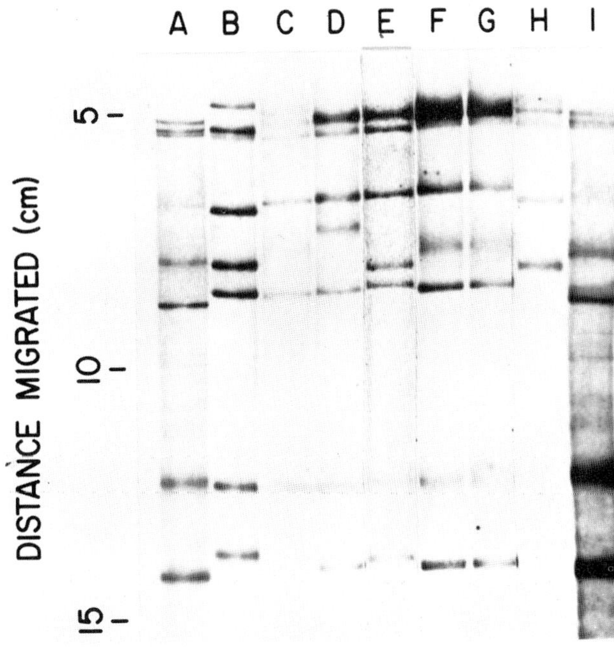


Fig. 1. Comparison of the RNA genome segments of human and swine influenza strains. Viral RNAs were iodinated and separated on a 65% formamide, 3% acrylamide gel. The reproduction shown is a composite of two exposures of the same gel. Lanes E and H, 5 days, all others, 1 day (2). Lane:

A, WSN/33 [HON1],
B, Sw/Iowa/15/30 [Hsw1 N1],
C, Sw/Wis/1/67 [Hsw1 N1],
D, Sw/Tn/1/75 [Hsw1 N1],
E, NJ/8/76 [Hsw1 N1],
F, Wis/263/76 [Hsw1 N1],
G, Sw/Wis/56/76 [Hsw1 N1],
H, Mem/110/75 [H3N2],
I, WSN/33 [HON1]

marked variation among these viruses involved band 5 which has been shown to code for the ribonucleoprotein (RNP) (5).

These studies show that swine influenza viruses can be transmitted to man and that there was considerable antigenic and genetic heterogeneity in the Hswl N1 influenza viruses from man and pigs. The biological significance of this heterogeneity remains to be elucidated.

III. Persistence of Hong Kong influenza virus variants in pigs. The A/Hong Kong/68 /H3N2/ influenza virus that has not been isolated from man for several years was recently isolated from pigs in Hong Kong (8). A/Victoria/3/75-like influenza viruses that are currently circulating in man were also isolated from pigs from Hong Kong and USA and serological evidence indicates infection of pigs with A/Victoria/3/75 in all countries tested. Both the A/Hong Kong/68 and the A/Victoria/75-like viruses transmitted readily from pig to pig in experimental studies.

## C. The Role of Ducks in the Ecology of Influenza Viruses

Recent studies have shown a high incidence of different influenza A viruses in both feral and domestic ducks (10). Thus, in studies on feral ducks in Canada, 19% of juvenile birds contained influenza viruses. All ducks from which viruses were isolated appeared healthy at the time of sampling. Antigenic characterization of the 111 influenza isolates revealed four hemagglutinin subtypes /Hav4, Hav5, Hav6, and Hav7/ and six neuraminidase subtypes /N1, N2, Neq2, Nav1, Nav2 and Nav5/ in various combinations (Table 2). In addition, several influenza viruses could not be classified with reference antisera and may possess novel antigens.

Table 2. Antigenic classification of influenza A viruses isolated from feral ducks in Canada

| Antigenic Characterization | | Number of virus isolates from | | |
|---|---|---|---|---|
| Hemagglutinin | Neuraminidase | Adults | Juveniles | Previous Isolates |
| Hav1 | Nav2 | 2 | 1 | A/turkey/Ore/71 |
| Hav4 | Nav1 | 3 | 12 | A/duck/Czech/56 |
| Hav4 | N1 | 0 | 3 | A/duck/Germany/210/67 |
| Hav4 | N(?) | 1 | 0 | None |
| Hav4 | Neq2 | 0 | 1 | A/Mynah/Mass/71 |
| Hav5 | N2 | 0 | 3 | A/turkey/Ont/4004/67 |
| Hav7 | Neq2 | 13 | 42 | A/duck/Ukraine/1/63 |
| Hav7 | Nav1 | 0 | 7 | A/duck/HK/228/76 |
| Hav7 | N1 | 0 | 1 | None |
| Hav7 | N2 | 0 | 1 | A/turkey/England/1/69 |
| H(?) | N1 | 1 | 11 | None |
| H(?) | Nav5 | 0 | 1 | None |
| H(?) | Nav1 | 0 | 1 | None |
| H(?) | N(?) | 0 | 2 | None |
| | | 20 | 86 | |

The isolation of many influenza viruses from the cloaca of feral avian species (1, 10, 12) - some of which are related to viruses from man - raises the question of the site of replication of these viruses and of the unlikely possibility of transmission of these avian viruses to man. Since A/duck/Memphis/546/74 /Hav3 Nav6/ influenza virus was isolated from the cloaca of mallard ducks (12), the question of how the virus reached the cloaca was studied.

Samples collected from ducks at hourly intervals after infection showed that virus was first detected in cloacal samples one hour after infection (Table 3,) but was never isolated from the blood, kidneys, spleen, or liver. On the third day after infection, virus was detected in the caecum and rectum, but not in the lungs or other organs. These experiments suggest that the A/duck/Memphis/546/74 influenza virus can pass through the digestive tract (despite the low pH in the gizzard), and replicate in the lower intestinal tract, without producing any signs of disease. Similar studies with mammalian influenza viruses in ducks showed that these viruses replicated in the upper respiratory tract of ducks, but not in the intestinal tract.

## D. Discussion

The isolation of swine influenza viruses /Hswl N1/ from pigs each month of the year shows that these viruses are continuously circulating in the pig population in USA. These viruses are antigenically and genetically heterogeneous and have the potential of transmission to man; the properties of the viruses and the ecological factors that will allow this to occur have not been resolved. Thus, swine influenza viruses may transmit relatively frequently to man but lack the properties that will permit them to cause an epidemic.

The isolation of A/Hong Kong/68-like influenza viruses from pigs in 1976 suggests that pigs may serve as a potential reservoir for future human pandemics as well as a possible source of genetic information for recombination between human and porcine strains of influenza virus.

The susceptibility of ducks to infection with a wide range of different avian influenza viruses, as well as with human influenza viruses, suggests that feral and domestic ducks may play an important role in the ecology of influenza A viruses. The present studies show that influenza A viruses isolated from the rectum of feral ducks replicate in the upper respiratory tract and also in the intestinal trace of ducks. Representative human influenza viruses replicate in the upper respiratory tract of ducks but not in the intestinal tract. In fecal material, the avian viruses retain infectivity for over 30 days at $4^{o}$ and for over 7 days at $20^{o}$. At low temperature ($4^{o}$) the virus retains infectivity for over 30 days when diluted in untreated water and for 4 days at $22^{o}$ (Webster, Yakhno, Hinshaw, Bean and Murti, unpublished results). The high concentration of duck influenza viruses in fecal material and the relatively long stability of the virus in untreated water offers a logical mechanism for transmission of avian influenza viruses from feral ducks to domestic avian and mammalian species. The susceptibility of ducks and pigs to infection with human influenza viruses suggests that these animals may play an important role in the ecology of influenza A viruses.

## E. Summary

Swine influenza viruses, Hswl N1, have been isolated from pigs in USA during each month of the year. These viruses are antigenically and genetically heterogeneous and some strains can infect man. The Hong Kong strains of human influenza

Table 3. Site of replication of A/Duck/Memphis/546/74 influenza virus in mallard ducks

| Days after Inoculation | Number of samples with virus/ Number of ducks inoculated | | | Virus titer $EID_{50}$/ml $log_{10}$ terms in the following organs | | | | | Signs of Disease |
|---|---|---|---|---|---|---|---|---|---|
| | Tracheal Swab | Cloacal Swab | Blood | Lung | Kidney, Liver, Spleen | Gizzard/pH/[b] | Caecum | Rectum | |
| 0 | 4/4 | 2/4[a] | 0/4 | | | | | | |
| 1 | 3/4 | 4/4 | 0/4 | | | | | | |
| 2 | 2/4 | 4/4 | 0/4 | 3.5* | <1.0 | <1.0 /2.9/ | ND** | 6.5 | Nil |
| 3 | 1/3 | 3/3 | 0/3 | <1.0 | <1.0 | <1.0 /4.4/ | 6.0 | 4.5 | Nil |
| 4 | 1/2 | 2/2 | 0/2 | <1.0 | <1.0 | <1.0 /4.0/ | 4.0 | 4.0 | Nil |
| 5 | 0/1 | 1/1 | 0/1 | <1.0 | <1.0 | <1.0 /3.5/ | ND | 3.5 | Nil |

Mallard ducks were infected with approximately $10^7$ $EID_{50}$ of A/duck/Memphis/546/74 in 0.5 ml by both oral and intratracheal inoculation.

a = virus detected on 1, 5, 7, 9 hours.
b = figures in brackets give the pH of the gizzard.

*Samples of virus isolated from each organ were identified in Hl tests with specific antiserum to A/duck/Mem/546/74.

**Not determined.

viruses persist in pigs. Studies in ducks show that many different influenza A viruses can be isolated from juvenile feral birds. Some of these viruses possess novel hemagglutinin and/or neuraminidase antigens. Influenza A viruses isolated from the rectum of feral ducks replicate in the upper respiratory tract and also in the intestinal tract of feral and domestic ducks. Representative human influenza viruses replicate in the upper respiratory tract of ducks but not in the intestinal tract. The susceptibility of ducks and pigs to infection with human influenza viruses suggests that these animals may play an important role in the ecology of influenza A viruses.

## F. Acknowledgements

This work was supported in part by Contract A1 52524 and Research Grant A1 08831 from the National Institute of Allergy and Infectious Diseases, Childhood Cancer Center Grant CA 08480 from the National Cancer Institute, and by ALSAC.

## G. References

1. Easterday, B.C.: Animal influenza. In: The Influenza Viruses and Influenza, edited by E.D. Kilbourne. Academic Press, New York, pp. 449-481 (1975).
2. Hinshaw, V.S., Bean, W.J.Jr., Webster, R.G., & Easterday, B.C.: The prevalence of influenza viruses in swine and the antigenic and genetic relatedness of influenza viruses from man and swine. Virology (In press, 1977).
3. Laver, W.G. & Webster, R.G.: Studies on the origin of pandemic influenza. II. Peptide maps of the light and heavy polypeptide chains from the hemagglutinin subunits of A2 influenza viruses isolated before and after the appearance of Hong Kong influenza. Virology 48, 445-455, (1972).
4. Morbidity and Mortality Weekly Report. U.S. Department of Health, Education and Welfare, 25, 47-48 (1976).
5. Ritchey, M.S., Palese, P., & Schulman, J.: Mapping of the influenza genome. III. Identification of genes coding for nucleoprotein, membrane protein and non-structural protein. J. Virol. 20, 307-313 (1976).
6. Schnurrenberger, P.R., Woods, G.T., & Martin, R.J.: Serologic evidence of human infection with swine influenza virus. Amer. Rev. Resp. Dis. 102, 356 (1970).
7. Seal, J.R., Sencer, D.J., & Meyer, H.M.: A status report on national immunization against influenza. J. Inf. Dis. 133, 715-720 (1976).
8. Shortridge, K.F., Webster, R.G., Butterfield, W.K., & Campbell, C.H.: Persistence of Hong Kong influenza virus variants in pigs. Science 196, 1454-1455 (1977).
9. Smith, T.F., Burgert, E.O., Jr., Dowdle, W.R., Noble, G.R., Campbell, R., & Van Scoy, R.E.: Isolation of swine influenza virus from autopsy lung tissues of man. New Eng. J. Med. 294, 708-710 (1976).
10. Webster, R.G.: Influenza viruses from avian and porcine sources and their possible role in the origin of human pandemic strains. Develop. biol. Standard, vol. 39, pp. 461-468, (1977).
11. Webster, R.G. & Laver, W.G.: Antigenic variations of influenza viruses. In: The Influenza Viruses and Influenza, pp. 209-314. Edited by E.D. Kilbourne, New York, Academic Press (1975).
12. Webster, R.G., Morita, M., Pridgen, C., & Tumova, B.: Ortho- and paramyxoviruses from migrating feral ducks: Characterization of a new group of influenza A viruses. J. Gen. Virol. 32, 217-225 (1976).

# Elements of Hyperendemic Transmission of Dengue Virus in the Caribbean

A.K. VENTURA

In this dissertation, I will examine the role of infants in the maintenance of endemic dengue in the Caribbean (1, 5). The studies to be outlined were conducted in Santo Domingo, Capital of the Dominican Republic, located on the southern coast of the Island of Hispaniola.

Ninety-eight percent of the cord sera collected from 915 deliveries in Santo Domingo possessed dengue 2 and/or 3 antibodies. The cord titers were often higher than their corresponding maternal titers; 38% were higher, 44% were equal and 18% were lower in titer. As depicted in Figure 1, the geometric mean titer for the cord sera was higher than the geometric mean titer for the maternal sera. Fractions derived by molecular exclusion chromatography in agarose gel columns from an arbitrary selection of cord-maternal serum pairs have indicated that most of the differences in antibody concentrations between cord and maternal sera were due to antibody differences in the IgG and not in the IgM or IgA fractions. Whether these higher IgG antibody titers in the fetal circulation confer better protection against *in utero* infections has yet to be determined (4).

Based on four-fold or greater changes in dengue titers in infants, estimates of the infection rates of children during the first year of life were calculated. Infection rates increased from 4% $\{\frac{10}{277}\}$ between birth and four months, to 9% $\{\frac{7}{80}\}$ between four and eight months, and up to 23% $\{\frac{7}{30}\}$ between eight and twelve months. Figure 2 traces the development of dengue antibody in infants below one year of age. Preliminary studies have shown that 6% of the children were born with elevated IgM concentrations and a few of them (approximately 1%) had IgM dengue antibodies, suggesting *in utero* dengue infections. Notwithstanding this, at birth the log 2 geometric mean dengue 2 HI titer for the neonates was 3.5. Thirty-three percent of the children with dengue antibodies at birth lost these antibodies by

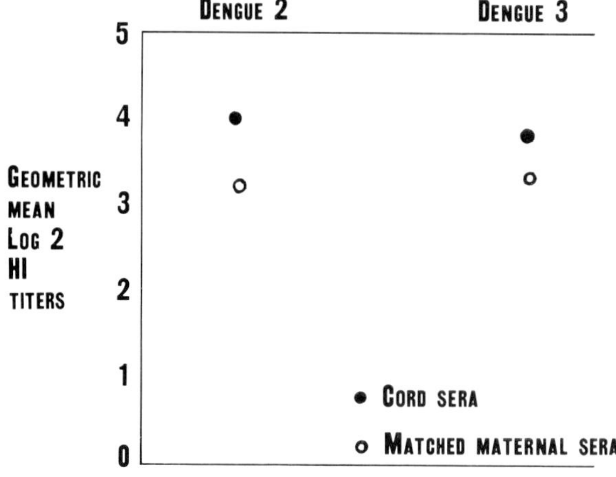


Fig. 1. Comparison of geometric mean HI titers between cord sera and matched maternal sera. HI = haemagglutination inhibition

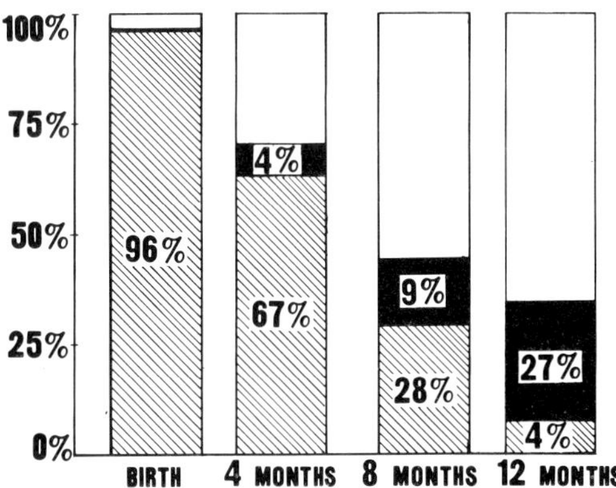


Fig. 2. Dengue antibody development in infants during the first year of life

maternal antibodies

actively acquired antibodies

no antibody

4 months of age, while 90% of the remainder sustained significant losses in titers during this period. Those with cord titers <1:40 fell to <1:10, and those with cord titers 1:640 dropped to 1:40 or less. At this stage, the log 2 geometric mean HI dengue 2 titer was 1.7 and only 4% of the infants could clearly be identified as having acquired antibody by infection. By 8 months, 72% had lost their maternal antibodies and 9% seemed to have been infected. At this juncture, the log 2 geometric mean dengue 2 HI titer had risen to 2.4. It appears that at most only 4% of the children could possibly have retained their maternal antibodies through to one year, while 23% appear to have sustained dengue infection between 8 and 12 months, and the geometric mean titer was now 2.7. These data signify an annual infection rate of 36% per annum or 3% per month, which is similar to the infection rate determined earlier in children under 10 years of age (1). These data indicate that children under one year of age in Santo Domingo were promptly infected with dengue viruses as soon as their maternal antibody protection waned.

It appears then, that infants who have just lost their maternal antibodies are the primary hosts for dengue transmission in the hyperendemic conditions in Hispaniola, since these children presumably can mount higher titer viremias than older children with pre-existing heterologous dengue antibodies or antibodies recalled by anamnestic responses. If this is so, then a target population has been identified for dengue control.

The infants included in this study experienced mild to moderate incidences of fever, cough, coriza and diarrhoea at varying times. Preliminary observations have revealed that children experiencing dengue infection between 4 and 12 months, suffer a worsening of these maladies, and many display classic dengue symptoms including typical maculopapular rash. This stands in contrast to dengue infections in 1 to 10 year old children. Most of the dengue infections in these children appear to be very mild or without obvious symptoms. This was true of dengue viremic children in Haiti (5), as well as those children which were infected during an eight weeks study in Santo Domingo (1).

In reviewing the circumstances which now exist in Hispaniola, circulation of at least two dengue serotypes (1, 5), repeated dengue infections at close intervals (1), high dengue infection rates (1), large populations with high reproductive rate, and high _Aedes aegypti_ indices, it is difficult to reconcile the absence of severe dengue in Hispaniola with the notion that secondary heterologous dengue infections are the bases for the manifestation of dengue-hemorrhagic-shock syndrome (3).

In conclusion, these studies have identified a target population for dengue control in the Caribbean; they have also provided information to indicate that dengue infections are rampant during the first year of life, and that these infections appear to have ill effects on children at this stage of life. Further, although circumstances in the Caribbean favour repeated dengue infections at close intervals, no case of dengue-shock-syndrome has yet been identified, thus challenging the tenets of the secondary dengue infection immunopathological theory for dengue-shock-syndrome.

Whether the advent of dengue 1 activity in Jamaica (2) may precipitate more serious secondary infections in the Caribbean is a matter of great concern and is being monitored closely.

## A. References


1. Ehrenkranz, N.J. & Ventura, A.K.: Endemic dengue virus infection in Hispaniola II. The dynamics of endemic dengue infection in children in Santo Domingo, D.R. J. Infect. Dis. submitted for publication (1977).
2. Moody, C., Dyer, A., Bowen-Wright, C., Basu, P., King, D., & Rose, E.: Dengue in Jamaica. CAREC Surveillance Report 7, 1-2, (1977).
3. Halstead, S.B.: Observations related to pathogenesis of dengue hemorrhagic fever VI hypothesis and discussion. Yale J. Biol. Med. 42, 350-361 (1970).
4. Ventura, A.K.: Concentration of low titered viral antibodies by the human placenta. Abstracts, 3rd International Congress for Virology, 223 (1975).
5. Ventura, A.K., & Ehrenkranz, N.J.: Endemic dengue virus infection in Hispaniola I, Haiti. J. Infect. Dis. 134, 436-441 (1976).

# Observations on the Survival of Pathogens in Water and Air at Ambient Temperatures and Relative Humidity

N.ST.G. HYSLOP

## A. Introduction

The environment is increasingly loaded with chemical and biological waste-products, and with pathogenic and non-pathogenic microorganisms of various types. Experimental data are cited to illustrate the varying rates of regression in viability of certain microorganisms of widely different classes under temperate conditions.

## B. Materials and Methods

I. Airborne organisms. Data on airborne organisms were obtained from stirred aerosols maintained at ambient pressure, at various temperatures and relative humidities (RH). Nearly all the results were derived from aerosols generated into an apparatus of about 25 litres capacity (4), except that experiments using poliovirus were performed on a smaller scale in a closed circuit apparatus (8).

II. Waterborne organisms. Experiments on the viability of microorganisms in water, sterilized water or sewage-slurry were performed in lightly cotton-plugged containers, held at ambient temperatures in shaded daylight.

## C. Results and Discussion

I. Bacteria in air. Experiments on longevity under selected conditions of temperature and RH revealed that bacteria, and even different strains of the same organism, exhibited a considerable degree of adaptation to the particular conditions under which they usually existed in the "host", though some organisms were able to survive for long periods in suboptimal environments. Differences between species were most marked among the *Mycoplasma* spp. tested; some strain differences were detected with all organisms.

Fig. 1 shows the differences in survival at 28$^{o}$C and 50% RH between a strain of *Mycoplasma mycoides var. bovis*, which caused pulmonary disease in cattle, and a strain of *Mycoplasma bovigenitalium* causing urogenital infection. Smaller differences in resistance to aerosolization were observed between *Mycoplasma mycoides var. bovis* and a pulmonary strain of *Mycoplasma mycoides var. capri* under the same conditions (Fig. 2). Complex interactions however, were recorded when the former organism was suspended in air at different RH values, and survival of the strain was shortest at about 80% RH (Fig.3). Furthermore, strains newly isolated from lung tissue tended to survive longer in the aerosol state than strains which had been passaged serially many times in broth culture media (Fig. 4). Differences were also observed between respiratory and genital *Mycoplasma* spp. of human origin (Figs. 5 & 6).

Three strains of *Brucella*, suspended in air at 25$^{o}$C and 59% RH (Fig. 7), and 2 strains of *Escherichia coli*, suspended at 20$^{o}$C and 63% RH (Fig. 8), survived for only relatively short periods (less than 3 hr). Nevertheless, such periods certainly might permit aerial spread of infection within buildings or at close

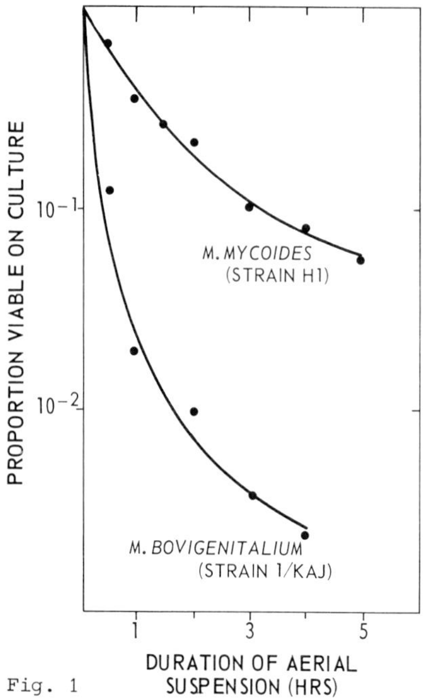


Fig. 1

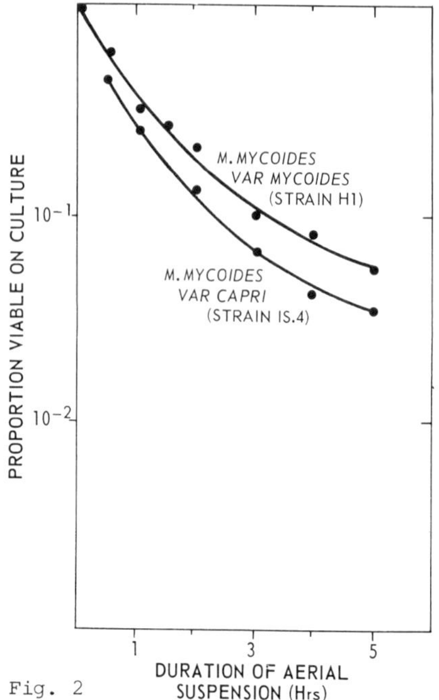


Fig. 2

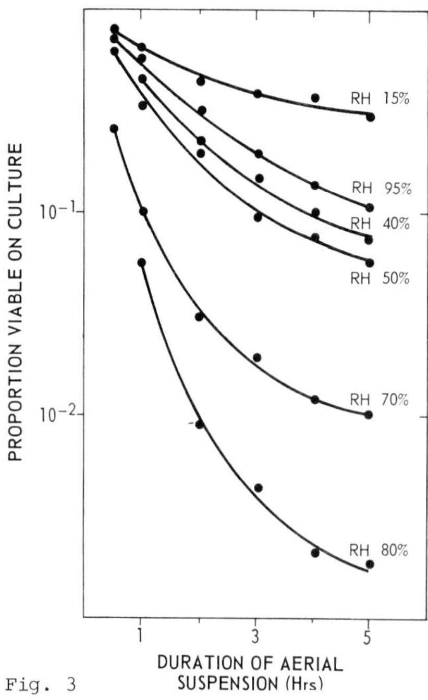


Fig. 3

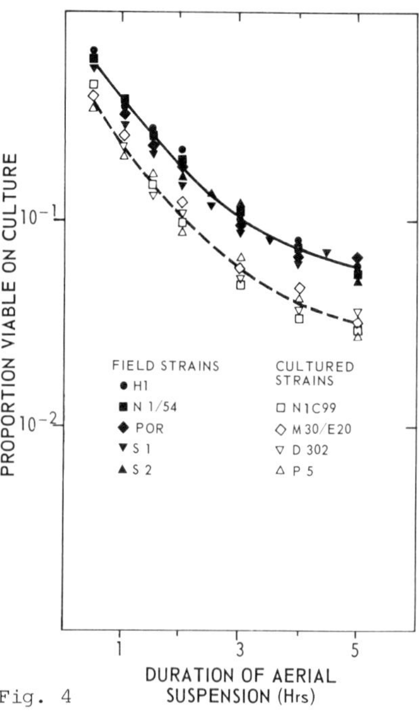


Fig. 4

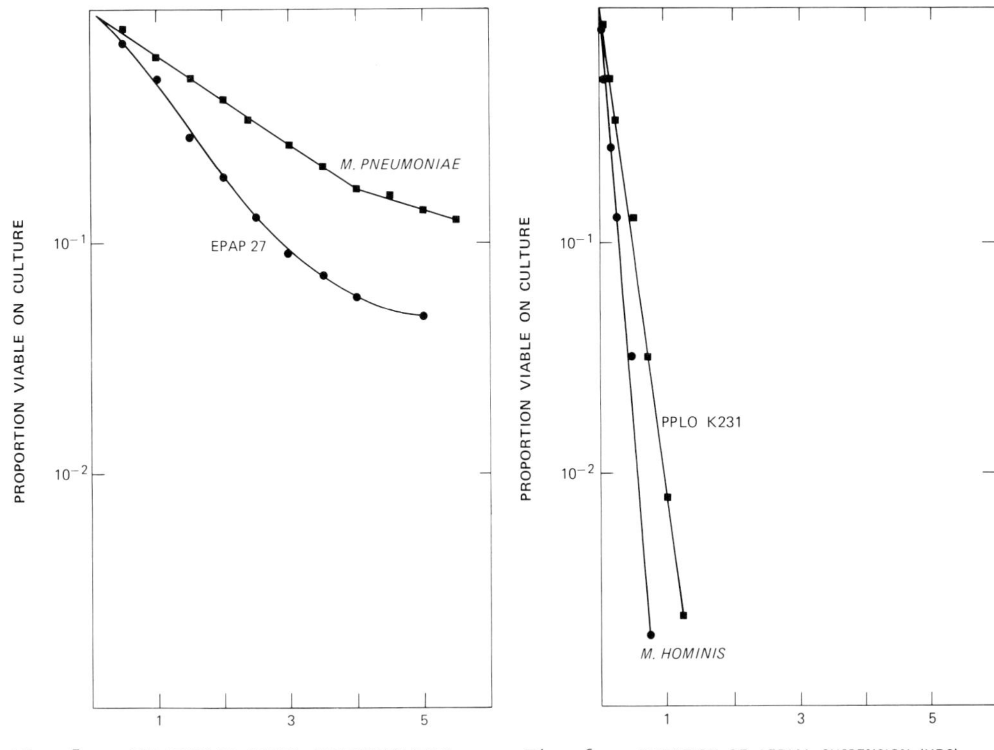


Fig. 5. Mean survival rates of Mycoplasma strains of human pulmonary origin in aerial suspension at 28°C and 50% RH in shaded daylight

Fig. 6. Mean survival rates of Mycoplasma strains of human genital origin in aerial suspension at 28°C and 50% RH in shaded daylight

---

◄ Fig. 1. Mean survival rates of Mycoplasma mycoides and Mycoplasma bovigenitalium in aerial suspension at 28°C and 50% RH in shaded daylight

Fig. 2. Survival of Mycoplasma mycoides var. mycoides and var. capri when suspended in the aerosol state at 28°C and 50% RH in shaded daylight

Fig. 3. Survival of Mycoplasma mycoides var. mycoides (strain H1) when suspended in the aerosol state at 28°C at different RH values in shaded daylight

Fig. 4. Survival of strains of Mycoplasma mycoides var. mycoides passaged in cattle or in culture when suspended in the aerosol state at 28°C and 50% RH in shaded daylight

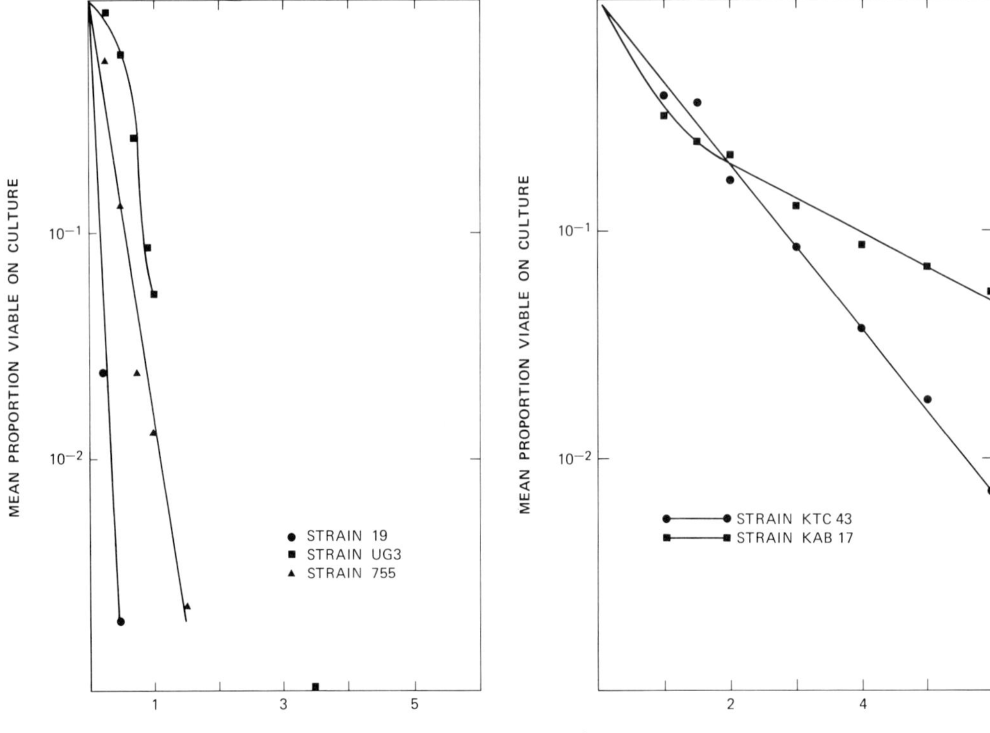


Fig. 7. Survival of Brucella abortus when suspended in the aerosol state at 25°C and 59% RH in shaded daylight

Fig. 8. Survival of E. coli when suspended in the aerosol state at 20°C and 63% RH in shaded daylight

quarters outdoors. It must be noted also that some strains of E. coli, not included in the present experiments, are known to be especially resistant to the effects of aerial suspension.

Although Salmonella typhimurium showed a fairly rapid loss of infectivity (Fig. 9), a significant proportion of the residual organisms were remarkably resistant, living organisms being recovered readily from the air for periods of at least 24 hr.

The aerial suspension of 2 pathogenic strains of Listeria monocytogenes, one of which was isolated from milk, revealed minor differences between the strains. After a moderate fall in the viable count, Listeria, strains from serovars I and IV persisted in the air for very long periods indeed (Fig. 10). Some infectivity remained after more than 10 days, though prolonged incubation of cultures of the collecting fluid from the air samples was necessary before good growth occurred; recovery of organisms was facilitated by "blind" subcultures.

II. Viruses in air. All of the viruses tested tended to be relatively short-lived in aerial suspension, though all were detectable for at least 30 minutes after aerosolization. Poliovirus, the viruses of foot-and-mouth disease (FMD) and rinderpest showed progressively decreasing resistance to desiccation (Fig. 13) though complex interactions between temperature and RH probably exist. Neverthe-

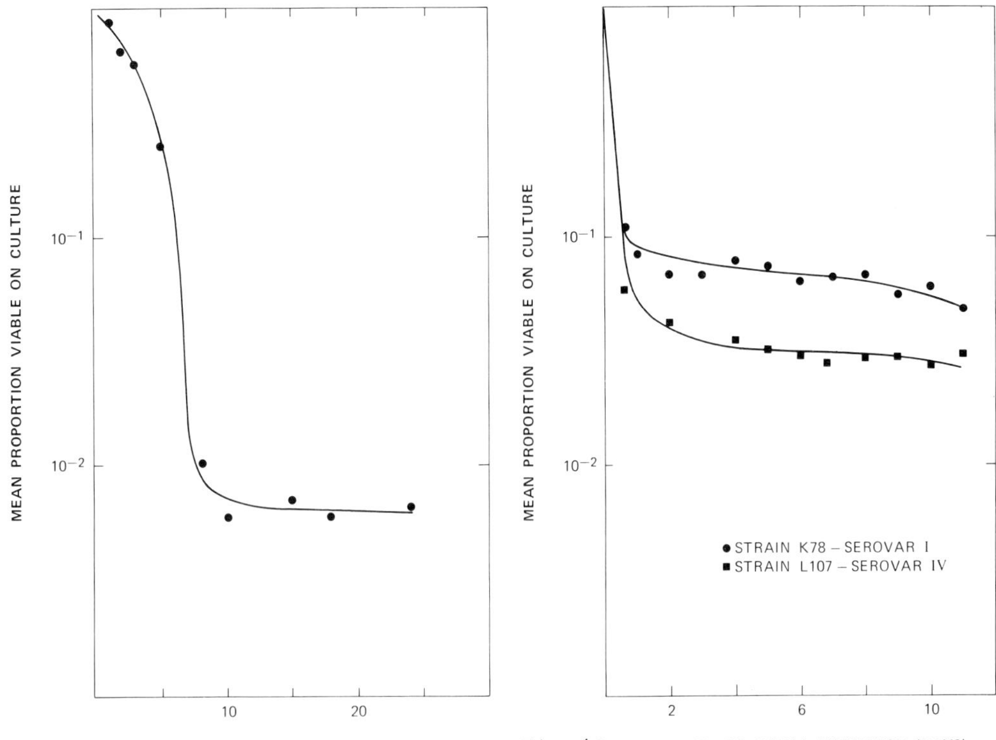


Fig. 9 Fig. 10

Fig. 9. Survival of Salmonella typhimurium when suspended in the aerosol state at 20°C - 26°C and 65% RH in shaded daylight or darkness

Fig. 10. Survival of Listeria monocytogenes when suspended in the aerosol state at 21°C - 28°C and c.50% RH in shaded daylight or darkness

less, the longevity in air of poliovirus and of FMD virus was sufficient to permit dissemination across distances of several miles at normal wind velocities. These results are generally consistent with observations made under field conditions.

Survival of airborne viruses depends principally on the virus type, particle size, pH, RH, altitude and insolation, though the effects of these determinants are modified by numerous other influences associated with physical, geographical and meteorological factors (6).

Lidwell and his associates have shown that dangers of cross-infection by various agencies exist even in air-conditioned hospitals. Furthermore, air-conditioners incorporating heat-exchangers of certain types may contaminate the air outside buildings (1).

III. Bacteria in water. A vast number of bacterial infections are predominantly food-or water-borne. Thus, contamination of the environment by "intestinal carriers" is of primary importance in the epidemiology of colibacillosis and of most Salmonella infections, although Salm. typhimurium and Salm. dublin appear also to be able to exist for significant periods in aerial suspension. Some Salmonella serotypes can exist in dried faeces for 3 years. Therefore, the increasing use of sludging on pastures as a fertilizer is particularly hazardous, not only be-

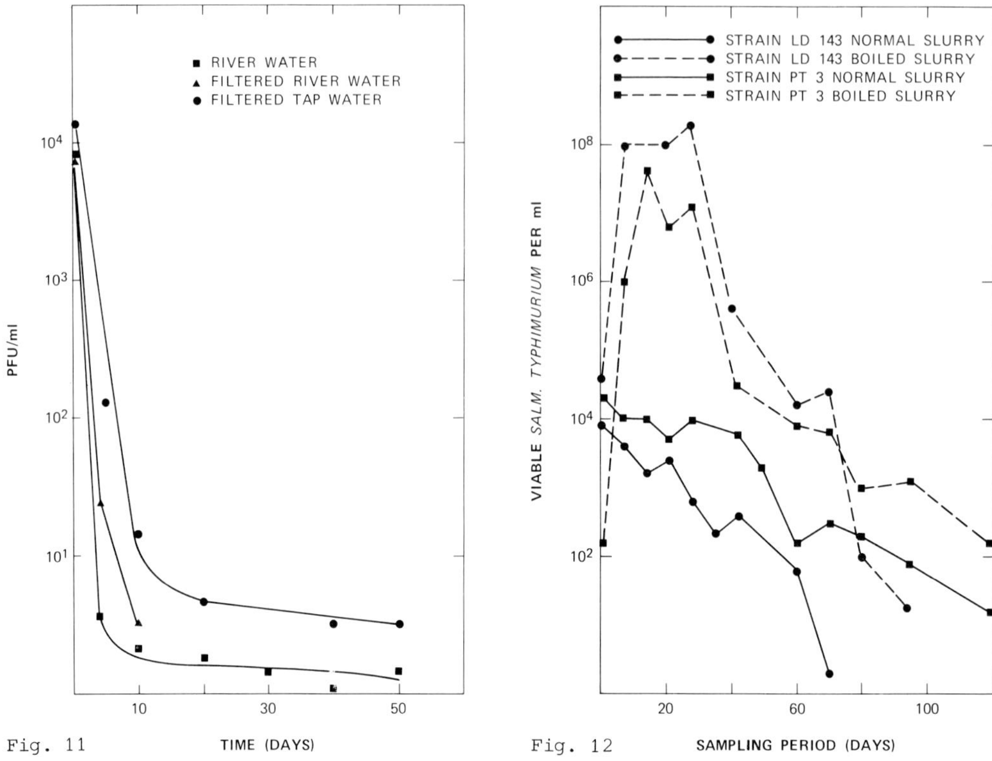


Fig. 11. Survival of poliovirus type I in river water, filtered river water or filtered tap water (hardness 123 ppm) at 22°C - 25°C

Fig. 12. Survival of Salmonella typhimurium in slurry (pH. 7.6, dry matter 3.5%) and after inoculation into boiled slurry at 18°C - 24°C

cause of direct contamination, but also because of the risk of creating aerosols which may be carried for quite long distances by prevailing winds.

The survival of 2 strains of Salmonella typhimurium in naturally infected liquid cattle slurry is shown in Fig. 12. When heat-sterilized slurry was seeded with a fairly heavy inoculum of the organisms, an initial growth phase occurred and was succeeded by a decline in infectivity which lasted for several months.

The importance of contamination of the environment by enteric organisms is increasing annually as human and animal populations proliferate. A concurrent tendency for concentration of human beings into cities, and of animals in large "economic units", exacerbates problems of sewage disposal. Consequently, epidemiologic studies reveal hazards associated with the exchange of pathogenic organisms between animals and man with increasing frequency (7, 3). In the recent report by Harbourne (3), Salmonella typhi B spread from a symptomless human carrier to the local sewage works, thence, via a stream carrying the effluent, to 5 of a herd of 59 milking cows and, finally, from the cows to farm workers on the same premises. It is easy to speculate how infection might have been disseminated further, especially since Salmonella spp. (and particularly Salm. typhimurium) have been recovered from milking equipment on infected farms. Harbourne also recorded the isolation of no less

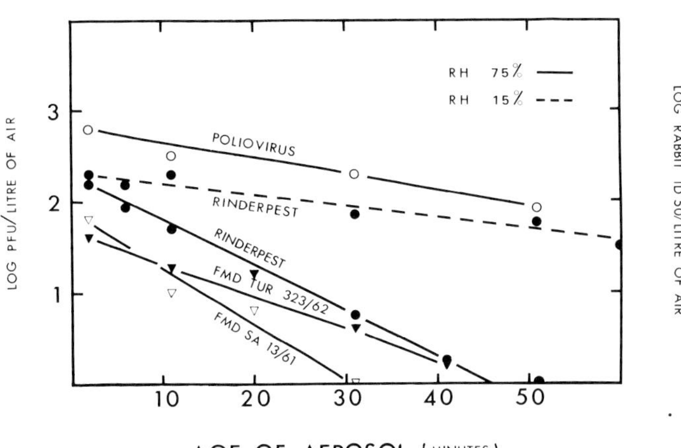


Fig. 13. Regression of infectivity of airborne viruses

than 316 cultures of Salmonella, of at least 29 different species, from the outfall of sewage works into local streams. The larger sewage installations tend to collect a greater variety of Salmonella spp. and of pathogenic organisms in general, whilst even modern sewage-handling systems do not act as highly effective "biological filters". Microorganisms may be discharged into the open air even when processing is completed inside a building. Salm. typhimurium can resist the effects of decomposition in sludge at 25°C for at least 8-12 weeks but, when sludge is sprayed onto pastures, at least 18 months should be allowed between spraying and grazing, because Salmonella spp. may persist in sludge spread on to grass for up to 72 weeks. Recent investigations in Canada show that many of the persistent Salmonella and coliform spp. remain for long periods in the top 15 cm of the soil. Pond water may contain viable Salm. dublin for at least 115 days after contamination (2).

Watercourses may also become contaminated by Brucella as the result of drainage of animal sewage; though the most frequent source is probably discarded foetal membranes from infected cows.

Some indication of the magnitude of problems in disposing of animal manure is provided by the realization that large N. American feedlots present an amount of waste equivalent to a city of 500,000 inhabitants. Concentration of dense human or animal populations, often with inadequate disposal of sewage, also leads to the build-up of heavy contamination of epiphytes and facultative pathogens, such as Listeria monocytogenes. This organism, frequently found in sludge used as a fertilizer, can multiply readily in damp or improperly cured silage. Thence, the organism may be carried to other parts of the silage by percolation of roof drainage or surface water, and it may be present also in fluids seeping from the silo (5). Listeria may proliferate further under ambient conditions of temperature and humidity, so that a wide variety of domestic and wild animals or birds develop clinical disease or become carriers and disseminate infection. The organism is transmissible from animals to man and we have shown milk to be a potent source of infection, with up to 36% of sero-positive milking cows excreting Listeria in the milk. The possibilities of airborne spread from liquids require further investigation, but results presented in Fig. 10 show that, despite an initial fall in viability, relatively little additional loss occurs during aerial suspension for periods up to 24 hr.

IV. Viruses in water. The longevity of viruses in water depends on virus type, pH, temperature, depth, solar radiation etc., but some enteroviruses (e.g. Cocksackie, polio- and echoviruses) adenoviruses, reoviruses, hepatitis viruses and possibly animal viruses, such as FMD virus, can exist in surface and river water for quite long periods. Poliovirus-type 1 may persist in water at 10°C for up to 60 days. Survival is usually prolonged by intra-cellular situation or by coating with faecal or other proteinaceous matter.

Our experiments on the viability of a limited number of strains of poliovirus (Type 1), and on the virus of foot-and-mouth disease, in river water, filtered river water and filtered tap water, showed that poliovirus may be isolated quite readily from contaminated water for at least a week (Fig. 11), and traces may be found for very much longer periods. FMD virus could not be recovered from neutral water at ambient temperatures for periods longer than six days.

Seawater is considerably more toxic than river water for the majority of microorganisms; this effect is caused partly by differences in pH, concentration of dissolved solids, oxygenation etc., but currently ill-defined factors also appear to play an important role. In temperate climates, coliform organisms but not anaerobes or heterotrophic aerobes are killed within 4 hr after discharge into the sea. Human and animal viruses seldom remain detectable for long periods in seawater. Nevertheless dumping of untreated sewage into coastal waters presents serious dangers as well as aesthetic problems.

## D. Conclusions

It must be concluded that increases in population with overcrowding will result in exposure of individuals to cumulative health hazards; precautions will depend on an improved knowledge of factors determining the survival of microorganisms. The relative importance of the various factors appears likely to be modified greatly by complex interactions of atmospheric conditions, and by species and strain differences in the genetic constitution of particular pathogenic organisms.

## E. Acknowledgement

The author wishes to acknowledge the valuable assistance of Mr. J. Ford and Mr. R.L.G. King.

## F. References

1. Beckert, J., Sinner, G.: Uber die Keimubertragung beim warme - und Feuchtigkeitsaustausch in Klimaanlagen Zbl. Bakt. Hyg. 1. Abt. Orig. B. 160, 473-498 (1975).
2. Gibson, E.A.: Ph.D. Thesis, University of London (1958).
3. Harbourne, J.F.: Intestinal infections of animals and man. Roy. Soc. Hlth. J. 97, 106-114 (1977).
4. Hyslop, N. St.G.: Observations on pathogenic organisms in the airborne state. Trop. Anim. Hlth. Prod. 4, 28-40 (1972).
5. Hyslop, N. St.G.: Epidemiologic and immunologic factors in Listeriosis. Proc. VIII Internat. Sympos. on Problems of Listeriosis, Leicester: Leicester University Press (1976) pp. 95-105.
6. Hyslop, N. St.G.: Aerial survival and contamination of the environment by airborne filterable organisms. In: Proc. Sympos. on Viruses in the Environment and their Potential Hazards. Burlington: International Commission and Canada Centre for Inland Waters. (In press 1977).

7. Jones, P.W., Matthews, P.R.J.: Examination of slurry from cattle for pathogenic bacteria. J. Hyg. Camb. 74, 57-64 (1975).
8. Thorne, H.V., Burrows, T.M.: Aerosol sampling methods for the virus of foot-and-mouth disease and measurement of virus penetration through aerosol filters. J. Hyg. Camb. 58, 409-414 (1960).

# Survival of Enterobacteria in Two Different Types of Sterile Soil

J. PAPAVASSILIOU and J. LEONARDOPOULOS

## A. Introduction

The different survival periods of enterobacteria in soil can offer interesting information about their ecology. The relatively short survival period of certain species, such as E. coli, supports the idea that its normal habitat is the intestinal tract of man and animals and therefore it constitutes a very good indicator of faecal pollution of water and food (3, 15). On the contrary other genera, Citrobacter, Klebsiella and especially Enterobacter have a much longer survival period and are frequently found in the environment and are considered to belong to its normal flora (11, 20). These bacteria can also be found in the faeces of man and other animals but it is not established if they can multiply in the intestine or are only passing through with water and food (14). Recent studies have shown that salmonellae are widely distributed in the environment and pose a potential threat to the health of humans and animals (4, 5). The wide distribution and the ease of recovering salmonellae suggest that these organisms may be a better indication of a health hazard than the coliforms (4). Some of these organisms, however, whose habitat was thought to be the intestinal tract, may be freeliving in nature (5). In this work we tested the survival period of 4 coliform and 4 Salmonella strains as well as the survival period of 1 Shigella and 2 Proteus strains in 2 different types of sterile soil.

## B. Materials and Methods

I. Types of soil. Samples from two soil types, one loam and one sandy-loam, were sieved (mesh 1.5 mm) and dried. Into each of 11 cylindrical glass containers was placed 100 g of loam soil and in another eleven, 100 g of sandy-loam. The containers were covered with aluminium foil and autoclaved for 1 h at 120$^{o}$C.

II. Strains: inoculation and survival. Samples of each soil were inoculated with 1 enterobacterium strain. The strains used were 4 coliforms (E. coli, Citrobacter, Klebsiella, Enterobacter), 4 salmonellae (S. typhi, S. paratyphi B, S. typhimurium, S. enteritidis), 2 Proteus (Pr. vulgaris, Pr. inconstans) and 1 Sh. flexneri. Each strain was grown for 24 h on a nutrient agar slant, washed three times with saline and resuspended in 10 ml of saline. Each soil sample was inoculated with 5 ml of the bacterial suspension; the number of bacteria added was estimated by the dilution spread plate technique using saline as diluent and MacConkey agar as the plating medium. The number of bacteria at inoculation varied from $10^6$ to $10^8$ per g of soil (Figs. 1 to 4). After inoculation the 22 glass containers were kept in the laboratory away from direct sunlight. The survival period of the strains was followed by estimating at intervals the number of bacteria per g of soil by removing after mixing 1 g of soil and suspending in 10 ml saline. Ten-fold dilutions in saline up to $10^{-6}$ or more were prepared and 0.1 ml of the appropriate dilutions inoculated in duplicate on MacConkey agar plates. Colonies

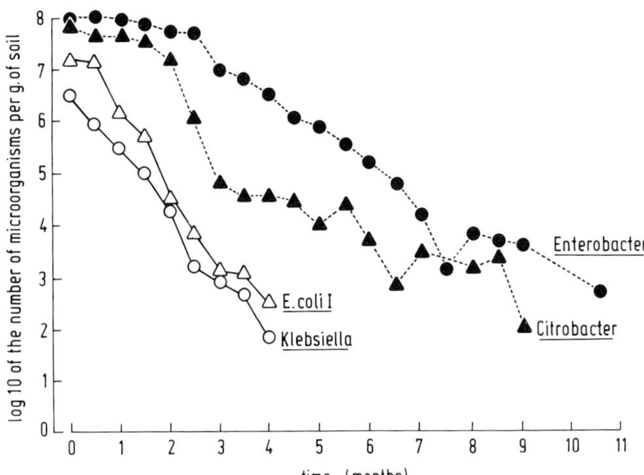


Fig. 1. Survival of coliform strains in sterile loam soil

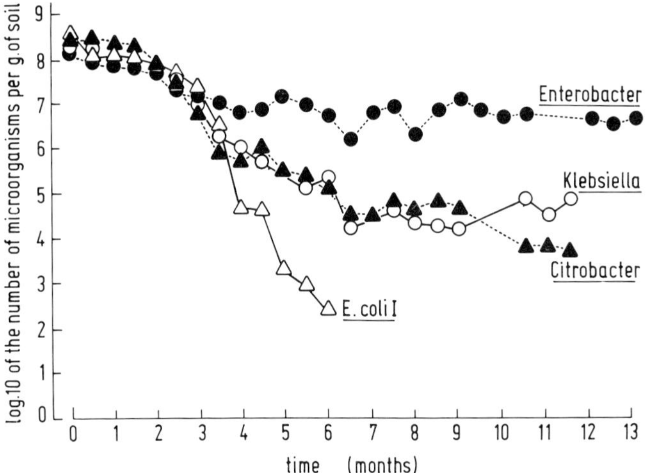


Fig. 2. Survival of coliform strains in sterile sandy-loam soil

were counted after 24 h of incubation at 37$^{o}$C. Further examination of a sample ceased when the number of bacteria per g of soil was found to be around $10^2$.

## C. Results and Discussion

These results are shown in figs. 1 to 4. In the loam soil the survival period was the same for E. coli and Klebsiella and considerably longer for Citrobacter and Enterobacter (Fig. 1). In the same soil S. typhi, Sh. flexneri and Pr. inconstans survived for less than 1 month and Pr. vulgaris for 39 days. S. paratyphi B, S. typhimurium and S. enteritidis survived for a period of several months (Fig. 3). In the sandy-loam soil the survival of the strains was generally longer, especially Citrobacter, Klebsiella and Enterobacter (Fig. 2). The survival period of S. paratyphi B, S. typhimurium and S. enteritidis was also longer in this type of soil though the increase in longevity was not as significant as for Citrobacter, Kleb-

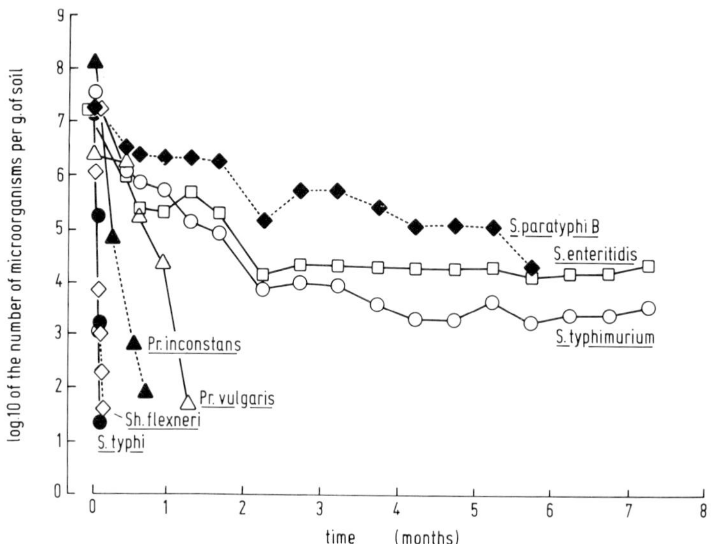


Fig. 3. Survival of Salmonella, Shigella and Proteus strains in sterile loam soil

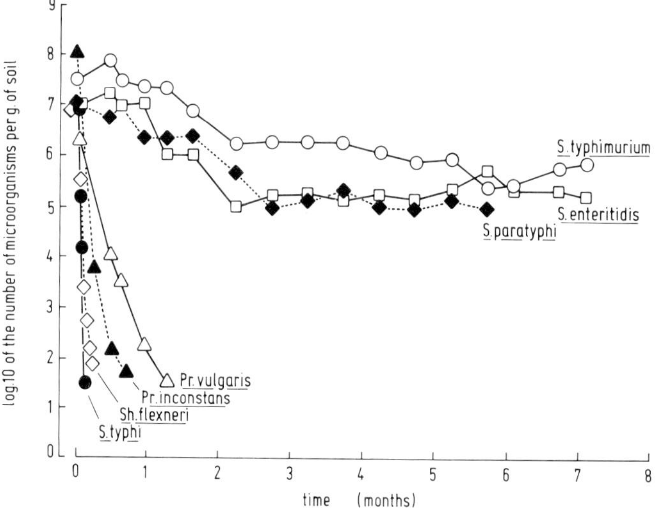


Fig. 4. Survival of Salmonella, Shigella and Proteus strains in sterile sandy-loam soil

siella and Enterobacter (Fig. 4). The survival time of S. typhi, Sh. flexneri, Pr. inconstans and Pr. vulgaris in the sandy-loam soil was the same as in the loam (Fig. 4).

From the results of this work it can be concluded that the survival of enterobacteria in soil depends on the bacterial species and on the type of soil. Some species and genera, such as E. coli and especially S. typhi, Sh. flexneri, Pr. inconstans and Pr. vulgaris cannot survive for a long period in the soil though others such as Enterobacter and some salmonellae can be found in constant numbers per g of soil for several months. The relatively short survival of E. coli in the soil is thought to be (8) due to their requirement for a high percentage of the available organic substances. What is usually left is not sufficient for multiplication and the number of bacteria is reduced quite rapidly. It is possible that the same explanation could also apply to the other short-living enterobacteria in soil. From the analysis of the 2 types of soil used in this work it was found that the sandy-loam had a higher content in organic matter and total nitrogen than the loam; the 2 types of soil were quite similar in most of their other properties. In the sandy-loam soil the survival period of enterobacteria was generally longer and could be attributed to the higher organic matter and total nitrogen content of the soil. The significance of these substances has also been stated in previous works (8, 10, 13). Our data on the longevity of some enterobacteria species in the soil are generally in accordance with the results of other workers. E. coli can survive for a few months in the soil but its longevity is significantly shorter than that of Enterobacter and Klebsiella (16, 6, 18, 9, 10). S. typhi dies out quickly, usually in a few days (13, 2, 1); S. paratyphi B (17), as well as S. typhimurium (7, 19, 12), can survive in soil for several months.

## D. References


1. Beard, P.J.: Longevity of Eberthella Typhose in Various Soils. Amer. J. Publ. Hlth. 30, 1077-1082 (1940).
2. Benedek, I.: The viability of typhoid bacillus in soil. Zh. Mikrobiol. 32, 61-65 (1961).
3. Buttiaux, R.: L'analyse bacteriologique des eaux de consomation, Flammarion, Paris, (1951).
4. Cherry, W.B., Hanks, J.B., Thomason, B.M., Murlin, A.M., Biddle, J.W., Croon, J.M., Salmonellae as an index of pollution of surface waters. Appl. Microbiol. 24, 334-340 (1972).
5. Cherry, W.B., Thomason, B.M., Gladden, J.B., Holsin, N., Murlin, A.M.: Detection of Salmonellae in foodstuffs, water and feces by immunofluorescence. Ann. N.Y. Acad. Sci. 254, 350-368 (1975).
6. Glath, H., Knoll, K.H., Makawai, A.A.M.: Die Lebensfähigkeit von Escherichia coli in verschiedenen Bodenarten. Z. Pflanzenernaehr. Dung. Bodenk. 100, 142-150 (1963).
7. Josland, S.W.: Aust. vet. J. 27, 264 (1951). Cited by Thomas, K.L., in Survival of Salmonella paratyphi B, Phage Type. 1 var. 6, in Soil. Monthly Bull. Min. Health. 26, 39-45 (1967).
8. Klein, D.A., Casida, L.E.: Escherichia coli die-out from normal soil as related to nutrient availability and the indigenous microflora. Canadian J. of Microbiol. 13, 1461-1470 (1970).
9. Leonardopoulos, J., Papavassiliou, J.: Über die Überlebensdauer von Escherichia in natürlich verunreigten Erdproben. Zbl. Bakt., I. Abt. Orig. 214, 79-85 (1970).
10. Leonardopoulos, J., Papavassiliou, J.: Über die Überlenbensdauer von Escherichiae in zwei verschiedenen sterilisierten Erdsorten. Zbl. Bakt. Hyg., I. Abt. Orig. A 218, 168-175 (1971).
11. Mackenzie, E.F.W., 33rd Ann. Rep. Metrop. Water Board, p.25. London (1938).
12. Mair, N.S., Ross, A.E.: Survival of Salmonella typhimurium in the soil. Monthly Bull. Min. Health. 19, 39-41 (1960).
13. Mallmann, W.L., Litsky, W.: Survival of selected Enteric Organisms in Various types of soil. Am. J. Publ. Health. 41, 38-44 (1951).
14. Papavassiliou, J.: Aerobacter (Enterobacter) cloaca in human and animal faeces. J. Bact. 85, 1176 (1963).

15. Rosebury, T.: Microorganisms Indigenous to Man, McGraw Hill Book Co. Inc., New York (1962).

16. Skinner, C.E., Murray, J.T.: The viability of B. coli and B. aerogenes in soil. J. Infect. Diseases, 38, 37-41 (1926).

17. Thomas, K.L.: Survival of Salmonella paratyphi B, Phage Type 1 var. 6, in soil. Monthly Bull. Minist. Health (London), 26, 39-45 (1967).

18. Waksman, S.A., Woodruff, H.B.: Survival of bacteria added to soil and the resultant modifications of the soil population. Soil Sci. 50, 421-427 (1940).

19. Watts, P.S., Wall, M.: Austr. vet. J. 28, 165 (1952). Cited by Thomas, K.L. in survival of Salmonella paratyphi B, Phage Type 1 var. 6 in Soil. Monthly Bull. Minist. Health (London), 26, 39-45 (1967).

20. Wilson, G.S., Miles, A.: Topley and Wilson's Principles of Bacteriology and Immunity, 5th Ed., E. Arnold, London, (1964).

# The Ecology of *Escherichia coli* Serotypes

K.A. BETTELHEIM

## A. Introduction

Kauffmann (9) established that strains of Escherichia coli from any source could be serologically typed. His antigenic scheme was based on three types of antigen: the somatic (O), the surfaces or capsular (K) and the flagellar (H) antigens. Currently most serotyping studies are based on the 'O' and 'H' antigens only. Although the 'K' antigens are still studied and sometimes reported, no internationally acceptable method of identifying them is available. To establish such a method is important because it has been demonstrated that some 'K' antigens are chemically part of the 'O' antigen complex (12). In this study, however, only 'OH' serotypes are reported. As there were available 163 'O' and 55 'H' antisera, theoretically it would be possible to obtain 8965 serotypes.

## B. Materials and Methods

Standard methods of serotyping have been employed to type strains of E. coli from a variety of sources (Table 1).

## C. Results and Discussion

The 20,700 strains studied over 15 years could be serologically subdivided as shown in Table 2 into 1173 serologically identifiable types. There were additionally identified strains with complex antigens such as: 05,065.H4; 091,0114.H16; 0129,0147.H32; 022.H5,H40. For some studies it was found necessary to prepare sera against unidentifiable types. Such types are thus serologically distinguishable, but have not yet been internationally accepted. These comprised an additional 36 different serotypes making 1209 serotypes identified.

It is impossible to list here the source of these 1209 serologically distinguishable types of E. coli. It is only possible to discuss some of the more important groups.

Since certain 'O' groups have been associated for many years with infantile gastroenteritis (IGE) it is interesting to examine these types in greater detail. Table 2 shows that there appear to be certain 'H' types associated with the 'O' types commonly isolated from babies with gastroenteritis. These tend to predominate in environments close to neonates and are often associated with IGE, whereas others appear widely distributed.

Of these 'O' groups only six were isolated from animal sources: 026.H32; 026.H42; 086.H10; 0111.H10; 0114.H0; and 0114.H10. It is noteworthy that the H10 antigen was carried by three of these 'O' groups and that the 0111.H10 serotype was not found among any of the human sources.

A number of 'O' groups have been commonly associated with urinary tract infections (UTI) in humans. The general view is that these are derived from the gastrointestinal tract of the patients and thus they represent the normal faecal flora of

Table 1. Sources of *E. coli* studied

| Sources of *E. coli* | No. of strains (x $10^{-2}$) | Reference |
|---|---|---|
| Healthy babies or babies with diarrhoea in UK | 16 | (1) |
| Patients with urinary tract infections and related conditions in UK | 7 | (1), (6), (8) |
| Patients with urinary tract infections in Hong Kong | 2 | (16) |
| Faeces of healthy adults in UK | 32 | (3). (4), (14) |
| Faeces of students from many parts of the world studying in Dublin. Specimens taken in country of origin and Dublin | 50 | (11) |
| Faeces of mothers and their babies in a maternity ward and ward environment in London | 80 | (7), (10) |
| Faeces and rectal swabs of domestic animals, meat and butchering equipment | 20 | (2), (5), (15) |
| TOTAL | 207 | |

Table 2. Distribution of serotypes of *E. coli* carrying 'O' antigens commonly associated with IGE

| 'O' Group | No. of 'H' types associated with | | | |
|---|---|---|---|---|
| | Babies with diarrhoea only | Healthy babies and babies with diarrhoea only | Babies with diarrhoea and other sources | Sources unconnected with babies |
| 026 | 4 | 1 | 1 | 2 |
| 055 | 9 | 1 | 1 | 0 |
| 086 | 8 | 0 | 3 | 5 |
| 0111 | 7 | 3 | 1 | 1 |
| 0114 | 1 | 0 | 3 | 2 |
| 0119 | 3 | 0 | 1 | 1 |
| 0125 | 5 | 0 | 2 | 2 |
| 0126 | 2 | 3 | 1 | 0 |
| 0127 | 7 | 2 | 1 | 4 |
| 0128 | 7 | 1 | 3 | 0 |
| 0142 | 0 | 0 | 1 | 1 |

man. In Table 3 the distribution of the different 'H' antigens associated with these 'O' groups is presented.

It is noticeable that despite the relative preponderance of strains of human material, twelve 'OH' serotypes, O1.H5, O2.H8, O2.H21, O2.H27, O4.H8, O4.H16, O4.H37, O6.H45, O7.H19, O8.H11, O8.H28 and O8.H39 belonging to these "common human" 'O' groups were represented only in animal sources. Also strains with the serological structure O1.H4, O4, H5, O6.H1, O7.H6, O75.H5, some of the most common isolates from human materials, were not found in any material from animals.

In a recent study, (13), the serological structures of 106 enterotoxigenic strains from children and adults from many parts of the world were examined. Although the strains were not fully representative they noted that enterotoxigenicity appeared restricted to certain 'OH' serotypes, which were not necessarily the most common. They described only 26 different fully 'OH' typeable strains; of these only 11 were represented in the studies reported here. The sources of these are given in Table 4. Strains of 'O' group 6 shown to be toxigenic (Table 4) were not the common O6.H1 or O6.H31, but O6.H16, which was relatively rare in this study. The same seems to apply to most of these serotypes.

Although over 20,000 strains of E. coli from a variety of sources were serotyped and 1173 serologically distinguishable types could be identified, it is difficult to draw many definite conclusions. It appears that certain serotypes predominate in animal faeces, while others occur in human faeces and yet others may cause human infections. Nevertheless there is considerable overlap. Quantitative methods of assessing the roles of different serotypes in the various ecological situations will have to be developed. Until then only tentative suggestions can be made.

D. Acknowledgement: Published with the authority of the Director-General of Health, Department of Health, Wellington, New Zealand.

Table 3. Distribution of serotypes of E. coli carrying 'O' antigens commonly associated with human urinary tract infections

| 'O' Group | No. of 'H' types associated with | | | | | | |
|---|---|---|---|---|---|---|---|
| | Urinary tract infections Only | and other human sources | and animal sources | and human and animal sources | Other human sources | Human and animal sources | Animal sources only |
| O1 | 4 | 2 | 1 | 2 | 3 | 5 | 1 |
| O2 | 3 | 1 | 0 | 4 | 10 | 1 | 3 |
| O4 | 1 | 3 | 0 | 1 | 4 | 1 | 3 |
| O6 | 2 | 4 | 1 | 0 | 5 | 0 | 1 |
| O7 | 1 | 1 | 0 | 1 | 8 | 1 | 1 |
| O8 | 1 | 4 | 0 | 2 | 7 | 3 | 3 |
| O75 | 1 | 1 | 1 | 0 | 10 | 1 | 0 |

Table 4. Sources in this study of 'OH' serotypes shown to be enterotoxigenic by Ørskov et al. (16)

| Serotype | No. of strains | Sources in this study |
|---|---|---|
| 02.H5 | 1 | Healthy adults, babies with diarrhoea; Urinary tract infection; cattle, meat. |
| 06.H16 | 7 | Urinary tract infection; poultry, meat. |
| 08.H9 | 8 | Healthy adults, babies with diarrhoea; Urinary tract infection, introital swab; cattle, meat. |
| 015.H11 | 11 | Babies with diarrhoea; Urinary tract infection. |
| 021.H21 | 1 | Healthy adults; Urinary tract infection; cattle. |
| 042.H37 | 1 | Healthy babies; Urinary tract infection; poultry. |
| 085.H7 | 1 | Healthy adults. |
| 0109.H21 | 1 | Healthy adults; cattle. |
| 0128.H7 | 3 | Babies with diarrhoea. |
| 0128.H21 | 2 | Healthy adults, babies with diarrhoea. |
| 0163.H19 | 1 | Healthy adults; cattle. |

## E. References


1. Bettelheim, K. A. Investigations into the pathogenicity of Escherichia coli in human infections. PhD Thesis, University of London. (1968).
2. Bettelheim, K. A., Bushrod, F. M., Chandler, M. E., Cook, E. M., O'Farrell, S., Shooter, R. A.: Escherichia coli serotype distribution in man and animals. J. Hyg. Camb., 73: 467-471 (1974).
3. Bettelheim, K. A., Cooke, M. E., O'Farrell, S., Shooter, R. A.: The effect of diet on the intestinal Escherichia coli. J. Hyg. Camb. 77, 43-45 (1977).
4. Bettelheim, K. A., Faiers, M., Shooter, R. A.: Serotypes of Escherichia coli in normal stools. The Lancet. ii: 1224-1226.
5. Bettelheim, K. A., Ismail, N., Shinebaum, R., Shooter, R. A., Moorehouse, E., Farrell, W.: The distribution of serotypes of Escherichia coli in cow-pats and other animal material compared with serotypes of E. coli isolated from human sources. J. Hyg. Camb. 76: 403-406 (1976).
6. Bettelheim, K. A., Taylor, J.: A study of Escherichia coli isolated from chronic urinary infection. J. med. Microbiol. 2: 225-236 (1969).
7. Bettelheim, K. A., Teoh-Chan, C. H., Chandler, M. E., O'Farrell, S., Rahamin, L., Shaw, E., Shooter, R. A.: Further studies of Escherichia coli in babies after normal delivery. J. Hyg. Camb. 73: 277-285 (1974).
8. Brooks, H. L. Properties of Escherichia coli in recurrent urinary tract infection. PhD Thesis, University of London. (1977).
9. Kauffman, F. Zur Serologie der Coli-Gruppe. Acta path. microbiol. Scand. 21: 20-45 (1944).
10. Lennox-King, S.M.J. Escherichia coli and the colonization of babies delivered by caesarian section. PhD Thesis, University of London (1976).
11. Majed, N. I. Escherichia coli as related to travellers' diarrhoea and other ecological problems. M. Phil Thesis, University of London (1976).
12. Ørskov, F., and Ørskov, I. Immunoelectrophoretic patterns of extracts from Escherichia coli test strains 01 to 0157. Examinations in homologous OK sera. Acta path. microbiol. Scand. B80: 905-910 (1972).

13. Orskov, F., Orskov, I., Evans, D. J., Jnr., Sack, R. B., Sack, D. A., Wadstrom, T. Special *Escherichia coli* serotypes among enterotoxigenic strains from diarrhoea in adults and children. Med. Microbiol. Immunol., 162, 73-78 (1976).

14. Shooter, R. A., Bettelheim, K. A., Lennox-King, S.M.J., O'Farrell, S.. *Escherichia coli* serotypes in the faeces of healthy adults over a period of several months. J. Hyg. Camb. 78: 95-98 (1977).

15. Shooter, R. A., Cooke, M. E., O'Farrell, S., Bettelheim, K. A., Chandler, M. E., Bushrod, F.: The isolation of *Escherichia coli* from a poultry packing station and an abattoir. J. Hyg. Camb. 73: 245-247 (1974).

16. Wong, W. T., Bettelheim, K. A. Serotypes of *Escherichia coli* from urinary tract infections in Hong Kong. Zbl. Bakt. Hyg., I. Abt. Orig. A, Z36: 481-486 (1976).

# An Ecological Profile of *Bacteroides nodosus*, the Chief Causative Agent of Footrot in Sheep

T.M. SKERMAN

## A. Introduction

Footrot is a contagious, indirectly-transmitted disease of the hoof epidermis in sheep, cattle, deer and goats. In foot lesions, Bacteroides nodosus, an obligate parasite of the avascular, uncornified tissue, is associated with Fusobacterium necrophorum, an organism of endogenous origin. Both these microorganisms are Gram-negative, non-sporeforming anaerobes, and both are implicated in the pathogenesis of the disease (3).

Substantial progress in understanding the aetiology, pathology and control of footrot has been made in the last 10 years (4). Because pathological changes are confined to the integument, and because afflicted animals do not benefit from disease-induced immunity, an unexpected outcome from this research was the demonstration that B. nodosus vaccines (bacterins) had both prophylactic and therapeutic effects on footrot in sheep (2). Field trials have indicated, however, that although these preparations may significantly reduce the incidence and severity of footrot infection in flocks, vaccine-induced immunity is of limited duration and degree (6).

Notwithstanding these advances, B. nodosus remains a relatively poorly-characterized species; little is known of its nutrition and metabolism, and the bacterial determinants of its pathogenicity and virulence have yet to be clearly identified. As a corollary, it follows that definitive studies of the structure, physiology and genetics of B. nodosus might not only clarify our understanding of the host-parasite relationships of this organism, but also yield guidelines for the development of more effective, better-defined immunizing preparations than at present seems attainable.

This paper refers briefly to a few attributes of B. nodosus that may have a bearing on mechanisms by which infection is established within the target tissue of the ovine hoof.

## B. Infectivity

Footrot transmission from clinically diseased to susceptible sheep is governed initially by effective numbers of B. nodosus discharged from foot lesions surviving in the external environment. Access to the new host then requires that the bacteria become fixed to the interdigital skin and that thereafter they multiply in an essentially unrestricted manner during the acute phase of invasion. It is, however, only following prolonged water maceration and antecedent infection of the superficial skin by F. necrophorum that colonization by B. nodosus can be observed (3).

The numerous pili borne by bacteria from certain B. nodosus cultures (7) have attracted interest in view of their resemblance to those filamentous appendages of other microbial species which are thought to be involved in the adherence of virulent bacteria to host mucosal surfaces. Recent studies at Wallaceville confirmed that B. nodosus cultures of piliated bacteria were substantially more virulent for sheep infected artificially than those containing non-piliated organisms. In addition, sheep vaccinated with purified preparations of B. nodosus pili were protected from challenge footrot infection, but it has not yet been established whether such protection is predicated upon restricted attachment of B. nodosus, or multiplication of the organisms, or both. Evidently, the pili of B. nodosus do not promote attachment to epithelial cells in vitro (8), hence their role in the invasiveness of this organism is not clear.

## C. Pathogenesis

Histopathological data reveal a special affinity between B. nodosus and living epidermal cells of the hoof (3). Major nutrient requirements in vitro - notably arginine and serine (5) - might therefore reflect an organotropic relationship with keratinaceous proteins which have been reported to have a relatively high content of these amino acids.

B. nodosus produces no exotoxin, and early workers ascribed a pathogenic role to its proteinase activity (1). Yet there was no histological evidence for a direct bacterial attack on the horn itself, and the severe inflammation and tissue damage characteristic of footrot is believed to be due principally to F. necrophorum which penetrates the epidermis once B. nodosus has become established (3). However, these findings do not preclude the possibility that either the proteinase, or some other metabolite of B. nodosus, exerts some aggressin-like or inflammatory action in vivo.

## D. Genetic Variation

We are currently studying in vitro changes of a B. nodosus isolate in regard to structural, physiological and antigenic properties. On primary isolation, colonies of this strain had a beaded appearance and contained organisms that were characterized as being virulent for sheep, possessed both K agglutinogen and numerous pili, and were able to induce protective immunity. Organisms from two other colony types (mucoid and smooth) from the same strain were usually avirulent, did not possess K agglutinogen, bore few pili, and were only weakly immunogenic. Cultures forming beaded colonies could be passaged on agar medium repeatedly. However, when liquid medium was inoculated, serial transfer of the resultant cultures tended to yield organisms producing firstly mucoid colonies, and then after further subculturing, smooth colonies.

The genetic bases of these changes induced in liquid medium have not been examined. Perhaps under these conditions a series of irreversible loss mutations are progressively selected for by some medium component(s) which, in the corresponding agar formulation, are either sequestered or neutralized. Whatever the explanation, genetic studies of B. nodosus are likely to be essential to provide for the conservation of protective antigens in cultures for vaccine production, and to understand the epidemiological significance of variations in the virulence of strains as expressed by benign and progressive forms of footrot disease.

## E. References


1. Beveridge, W.I.B.: Foot-rot in sheep: a transmissible disease due to infection with Fusiformis nodosus n. sp. Counc. scient. ind. Res. Melb. Bull. No. 140 (1941).

2. Egerton, J.R.: Successful vaccination of sheep against foot-rot. Aust. vet. J. 46, 114-115 (1970).

3. Egerton, J.R., Roberts, D.S., Parsonson, I.M.: The aetiology and pathogenesis of ovine foot-rot. I. A histological study of the bacterial invasion. J. comp. Path. 79, 207-227 (1969).

4. Schmitz, J.A., Gradin, J.L.: Ovine foot root: a review. Proc. ann. mtng. U.S. anim. Hlth. Assn. 78, 385-392 (1974).

5. Skerman, T.M.: Determination of some in vitro growth requirements of Bacteroides nodosus. J. gen. Microbiol. 87, 107-119 (1975).

6. Skerman, T.M., Cairney, I.M.: Experimental observations on prophylactic and therapeutic vaccination against foot-rot in sheep. N.Z. vet. J. 20, 205-211, (1972).

7. Stewart, D.J.: An electron microscopic study of Fusiformis nodosus. Res. vet. Sci. 14, 132-134 (1973).

8. Stewart, D.J.: Foot-rot of sheep. In: Ann. Rept. 1975. Div. Anim. Hlth. C.S.I.R.O., Australia, p.46.

# The Ecology of *Pithomyces chartarum* and the Control of Facial Eczema

J.N. PARLE and M.E. DI MENNA

Pithomyces chartarum (Berk. & Curt.) M.B. Ellis, the imperfect mold which causes facial eczema, a mycotoxicosis of sheep and cattle, is saprophytic (1), nutritionally unexacting (27), and world wide in distribution (12, 14, 16). The countries in which facial eczema occurs, New Zealand (13), Australia (15), South Africa (17) and possibly Uruguay and Brazil, are those in which toxic levels of the fungus are reached at sites to which grazing animals have access. Although P. chartarum has been found on all types of dead vegetable material, conditions under which it sporulates freely enough to be dangerous occur only in pasture under climatic conditions which are reached most often in the North Island of New Zealand. It is only in countries with a similar climate and in which animals are fed almost exclusively on pasture that facial eczema occurs or seems likely to occur.

Facial eczema was recognised in New Zealand about 70 years ago but it is less than 20 years since Thornton and Percival (31) found its cause. The toxin responsible, sporidesmin which was characterised and named when P. chartarum was a synonym of Sporidesmium bakeri Syd. (34), causes inflammation and blockage of the bile duct. Retained bile damages the liver and spills back into the bloodstream. Phylloerythrin in the bile gives rise to the photosensitization which causes the external symptoms. Because of the process of the disease, symptoms are not seen until 10-20 days after ingestion of the toxin, a time lag which hindered workers looking for the cause.

The discovery of the causal agent allowed rational methods of control of the disease to be tested and explained why some of the empirical methods used had been effective.

## A. Control by Control of Substrate

Because P. chartarum was a saprophyte, it was hoped that reduction of dead material in pasture would reduce spore numbers sufficiently to control the disease (2). This did not prove effective. Pastures can be made more dangerous by topping and allowing the dead material to lie but, if conditions are favourable for growth of the fungus, they cannot be made innocuous. There is always enough dead material, often attached to leaf sheaths, to support dangerous levels. Grazing management which keeps dead material to a minimum and also uses pasture most efficiently, causes animals to eat to the base of pasture where the fungus grows (3).

It was also hoped that adjustment of the species composition of pasture might give swards unfavourable for the growth of P. chartarum but, once again, any reduction in numbers was insufficient for control. It is only in crops such as maize and brassicas, in which leaves are held above the soil surface, that P. chartarum does not reach dangerous levels. Facial eczema has increased in prevalance in New Zealand, not because the climate has changed, but because stocking rates have increased and improved pastures contain species susceptible to summer drought.

Control of facial eczema therefore depends upon recognition and avoidance of dangerous pasture.

B. Control by Avoiding Dangerous Pasture

Sporidesmin is formed when P. chartarum spores. The amount of toxin produced in culture by New Zealand isolates was proportional to the number of spores formed provided that the incubation temperature did not much exceed 20°C (28, 29, 11, 10, 19). The number of spores produced is proportional to the heat input when humidity is 100%. Although P. chartarum grows at RH values as low as 60%, it only spores freely at 100% (1). Fig. 1 shows the time taken by 20 New Zealand isolates to spore when grown on potato carrot agar plates at four temperatures between 10 and 27°C. Sporulation time varied with isolate and increased as the temperature was reduced.

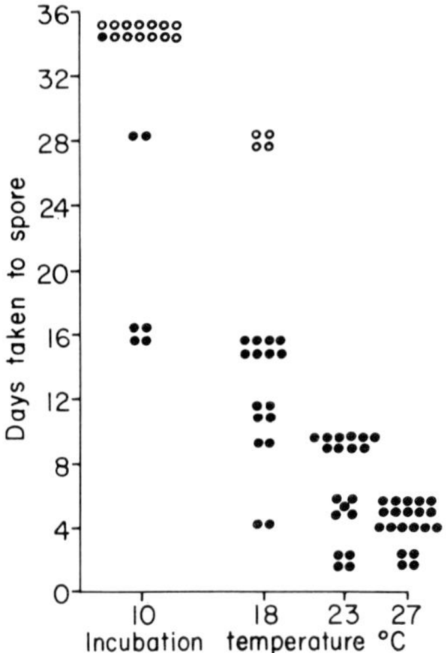


Fig. 1. Sporulation times of 20 isolates of Pithomyces chartarum at four incubation temperatures

• I strain sporing
○ I strain not sporing, discarded

C. Use of Climatological Observations

The relationship between moisture, temperature and facial eczema outbreaks was recognised as early as 1940 (8). In 1959, before the cause of the disease was known, the so-called danger periods which preceded all outbreaks were defined as times when rainfall of 4 mm or more was accompanied by at least two nights on which grass minimum temperatures did not fall below 12°C (21). All outbreaks were preceded by danger periods but, particularly early in the season, not all danger periods were followed by outbreaks.

This information (21) was incorporated into a predictor (9), designed to allow farmers who owned a rain gauge and a grass minimum thermometer to assess facial eczema danger on their own farms, but the predictor was not precise enough to be useful (18).

D. Use of Spore Counts

Before it was known that P. chartarum caused facial eczema, the toxicity of pasture could be measured only by feeding trials using susceptible animals, the most valid of assay methods but too slow to be used to protect stock in the field. The rate at which P. chartarum can increase in pasture is such that daily assessment of facial eczema danger is essential. As yet, chemical assay of sporidesmin in pasture is impractical and the quick and simple but indirect measure of toxin by spore counts in pasture remains the measure used.

Two factors may affect this measure. In the field, destruction of spores is slower than destruction of toxin. Sporidesmin, leached from the spore is rapidly destroyed by ultra-violet light (6, 7), and the half life of sporidesmin in spores grown in culture and exposed to the weather is about one day. In pasture, spore numbers equal production rate minus destruction rate. Production is dependent upon temperature and humidity, but the causes of destruction, germination, ingestion by herbivores and soil invertebrates, removal by rain and wind, and disruption by osmotic effects, are easier to list than measure.

Whilst all New Zealand isolates have produced toxin in proportion to sporulation, strains have been recovered in South Africa and the United States which spored freely but produced no measurable toxin (Marasas, personal communication, 35). It is likely that such strains occur in New Zealand but they appear uncommon enough not to affect the hypothesis that, in this country, production of spores equals production of toxin.

In the summer of 1975 a study was made of molds, particularly *P. chartarum* on leaves of ryegrass-white clover pasture on a plot 100 m from Rukuhia (Hamilton) Soil Research Climatological Station. Leaf samples were washed with sterile water and *P. chartarum* spore counts made daily (18) and culture of total moulds twice weekly (20). Fig. 2 shows the daily rainfall and the grass minimum temperatures for January-March with the *P. chartarum* counts obtained by the direct method (= live plus dead spores) and by culture (= viable propagules).

After facial eczema danger periods, one in the first half of January and one at the beginning of March, there were peaks in both *P. chartarum* counts. In early February, which was dry with low night temperatures, numbers of viable propagules dropped more than did numbers of spores, suggesting that many of the spores were dead. A rise in viable propagules followed some high grass minimum temperatures in mid-February, although appreciable rain did not fall until February 24. It is

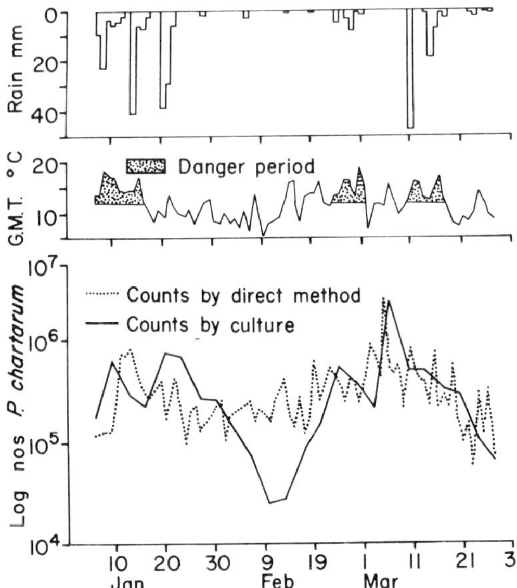


Fig. 2. *Pithomyces chartarum* counts on pasture leaves, rainfall and grass minimum temperatures, Rukuhia, summer, 1975

likely that the first part of this rise was due to mycelial extension rather than sporulation.

Spore counts are not an entirely reliable measure of pasture toxicity but are more reliable than climatological observations and, because sporidesmin is destroyed faster than spores, errors are on the side of safety. Since the early 1960s, district warnings of facial eczema danger have been issued by local warning committees, usually on the basis of spore counts made at several sites, often those known to be prone to facial eczema. The disadvantage of general warnings is that spore numbers change not only with time but with distance. Even if rainfall is evenly distributed throughout a district, spore numbers vary with terrain as terrain affects temperature. Facial eczema outbreaks were observed to be restricted to the north and west slopes of a volcanic cone (8) and it was confirmed (18) that the greater warmth of sunny slopes or the sheltered sides of hedges was reflected in greater spore numbers.

Warnings issued to protect stock on all farms caused many farmers to take needless precautions. Increase in the number of sites at which spore counts were made improved district warnings, but it was concluded that, to eliminate facial eczema outbreaks, spore counts on all farms were needed (18). A pilot scheme to teach farmers to make spore counts was introduced and, when it proved workable, was adopted extensively.

By this means a farmer can use what naturally safe pasture he has or, if he has none, can use stored fodder or fungicide sprayed pasture as sparingly as possible.

## E. Control by Fungicides

Early attempts to control *P. chartarum* in pasture by application of fatty acids (32) or other available fungicides (24) were unsuccessful. In 1967 thiabendazole and, later, other substituted benzimidazoles were found to reduce spore numbers and to protect grazing stock for up to six weeks after application at rates of 140 g $ha^{-1}$ or more (4, 5, 23, 25, 30). These fungicides, which prevent growth at levels as low as 0.2 ppm, appear to be effective because of their high activity. Spray applied to pasture can only reach the site at which *P. chartarum* grows by moving down the leaf in a moisture film and it is likely that only small amounts of fungicide applied reach the base of the plant. The period over which control can be achieved is related to the systemic nature of the fungicides which are taken up by the pasture within 3 days and appear to be released again over an extended period.

There is no evidence that *P. chartarum* becomes resistant to substituted benzimidazoles (24), perhaps because they block a pathway to vitamin B12 synthesis to which the fungus has no alternative (33). Where it has been possible to investigate cases of breakdown in control in the field, they have been found due to heavy rain immediately after fungicide application, before it had become systemic (25), or to challenge by very high rises in spore numbers (22, 26). No rate of fungicide used has eliminated *P. chartarum* from pasture but has only reduced spore numbers by a greater or lesser percentage. When the percentage reduction still does not bring spores below the danger level, facial eczema outbreaks occur in stock.

*P. chartarum* was once an obscure species of interest only to taxonomists. If it had not been toxigenic, its ability to proliferate extensively although ephemerally in pasture might have passed unnoticed. Because it is nutritonally undemanding and has a simple life cycle it has not been possible to control its growth sufficiently by any means but fungicide application. The understanding of its ecology has however made prediction of its growth possible and has reduced the control of facial eczema, once uncertain, to a piece of farm management.

## F. References

1. Brook, P.J.: Ecology of the fungus Pithomyces chartarum (Berk. & Curt.) M.B. Ellis in pasture in relation to facial eczema disease of sheep. N.Z. J. agric. Res. 6, 147-228 (1963).
2. Brook, P.J., Mutch, G.V.: Field control of facial eczema of sheep. N.Z. J. agric. Res. 7, 138-145 (1964).
3. Campbell, A.G.: Recent advances in the control of facial eczema. Proc. Internat. Grass. Conf. 774-777 (1970).
4. Campbell, A.G., Sinclair, D.P.: Control of facial eczema in lambs by the use of fungicidal sprays. Proc. Ruakura Farm. Conf. 148-155 (1968).
5. Campbell, A.G., Sinclair, D.P., Parle, J.N.: Control of facial eczema by fungicides. Proc. Ruakura Farm. Conf. 96-102 (1971).
6. Clare, N.T., Gumbley, J.M: Some factors which may affect the toxicity of spores of Pithomyces chartarum (Berk. & Curt.) M.B. Ellis collected from pasture. N.Z. J. agric. Res. 5, 36-42 (1962).
7. Clare, N.T., Mortimer, P.H.: The effect of mercury arc radiation and sunlight on the toxicity of water solutions of sporidesmin to rabbits. N.Z.J. agric. Res. 7, 258-263 (1964).
8. Cooper, E.R., Walker, D.: Climatic factors relating to outbreaks of facial eczema in New Zealand. N.Z.J. Sci. Tech. A22, 30-41 (1940).
9. Crawley, W.E., Woolford, M.W.: Predicting facial eczema danger periods. Proc. Ruakura Farm. Conf. 15-21 (1965).
10. Davison, S., Marbrook, J.: The effect of temperature on the toxicity of spores of Pithomyces chartarum (Berk. & Curt.) M.B. Ellis. N.Z.J. agric. Res. 8, 126-130 (1965).
11. Dingley, J.M., Done, J., Taylor, A., Russell, D.W.: The production of sporidesmin and sporidesmolides by wild isolates of Pithomyces chartarum in surface and in submerged culture. J. gen. Microbiol. 29, 127-135 (1962).
12. Ellis, M.B.: Dematiaceous hyphomycetes. 1. Mycol. Pap. No.76. Commonwealth Mycological Institute, Kew, Surrey, 1960, p.7
13. Gilruth, J.A.: Facial eczema. In: Report N.Z. Dep. Agric. Appendix VII, Vet. Div. 188-191 (1908).
14. Gregory, P.H., Lacey, M.E.: The discovery of Pithomyces chartarum in Britain. Trans. Brit. Mycol. Soc. 47, 25-30 (1964).
15. Hore, D.E.: Facial eczema. Aust. vet. J. 36, 172-176 (1960).
16. Hughes, S.J.: Fungi from the Gold Coast II. Mycol. Pap. No.50. University College of the Gold Coast Publications Board and The Commonwealth Mycological Institute, Kew, Surrey. p. 66 (1953).
17. Marasas, W.F.O., Adelaar, T.F., Kellerman, T.S., Minne, J.A., van Rensburg, I. B.J., Burroughs, G.W.: First report of facial eczema of sheep in South Africa. Onderstepoort J. vet. Res. 39, 107-112 (1972).
18. Menna, M.E. di, Bailey, J.R.: Pithomyces chartarum spore counts in pasture. N.Z. J. agric. Res. 16, 343-351 (1973).
19. Menna, M.E. di, Campbell, J., Mortimer, P.H.: Sporidesmin production and sporulation in Pithomyces chartarum. J. gen. Microbiol. 61, 87-96 (1970).
20. Menna, M.E. di, Parle, J.N.: Moulds on the leaves of ryegrass and white clover. N.Z.J. agric. Res. 13, 51-68 (1970).
21. Mitchell, K.J., Walshe, T.O., Robertson, N.G.: Weather conditions associated with outbreaks of facial eczema. N.Z.J. agric. Res. 2, 584-604 (1959).
22. Parle, J.N.: Facial eczema. 1. Problems in control with fungicides. Proc. Ruakura Farm. Conf. 46-49 (1973).

23. Parle, J.N., di Menna, M.E.: Fungicides and control of Pithomyces chartarum. Proc. Ruakura Farm. Conf. 141-147 (1968).

24. Parle, J.N., di Menna, M.E.: Fungicides and the control of Pithomyces chartarum. I. Laboratory trials. N.Z.J. agric. Res. 15, 48-53 (1972a).

25. Parle, J.N., di Menna, M.E.: Fungicides and the control of Pithomyces chartarum. II. Field Trials. N.Z.J. agric. Res. 15, 54-63 (1972b).

26. Parle, J.N., di Menna, M.E., Neville, F.J.: An examination of factors affecting the field control of Pithomyces chartarum by fungicides. In press.

27. Ross, D.J.: A study of the physiology of Sporidesmium bakeri Syd. I. Nutrition. N.Z.J. Sci. 3, 15-25 (1960).

28. Ross, D.J.: A study of the physiology of Pithomyces chartarum (Berk. & Curt.) M.B. Ellis. 4. The influence of temperature and period of incubation on production of sporidesmin on a potato-carrot extract medium. N.Z.J. Sci. 5, 246-252 (1962).

29. Ross, D.J., Thornton, R.H.: A study of the physiology of Pithomyces chartarum (Berk. & Curt.) M.B. Ellis. 3. Production of toxin on some laboratory media. N.Z.J. Sci. 5, 165-183 (1962).

30. Sinclair, D.P., Parle, J.N.: Thiabendazole and facial eczema. Proc. Ruakura Farm. Conf. 160-169 (1967).

31. Thornton, R.H., Percival, J.C.: A hepatotoxin from Sporidesmium bakeri capable of producing facial eczema diseases in sheep. Nature 183, 63 (1959).

32. Thornton, R.H., Taylor, W.B.: Anti-fungal activity of fatty acids to Pithomyces chartarum (Berk.& Curt.) M.B. Ellis in field trials. N.Z.J. agric. Res. 6, 329-342 (1963).

33. Stutzenberger, F.J., Parle, J.N.: Effect of 2-substituted benzimidazoles on the fungus Pithomyces chartarum. J. gen. Microbiol. 76, 197-209 (1973).

34. Synge, R.L.M., White, E.P.: Sporidesmin: A substance from Sporidesmium bakeri capable of producing facial eczema diseases in sheep. Chem. Ind. 1546-1547 (1959).

35. Ueno, K., Giam, C.S., Taber, W.A.: Absence of sporidesmin in a Texas isolate of Pithomyces chartarum. Mycologia 66, 360-362 (1974).

# Balloon Print Replica Plating, a New Technique for Isolating and Enumerating Microorganisms on Skin

R. MICHITSCH, G. STOTZKY, W.D. ROSENZWEIG, and P. DELLA-LATTA

## A. Introduction

Several methods exist for quantifying bacteria on skin surfaces (8). In addition to the contact plate (6) and the plastic tape (7) procedures, skin can be sampled with velvet pads, which are either pressed onto agar plates (2) or rinsed leading to greater bacterial recovery than with the imprint method (4). A gauze-capillarity technique (1) and a moist swab culture method (3) have been used to sample from burn wounds. Using either dilution plating or direct transfer of microorganisms from the skin to an agar plate, slow growing organisms and those present in low numbers may go undetected.

A method using a balloon as a sampling tool (5) for detecting slow growing bacteria in the presence of fast growing bacteria on plant surfaces enabled a broader range of organisms to be detected, as inflating the balloon after sampling and before replicating to agar increased the distance between microorganisms and provided those that grew slowly a better chance for survival and of developing a colony.

This balloon print technique has been employed in the quantitative and qualitative sampling of microorganisms on several different surface sites of the human body, and the results have been compared to sampling by contact plates.

## B. Materials and Methods

A contract plate (60 mm diam) was pressed against a specific body site, and, while being appressed, the circumference of the contact plate was traced on the site. After sampling with the contact plate, an uninflated, autoclaved balloon, with the circumference of a contract plate previously traced onto it, was pressed against the same delineated site, removed, and inflated so that the area that had made contact with the skin was equal to the area of a 100 mm diam Petri dish. The balloon was then rolled on to a 100 mm plate, being careful to keep the center of the balloon in a fixed position to avoid multiple inoculations. The procedure was repeated with a second balloon, which was inflated to equal the area of a 150 mm diam Petri dish. There was a 15 min interim between each sampling. The time of contact on the skin for balloons and contact plate was 4 sec. All plates were incubated for 48 h at 37±2 C.

Nine body sites on 5 individuals (3 males and 2 females) were sampled, and replicate inoculations made onto Bacto-Tryptic Soy agar (TSA) and to TSA containing 5% sheep blood (kindly provided by Scott Labs). The contact plates and Petri dishes were filled to the rim with agar. The sampling sequence was: contact plate, small balloon, large balloon.

After incubation, colonies on the plates were counted and scored on morphological characteristics (color, size, shape, texture). Two representatives of each colonial type, on each plate, were Gram stained. Gram-negative bacilli were subcultured on Eosin-Methylene Blue agar, and the resultant colonies scored on morphological appearance. Gram-positive bacilli were scored on colonial morphology on the initial isolation plates. Gram-positive cocci were subcultured on Mannitol Salt (MS) agar. Those growing on MS agar were designated as either mannitol-fermenting staphylococci or non-mannitol-fermenting staphylococci. Gram-positive cocci that did not grow on MS agar were scored on their colonial appearance on TSA or blood agar. Fungal colonies were scored as being either mold or yeast.

Preliminary studies were conducted to determine the affinity of rubber balloons for microorganisms. A 60 mm diam contact plate containing TSA, with one-half covered by sterile gauze, was pressed against the chest area, and an uninflated, sterile balloon was then pressed on the other half of the area previously untouched by the contact plate. The uninflated balloon was then replicated to a TSA plate. After incubation for 48 h at 37±2 C, both contact and balloon print plates were scored for total colonies.

To be certain that the balloons were not contaminated during expansion, 3 autoclaved balloons were inflated and rolled onto Petri dishes containing TSA.

All handling of balloons and plates was with sterile gloves to minimize contamination.

## C. Results

In the preliminary studies, an average of 27 colonies was enumerated from contact plates pressed against half of the sampling area on the chest, and an average of 25 colonies was transferred to TSA plates by the uninflated balloons pressed against the other half of the circular area. The rubber balloons, therefore, appeared to have an affinity for microorganisms on the skin equal to that of contact plates. Contamination was not a problem in this technique, as no growth developed on any of the plates replicated with inflated sterile balloons.

The balloon print technique gave higher total colony counts than did the contact plate method for all body sites tested(Fig.1). The large plates to which the balloons were replicated showed higher total colony counts per plate than did the small plates. An average of 23 colonies was isolated on the contact plates for all sites sampled,

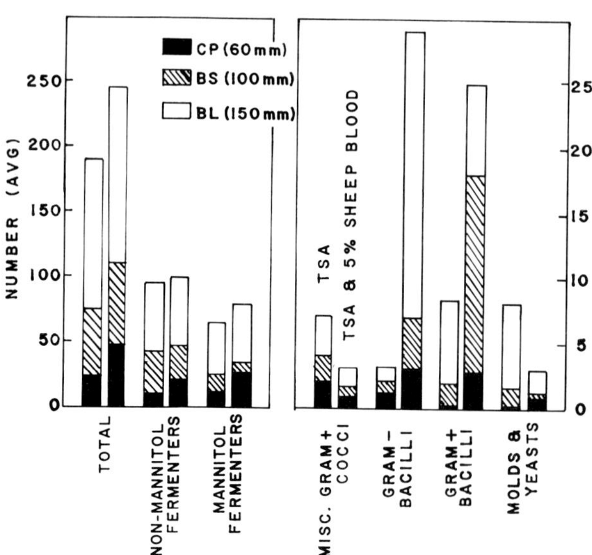


Fig. 1. Number of organisms enumerated from skin by the contact plate (CP) and by balloon replicas to small (BS) and large (BL) Petri plates. Average of 9 body sites on 5 individuals (3 males and 2 females). Media used were Tryptic Soy Agar (TSA) and TSA containing 5% sheep blood

whereas the average was 75 on the small plates and 187 on the large plates, using TSA as the isolation medium. When TSA containing sheep blood was the isolation medium, the average colony counts were 47 for the contact plates, 109 for the small plates, and 247 for the large plates.

In the 42 cases where the same type of organism (e.g., mannitol-fermenting staphylococci, Gram-negative bacilli) was isolated from a particular body site on all three plates, 34 of the 42 isolations (81.0%) showed an increase in total numbers on the small plates as compared to the contact plates, and 34 isolates also showed an increase on the large plates when compared to the small plates.

Twenty isolates present on small plates and 25 isolates present on large plates were not observed on the corresponding contact plates. Conversely, only 8 isolates (6 bacteria and 2 yeasts) not detected on either small or large plates were observed on the corresponding contact plates.

Sixty-two of the 78 total isolations from all body sites (79.5%) showed increases in colony counts on large plates over small plates, while 9 of the 78 isolates (11.5%) showed either a decrease in colony counts or the absence of a specific isolate. Nine percent of the 78 isolates showed no change in counts between the small and large plates.

## D. Discussion

The small plates to which the balloons were replicated had an average of approximately 3.2 (on TSA) and 2.3 times (on TSA plus blood) more colonies than detected on the contact plates for all sites sampled. The large plates had a further average increase of 2.5 (on TSA) and 2.3 times (on TSA plus blood) the total colony numbers noted on the small plates. Inasmuch as the number of colonies replicated by uninflated balloons was similar to the numbers detected by contact plates from the same sample sites, the balloons did not have a greater or lesser affinity for microorganisms than did the contact plates. Multiple inoculations were probably also not responsible for the results, as care was taken to avoid this while rolling the expanded balloon onto the agar plate.

Disruption of microcolonies, transferred from the skin to the balloon, by inflation of the balloon might have been responsible, in some cases, for the increases in total numbers of colonies per plate with the balloon print technique. However, the increase in the types of microbes isolated with the balloon print replica plating method cannot be explained by disruption of microcolonies. Expansion of the balloons, which provided a greater distance between microorganisms on the replica plates, probably was responsible for the increase in both numbers and types of organisms isolated with the balloon technique. As the distance between microorganisms was increased, antagonism (e.g., competition, amensalism) between neighbouring microorganisms was probably decreased or eliminated and slow growing organisms were not overgrown by faster growing organisms. By alleviating these effects, organisms normally incapable of developing into colonies on standard contact plates were detected.

The numbers of organisms enumerated on both the small and large plates probably underestimated the total numbers capable of being detected, as the order of sampling from a specific site was purposely contact plate, small plate, and large plate. With each successive sampling of an area, the total number of organisms remaining on the skin probably decreased. However, the results consistently showed that the greater the area on which the microorganisms were cultured after sampling, the greater was the number and variety recovered.

The higher colony counts obtained on TSA supplemented with 5% sheep blood than on TSA alone were probably because fastidious organisms were able to grow on the enriched medium.

The results of these studies indicate that the balloon print replica plating technique provides an inexpensive, simple, and accurate method for enumerating microorganisms on skin. Furthermore, the flexibility of the balloon enables sampling skin on uneven areas of the body. This technique should be useful in obtaining a better understanding of the activity, ecology, and population dynamics of microorganisms on skin.

## E. References


1. Brentano, L., Gravens, D.J.: A method for the quantitation of bacteria in burn wounds. Appl. Microbiol. 15, 670-671 (1967).
2. Holt, R.J.: Aerobic bacterial counts on human skin after bathing. J. Med. Microbiol. 4, 319-327 (1971).
3. Georgiade, N.G., Lucas, M.C., O'Fallon, W.M., Osterhout, S.: A comparison of methods for the quantitation of bacteria in burn wounds. I. Experimental evaluation. Am. J. Clin. Pathol. 53, 35-39 (1970).
4. Raahave, D.: New technique for quantitative bacteriological sampling of wounds by velvet pads: Clinical sampling trial. J. Clin. Microbiol. 2, 277-280 (1975).
5. Rusch, V., Leben, C.: Epiphytic microflora: The balloon print isolation technique. Can. J. Microbiol. 14, 486-487 (1968).
6. Ulrich, J.A.: Dynamics of bacterial skin population. In: Skin Bacteria and Their Role in Infection. Maibach, H.I., Hildick-Smith, G. (ed.). New York: McGraw-Hill, pp. 219-234 (1965).
7. Updegraff, D.M.: A cultural method of quantitatively studying the microorganisms in the skin. J. Invest. Dermatol. 43, 129-137 (1964).
8. Williamson, P.: Quantitative estimation of cutaneous bacteria. In: Skin Bacteria and Their Role in Infection. Maibach, H.I., Hildick-Smith, G. (ed.) New York: McGraw-Hill, pp. 3-11 (1965).

# Immunological Aspects of Environmental Contamination

N.ST.G. HYSLOP

In mammals, immunity results from an extremely complex series of interacting humoral, cell-mediated and "non-specific" processes. Additional modifications of basal immunity are caused by diet and stress both of which may be related to the environment, so that the environment often exerts a variety of important "nurturing" or "stressor" effects (6).

For more than 80 years, it has been recognized that adult human beings, indeed most animals, may possess "natural" antibodies against a wide range of pathogenic organisms, although they had remained free from the diseases caused by those organisms. The recently developed highly sensitive tests (particularly those involving observations of dynamic inhibition) reveal that such antibodies are often biochemically and biologically indistinguishable from the antibodies evoked by recovery from clinical disease or by vaccination.

Thus, when samples of the sera of 572 cattle, sheep, pigs and horses, which had been maintained wholly under extensive conditions of husbandry, were tested for antibodies against certain Salmonella species immediately before slaughter, no less than 380 samples proved to be positive. Table 1 shows that, although the animals had been maintained in the open air for their entire lives, no less than 63.0% of cattle, 58.7% of sheep, 76.6% of pigs and 57.0% of horses had low but significant levels of antibody. Cross-absorption tests on many of the sera which were active against Salm. dublin, Salm. typhimurium or Salm. cholerae-suis indicated a high degree of specificity against a single species. In other experiments, even greater proportions of apparently healthy animals have been found to possess antibodies against Salmonella spp. In many cases, however, the animals had been concentrated and housed for part of their lives The antibodies directed against viruses are equally indistinguishable, and Figure 1 demonstrates that, provided the same amounts of immunoglobulin are employed, the rates of neutralization of infectious bovine rhinotracheitis (IBR) virus by "natural" antibodies or by "convalescent" antibodies are identical.

A progressive increase in antibody gamma-globulins occurs during early life in both man and domestic animals, and then declines later as senile changes appear. Nevertheless, groups of healthy people show significant age-, sex- and race-associated variations in the mean titres of the serum immunoglobulins IgM, IgG and IgA. Individual variations within the groups may be mediated partly by physiological factors. Nevertheless, it may be possible to evaluate altered immunoglobulin concentrations in individuals by the application of statistical methods (2). Not all of the total serum immunoglobulin can be shown to possess antibody activity directed against any known pathogenic organism, though part may result from past experience of commensal species, allergenic substances of various types, and even dietary sources of antigen. It has been postulated however, that some degree of immunity may be acquired against diseases caused by certain encapsulated bacteria, at least partly as a result of contact with cross-reacting antigens of non-pathogenic bacteria.

Table 1. Proportion of serum samples from healthy animals, maintained under extensive conditions of husbandry, which contained 'H' or 'O' agglutinins to *Salmonella* species *

| Animal Species | No. Tested | Titre | Agglutinins detected against *Salmonella* species: | | | | | | |
|---|---|---|---|---|---|---|---|---|---|
| | | | *typhimurium* | *dublin* | *choleraesuis* | *newport* | *thompson* | *zanzibar* | *kiambu* |
| Cattle | 208 | 1/16 | 16.4 | 23.6 | 0 | 0.9 | 0 | 0 | 0.5 |
| | | 1/32 | 6.7 | 14.9 | 0 | 0 | 0 | 0 | 0 |
| Sheep | 160 | 1/16 | 31.8 | 15.0 | 1.9 | 6.9 | 0 | - | 0 |
| | | 1/32 | 2.5 | 0.6 | 0 | 0 | 0 | - | 0 |
| Pigs | 197 | 1/16 | 9.6 | 2.5 | 31.5 | 22.3 | 1.5 | 0 | 0 |
| | | 1/32 | 0 | 0 | 4.6 | 3.6 | 0 | 1.0 | 0 |
| Horses | 7 | 1/16 | 14.2 | 42.8 | 0 | 0 | 0 | - | - |

* - Samples collected from randomly selected animals at Athi River or Uplands Abbatoirs; Cultures of intestinal lymph nodes at time of slaughter were all negative.

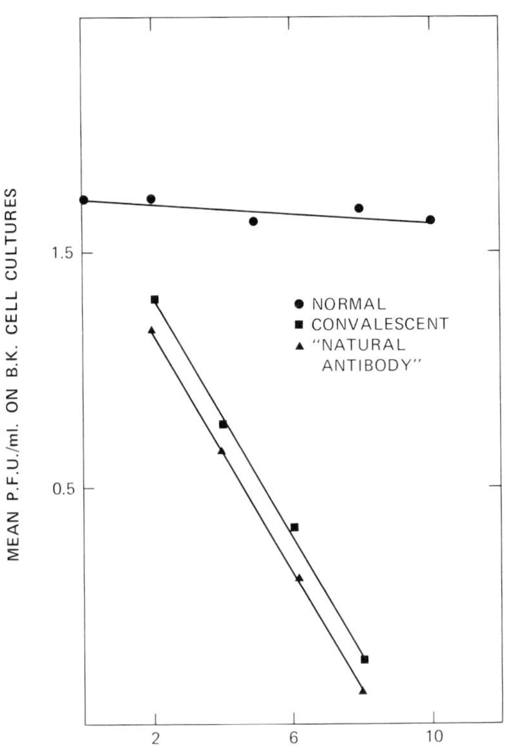


Fig. 1. Neutralization kinetics of IBR virus (Strain Okehampton 1) - by equivalent concentrations of pooled bovine IgG from normal steers, convalescent steers, or steers with "natural antibody"

In the immature subject, part of these circulating antibodies may be of maternal origin, but in adults their incidence must generally reflect the prevalence of the causal organism in the particular environment. Thus, our tests on sera and lymph nodes of healthy pigs, cattle, sheep and horses revealed that up to 70 per cent possessed antibody against Salmonella spp. Similarly, in man, a gradual increase occurs throughout life in antibody directed against a wide range of organisms, including Pseudomonas aeruginosa (4, 12). In widely separated populations and in different vocational groups, a high proportion of individuals may possess antibodies against Listeria monocytogenes (14), and Listeria can be isolated frequently from apparently healthy persons and animals (10, 7).

In no case is it necessary to assume that a clinically overt infection may have occurred when "natural" antibody is detectable, especially when it is recalled that experiments on oral immunization have revealed that antigenic components of living or dead organisms are capable of passing across the mucosa of the alimentary tract and evoking circulating antibody. It is of interest to note that the formation of antigen-antibody complexes in the intestinal mucosa may retard the passage of the homologous antigen, but may also impair the integrity of the mucosa (presumably by a mild inflammatory reaction) sufficiently to enhance the passage of heterologous antigens or macro-molecules (1). Stimulation of complex sequences, resulting in the production of circulating antibodies in varying amounts, also occurs after the inhalation of certain microorganisms or other antigens. In some species at least, the action of "natural" antibodies may be supplemented by virus-specific neutralizing substances of low molecular weight, which are quite distinct from the series of true antibodies found amongst the immunoglobulins of mammalian sera (13).

In addition to causing those conditions which arise as a result of the presence of true pathogens, continuing contamination of the environment with a wide range of pollens, bacteria, molds and other proteinaceous substances leads to widespread sensitization not only of human beings, but also of their domestic animals.

The vast scale on which microbial contamination of the environment may occur is well illustrated by the estimate that 50 million Salmonella organisms enter the river Rhine each second (11). Other rivers are probably equally contaminated with Salmonella and other organisms. These organisms are derived from man, other mammals, birds, amphibia and insects, and are present in food products, sewage, industrial effluents and groundwater. The carriage of microorganisms by airstreams is discussed elsewhere (8).

Clearly, heavy contamination of the environment may be of little immediate significance if virtually all members of the local host population possess either maternal antibodies or immunity arising from clinical or subclinical infection, and if stress factors are controlled adequately. The existence, however, of heavy "background" contamination may become vitally important if stresses increase, or if the population is constantly changing, as in barracks and in animal stockyards. Not only may the ostensibly healthy arrivals rapidly develop severe clinical disease after entering the infected environment, but they may also introduce new and invasive strains of organisms into a previously apparently resistant host population.

Alternatively, it has been shown that the introduction of low levels of infection into semi-immune populations may lead to the establishment of inapparent infection or aberrent clinical signs: atypical responses may occur even with foot and mouth disease (9,3) or with other diseases normally considered to be of high virulence. Furthermore, repeated passage in the presence of sub-inhibitory levels of antibody may result in the emergence of completely different strains of virus (5, 9).

A particularly important aspect of immune responses, acquired as a result of contamination of the environment, lies in their interference with diagnostic serological reactions. For example, certain fungal species contain substances which produce precipitin reactions similar to those of C-reactive protein of the completely unrelated pneumococcus. Cross-reactions between antisera and widely disparate types of bacteria may be caused apparently by fortuitous spatial similarities of protein fragments or, more frequently, by possession of common polysaccharide moieties. Thus, in the past, various investigators have shown that Salmonella serotypes may possess serological cross-reactivity with some species of Brucella, Pasteurella, Vibrio, Yersinia, Rickettsia and Klebsiella; Klebsiella serotypes may show similar cross-reactions with pneumococcus. Recently, further cross-reactions have been encountered between Salmonella species, including Salm. cholerae-suis, and some of the numerous species of the yeasts Candida and Torulopsis.

Whether immunological cross-reactivity may be regarded as a bane or a benefit remains a matter of opinion. Certainly, some diagnostic tests may be rendered difficult to interpret. Having regard, however, to the vast numbers of pathogenic organisms (and facultative or opportunistic pathogens), which spill into the biosphere from its ever-increasing human and other animal populations, one must be grateful for the opportunity to build up some degree of immunity (even if only a partial one), against a wide range of diseases, merely by encountering very small numbers of living, attenuated, dead or antigenically related microorganisms which are present in the general environment.

## A. Acknowledgement

The author is indebted to Dr. A. Ginsberg for advice and assistance with the observations recorded in Table 1.

## B. References

1. Brandtzaeg, P., Tolo, K.: Mucosal penetratability enhanced by serum derived antibodies. Nature, Lond. 266, 262-263 (1977).
2. Buckley, C.E., Dorsey, F.C.: Serum immunoglobulin levels throughout the life span of healthy man. Ann. int. Med. 75, 673-682 (1971).
3. Fagg, R.H., Hyslop, N. St. G.: Isolation of a varient strain of foot and mouth disease virus (Type O) during pąssage in partly immunized cattle. J. Hyg. Camb. 64, 397-404 (1966).
4. Gaines, S., Landy, M.J.: Prevalence of antibody to *Pseudomonas aeruginosa*. J. Bact. 69, 628-633 (1958).
5. Hyslop, N. St. G.: Isolation of variants from strains of foot-and-mouth disease virus propagated in cell cultures containing anti-viral sera. J. gen. Microbiol. 41 135-142 (1965).
6. Hyslop, N. St. G.: Productivity, disease and climate. Int. J. Biometeor. 17, 107-108 (1973).
7. Hyslop, N. St. G.: Epidemiologic and immunologic factors in Listeriosis. In: Problems of Listeriosis. Woodbine, M. (ed.) Leicester: Leicester University Press (1975), pp. 94-105.
8. Hyslop, N. St. G.: Observations on the survival of pathogens in water and air at ambient temperatures and relative humidity. These proceedings (1977)
9. Hyslop, N. St. G., Fagg, R.H.: Isolation of variants during passage of a strain of foot-and-mouth disease virus in partly immunized cattle. J. Hyg. Camb. 63, 357-368 (1965).
10. Kampelmacher, E.H., van NoorleJansen, L.M.: Further studies on the isolation of *L. monocytogenes* in clinically healthy individuals Zbl. Bakt. Hyg. I. Abt. Orig. A221, 70-77 (1972).
11. Kampelmacher, E.H., van NoorleJansen, L.M.: Comparative studies on the isolation of *Salmonella* from effluents. Zbl. Bakt. Hyg. I. Abt. Orig. B157, 71-77 (1973)
12. Kefalides, M.: The role of infection in mortality from severe burns. N. Eng. J. Med. 267, 317-323 (1962).
13. Levy, T.A., Ihle, J.N., Oleszko, O., Barnes, R.D.: Virus specific neutralization by a soluble non-immunoglobulin factor. Proc. Nat. Acad. Sci. U.S. A. 72, 5071-5075 (1975).
14. Osebold, J.W., Sawyer, M.T.: Agglutinating antibodies for *Listeria monocytogenes* in human serum. J. Bacteriol. 70, 350-351 (1955).

# Microorganisms and the Gastro Intestinal Tract

## Gastrointestinal Microecology: One Opinion

D.C. SAVAGE

### A. Introduction

Principles of ecology apply to the microbial inhabitants of the alimentary canal as surely as they do to trees in a forest (1). Nevertheless, for most animal types including humans, the gastrointestinal microbial ecosystem is not understood. Indeed, for most animal types, the composition of microbial communities inhabiting the gastrointestinal tract is described poorly or not at all. Moreover, little is known about the multivariant factors regulating the population levels and localization of the microbial communities. Even less is known about the nutrition, physiology and genetics of the microbes and their interactions with each other and their animal hosts (10).

Indigenous microbes in the gastrointestinal tract influence profoundly the lives of their hosts (4) and research in the field should be encouraged. In my opinion, however, such research is likely to be most successful in the hands of investigators who understand that microbes indigenous to the gastrointestinal ecosystem occupy particular habitats in that system, and function most actively and typically when present in those habitats. Such investigators cannot be satisfied to isolate a microorganism from a fecal sample and call it "indigenous" without determining whether or not it satisfies criteria for autochthony in the system (10). Moreover, such persons are reluctant to attempt to assess the impact on host physiology or immunology of a particular indigenous microbial species or group without first learning about the ecology of the organisms. If at all possible, such investigators strive to perform their experiments so as to learn how the microbes live and influence their host while residing in their native habitats. In support of these opinions, I review some pertinent literature and discuss some recent findings from my laboratory.

### B. The Ecosystem - Communities, Habitats and Niches

In animal species examined appropriately, communities of indigenous microorganisms may occupy habitats distributed vertically and horizontally in the gastrointestinal canal (10). The habitats can be distributed vertically by being located in any region of the tract from esophagus to anus. They can be distributed horizontally by being located in the lumen, on the epithelial surfaces or in the crypts in the same area of the tract (10). In any of these habitats, the microbes may associate intimately with, and even attach physically to, a surface (7, 9). In the lumen they may attach to particles of digesta (2). On the epithelial surface or in the crypts, they may attach to epithelial cells themselves (9). Such interface phenomena (6), undoubtedly are important in stabilizing the system (10).

The population levels of microbes occupying habitats in the large bowel may be enormous, exceeding $1 \times 10^{11}$ organism per g of material (5, 10). By contrast, the levels of the microbes in communities in habitats in the stomach or small intestine are usually $1 \times 10^{8-9}$ organisms per g or less (10). Because their population levels are so small, microbes occupying such distal habitats may not be detected at high population levels in feces. Thus, by comparison with colonic or cecal microbes, the organisms from habitats above the large bowel are often not regarded as important components of the gastrointestinal microbiota. Such a view may be erroneous. Even though their population levels may be small, microbes in the stomach and small intestine may be better located than are cecal and colonic microbes, to have an impact on the physiological, particularly the digestive, processes of their hosts. Thus, guesses about their importance in the system based upon their population levels in feces could be misleading.

More importantly, perhaps, any of the nutritional or biochemical activities of microbes native to habitats above the large bowel, if measured in feces, might be underestimated or even go undetected in the background of activity generated by the billions of microbes deriving from the large bowel. Thus, not only their population levels, but also the niches occupied by the organisms in their natural habitats might be misinterpreted in experiments involving fecal samples.

Similarly, experiments may be misleading when they involve tests with pure cultures of particular microbial types. Such experiments can never reveal the habitat, and composition of the community in the habitat, and at best can be only suggestive about the niches of the microbes *in vivo*. Thus, as with experimentation involving fecal samples, studies with pure cultures may be misleading. That is not to say, of course, that pure cultures should not be used in studies of the nutrition, physiology and genetics of indigenous microbes. It is to say, however, that information derived in such a way may reveal little about the behavior of the microbes occupying their habitats and niches in their natural hosts.

## C. Factors Regulating Population Levels and Localization of Microbial Communities

As noted, the factors dictating where indigenous microbes localize and multiply in the gastrointestinal tract are at best poorly understood (10). Such factors can be environmental, or derive from activities of the microbes in modifying their own environment. Whatever their origin, the factors dictate where communities of particular microbial species localize during succession in infants. In addition, they undoubtedly stabilize the composition of the climax community during adulthood. The factors may never be understood well unless the microbes are studied in some fashion while occupying their native habitats and niches.

## D. A Case in Point

We have studied some properties of an indigenous microbe growing both *in vitro* in culture media and in its native habitat. Results from experiments with the microbe growing *in vivo* have often surprised us when our hypotheses were based upon findings from experiments with cells grown *in vitro*. The microbe in the experiments is an indigenous yeast, *Torulopsis pintolopesii*, that colonizes the epithelium of the glandular portion of the stomachs of adult mice and rats raised under conventional conditions (8). During succession of the gastrointestinal microbiota, *T. pintolopesii* colonizes its habitat only after the infants are weaned (8). It is usually not present in specific pathogen-free mice from certain suppliers. Presumably, it is never present in the environment of such animals as long as they are maintained under conditions that keep them free of their own pathogens. *T. pintolopesii* is not a pathogen for rodents, however, and does not harm them even when injected parenterally in large numbers by various routes (Savage, unpublished observations).

Isolated in pure culture, T. pintolopesii can grow either aerobically or anaerobically. Growing aerobically in media used in our laboratory, the cells contain cytochrome B, can be shown to have a KCN-sensitive system for taking-up $O_2$ and divide with an average doubling time of about 98 min. Growing anaerobically, the cells have no KCN-sensitive system for assimilating $O_2$, produce ethanol as a main end-product and require about 164 min. to divide (Artwohl, Energy-yielding metabolism of indigenous yeast found in the stomach of mice. Thesis for the M.A. degree, University of Illinois, 1977). Thus, the microbe undoubtedly derives energy, when growing aerobically, mainly from respiratory metabolism with $O_2$ as terminal electron acceptor, and when growing anaerobically, from fermentative metabolism. This concept was reinforced when we isolated from agar medium inoculated with the organism, a petite mutant strain that behaved when growing aerobically just as did the parental wild-type strain growing anaerobically.

Since the microbe is able to grow in vitro at a faster rate in an atmosphere containing $O_2$ than in an anaerobic atmosphere, we predicted that the native environment of the microbe (on the surface of gastric secreting mucosa) must contain some oxygen. We tested that hypothesis in two ways.

First, we monoassociated numerous germfree mice with the microbe, which, of course, colonized the gastric epithelium. Then we harvested the cells of the microbe from that surface and attempted to detect in them KCN-sensitive $O_2$-uptake. The cells could not be shown to have such a system (Artwohl, 1977). In addition, we monoassociated germfree mice with either the petite mutant or the parental wild-type strain by inserting a known amount of yeast cells in the stomach. At various

Table 1. Yeast Populations in Stomachs of Gnotobiotic Mice Monoassociated with T. pintolopesii strains [a]

| Time (Hr) | Parent[b] | Petite[b] | Control[c] |
|---|---|---|---|
| 24 | $4 \times 10^4$ [d] | $6 \times 10^4$ | NC |
| 48 | $4 \times 10^5$ | $8 \times 10^5$ | NC |
| 72 | $3 \times 10^6$ | $4 \times 10^6$ | NC |
| 120 | $1 \times 10^7$ | $3 \times 10^7$ | NC |
| 240 | $2 \times 10^7$ | $2 \times 10^7$ | NC |

[a] From: Artwohl, J.E., & Savage, D.C. Ecological determinants in microbial colonization of the murine gastrointestinal tract: Carbon and nitrogen nutrition and energy-yielding metabolism of Torulopsis pintolopesii. In preparation (1977).

[b] Germfree mice were monoassociated with either a wild-type strain (parent) or a spontaneous mutant with petite colonial form and fermentative metabolism (unable to utilize oxygen as terminal electron acceptor) derived from the facultative wild-type strain.

[c] Germfree mice.

[d] Number of viable yeast/g of wet tissue.

times thereafter, for several days, we sacrificed animals and estimated the population levels of viable yeast cells in their stomachs (Table 1). Both the petite mutant and wild-type strains colonized the stomach at essentially the same rates and to the same final population levels. The mutant did not revert to the wild-type in the stomach for the full period of the experiment.

We concluded from these findings that the microbe need not utilize $O_2$ to occupy its native niche. We could not conclude, however, in spite of our expectations, that $O_2$ was present in the environment. We do not yet understand, therefore, how its ability to respire aerobically *in vitro* is advantageous to *T. pintolopesii* in its native niche.

Similarly, we had some surprises when we began to study the mechanisms by which the microbe remains in close association with the epithelium of the gastric mucosa (9). In preparations of such mucosa examined in the scanning electron microscope, we could see that the organism appears to attach physically to the epithelium (Fig. 1). Moreover, we knew that it could colonize the keratinized, stratified squamous epithelium of the nonsecreting portion of the stomach in mice and rats treated with antibiotics to remove the indigenous lactic acid bacteria that normally colonize that surface (11). Lactobacilli are known to attach specifically to such cornified epithelium, possibly via acidic mucopolysaccharides (3). We speculated that the yeast attaches physically to the gastric epithelia, and does so by a mechanism similar to that mediating attachment of the lactic acid bacteria.

To test these hypotheses, we developed a system for assaying the numbers of yeast cells attaching to mucosal tissues exposed to the cells *in vitro*. To our surprise, in that assay system the microbe attached to tissue removed from any region of the murine gastrointestinal tract (Table 2). By contrast, it colonized and formed visible layers only on the epithelia of the secreting and nonsecreting

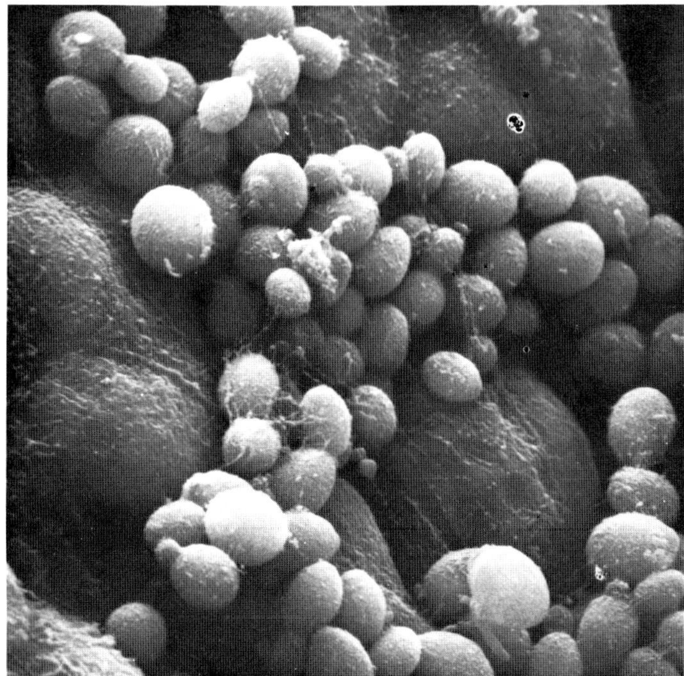

Fig. 1. Scanning electron micrograph showing *T. pintolopesii* on the surface of the secreting gastric epithelium in a monoassociated gnotobiotic mouse. X 2,295

Table 2. Attachment *in vitro* of *T. pintolopesii* to Epithelia from the Digestive Tracts of Mice [a]

| Area of Tract | SPF[b] | GF[b] |
|---|---|---|
| Stomach - nonsecreting | 0.6[c] | 2.0 |
| - secreting | 1.0 | 5.0 |
| Small Intestine - upper | 0.5 | 2.5 |
| - middle | 2.6 | 2.5 |
| - lower | 2.4 | 1.5 |
| Large Intestine - cecum | 0.6 | 4.5 |
| - colon | 0.8 | 2.5 |

[a] From: Suegara, N., & Savage, D.C. Ecological determinants in microbial colonization of the murine gastrointestinal tract: Adherence of *Torulopsis pintolopesii* to gastric epithelia. In preparation (1977).

[b] SPF: specific pathogen-free mice; GF: germfree mice.

[c] Number of yeast cells (x $10^{-5}$)/mg wet tissue. The yeast cells were labeled with $^3$H-methionine. Weighed pieces of mucosa were exposed to the labeled cells, and then washed vigorously and digested. The number of yeast cells adhering to the tissue was assessed by liquid scintillation spectrometry.

Table 3. Layers of *T. Pintolopesii* on Epithelial Surfaces in the Digestive Tracts of Associated Mice[a]

| Area | SPF[b] | GF[b] |
|---|---|---|
| Stomach - nonsecreting | 0/5[c] | 5/5 |
| - secreting | 5/5 | 5/5 |
| Small Intestine - upper | 0/5 | 0/5 |
| - middle | 0/5 | 0/5 |
| - lower | 0/5 | 0/5 |
| Large Intestine - cecum | 0/5 | 0/5 |
| - colon | 0/5 | 0/5 |

[a] From: Suegara, N., & Savage, D.C. Ecological determinants in microbial colonization of the murine gastrointestinal tract: Adherence of *Torulopsis pintolopesii* to gastric epithelia. In preparation (1977).

[b] SPF: Specific pathogen-free mice; GF: Germfree mice.

[c] Number of animals with layers/number examined. Yeast layers on epithelial surfaces in Gram-stained histological sections of mucosa were detected by light microscopy.

portions of the stomachs of monoassociated gnotobiotic (ex-germfree) mice and the secreting portion of the stomachs of SPF mice (Table 3). SPF mice of the type used are normally free of the yeast, but have indigenous lactobacilli colonizing the nonsecreting epithelium of their stomachs. Thus, the yeast could attach in vitro to any gastrointestinal epithelium, but could colonize in vivo only the surfaces of its native habitat.

We concluded from these findings that the ability of the microbe to attach to epithelia is only one factor influencing its ability to colonize its habitat. We might also have concluded, however, that in observing the microbe attaching to epithelia from all areas of the tract in vitro, we are observing an artifact that is telling us nothing of the way the microbe functions in vivo. These and other such considerations made from findings from studies involving indigenous microbes other than T. pintolopesii have made us endeavor to examine properties of all microbes of the gastrointestinal microbiota as they occupy their native habitats. We are convinced that such microbes function most actively and typically, and more readily reveal their ecology, when studied under such conditions than when studied in any other way.

## E. Acknowledgements

I am grateful to the National Institute of Allergy and Infectious Diseases (AI 11858) and the National Dairy Council for supporting my research.

## F. References

1. Alexander, M. Microbial Ecology, pp. 3-93. New York, 1971, Wiley.
2. Akin, D.E. Ultrastructure of rumen bacterial attachment to forage cell walls. Appl. Environ. Microbiol. 31, 562-568 (1976).
3. Fuller, R. Nature of the determinant responsible for the adhesion of lactobacilli to chicken crop epithelial cells. J. Gen. Microbiol. 87, 245-250 (1975).
4. Gordon, H.A., & Pesti, L. The gnotobiotic animal as a tool in the study of host microbial relationships. Bacteriol. Rev. 35, 390-429 (1971).
5. Holdeman, L.V., Good, I.J., & Moore, W.E.C. Human fecal flora: variations in bacterial composition within individuals and a possible effect of emotional stress. Appl. Environ. Microbiol. 31, 359-375 (1976).
6. Marshall, H.C. Interfaces in Microbial Ecology, pp. 1-156, Cambridge, MA and London, 1976, Harvard Univ. Press.
7. Savage, D.C. Associations and physiological interactions of indigenous microorganisms and gastrointestinal epithelia. Amer. J. Clin. Nutr. 25, 1372-1379 (1972).
8. Savage, D.C. & Dubos, R.J. Localization of indigenous yeast in the murine stomach. J. Bacteriol. 94, 1811-1816 (1967).
9. Savage, D.C. Indigenous microorganisms associating with mucosal epithelia in the gastrointestinal ecosystem. In: Microbiology-1975 (ed. D. Schlessinger), pp. 120-123. Washington, D.C., 1975, American Society for Microbiology.
10. Savage, D.C. Microbial ecology of the gastrointestinal tract. Ann. Rev. Microbiol. 31, 107-133 (1977).
11. Savage, D.C. Microbial interference between indigenous yeast and lactobacilli in the rodent stomach. J. Bacteriol. 98, 1278-1283 (1969).

# Quantitation of Autochthonous Bacteria in Rat Ileum by Scanning Electron Microscopy and Transect Line Analysis

C.D. GARLAND, A.E. STARK, A. LEE, and M.R. DICKSON

## A. Introduction

Within the gastro-intestinal tract is a complex ecosystem in which the epithelial surfaces represent the primary interface between the host and numerous components of the gut lumen. Of particular interest are the observations of dense bacterial populations specifically associated with this interface. It seems important to gain an understanding of the spatial distribution of these bacteria on the epithelial surface, since it may be related to the role(s) they play in the gut ecosystem.

Consequently we devised a method for quantitating epithelium-associated bacteria in situ. A population of filamentous bacteria colonizing the villous epithelium in normal rat ileum was chosen in order to test this quantitative method. Chase and Erlandsen (1) have described the distinct morphology of these organisms but all attempts to culture them have failed. To observe these bacteria in their villous habitat scanning electron microscopy (SEM) was used, and to count them rigorous criteria based on transect line analysis were adopted.

## B. Materials and Methods

I. Preparation of tissue for SEM. Young adult rats of either sex were anaesthetized with ether and killed by cervical dislocation. A specific area of the ileum, 2.5 to 3.5 cm proximal to the ileo-caecal junction, was removed and washed vigorously three times in saline. This area was cut into four pieces (sites) and fixed in Karnovsky's fixative pH 7.2 (2) containing 0.05% ruthenium red (3) for 3 hrs. at room temperature. Tissue specimens were dehydrated through an alcohol and acetone series, critical point dried, mounted on stubs with silver paint and then coated with gold. They were examined in a Cambridge S4-10 Stereoscan (SEM) at 10KV.

II. Transect Line Analysis. A grid system of 23 parallel transect lines, each 5 mm apart, was drawn on a clear acetate sheet (Fig. 1A). The image of the epithelial surface of a randomly selected villus was obtained on the SEM screen (Fig. 1B). The transect line grid was placed over the screen (Fig. 1C) and the display area of the screen was reduced to produce transect lines of 40 μm (Fig. 1D) at constant magnification (750X). Five transect lines were randomly selected and the number of intersections the filamentous bacteria made per selected transect line was counted. A representative sample of the bacterial population was obtained by counting 5 transects per villus (villus count), 5 villi per site (site count), 4 parallel sites in the ileum (animal count), and the ilea of 15 animals.

## C. Results and Discussion

I. Analysis of transect line data. The frequency distribution of transect line counts obtained at different levels is shown in Fig. 2.Statistical analysis of the data provided the following components of variance (expressed per transect line):

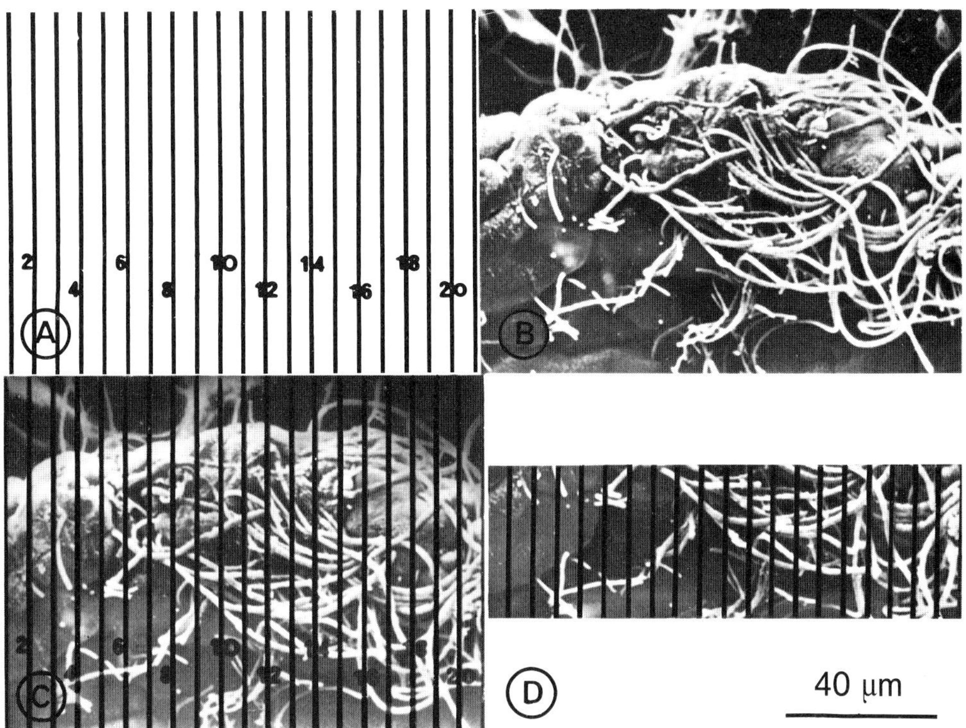


Fig. 1 A-D. Transect line analysis. (A) Grid system, (B) epithelial surface, (C) grid overlaid on epithelial surface, (D) reduced display screen area (transect lines ≡ 40 µm)

$\sigma_t^2$ = 7.850 (transect to transect variance): $\sigma_v^2$ = 1.118 (villus to villus variance): $\sigma_s^2$ = 0.742 (site to site variance): $\sigma_r^2$ = 0.394 (rat to rat variance): $\mu$ = 3.75 (mean).

The largest component of variance occurred at the transect to transect level, as reflected in the pronounced skew of the transect frequency distribution (Fig. 2A). The smallest component of variance occurred at the rat to rat level, as reflected in the near normal distribution of the animal frequency distribution (Fig. 2D).

II. Distribution of bacterial population. At the site to site level, there was no significant difference between site counts (F test, $p>0.05$). This result indicates that the organisms are fairly uniformly distributed within the four sites of the ileum examined, even though the variability at the villus to villus and transect to transect level was marked. There was a significant difference between animal counts (F test, $0.05>p>0.01$), indicating that some animal to animal variation does occur, as might be expected.

III. Accuracy of *in situ* quantitation. In total, 1500 transect lines on 300 villi in 60 sites in the ilea of 15 animals were counted. Analysis of the variance data derived from this sample produced a standard error of the mean of 0.219, approximately 6% of the mean (3.75). In view of the accuracy of the standard error of the mean, it is suggested that the *in situ* quantitative method is both rigorous

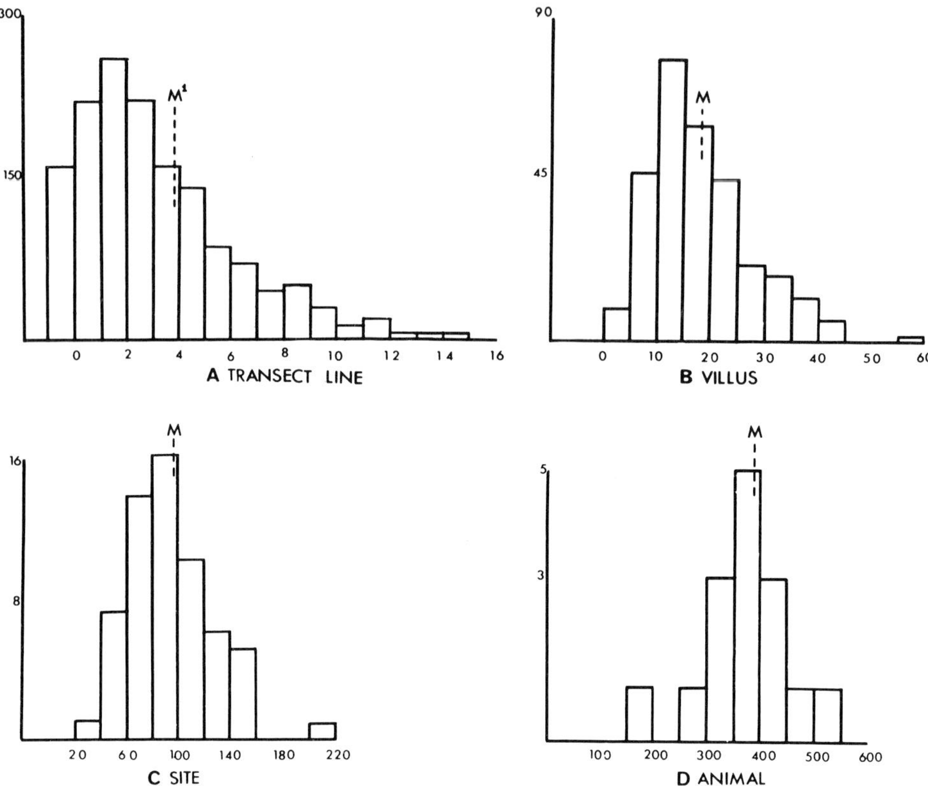


Fig. 2 A-D. Frequency distribution of transect counts at different levels ($M^1$ = mean)

and reproducible. It should be added that the sampling technique is arbitrary and can be devised to suit particular experimental conditions.

D. Applications

I. Stress factors. Since the variability of the bacterial population in our model has been quantitated, the effect of stress factors on the numerical stability of that population can be assessed. Stress factors which might include diet, starvation, age, antibiotics, or infection, can be assessed. It has been postulated that layers of epithelium-associated bacteria promote resistance to infection by physically blocking pathogens from access to the tissue surface of the host (4). Using our method, we can visualize and quantitate *in situ* the role filaments may play in preventing infection, and also quantitate the role of attachment and proliferation in the pathogenesis of the infectious organism.

II. Epithelium associated bacteria. The *in situ* quantitative technique would be useful for examining other bacterial populations which colonize epithelial surfaces, especially those bacteria which have not as yet been cultured.

III. Preparative methods for EM. A wide range of EM preparative methods has been reported, and a considerable lack of standardization exists. At present, by using this quantitative method as a basis for comparison, we are assessing commonly used aldehyde fixatives, with and without heavy metal stains. Preliminary results indicate that fixative composition profoundly influences the preservation of high numbers of bacteria associated with epithelium.

## E. References

1. Chase, D.G. & Erlandsen, S.L.: Evidence for a Complex Life Cycle and Endospore Formation in the Attached, Filamentous, Segmented Bacterium from Murine Ileum. J. Bacteriol. 127: 572-583 (1976).
2. Karnovsky, M.J.: A Formaldehyde-Glutaraldehyde Fixative of High Osmolality for use in Electron Microscopy. J. Cell. Biol. 27: 137A-138A (1965).
3. Luft, J.H.: Ruthenium Red and Violet. II. Fine Structural Localization in Tissues. Anat. Rec. 171: 369-416 (1971).
4. Savage, D.C. and McAllister, J.S.: Microbial Interactions at Body Surfaces and Resistance to Infectious Disease. In: Resistance to Infectious Disease. Dunlop, R.H. & Moon, H.W. (ed.). Saskatoon Modern Press, pp. 113-127 (1970).

# Fluorescence Microscopy of Gut Microbes

R.T.J. CLARKE and G.E. NAYLOR

## A. Introduction

New methods are needed for the identification and enumeration of microbes in their specific niches in the gut and for quantification of their activities. This paper describes preliminary studies on rumen ciliate protozoa and bacteria using acridine orange and immunofluorescence.

## B. Materials and Methods

Protozoa were stained by adding an equal volume of 0.006% (w/v) aqueous acridine orange to rumen contents. Pieces of plant material were immersed in a solution of acridine orange (0.1%, w/v, in 0.1M phosphate buffer, pH8) for 10 min, rinsed in phosphate buffered saline (1), blotted to remove excess liquid, and mounted in immersion oil. Fluorescent conjugates were prepared with fluorescein isothiocyanate from antisera produced in rabbits (1). Rhodamine Bovine Albumin (BBL) was used as counterstain.

## C. Results and Discussion

### I. Rumen Protozoa

The identification in rumen contents of some ciliate protozoa, particularly entodinia, is difficult because of their small size and lack of easily discernible internal structure and because they are easily obscured by plant material. In *Entodinium* spp. differentiation is based on size, shape, and position of the two nuclei. Addition of acridine orange (AO) to rumen contents and examination with blue (420 nm) light renders ciliates easily visible among plant fragments and stains the nuclei.

### II. Bacteria on Surfaces

1. *Acridine orange*. A major limiting factor in studying bacterial colonisation of food particles or epithelium in the gut has been the inability to see bacteria clearly by light microscopy. The use of AO (2) overcomes this difficulty. By use of a microscope equipped with transmitted and incident illumination and blue and white light, bacteria on plant fragments and on gut epithelium are readily visible at magnifications up to 1200X. Even fragments of tissue too fragile for SEM (scanning electron microscopy) can be examined satisfactorily. The following observations were made by studying plant material removed from the rumen and lower gut of sheep.

a) Colonisation of dried and fresh feed material was rapid, large lumps of bacteria adhering within 20 min.

b) Entodiniomorph protozoa of several genera also adhered rapidly to dried and fresh feedstuffs.

c) The cuticle of clover and grass leaves became heavily colonised, especially by large rumen bacteria, (*Oscillospira*, *Lampropedia*, and ovals) none of which have been grown in pure culture. The concentrations of Oscillospira and Lampropedia on leaves were considerably higher than in the rumen fluid. The surface of green material in the rumen may constitute a specific niche for these bacteria.

d) Where damage to the plant was sufficient, bacteria rapidly invaded cells and structural elements of phloem and xylem. Some cells soon appeared filled with bacteria and many contained bacteria of only one morphological type. Digestion of the cell walls proceeded from within the cell.

e) Fragments of tissue were still heavily colonised when they reached the abomasum. Plant material in the large intestine and caecum was also heavily colonised.

2. Immunofluorescence. During development of a technique for identifying and counting "free" bacteria in rumen contents, known numbers of *Streptococcus bovis* ($10^6$ - $10^9$, estimated by direct count) were suspended in saline, rumen fluid (RF) clarified by centrifugation, unmodified RF, or human small intestine fluid (IF). The streptococci in these preparations were counted on black membrane filters after staining with specific and control conjugates and with a counterstain. Counts of added cells were within the range normally obtained with direct counts: saline, 88-118%; clear RF, 113%; RF, 88-118%; IF, 87-100%. Thus it should be possible to count most bacteria in gut fluid in spite of the large amounts of protein present.

Immunofluorescent techniques were applied to bacteria on clover leaf removed from rumen, or damaged and (a) suspended in the rumen in a nylon bag, or (b) incubated with cultures of bacteria. With a conjugated serum highly specific for *Lachnospira multiparus*, this pectin-degrading rumen bacterium was shown to adhere to the adaxial surface of clover leaves in the hallows between the epidermal cells. Long mono-filaments of cells were seen lying along the cell boundaries beneath the cuticle after staining with AO, and were identified with a specific antiserum. The observations were made with leaf suspended in cultures of *Lachnospira*. Studies are being made with material from the rumen, and with conjugates for other fibre-degrading bacteria.

## D. References

1. Cherry, W.B.: Fluorescent-antibody techniques. In: Manual of Clinical Microbiology. Blair, J.E., Lennette, E.H., & Truant, J.P. (eds). Bethesda: American Society for Microbiology, pp. 693-704 (1970).

2. Francisco, D.E., Mah, R.A. & Rabin, A.C.: Acridine orange-epifluorescence technique for counting bacteria in natural waters. Trans. Amer. Micros. Soc. 92, 416-421 (1973).

# Microbial Flora of Australian Termites and Their Significance

M.L. EUTICK, R.W. O'BRIEN, and M. SLAYTOR

## A. Introduction

Most studies of the bacterial flora of the termite gut have centred around attempts to isolate cellulose-degrading, or nitrogen-fixing organisms, with few attempts to examine the whole gut population (3). We have surveyed the gut bacteria of nine species of Australian termites and have studied the effects of antibiotics on the gut flora and the life span of *Nasutitermes exitiosus* hoping to gain an insight into the symbiosis between termites and their bacteria.

## B. Materials and Methods

I. *Termites*. The termites used are listed in Table 2.

II. *Isolation of gut bacteria*. The termites were surface sterilized (7) and the gut of five excised under sterile conditions, homogenized in Krebs-Ringer buffer (pH 7.4), serially diluted and spread on reinforced clostridial medium plates (Oxoid). The plates were incubated under anaerobic and aerobic conditions and the bacterial colonies counted. Gut contents were also spread on N-deficient agar plates (glucose as the carbon source) (5) and incubated under nitrogen, and inoculated into cellulose media (1). All incubations were at 30$^{o}$C.

III. *Morphological and biochemical characteristics of gut bacteria*. All tests were carried out at 30$^{o}$C according to Skerman (8), except for glucose oxidation and fermentation (2) and gluconate utilization (6).

IV. *Treatment of N. exitiosus with antibiotics, acid fuchsin and oxygen*. Sterile wood impregnated with antibiotics (Table 3), pimaricin, and acid fuchsin was placed in a series of micronests (4) containing 90 workers and 10 soldiers. Micronests (containing termites and untreated wood) were exposed to oxygen at atmospheric pressure for 2 h. At intervals one nest of each series was broken up to: (a) count the live termites; (b) count the gut bacteria; and (c) examine the gut microscopically for live spirochetes.

## C. Results

The bacteria isolated from the gut of the termites were identified as *Enterobacter*, *Staphylococcus*, *Streptococcus*, *Bacillus* and *Flavobacterium* (Table 1). Their distribution among the termites is shown in Table 2. The data in Table 1 presents only those tests which are characteristic of the different genera and not all of the tests which were carried out. *Streptococcus* was further characterized on the basis of forming chains of cells and being homofermentative; *Staphylococcus* on the basis of forming packets of cells, and *Flavobacterium* on the basis of forming water-insoluble yellow pigments. Cellulose-degrading, or strictly anaerobic organisms were not detected. All the *Enterobacter* strains grew on N-deficient medium under nitrogen in the presence of sulfuric acid. *N. exitiosus* and *N. walkeri* did not yield any organisms capable of growth on N-deficient medium. The major bacterium of each termite was present in numbers approximating 1 to 2 x $10^7$ cells per ml of gut, with the minor species falling to as low as $10^5$ cells per ml of gut.

Table 1. Characteristics of termite gut bacteria

| Characteristic | [a] group | | | | |
|---|---|---|---|---|---|
| | 1 | 2 | 3 | 4 | 5 |
| Gram reaction | [b] - | [b] + | + | + | - |
| Endospores | - | - | - | + | - |
| Motility | + | - | - | + | + |
| Growth on citrate | + | - | - | - | - |
| gluconate | + | - | [b] +/- | +/- | + |
| malonate | + | - | - | +/- | + |
| cyanide broth | + | - | - | +/- | + |
| Methyl red | - | +/- | + | +/- | - |
| Voges-Proskauer | + | +/- | - | - | - |
| Catalase | + | + | - | + | + |
| Benzidine | + | + | - | + | + |
| Kovac's oxidase | - | - | +/- | + | - |
| Nitrate reduction | + | - | - | +/- | + |
| Acid and gas from: | | | | | |
| glucose, anaerobic | [c] AG | A | A | A/- | - |
| glucose, aerobic | AG | A | A | A/- | - |

[a] Group 1, Enterobacter; 2, Staphylococcus; 3, Streptococcus; 4, Bacillus; 5, Flavobacterium.

[b] Minus sign, negative reaction or no growth; plus sign positive reaction or growth; +/-, some strains positive, some negative.

[c] A, acid; G, gas.

The effects of antibiotics, the fungicide pimaricin, acid fuchsin and oxygen on the gut flora and the life span of N. exitiosus are shown in Table 3. Antibiotics which killed both the bacteria and the spirochetes (ampicillin, tetracycline) reduced the life span of the termite from about 250 to 13 days. Death of the whole colony rapidly followed death of half of the colony. Metronidazole and oxygen, which killed only the spirochetes, also reduced the life span to 13-22 days. The gut bacteria were unaffected by these treatments and their concentration remained in the range observed with untreated termites (1-2.5 x $10^5$ per gut). In treatments where the bacteria and spirochetes were unaffected (penicillin, pimaricin, acid fuchsin) the life span of the termite was also unaffected. Control experiments indicated that the antibiotics were not affecting the termites themselves.

Table 2. Distribution of bacteria among termite species

| Termite Family | Species | Group[a] 1 | 2 | 3 | 4 | 5 |
|---|---|---|---|---|---|---|
| Mastotermitidae | Mastotermes darwiniensis | [b]+ | | [b]+++ | | + |
| Kalotermitidae | Cryptotermes primus | + | | +++ | | |
| Rhinotermitidae | Heterotermes ferox | +++ | | + | | |
| | Coptotermes acinaciformis | +++ | | | + | |
| | Coptotermes lacteus | +++ | | + | | |
| | Schedorhinotermes intermedius intermedius | +++ | | + | + | |
| Termitidae | Nasutitermes exitiosus | | +++ | | | |
| | Nasutitermes gravelolus | + | +++ | | | |
| | Nasutitermes walkeri | | +++ | | | |

[a] Group 1, Enterobacter; 2, Staphylococcus; 3, Streptococcus; 4, Bacillus; 5, Flavobacterium.

[b] +++, major gut bacterium; +, minor gut bacterium.

Table 3. Effect of antibiotics, pimaricin, acid fuchsin and oxygen on the gut flora and life span of N. exitiosus

| Antibiotic | Flora surviving | Life span of N. exitiosus (days) 50% mortality | 100% mortality |
|---|---|---|---|
| None | [a]B,S,F | - | 250 |
| Penicillin | B,S,F | > 180 | >200 |
| Ampicillin | F | 11 | 13 |
| Tetracycline | none | 11 | 13 |
| Metronidazole | B F | 11 | 13 |
| Oxygen | B F | 14 | 22 |
| Pimaricin | B,S | - | 275 |
| Acid fuchsin | B,S,F | - | 210 |

[a] B, bacteria; S, spirochetes; F, fungi.

## D. Discussion

A correlation appears to exist between the major gut bacterium and the family to which the termite belongs (Table 2). Thus Streptococcus was the major bacterium of the two lowest families (Mastotermitidae and Kalotermitidae), whereas Enterobacter predominated in the Rhinotermitidae, and Staphylococcus in the Termitidae. Putative nitrogen-fixing Enterobacter spp. were isolated from seven of the termites and thus a role seems to exist for these bacteria; the role of the other bacteria was not discovered. The flora of C. acinaciformis seems to be constant, irrespective of its geographical location, since this species collected from three widely separated areas always had Enterobacter as the major bacterium.

Feeding of N. exitiosus (a higher termite devoid of protozoa) with metronidazole, or subjecting it to pure oxygen, demonstrated that this species is dependent for survival on its gut spirochetes and that the other bacteria are unable to sustain it. Failure of pimaricin to shorten the life span of N. exitiosus showed that the gut fungi are not essential. The spirochetes were not affected by acid fuchsin.

## E. Acknowledgments

We thank Dr. J.A.L. Watson, CSIRO Division of Entomology, Canberra, for providing and identifying termites; Dr. L.I. Sly, Department of Microbiology, University of Queensland, for assistance in identifying bacteria. M.L.E. was the recipient of an Australian Government Postgraduate Studentship.

## F. References

1. Aaronson, S.: Experimental Microbial Ecology. New York-London: Academic, p. 80 (1970).
2. Baird-Parker, A.C.: Methods for classifying staphylococci and micrococci. In: Identification Methods for Microbiologists. Gibbs, B.M. & Skinner, F.A. (ed.). London-New York: Academic, pp. 59-64 (1966).
3. Breznak, J.A.: Symbiotic relationships between termites and their intestinal microbiota. In: Society for Experimental Biology Symposium XXIX. Jennings, D.H. & Lee, D.L. (ed.). Cambridge: University Press, pp. 559-580 (1975).
4. Eutick, M.L., O'Brien, R.W., Slaytor, M.: Aerobic state of gut of Nasutitermes exitiosus and Coptotermes lacteus, high and low caste termites. J. Insect Physiol. 22, 1377-1380 (1976).
5. Hino, S., Wilson, P.W.: Nitrogen fixation by a facultative bacillus. J. Bacteriol. 75, 403-408 (1958).
6. Holding, A.J., Collee, J.G.: Routine biochemical tests. In: Methods in Microbiology Vol. 6A. Norris, J.R. & Ribbons, D.W. (ed.). London-New York: Academic, pp. 2-32 (1971).
7. Mannesmann, R.: A comparison between cellulolytic bacteria of the termites Coptotermes formosanus Shiraki and Reticulitermes virginicus (Banks.) Int. Biodeterior. Bull. 8, 104-111.
8. Skerman, V.B.D.: A Guide to the Identification of the Genera of Bacteria, 2nd ed. Baltimore: Williams and Wilkins. (1967).

# Variations in the Intestinal Microflora of Salmonid Fishes

T. J. TRUST, J. I. MacINNES, and K. H. BARTLETT

## A. Introduction

There has been a long-standing controversy over the nature of the gastrointestinal microflora of fish. Some workers have reported significant bacterial numbers in feral (or free-living) fish (5), whereas others have failed to demonstrate any viable bacteria (1, 2).

This study was undertaken to clarify this apparent conflict. Both feral fish and fish grown under defined culture conditions were examined. The use of this control group provided a simple model free from the many undefined environmental variables associated with the feral fish.

## B. Materials and Methods

Feral fish were obtained from 30 different locations in British Columbia. The fish examined included Kokanee salmon, brook, golden, cutthroat, and rainbow trout. The control fish were housed at 10-13 C in hemispherical plastic tanks supplied with sterile well water at 200 ml $min^{-1}$. Oxygen levels were 7-9 mg $l^{-1}$. Tank water was exchanged every 4 h with a central siphon. The average bacterial load of the tank water was $5 \times 10^3$ viable mesophilic aerobes $ml^{-1}$. Fish were fed twice daily with dry pelleted diets which generally contained an average of $5 \times 10^3$ aerobic and facultative mesophilic bacteria $g^{-1}$ dry weight. Fish were sampled before the morning feeding.

Aseptic surgery was employed for the removal of gastrointestinal tracts. The tract was either examined in its entirety or divided into 3 sections: stomach and pyloric caeca, midgut, and hindgut. In some cases lumen contents and tissue were examined separately. After removal of the lumen contents tissues were washed 3 times in isotonic saline to remove any adherent matter. The qualitative and quantitative examinations were similar to those previously described (4). Isolates identified as *Aeromonas* species were more fully characterized by their reactions to 32 biochemical tests and their sensitivities to 22 antibacterials. Aeromonads were grouped by cluster analysis (3).

## C. Results and Discussion

The results (Table 1) demonstrate the significant bacterial populations in the gastrointestinal tract of feral salmonids. All feral fish (including spawners) had countable populations. Typically the number of bacteria increased from the anterior to the posterior regions of the tract. The magnitude of the populations found in these fish indicates that bacterial multiplication occurs in the digestive tract. In the midgut and hindgut, most of the facultative anaerobes were in the lumen contents.

In the controlled environment fish, it was not always possible to demonstrate countable bacteria. Over a 24 month period, 340 of the 400 fish examined (10.5 -

Table 1. Bacterial numbers in the alimentary tract of salmonids

| Samples | Fish | Incubation[a] | No. of samples | No. of viable bacteria $g^{-1}$ (wet weight) Average | Range |
|---|---|---|---|---|---|
| Total tract + contents | feral | $O_2$[b] | 22 | $1.9 \times 10^7$ | $2.9 \times 10^3 - 1.2 \times 10^8$ |
| | control | An[c] | 22 | $1.4 \times 10^7$ | $1.6 \times 10^3 - 1.5 \times 10^8$ |
| Stomach and pyloric caeca + contents | feral | $O_2$ | 69 | $4.5 \times 10^5$ | $1.0 \times 10^2 - 2.5 \times 10^6$ |
| | control | $O_2$ | 25 | $1.2 \times 10^3$ | TFC[d] $- 7.0 \times 10^3$ |
| Midgut + contents | feral | $O_2$ | 68 | $2.2 \times 10^7$ | $1.0 \times 10^2 - 2.5 \times 10^8$ |
| | control | $O_2$ | 25 | $5.2 \times 10^4$ | TFC $- 3.4 \times 10^5$ |
| Midgut tissue | control | $O_2$ | 26 | 240 | TFC $- 4.6 \times 10^3$ |
| Hindgut + contents | feral | $O_2$ | 69 | $9.2 \times 10^7$ | $8.0 \times 10^2 - 9.0 \times 10^8$ |
| | control | $O_2$ | 26 | $3.3 \times 10^5$ | TFC $- 1.3 \times 10^6$ |
| Hindgut tissue | control | $O_2$ | 26 | 240 | TFC $- 4.1 \times 10^3$ |

a Trypticase Soy agar (BBL), 30 C, 48 h.
b $O_2$ is aerobic.
c An is anaerobic.
d TFC is too few to count.

950 g) did not yield countable bacteria with the culture techniques employed. It would appear that salmonid fishes are able to exist in a non-sterile environment without a nutritionally nonfastidious gut microflora. This is in contrast to the complex microflora that is always associated with the gut of homeotherms.

A wide variety of bacterial species were isolated from the feral fish whereas the control fish contained a very limited flora (Table 1). The predominant bacteria isolated from both groups belonged to the *Aeromonas hydrophila* complex. In many of the control fish these were the only bacteria present in countable numbers. Cluster analysis of these aeromonads revealed two groups: the first was characterized by the ability to hydrolyze esculin, grow at 37 C, ferment sorbose and sorbitol, but was unable to ferment cellobiose, and gave a negative methyl red reaction.

These aeromonads were most readily isolated from the hindgut; of a total of 91 isolations, 54 were from the hindgut, 22 were from the midgut and 15 were from the stomach-pyloric caeca. In the midgut and hindgut, *A. hydrophila* was most readily isolated from the lumen; of a total of 112 isolations, 99 were from the lumen contents, while 13 were from the washed mucus-mucosal tissue section. It should be noted that while a large number of the diets fed to the control group have been sampled, *A. hydrophila* has not been isolated. Apparently *A. hydrophila*

Table 2. Frequency of isolation of bacterial strains from the alimentary tract of salmonids

| Genus | Feral[a] (%) | Control[b] (%) |
|---|---|---|
| Achromobacter/Alcaligenes | 2 | |
| Acinetobacter | 12 | 15 |
| Aeromonas | 13 | 72 |
| Bacillus | 1 | |
| Coryneforms | 1 | |
| Clostridium | 1 | |
| Enterobacter | 36 | |
| Escherichia | 1 | |
| Micrococcus | 6 | |
| Pseudomonas | 21 | 13 |
| Serratia | 1 | |
| Streptococcus | 1 | |
| Vibrio | 3 | |
| Xanthomonas | 1 | |

[a] 670 isolates.
[b] 458 isolates.

is selected by the fish, but other species, e.g. Enterobacter, which was isolated only from feral fish, simply reflect the environment of the fish.

The commensal population in the salmonid gut is markedly different from that found in homeotherms. The salmonid appears to lack the large numbers of non-fastidious strict anaerobes commonly isolated from homeotherms. The controlled environment salmonids did not yield such anaerobes even after extensive examination using conventional anaerobic jar techniques and an anaerobic cabinet with a wide variety of prereduced culture media. Microscopic examination of the mucus layer in the gastrointestinal tract did not reveal the large population of bacteria seen in similar preparations from homeotherms. However, small numbers of spiral and spirochaetal organisms (Fig. 1) were demonstrated in sealed saline wet mounts of mucus-mucosal scrapings after room temperature incubation for 4 to 8 h. These unusual bacteria were seen in the pyloric caecal region of both the control trout and marine salmon, but have not been cultured. Such organisms have not been previously described in fish, and appear to be attached to or associated with mucosal tissue cells, and may represent intracellular forms.

The study has provided the following model for the freshwater salmonid gastrointestinal microflora. While feral fish normally contain significant bacterial numbers in their gastrointestinal tract, the salmonid can mature over a considerable weight span without a gastrointestinal microflora. When present, the microflora consists of aerobes and facultative anaerobes, many of which reflect the environment of the fish. The predominant species apparently selected by the fish are of the Aeromonas hydrophila complex. The bacterial numbers increase along the tract, and are associated with the lumen contents. The large numbers of readily cultured anaerobes associated with the homeothermic gut are absent.

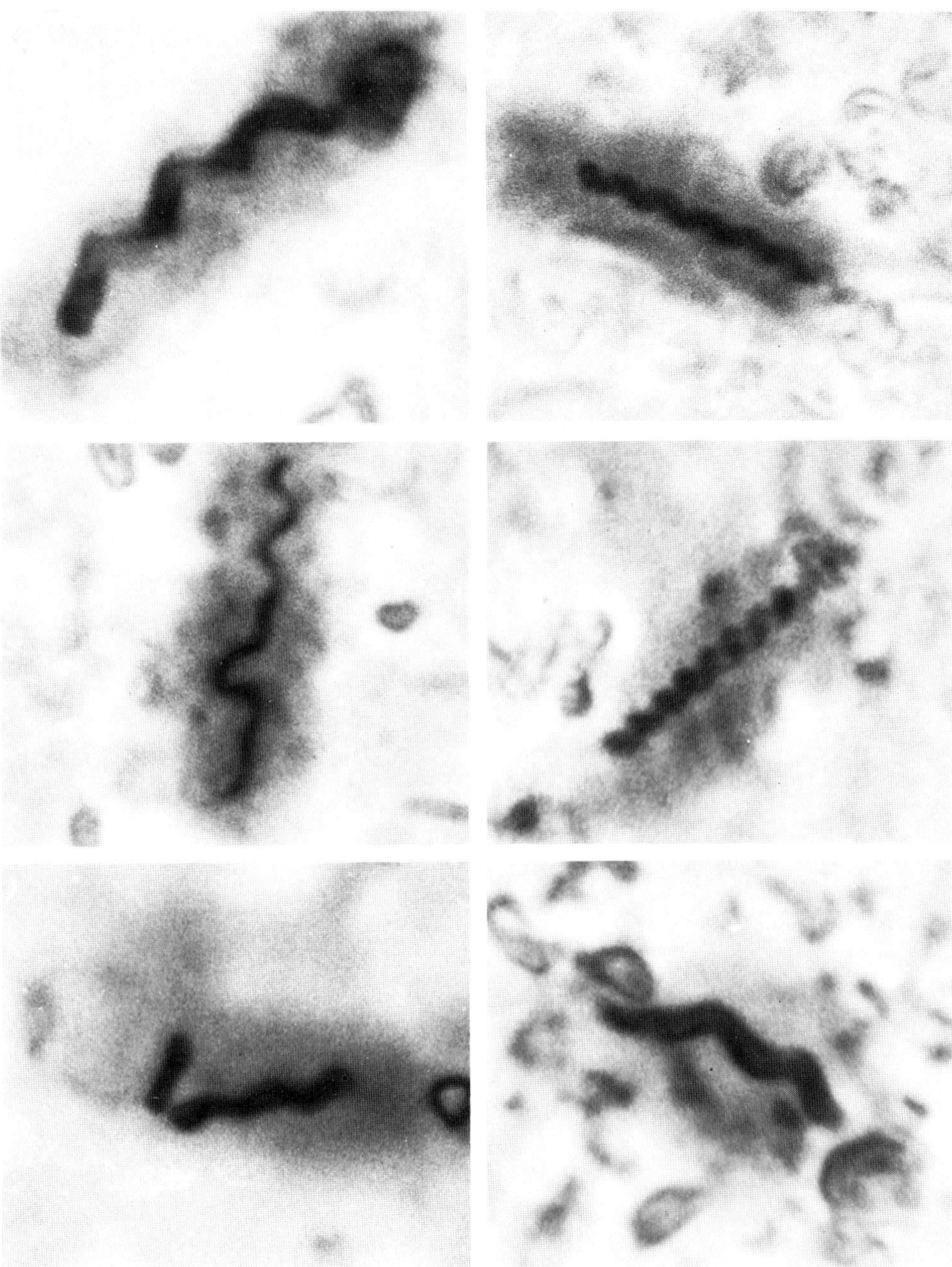

Fig. 1. Spiral and spirochaetal organisms in sealed saline wet mounts of intestinal mucus-mucosal scrapings from trout. X 10,000

## D. References

1. Blake, I.: Some observations on the bacterial flora of the alimentary tract of certain Salmonidae. In: Final Report of the Furunculosis Committee. App. B.H.M. Stationary Office, Edinburgh, pp. 60-67.
2. Margolis, L.: The effect of fasting on the bacterial flora of the intestine of fish. J. Fish. Res. Board Can. 10, 62-63 (1953).
3. Sneath, P.H.A. & Sokal, R.R.: Numerical taxonomy. San Francisco: W.H. Freeman and Co., 1973, p.226.
4. Trust, T.J.: Bacteria associated with the gills of salmonid fishes in freshwater. J. appl. Bact. 38, 225-233 (1975).
5. Yoshimizu, M. & Kimura, T.: Study on the intestinal microflora of salmonids. Fish Pathology 10, 243-259 (1976).

# Colony Incompatibility Among Strains of *Salmonella*

K. A. BETTELHEIM

## A. Introduction

Although the incompatibility reaction between strains of *Proteus*, known as the Dienes phenomenon, has been described as long ago as 1946 (4), and extensively reviewed and discussed since then (3,5,6), it has only recently been demonstrated that a similar reaction occurs among other strains of Enterobacteriaceae (1). As only *Proteus* strains but not other organisms swarm on normally constituted agar, the concentration of agar must be reduced in order to demonstrate the incompatibility reaction. This simple procedure will permit a study of this phenomenon among bacterial groups other than *Proteus*. Strains of *Salmonella* can be grouped by a variety of methods and often a number of these are used simultaneously. This study reports on a preliminary investigation using this technique on strains of *Salmonella* isolated during the last few months in New Zealand.

## B. Materials and Methods

I. Bacterial strains. As a reference strain in a number of studies *Escherichia coli* K12 was used. All other strains were recent isolates of *Salmonella* from patients or food in New Zealand. They were all serotyped and phage-typed by standard methods at the National Health Institute, Wellington.

II. Cultural methods. The medium used contained Bacto Tryptone (Difco) 1%, sodium chloride 0.5%, and Bacto agar (Difco) 0.35%. In Petri dishes the medium must not be inverted during incubation. The agar surface was dried by leaving it at $4^{o}C$ for two to three days. Strains in which full motility had been induced by passage through semi-solid medium were inoculated on either side of a plate. In all studies controls of identical strains inoculated at both sides were included. The dishes were incubated the right way up in a $37^{o}C$ incubator for four days and examined after three and four days, under oblique illumination.

## C. Results

Figure 1 shows a typical reaction of two identical strains merging with one another; sometimes small gaps are formed in the centre where presumably the medium was depleted before growth reached that area. Similar effects were also found when some pairs of different strains were tested. Nevertheless other pairs of strains when tested against each other showed varying degrees of incompatibility (Figs. 2-4).

Two strains of *Salm. typhimurium* were compared. Strain A (Phage-type 1) was isolated in Hamilton and Strain B (Phage-type 179) in Auckland. The results of testing them against each other in all combinations in both H phases is given in Table 1. The strong incompatibility reaction observed is also shown in Figure 4.

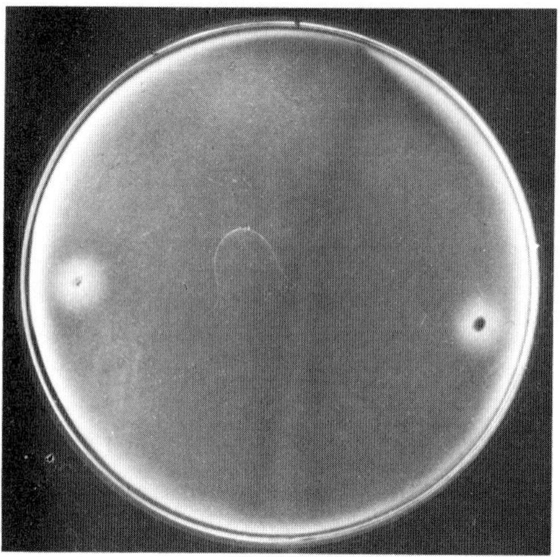

Fig. 1

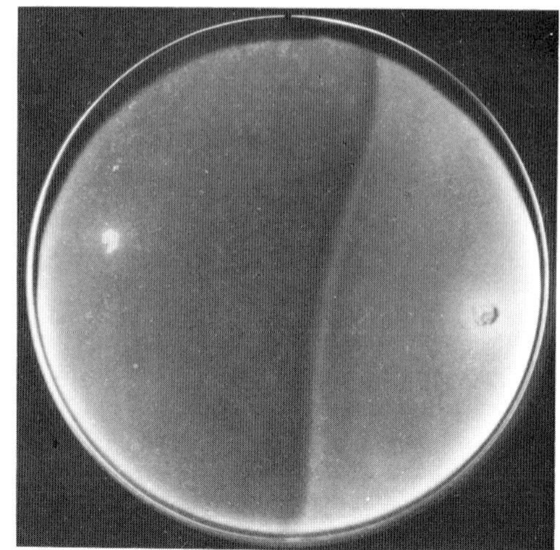

Fig. 2

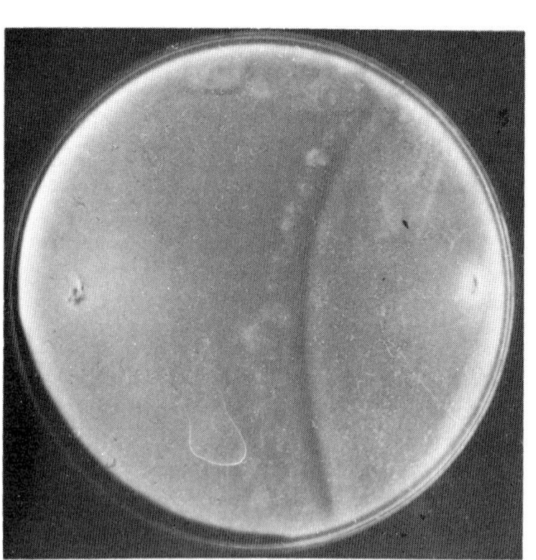

Fig. 3

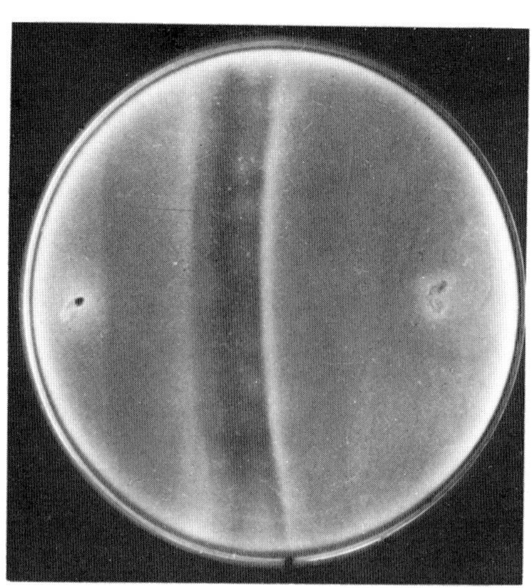

Fig. 4

Fig. 1. Lack of interaction when two strains of *E. coli* K12 are inoculated at either end of a Petri dish containing soft agar

Fig. 2. Slight interaction between *E. coli* K12 and a strain of *Salm. typhimurium*, indicated by a bunching up where they meet

Fig. 3. Stronger interaction with demarcation gap formed between a strain of *Salm. anatum* and a strain of *Salm. typhimurium*

Fig. 4. Very strong interaction between two strains of *Salm. typhimurium*, phage-types 1 and 179

Table 1. Interaction between two strains of *Salmonella typhimurium*, phage-types 1 and 179

| Strain | Phage - type 1 | | Phage - type 179 | |
|---|---|---|---|---|
| | Phase I | Phase II | Phase I | Phase II |
| Phage-type 1 | | | | |
| Phase I | No reaction | No reaction | Strong reaction | Strong reaction |
| Phase II | | No reaction | Strong reaction | Strong reaction |
| Phage-type 179 | | | | |
| Phase I | | | No reaction | No reaction |
| Phase II | | | | No reaction |

## D. Discussion

This preliminary study is intended to show the type of incompatibility reactions that can be observed. There is at present no adequate explanation for the Dienes phenomenon among strains of *Proteus*. It may be the type of reaction commonly associated with Fungi, particularly as it has recently been demonstrated that deoxyribonuclease is produced in the demarcation line (2).

Whether the incompatibility observed among the two strains of *Salm. typhimurium* is related to their carriage of different phages is being investigated by determining the patterns of incompatibility between different phage-types. This may also help to determine the ecological significance of different phage-types, and may determine which strains will survive and which will not in a given habitat.

## E. Acknowledgements

The photography of Mr. G. Bruggemans of the Clinical Photography Department of Wellington Hospital is gratefully acknowledged.

Published with the authority of the Director-General of Health, Department of Health, Wellington, New Zealand.

## F. References

1. Bettelheim, K.A., Carlile, M.J.: Colony incompatibility in bacteria. Nature, London., 264, No. 5588, 757-758, (1976).
2. Chambers, T.J.: Deoxyribonuclease release by incompatible strains of Proteus. Zbl. Bakt. Hyg. I Orig., 233, 228-231, (1975).
3. Coetzee, J.N.: Genetics of the Proteus group. Ann. Rev. Microbiol., 26, 23-54. (1972).
4. Dienes, L.: Reproductive processes in *Proteus* cultures. Proc. Soc. exp. Biol. Med., 63, 256-270, (1946).
5. Sourek, J.: On some findings concerning Dienes phenomena in swarming *Proteus* strains. Zbl. Bakt. Hyg. I Orig., 208, 419-427, (1968).
6. Sturdza, S.A.: A propose de l'utilisation du phenomene de Dienes (ligne de demarcation des *Proteus*) en practique bacteriologique. Arch. Roum. Path. exp. Microbiol., 28, 503-518, (1969).

# Gut Microbiology

R.E. HUNGATE

This is a vast subject for a brief presentation. Many facets of gut microbiology cannot be touched upon, and limited space prevents acknowledgment of the many investigators whose work is the basis for the generalizations presented.

## A. Models of Mammalian Gut Ecosystems

Mammals can be classed as (a), those using foods, usually of animal origin, readily digested by the mammal, and consisting largely of protein, fat and carbohydrate (often referred to jointly as concentrates), and (b), those using foods of plant origin, containing relatively small amounts of protein, fats and digestible carbohydrates, but with a much larger proportion of plant fiber undigestible by the mammal, with cellulose and hemicellulose as its chief components.

When foods consist chiefly of concentrates, digestible by both mammal and microbe, the nutritional relationship is clearly one of competition. The microbe, with its rapid growth, would seem to have advantage over the animal, but two features of mammalian digestion hold microbial development in check (a), high acidity in the stomach is lethal to most microbes, and (b), after alkaline secretions in the duodenum restore a favorable pH, plug flow along the intestine carries the digesta to the site of digestion and absorption by host enzymes before the gut microbes have a chance to develop. The importance of plug flow can be seen from the exceptional case of the protozoan, Giardia, which can attach to the intestinal epithelium and resist removal by flow. It can become exceedingly numerous and deleteriously affect the mammal by absorbing nutrients from the small intestine.

The competition model is presumably characteristic of the mammalian orders Carnivora, Pinnipedia, Insectivora, Chiroptera, carnivorous marsupials and termitophagous orders, though not all have been studied in this regard. All these would be expected to have a highly acid first stomach in which the microbes are eliminated, allowing host enzymes to digest the various concentrate foods and absorb the sugars, amino acids and peptones, with little loss to the microbes.

A cooperative relationship predominates in herbivores whose food contains abundant plant material high in carbohydrates indigestible by mammalian enzymes. The microbial fermentation chamber precedes the acid stomach. It is kept at a near neutral pH, and is populated by myriads of microbes growing in the absence of oxygen and fermenting carbohydrates to form new cells and give off $CO_2$, $H_2$ or $CH_4$, and short-chain volatile fatty acids as waste products. The acids are absorbed and oxidized by the mammal. This model is best represented in ruminants. Their fermentation chamber, the rumen, is offset from the direct line of food passage. But the model fits also the herbivorous marsupials, the leaf-eating monkeys, and possibly some other mammals in which the fermentation chamber is not offset but is an enlarged and partially separated proximal portion of the stomach, leading into the acid pyloric end, but somewhat separated by a constriction.

The caecal model is a combination of competition and cooperation. Mammals with a well developed caecum have an acid stomach in which the bacteria are killed.

Host enzymes break down the concentrates, which are absorbed, and the materials indigestible by the host are fermented in the enlarged caecum and upper colon. The microbial cell bodies cannot be digested because they are formed distal to the host enzymes, and in this respect the caecal model appears less effective than that of the ruminant. The lagomorphs and many of the rodents overcome this handicap by consuming their feces (coprophagy), thus salvaging as food the microbial cell bodies as well as the fermentation acids.

There is at least some caecal and colonic microbial development in all mammals, even in carnivores, with undigested food residues, sloughed epithelial cells and mucoid secretions as substrates. When concentrates and relatively digestible plant materials are both consumed, as in omnivores, the caecum and colon provide more fermentation acids than are produced in carnivores, and as will be discussed later, this may be important in human colonic disease.

## B. Microbial Interrelationships

Rumen studies have disclosed innumerable interrelationships between individual microbial species, and it is highly probable that similar phenomena occur to varying extents in the other nutritional models. Nearly all these interrelationships have a nutritional basis, one species giving off materials used by other species. Two types of materials can be distinguished, (a), primary substrate formed outside the cell and (b) waste products formed within the cell during growth and maintenance.

I. Primary substrate leakage. Insoluble plant carbohydrates such as cellulose and hemicellulose are digested by extracellular enzymes elaborated by some but not all gut microbes. The fiber digesters constitute only ca. 5% of the total microflora, even though the digestible fiber is about half of the available substrate. A large fraction of the soluble products of fiber digestion are captured by non-fiber digesters.

This phenomenon is well illustrated by the difference between cellulose digestion in liquid cultures and in cooled melted agar roll tubes inoculated with parallel serial dilutions of rumen contents. Little cellulose digestion occurs in lower dilution tubes of agar medium; the non-fibrolytic bacteria are so abundant that they absorb all sugars produced, and the fibrolytic colony never reaches visible size. Only in higher dilutions, with fewer non-fibrolytic bacteria, can small fibrolytic colonies be detected by the clearing of cellulose around them. A rough estimate suggest that the fibrolytic bacteria, about 5% of the total, may support up to six times their number of non-fiber digesters.

II. Waste product utilization. 1. Protein metabolism. Proteins are rapidly degraded in the rumen, and ammonia is formed, along with the $CO_2$, $H_2$ and various acids of the carbohydrate fermentation. In addition, isovaleric, isobutyric, 2-methylbutyric and phenylacetic acid are formed, and are used by other species for amino acid synthesis. Most of the rumen bacteria can use ammonia, and many require it. This is the basis for the common practice of feeding urea as a substitute for about one-third of the ruminant nitrogen requirement.

Why cannot urea supply all the required nitrogen? This is a challenging problem. In theory it seems possible to devise a ration of carbohydrate, ammonia and balanced minerals that would support a rumen fermentation of sufficient magnitude to feed the host. This goal has not been reached, though animals have been maintained on such a diet without growth or increase in weight. Lack of organic growth factors may explain the slowed fermentation. The nature of the influence of growth factors on the overall fermentation rate is important from both theoretical and practical standpoints. Hopefully, future research on cooperative mammal-microbe ecosystems may

Table 1. Types of Mammal-Microbe Nutritional Interrelationships

| Model | Examples | Characteristic Food | Microbiota | Type of gas | Microbes digested | Acids oxidized for host energy | Coprophagy |
|---|---|---|---|---|---|---|---|
| Competition | Carnivora Pinnepedia insectivores | chiefly animal protein carboh. and fats little fiber | bacteria in colon | $H_2$ or $CH_4$ | no | very little | rare |
| | omnivores humans | chiefly animal and conc. plant foods little fiber | bacteria in colon | $H_2$ or $CH_4$ | no | very little | no |
| Cooperation offset ferm. chamber | ruminants tylopods hippopotamus | plant bodies including fiber | bacteria and protozoa | $CH_4$ | yes | yes | no |
| in-line ferm. chamber | herbivorous marsupials | plant bodies including fiber | bacteria and protozoa | $H_2$ or $CH_4$ | yes | yes | no |
| | leaf-eating monkeys | plant bodies with some fiber | bacteria | $H_2$ or $CH_4$ | yes | yes | no |
| Caecal model (competition & cooperation) | Perissodactyla Proposcidia | plant bodies with concentrated plant foods and fiber | bacteria and protozoa | $CH_4$ | no | yes | no |
| | Rodentia lagomorphs | plant bodies with concentrated plant foods and fiber | bacteria, a few proto-zoa | $H_2$ or $CH_4$ | yes | yes | yes |

disclose an explanation for the reduced carbohydrate fermentation in the absence of organic growth factors including proteins, so that feeding practices can more fully exploit the microbial potential for de novo synthesis of the total host protein and energy needs.

2. Carbohydrate metabolism. In all normal gut ecosystems acetic acid, $CO_2$, and $H_2$ or $CH_4$, are major fermentation products from carbohydrates, with propionic and butyric acids also formed. The methane and hydrogen are the normal metabolic sinks for disposal of electrons removed in the anaerobic oxidation of triose. Ethanol and lactate, because they tie up acetyl-CoA, are rarely important repositories for hydrogen in ecosystems with limiting soluble carbohydrate. This is ascribed to the lower yield of adenosine triphosphate (ATP) per triose, when pyruvate cannot be converted to acetate. Four electrons are removed in the anaerobic oxidation of triose to acetate and $CO_2$. If these electrons can be disposed of as $H_2$ or $CH_4$ the acetyl-CoA can yield ATP and acetic acid. If electrons cannot be converted to $H_2$ or $CH_4$ they must be disposed of by reduction of pyruvate and acetyl-CoA to lactate and ethanol, respectively, losing the potential ATP available when $H_2$ and $CH_4$ are formed.

III. <u>Other habitats</u>. Leakage of primary substrate and utilization of waste products must be important in many microbial ecosystems. The close association of specific bacteria to the gut epithelium, disclosed in the mouse and rat, the rumen, and the termite, is most easily accounted for on this basis. The production of methane in some termites, in bird caecae and in the human mouth indicates interspecies transfer of $H_2$ to methanogenic bacteria also in these habitats.

The numerous examples of spirochetes and bacteria clinging to the surface of so many termite protozoa must similarly indicate some nutrient transfer. Attention to this problem may provide information on which successful methods of cultivation for intimate symbiotes can be developed.

## C. Regulation

The soluble sugars in most forages are insufficient to support maximal fermentation, and the fiber, though abundant, is slowly digested, perhaps because the fibrolytic enzymes must be adsorbed to the limited areas of the fiber surface. The concentration of soluble carbohydrate limits the normal rumen fermentation rate. Efficiency in producing 2 ATP per triose is advantageous when substrate is limiting, and can explain why the predominant species form acetate.

If sugar or readily digested starch is added in excess, the normal rumen ecosystem is catastrophically changed. Lactic acid bacteria grow almost explosively, producing so much acid that within a few hours the pH drops as low as 4.0. Rumen motility ceases, and the animal may die within 24 hours. This illustrates that the critical competitive feature in these anaerobic ecosystems is not maximal ATP production per triose molecule, but rather maximal ATP production per unit time. As long as soluble sugar substrate is limiting, maximal ATP/molecule of substrate is equivalent to maximal ATP/unit time. With excess sugar the cells producing lactic acid, though obtaining only one ATP/triose, are able to metabolize the sugar molecules more rapidly and produce more ATP per unit time than can the predominant biota adapted to more efficient use of limited substrate.

On the basis that work is done in the biochemical reactions forming ATP, the critical factor in the anaerobic competition has the dimensions of work per unit time, and since work/time is power, it must be concluded that the power at the disposal of an organism is a crucial factor in its competitive success, whether man or microbe.

If inhibitors of methanogenesis are added to a rumen fermentation, less $CH_4$ is produced. Copious $H_2$ is formed, though less in amount than the quantity needed to account for the diminished $CH_4$ production. The accumulated $H_2$ retards the formation of $H_2$ from reduced pyridine nucleotides, and they are used in the production of propionate. This increase in propionate aids the host if its carbohydrate supply is limiting, as in milk producing cows, but it has not been conclusively shown in feed-lot animals that an increase in host protein parallels the increase in rumen propionate.

Host regulation of the rumen ecosystem occasionally fails disastrously when fresh legumes are consumed. The digesta retain tiny bubbles of fermentation gas and increase in volume as does bread dough, causing elevated pressure sufficient to cut off respiration and circulation, killing the animal. It is probable that microbes play a role in this sudden change in the nature of rumen digesta, but though many hopeful hypotheses have been developed, there is no general agreement on the cause of legume bloat. It remains one of the enigmas of the rumen microbial ecosystem, challenging future investigators.

## D. Present Interests in Gut Microbiology

I. Human gut ecosystem. Increased knowledge of intestinal anaerobes has stimulated a surge of interest in their role in human disease. Anaerobic pathogens failing to grow with older methods can now be easily cultured. The etiology of many deep seated infections has been explained by the isolation of anaerobes normally resident in the gut, but capable of opportunistic invasion of tissue habitats. Methods for clinical culture, and identification of the antibiotic susceptibility of these organisms have provided the physician with new insight into the rational treatment of pathologic conditions for which the etiological agent was previously undetected.

The other currently interesting aspect of human gut microbiology is the role of gut bacteria in the etiology of colonic cancer and other alimentary disorders. In contrast to infections a microbial role in colonic cancer must be subtle, of long-term duration, and more difficult to analyze for cause and effect relationships.

A renewed interest in the importance of fiber in human nutrition has been sparked by studies on primitive peoples consuming much fiber, and exhibiting a low incidence of colonic disorder, including cancer. Evidence is limited to epidemiological comparisons of the incidence of the colonic disease in some fossil fuel-dependent and fossil fuel-independent nations. Statistically significant correlations between colonic cancer and low-fiber diets are interpreted as evidence for an importance of dietary fiber in reducing colonic disease.

Attempts to demonstrate significant differences in the fecal microbiota of individuals in fossil-fuel dependent and independent cultures have been largely negative. Food differences do not markedly change the composition of the fecal flora, though there may be small but significant changes, undetected because of the variability inherent in attempts to identify the numbers and kinds of all the microbes in the teeming ecosystem. Since it is almost axiomatic in microbial ecology that a difference in micro flora accompanies a difference in substrate, the above noted similarities of fecal flora when diverse foods are consumed indicates that the foods consumed are not the chief substrate for the colonic flora. Presumably the earlier mentioned epithelial cells sloughed off in the small intestine, plus mucoid and other secretions, are the substrate supporting the fecal flora. These materials do not vary according to type of food consumed. Food differences, particularly in fiber content, must exert minor effects on the nature of the colonic population, and future studies should be directed toward the discovery of those species important in colonic diseases.

Several possible microbial roles in the etiology of colonic disease have been postulated. The volatile fatty acids, particularly propionic and butyric acid, are essential for the normal development of the caecal and ruminal wall in rodents and sheep, respectively, and may be similarly important in man. This is the simplest hypothesis to explain how fermentation of fiber in the colon could affect the health of this organ.

Another hypothesis is that fiber fermentation diminishes the toxicity of ammonia. Any net acid production from fiber would diminish the concentration of $NH_3$ by converting some of it to $NH_4^+$. Cell membranes are extremely permeable to ammonia, and chronic ammonia toxicity may be a factor in colonic disease. It seems improbable that an anticarcinogenic effect of fiber could be exerted simply by stimulating the rate of passage (a well established effect of dietary fiber). An effect from its fermentation is a more probable hypothesis, deserving of careful study.

Other factors than food fiber must also be taken into account in interpreting the results of epidemiological comparisons of colonic cancer in human populations. Increased longevity, higher dietary fat and protein and a generally higher plane of nutrition may increase colonic disease.

II. The rumen. An anticoccidial drug, monesin, increases the efficiency of food utilization for body weight gain in ruminants. Equivalent gain is obtained with less feed. The mechanism is unknown. A possible explanation is that the drug inhibits not only coccidia, but also the rumen protozoa. The protozoa feed chiefly on bacteria to obtain their protein supply, thus constituting an additional link in the food chain from bacteria to mammal. Since each predatory link in a food chain contains less protein than the food it consumes, the quantity of protozoal protein is less than the quantity of bacterial protein consumed. Although the quantity is less, a superiority of protozoal protein over bacterial protein has been thought to be responsible for the better growth of faunated than of defaunated lambs. Further study is needed to show whether this better quality of protozoal protein is sufficient to overcome its diminution in quantity, as compared to the bacterial protein consumed by the protozoa.

III. Other ecosystems. The microbial ecosystem of the human mouth is currently receiving much attention in relation to dental caries. A previously unsuspected abundance of obligately anoxic bacteria has been demonstrated, and, as in so many ecosystems, complicated interrelationships between microbial types affect the health of teeth.

The termite gut has always excited interest because of the wood-digesting ability of these insects due to their gut microbiota. A capacity for dinitrogen fixation has been demonstrated in some of these groups, though for some termite species it is not sufficient to account for a significant fraction of the nitrogen assimilated.

The termite gut ecosystem is so inherently interesting that one can anticipate an acceleration of research to understand it.

In all these ecosystems new techniques are being developed for direct study of the microbes in the habitat. Scanning electron microscopy and fluorescent antibody techniques are particularly valuable for detecting spatial relationships in the relatively undisturbed ecosystem.

## E. Future Trends

Several promising lines of investigation have already been mentioned. Additional future trends may be envisioned as follows:

I. Antibiotic resistance factors. Current studies indicate that the feeding of antibiotics to improve the growth domestic animals results in an increased incidence of antibiotic resistance factors in the microbial population. The resistance factors are acquired also by the microorganisms in the gut of humans associated with the animals. A more thorough study of the mechanism of the antibiotic effect on growth is needed for development of alternative feed supplements promoting the growth of domestic animals, to avoid building up resistance factors in humans.

II. Gnotobiotic studies. Continued study of gnotobiotic mammals should provide increasingly precise information on the relationships of gut microbes to the host and on the interrelationships between the microbes themselves. The extensive investigations on gnotobiotic mice and other mammals are being increasingly extended to ruminants and other mammals, and provide a means for discovering new microbe mammal relationships, as well as confirming hypotheses already formulated.

III. Analysis of the human gut ecosystem. The human ecosystem is almost certain to receive further study, much of it in relation to health, but hopefully some attention will be devoted to the understanding of the human gut ecosystem without regard to its practical importance. Basic ecological studies can provide new insight into the factors influencing many pathological states, including colonic diseases. Development of techniques for the quantitative description of rates of microbial growth in various part of the human gut, coupled with identification of the substrates used, the influence of rate of passage, secretion of enzymes, and absorption of digestion products, can provide a foundation for the understanding of disease. As an example, excess lactic acid production in man may account for some of the symptoms associated with indigestion.

IV. Increased supply of digestible plant fiber. Most of the carbohydrate in plant bodies is not digestible, because of lignification and other processes. One of the pressing needs is to increase the digestibility of plant fiber in order to exploit more fully those mammals exhibiting the cooperative model of microbial interaction. Particularly in tropical countries research on development of new forages is much needed. The great productivity of moist tropical countries is largely unexploited at present. Lack of digestible fiber limits utilization of the cooperative mammal-microbe ecosystems to provide much needed protein food. Mammals with cooperative microbes occur in tropical countries, and further research toward their domestication may increase the supply of food.

These are only a few of the potentials for future research in gut microbiology. They illustrate the opportunities to utilize microbes for the improvement of human life, but many additional gut microbial ecosystems are potentially important and need study. Each has its own peculiarities which tax the imagination and ingenuity of the ecologist studying it, and will be well rewarding when understood.

# Microbial Insecticides

## Viral and Bacterial Insecticides: Safety Considerations

E. KURSTAK

### A. Introduction

The use of microbial insecticides has been proposed for the control of insects important in agriculture and veterinary and public health. Baculoviruses and Bacillus thuringiensis have been considered seriously as potential insecticides and as an alternative to chemical pesticides, some of which can no longer be used because of their toxicity.

It is considered unlikely that microbial insecticides could entirely replace chemical insecticides in all situations in which vectors and pests are susceptible to microorganisms. The use of insect viruses, bacteria, protozoa, fungi or rickettsia however, in some ecosystems may be sufficient to keep pest populations at acceptable economic and medical levels.

Until recently, the safety considerations and possible hazards in the use of microbial insecticides, have not been sufficiently stressed (2).

### B. Current Status

While at present there is no direct evidence that any microorganism proposed as an insecticide can infect man or other vertebrates, in many cases satisfactory investigations on safety are lacking.

Whenever microbial insecticides are used for the control of pests and vectors of disease agents, man and other non-target animals will be exposed. The degree of exposure depends on the nature and scale of the operation, the method used, and the geographical area. The risks to animals exposed within the treated area will depend on the specificity of the microorganism used. Risks to non-target species outside the treated area depend mainly on the possibility of dissemination by wind, rain, predators and their parasites. It is necessary to take into account in testing exposed species, both invertebrates and vertebrates, for their susceptibility to infection, such parameters as body temperature, the age of the animals, the dose of the insecticide in question, the duration of the assay, and the passage level in the case of viruses. Other risks for vertebrates such as oncogenicity, teratogenicity, possible genetic recombination between viruses and mutations should be considered. It is well known in medical and veterinary virology that the passage of a virus in any host by unnatural means or the infection of an unfamiliar host may result in the selection of a mutant virus with different properties.

During the manufacture, formulation, packaging, field preparation and application of microbial insecticides, possible risks to man may arise; exposure of the respiratory and gastrointestinal tract, the skin and eyes may lead to infective, allergic or toxic effects.

The hazards presented by exposure to microbial insecticides, particularly with respect to infection by viruses, can only be assessed by very extensive testing. A potentially serious risk must be presumed to exist if a virus can initiate any sort of infection in any species of vertebrate. The end result may be difficult to assess rapidly because of the possibility of the adaptation and spreading of the virus to other animals. Special attention should be paid to infective hazards enhanced by immunodepression, malnutrition and certain diseases, such as malaria and leukaemia. In some parts of the world, the potential susceptibility of man and other animals to microbial insecticides is expected to differ.

Allergic reactions to microbial insecticides are possible and detectable after several exposures and may be caused by the microorganism itself, contaminating microorganisms, the insect host material or any allergenic material added to the formulation. Efforts should be made to ensure that microbial insecticide preparations proposed for field use do not contain allergens. It is known that allergic individuals often suffer asthma or pulmonary fibrosis.

Toxic effects are possible from heavy contamination of microbial insecticides. Testing for acute and chronic toxicity should be mandatory before field use and precautions are necessary to protect workers manufacturing such insecticides. Harmful effects due to accumulation of microbial insecticides in the environment, similar to that reported for organochlorine pesticides, is not evident so far.

Safety considerations should also include protection of non-target invertebrates (parasites and predators) in the ecosystem. Exhaustive testing of the specificity of microbial insecticides is required.

## C. Viral Insecticides

It is clear that Baculoviruses which induce polyhedrosis and granulosis diseases in insects are the main live viral insecticides.

In general, at the present time only nuclear polyhedrosis viruses and granulosis viruses are recommended as potential and relatively safe viral insecticides (1, 4). All Baculovirus insecticide candidates must be tested for their specificity for harmful insects and for their safety for vertebrates including man.

In the U.S.A. and Canada, guidelines to ensure the safe use of viruses for pest control have been established. Recommendations on the use of viruses for the control of insect pests and disease vectors are also available from the WHO/FAO experts' report (6). The main recommendations are: (i) the viruses to be used as insecticide should be well characterized; (ii) the use of any viral-insecticide must be conditional on satisfactory evidence that it is safe to man, domestic and wild animals, beneficial arthropods and plants; (iii) appropriate measures must be taken to protect workers involved in the various stages of production and use of viral insecticides; (iv) the possible relationship between insect viruses and vertebrate (including man) viruses should be investigated urgently; (v) studies should be performed on the mutability of insect viruses and on the problem of the change in the susceptibility of the host to the virus infection.

The safety tests for nuclear polyhedrosis viruses (NPV) conducted in the U.S.A. and Canada were reasonably satisfactory, and three of them, the *Heliothis*, *Neodiprion* and *Trichoplusia* NPV were produced industrially and marketed as insecticides (1). However, the U.S. Environmental Protection Agency registered only the *Heliothis zea* and *Orgyia pseudottsuga* NPV. Recent findings indicate that in vitro Baculoviruses can infect not only cells of poikilothermic but also of certain warm-blooded animals (5) strongly suggesting that safety tests for Baculovirus insecticides should not be discontinued.

Additional safety tests are necessary for cytoplasmic polyhedrosis viruses and Entomopoxviruses. These viruses are classified in the families of Reoviridae and Poxviridae, respectively, and include viruses causing well-known diseases in man and animals. A similar situation exists with respect to small DNA and RNA viruses of invertebrates, such as Densonucleosis viruses, acute bee paralysis virus, cricket paralysis virus, Nodamura and Gonometa viruses, which were incorporated into the Parvoviridae and Picornaviridae families. Serological reactions between certain insect pathogenic viruses and mammalian sera, including those from laboratory workers (3), increased the interest in the safety of these viruses.

## D. Bacterial Insecticides

I. Bacillus thuringiensis and Bacillus popilliae. These two bacteria were recommended as agents for the biological control of insect pests (1).

After safety studies on game animals and human volunteers, B. thuringiensis was accepted as an insecticide by the U.S. Food and Drug Administration. Today, mainly B. thuringiensis var. kurstaki and B. thuringiensis var. thuringiensis strains are used in the U.S.A. and Japan for commercial production.

Toxins of B. thuringiensis at high concentrations, which are never likely to occur in field use, are toxic for non-target insects, such as honey-bees. The problem of the destruction of non-target insects by biological insecticides requires more investigations. The safety of bacterial insecticides must also be conducted for game birds, commercial fish and shellfish.

B. popilliae was used several years ago in the U.S.A. to control Japanese beetle. Restricted safety tests, at that time, did not demonstrate adverse effects on birds. The argument that these bacteria do not grow at the temperature of vertebrates permitted the use of B. popilliae, milky-disease agent, in the field. However, the efficiency of the control of insects by spores of B. popilliae was poor. For this reason, this live insecticide is no longer used on a large scale and, therefore, has not been submitted for additional tests on its safety.

## E. Conclusions

Compared to chemical insecticides, certain viruses and bacteria present some advantages as insecticides. The problem of safety of these live microbial insecticides was stressed only recently and investigations were started only during the last few years.

At present, no complete information is available to declare such insecticides without any danger for warmblooded animals and man. For this reason, research workers should intensify studies on new microbial insecticides and on their efficiency and safety. For the diagnosis of infections the use of modern methodology is essential.

## F. Acknowledgements

Work sponsored by the National Research Council of Canada.

## G. References

1. Heimpel, A.M.: Safety of Insect Pathogens for Man and Vertebrates. In: Microbial Control of Insects and Mites (H.D. Burges & N.W. Hussey, eds). Academic Press, New York, 461-489 (1971).
2. Kurstak, E.: Répercussions éventuelles de la lutte microbiologique sur les vertébrés. Cas des insecticides biologiques à base de virus. Ann. Parasitol. Humaine et Comparée, Paris, 46 (3 bis), 277-288 (1971).

3. Longworth, J.F., Robertson, J.S., Tinsley, T.W., Rowlands, D.J., & Brown, F.: Reactions between an insect picornavirus and naturally occurring IgM antibodies in several mammalian species. Nature, 242, 314-316 (1973).
4. Maksymiuk, B.: Pattern of use and safety aspects in application of insect viruses in agriculture and forestry. In: Baculoviruses for Insect Pest Control: Safety Considerations (M. Summers, R. Engler, L.A. Falcon, P. Vail, eds.). ASM Publications, Washington, pp. 123-128 (1975).
5. McIntosh, A.H. & Maramorosch, K.: Annual Report of Waksman Institute of Microbiology, in press (1977).
6. World Health Organization: The use of viruses for the control of insect pests and disease vectors. Tech. Rep. Series, 531, 3-48 (1973).

# An Ecological Approach to the Use of Insect Pathogens for Pest Control

J.F. LONGWORTH and J. KALMAKOFF

## A. Introduction

The purpose of this review is to point to the urgent need for basic research into the ecology of insect pathogens since in some instances it may lead to a successful alternative to using pathogenic organisms as "insecticides". We have recently outlined (5) three different ways of using insect pathogens in biological control systems; (a) the insecticidal approach, (b) limited application or release, and (c) the manipulation of pathogenic diseases already present in pest population.

## B. The Insecticidal Approach

The concept of using insect viruses as insecticides has dominated research in insect pathology in the last decade and there has been considerable emphasis upon the characterization and identification of these viruses particularly baculoviruses, as a prerequisite to demonstrating their safety.

An example of the insecticidal approach is the use of the baculovirus of *Heliothis zea* which was developed primarily to control this pest on cotton (3). Notwithstanding a high mortality of insect larvae after treatment of cash crops with this virus, sufficient can remain alive to cause significant damage. This is likely to occur with multivoltine pests, since overlapping generations or even an extended egg laying period could give rise to a larval population of widely different ages and susceptibility with a consequent need for frequent applications of virus within a season and from year to year. The insecticidal approach may be necessary with cash crops in particular, to quickly achieve a high degree of control either to prevent losses in yield or to minimise undesirable cosmetic damage. Such an approach does not take advantage of the ability of the virus to persist from season to season and cause spontaneous epizootics. Further, the insect host must be easily reared in the laboratory, preferably on an artificial diet, to produce the large quantities of virus which are required from year to year.

## C. The Limited Release Approach

This approach is different in concept from the insecticidal one since the aim is to reduce pest populations to an economically tolerable level. With pests of pastures or forests it is often possible to apply the virus as a single application at an appropriate time, leading to a stable host-pathogen complex which may persist for many years. It may also be possible to introduce the pathogen strategically into the population by means of contaminated baits, or by infected individuals, either larvae or adults, rather than by spraying.

I. *Limited application*. The microsporidan, *Nosema locustae*, is being developed in the U.S.A. for control of grasshoppers on rangelands, not as a conventional insecticide, but rather as a management tool for suppression of populations below economic thresholds (2).

One application of virus by conventional insecticidal techniques can bring about acceptable long-term control. A good example of this approach is the control of a jackpine sawfly in pine forests in Canada (7). Further, it seems it may be necessary to only use the newly registered nuclear polyhedrosis virus of the Douglas fir tussock moth periodically (2).

II. Strategic release. A classical example of strategic pathogen release is provided by the method used to apply the baculovirus of *Oryctes rhinoceros* in Western Samoa (6). In this case the virus was rapidly and effectively distributed via the adult beetles. Unlike most other baculoviruses the virus does not form protein inclusion bodies and is quickly inactivated outside the host. The host itself is not easily reared in quantity, and these factors combined with the fact that the host causes economic damage at low population density and breeds in protected situations precluded an insecticidal approach. A good understanding however, of the biology of the host and the ecology of the virus allowed a highly selective method of application in which mobile, long-lived adults became infected after being attracted to artificially contaminated breeding sites. Thereafter rapid and effective dispersal was achieved by these adults. From these initial foci of infestation, infected beetles transmitted virus via the faeces to larvae present in other breeding sites as well as to their own progeny. Further migration of these infected adults and of others developing from larvae infected at a late stage led to an epizootic wave of infection which spread from the initial centres at a rate of 3 km $month^{-1}$ (8). This substantially decreased the *Oryctes* population in the affected area and the coconut palms showed a marked improvement in condition.

## D. The Manipulation of Pathogenic Enzootic Viruses

In the examples described the pathogen has been applied to the pest population. The pathogen, however, may be sufficiently widespread in nature so that suitable management may enhance its effect. This approach is exemplified by the proposed management of *Wiseana* spp. in New Zealand. The soil-dwelling larvae of this univoltine insect is a widely distributed pest of pastures causing severe damage. Field surveys have shown that over 80% of "established" pastures (>5 years old) contain larvae infected with viruses primarily with nuclear polyhedrosis virus (4).

*Wiseana* causes serious damage in newly established pastures as oviposition, larval survival and development are favoured by the vigorous grass growth. Furthermore, cultivation buries deeply any virus which may have been present and prevents it from being encountered by the larvae. Dense larval populations develop in 2-3 years in these circumstances resulting in extensive pasture damage. In "established" pastures, the virus-infected larvae tend to die at the soil surface and stock animals then distribute the virus between adjacent paddocks. Also bird feces collected around and within such areas contain high concentrations of virus. The distribution of virus is sufficient to introduce foci of infection into the larval populations in new pasture and within 1-2 years an epizootic or an enzootic develops reducing pest numbers to a tolerable low level, at which they are maintained by the virus (1).

Certain pasture management practices have been suggested as a means of controlling *Wiseana* (1) these include oversowing affected areas, intensive rotational stocking, and haying older pastures. The advantages of such a system are that it eliminates the need to introduce large doses of virus into the ecosystem as well as being much cheaper and posing far fewer ecological hazards.

## E. Future Research in the Selective Use of Insect Pathogens

To use any insect pathogen to the best advantage calls for an inter-disciplinary effort to define: (a) the life cycle of the pest, (b) the factors responsible for the natural regulation of field populations of the pest, (d) the mechanism of spread of the pathogen in naturally occurring epizootics, (d) the economic threshold levels of the pest population, (e) the properties of the pathogens and (f) the effect of different management practices and climatic conditions on the induction of epizootics and the maintenance of enzootics. This knowledge may well reveal new opportunities for control and also decrease the possible risks both from a public health point of view and to non-target insects. This requires an increased research effort on the ecology of insect pathogens and more knowledge on the fundamental properties of the pathogens themselves.

## F. References


1. Crawford, A.M., Kalmakoff, J.: A host-virus interaction in a pasture habitat: Wiseana spp. (Lepidoptera: Hepialidae) and its baculoviruses. J. Invertebr. Pathol. 29, 81-87 (1977).
2. Harper, J.D.: Proc. 1st Int. Coll. on Invertebr. Pathol. 69, Kingston (1976).
3. Ignoffo, C.M., Chapman, A.J., Martin, D.F.: The nuclear polyhedrosis virus of Heliothis zea (Boddie) and Heliothis virescens (Fabricus). III. Effectiveness of the virus against field populations of Heliothis on cotton, corn and grain sorghum. J. Invertebr. Pathol. 7, 227-235 (1965).
4. Kalmakoff, J., Moore, S.C.: The ecology of nucleo-polyhedrosis virus in porina (Wiseana spp. Lepidoptera: Hepialidae). NZ. J. Agric. 133, 41-42 (1976).
5. Longworth, J.F., Kalmakoff, J.: Insect virus for biological control: An ecological approach. Intervirology 8, 68-72 (1977).
6. Marschall, K.J.: Introduction of a new virus disease of the coconut palm rhinoceros beetle in Western Samoa. Nature (London) 252, 288-289 (1970).
7. Smirnoff, W.A.: A virus disease of Neodiprion swainei Middleton. J. Insect. Pathol. 3, 29-46 (1961).
8. Young, E.C.: The epizootiology of two pathogens of the coconut palm rhinoceros beetle. J. Invertebr. Pathol. 24, 82-92 (1974).

# Microbial and Integrated Control of Vectors of Medical Importance

M. LAIRD

## A. Introduction

As invertebrate pathology was maturing, an anti-science preaching "small is good" was emerging as a reaction to the fact or fear of adverse health and environmental effects of the indiscriminate use of synthetic organic compounds. Such use of persistent pesticides, still-valuable as elements in integrated programs for vector control, was a particular target for environmentalists and the mass media. The latter, often seemingly unaware that insecticide resistance problems were seriously prejudicing the success of anti-vector efforts and that a search for alternative and supplementary means of control was already well-advanced, not infrequently contrasted what they represented as totally-bad chemical agents with totally-good biological ones. Actually, biocontrol agents must be employed just as responsibly as chemical pesticides. If they are conventional predators or parasitoids, searching prior analysis of the likely ecological consequences of their introduction to and establishment in a new area, should be undertaken (it has by no means always been, in the past). If they are novel pathogens or parasites, the application of which to vector control needs represents a practical outcome of invertebrate pathology, their health and environmental acceptability must be assured (31).

For microbial elements in future integrated control methodologies designed for the needs of specific vector-suppression situations, will seldom be employed on an introduction/establishment basis - although there will be exceptions to this generalization, notably where small and isolated islands are concerned (for example, the relative permanency of a comparatively small number of mosquito larval habitats favoured the modestly-successful use of *Coelomomyces* fungi by a single application against the filariasis vector, *Aedes polynesiensis*, on Nukunono atoll, Tokelau islands, nearly 20 years ago (13)). Usually, though, microbial biocontrol agents will have to be applied repetitively in similar fashion to chemical pesticides. This will call for mass-production and for the commercial availability of adequate quantities of effective and safe products based upon them. Such products, largely bacterial but now viral too, are already widely registered for use against economic pests. Their numbers and markets continue to expand, and they are currently the subject of much attention on the part of industry (23).

So far, though, only a single such product is becoming available for use against vectors. This (SKEETER DOOM: Fairfax Biological Laboratory Inc., Clinton Corners New York) is based upon the warm-water mosquito worm, *Romanomermis culicivorax*. Apparently motivated by misgivings about possibly small markets for highly specific microbial agents in the vector control field, and about a diversity of other factors ranging from non-patentability problems to questions of possible resistance and inadequate shelf-life, the commercial sector is otherwise maintaining a wait-and-see attitude about this paper's subject. This is doubly unfortunate in that without industrial participation further progress is likely to be limited by lack of funds and opportunities for those in the university/government sectors now well-advanced in that subject's more basic areas, while world needs for an expanded vector-control armament have become pressing. The enhanced human population of 2000 A.D. will face enormous risks of disease, famine and economic chaos likely to make vector control in developing countries an over-riding priority. To meet the need, goal-oriented research programs emphasizing team-work and joint ventures by the university/industrial/governmental sectors should _already_ be far advanced. Valuable time and opportunities are being lost, to profit from recent advances concerning such microorganisms as those briefly mentioned in the following section and more fully discussed elsewhere (14).

## B. Candidate Microbial Control Agents

I. _Entomopathogenic viruses_. Investigations of baculoviruses of crop and forest pests are far advanced, their safety prognosis is excellent (24), and commercial exploitation has already begun in the U.S.A. (e.g., Sandoz' ELCAR). A variety of these entities have been discovered in mosquitoes in recent years (4). These include Nuclear Polyhedrosis Viruses, which are as yet not known from blackflies, although Cytoplasmic Polyhedrosis Viruses now are (2, 3). Neither group has yet been reported from tsetse-flies (5), but these have received far less attention from invertebrate pathologists than the Culicidae or Simuliidae.

II. _Entomopathogenic bacteria_. Strains of _Bacillus thuringiensis_ (18) and _B. sphaericus_ (22) are showing promise against mosquitoes in laboratory tests, which in the case of the former species are yielding encouraging data for blackflies too (12). Being both mass-producible by available technology and broad-spectrum with regard to host-range, the industrial attraction here is obvious. It could well be so for _Bacterium mathisi_ of tsetse, too, but the culture of this entity was lost shortly after its discovery and initial favourable reports (19, 20). However, in the latter case close host-specificity might be an industrial non-incentive, and there would be special difficulties of application in the field (15).

III. _Entomopathogenic fungi_. Early in its biological control program and in co-sponsorship with New Zealand, the World Health Organization supported the first pilot project resulting in the establishment of _Coelomomyces_ fungi against mosquitoes in a region remote from that whence the pathogen was derived (13). Although many species of this genus had then been described and upwards of 25 (most of them from the Culicidae) are now known, it was only three years ago that cyclopoid copepods were first found to serve as an intermediate host (29, 30). The confirmations since published include one from New Zealand (17), and a similar claim has been made for ostracods (28). The incident serves as a timely reminder of the continuing need for basic studies of each candidate microbial agent under investigation from the practical standpoint. While _Coelomomyces_ can now be cultured _in vivo_ (7), relevant _in vitro_ work remains in its infancy (21).

The above and other fungal options for mosquitoes (_Culicinomyces_, _Lagenidium_), for these insects and blackflies (the already mass-producible and broad-spectrum _Metarrhizium anisopliae_, which is well-advanced as to safety-testing and has in fact a history of practical use), and for tabanids and tsetse, are being considered elsewhere (14).

IV. _Entomopathogenic protozoa_. The Microsporida, with field-testing against anophelines under way in Pakistan (1), and tetrahymenid ciliates of the Culicidae and Simuliidae - notably _Lambornella_ of tree-hole-utilizing mosquitoes (6) - are leading candidates. Safety considerations concerning microsporidans require more investigation, though (25, 27); and these organisms, although mass-producible _in_ vivo (26), have yet to be cultured in vitro. The latter is straight forward for tetrahymenids.

V. Entomopathogenic nematodes. One group of worms, the Mermithidae, is represented by species causing mortalities among each of the groups of vectors referred to herein. As already indicated, Romanomermis culicivorax is further advanced towards practical use than any other vector-control pathogen or parasite. Not only is its future promising for warm-water mosquito control, but also laboratory experiments with it against blackflies in North America (8) led to laboratory confirmation involving the onchocerciasis vector, Simulium damnosum, in West Africa (9) followed in 1977 by an as-yet unpublished field demonstration by Drs. Eder and James Hansen of its capacity to infect larvae of this blackfly, causing the downstream drift of attacked and penetrated examples, under actual riverine conditions.

## C. Summary and Conclusions

There is no lack of candidate microbial control agents of established efficacy against vectors, recent status reports being available for mosquitoes (16), blackflies (10) and tsetse (11). Their established or likely health and environmental acceptability should render them as attractive to industry as their established or likely mass-producibility, and the certainty of large markets for products based upon them; not only in those huge areas of the developing world subject to malaria, filariasis, onchocerciasis, African trypanosomiasis and other arthropod-borne diseases, but also in vast northern areas the full opening-up of which awaits practical integrated methodologies for biting-fly control. Without major industrial participation in relevant investigations at this stage, though, some of the questions of prime interest to industry, relating to such topics as shelf-life, hybridization vis-à-vis patentability, etc., may await answers for longer than necessary. For, in this area of enterprise, "small" is just not going to be good enough.

## D. References


1. Alger, N.E. Personal communication. (1977).
2. Bailey, C.H.: Field and laboratory observations on a cytoplasmic polyhedrosis virus of blackflies (Diptera: Simuliidae). J. Invertebr. Pathol., 29, 69-73 (1977).
3. Bailey, C.H., Shapiro, M. & Granados, R.: A cytoplasmic polyhedrosis virus from the larval blackflies Cnephia mutata and Prosimulium mixtum (Diptera: Simuliidae). J. Invertebr. Pathol., 25, 273-274 (1975).
4. Bay, E.C., Berg, C.O., Chapman, H.C. & Legner, E.F.: Biological control of medical and veterinary pests. In: Theory and Practice of Biological Control. Academic Press, Inc., 1976, pp. 457-479.
5. Briggs, J.D., Riordan, K., Touré, S.M., & Nolan, R.A.: Pathology and nematode parasitism. In: TSETSE - The Future for Biological Methods in Integrated Control. Laird, M. (ed.). International Development Research Centre, Ottawa, 1977. pp. 75-88.
6. Corliss, J.O. & Coats, D.W.: A new cuticular cyst-producting tetrahymenid ciliate, Lambornella clarki n. sp., and the current status of ciliatosis in culicine mosquitoes. Trans. Amer. Micros. Soc., 95 (4), 725-739 (1976).
7. Couch, J.N.: Mass production of Coelomomyces, a fungus that kills mosquitoes. Proc. Nat. Acad. Sci. USA, 69 (8), 2043-2047 (1972).
8. Finney, J.R.: The penetration of three simuliid species by the nematode Reesimermis nielseni. Bull. Wld Hlth Org., 52 (2), 235, (1975).

9. Hansen, E.L. & Hansen, J.W.: Parasitism of Simulium damnosum by Romanomermis culicivorax. IRCS Medical Science, 4, 508 (1976).

10. IDRC. (Laird, M. ed.) Preventing Onchocerciasis through Blackfly Control. International Development Research Centre, Ottawa. 11 pp. (1976).

11. IDRC. (Laird, M. ed.) TSETSE - The Future for Biological Methods in Integrated Control. International Development Research Centre, Ottawa. 220 pp. (1977).

12. Lacey, L.A. & Mulla, M.S.: Evaluation of Bacillus thuringiensis as a biocide of blackfly larvae (Diptera: Simuliidae). J. Invertebr. Pathol., 30, 46-49 (1977).

13. Laird, M.: A coral island experiment: a new approach to mosquito control. WHO Chronicle, 21, 18-26 (1967).

14. Laird, M.: Microbial and integrated control of vectors of medical importance. In: Proceedings of Symposium on Ecology of Microbial Insècticides, Dunedin, New Zealand, August 1977, (Kurstak, E. and Maramorosch, K. eds.), Academic Press.

15. Laird, M. & Simmonds, F.J.: Biocontrol prospects in future integrated tsetse control programs. In: TSETSE - The Future for Biological Methods in Integrated Control. Laird, M. (ed.). International Development Research Centre, Ottawa, pp. 177-186 (1977).

16. NAS. Mosquito Control: Some Perspectives for Developing Countries. National Academy of Sciences, Washington, D.C. 63 pp. (1973).

17. Pillai, J.W., Wong, T.L. & Dodgshun, T.J.: Copepods as essential hosts for the development of a Coelomomyces parasitizing mosquito larvae. J. Med. Ent., 13 (1), 49-50 (1976).

18. Reeves, E.L. & Garcia, C.: Pathogenicity of bicrystalliferous Bacillus isolate for Aedes aegypti and other aedine mosquito larvae. Proc. 4th Int. Colloq. Insect Pathol., College Park, Maryland, pp. 219-228 (25-28 August 1970).

19. Roubaud, E. & Treillard, M.: Un coccobacille pathogène pour les mouches tsétsés. C.R. Acad. Sci (Paris), 201, 304-306 (Rev. appl. Ent. B, 23, 242) (1942).

20. Roubaud, E. & Treillard, M.: Infection expérimentale de Glossina palpalis par un coccobacille pathogène pour les mouches. Bull. Soc. Path. Exot., 29, 145-147 (1936).

21. Shapiro, M. & Roberts, D.W.: Growth of Coelomomyces psorophorae mycelium in vitro. J. Invertebr. Pathol., 27, 399-402 (1976).

22. Singer, S.: Isolation and development of bacterial pathogens of vectors. In: Biological Regulation of Vectors: The Saprophytic and Aerobic Bacteria and Fungi. (Briggs, J.D. ed). U.S. National Institutes of Health, pp.3-17 (1977)

23. SRI. New, Innovative Pesticides: An Evaluation of Incentives and Disincentives for Commercial Development by Industry. Stanford Research Institute, Menlo Park, California. 1977 xiv + 318pp. (1977).

24. Summers, M., Engler, R., Falcon, L.A., Vail, P. (Editors). Baculoviruses for Insect Pest Control: Safety Considerations. American Society for Microbiology, Washington, D.C. 1975. 186 pp.

25. Undeen, A.H. & Alger, N.E.: Nosema algerae: Infection of the white mouse by a mosquito parasite. Experimental Parasitology 40, 86-88 (1976).

26. Undeen, A.H. & Maddox, J.V.: The infection or nonmosquito hosts by injection with spores of the microsporidan Nosema algerae. J. Invertebr. Pathol., 22, 258-265 (1973).

27. Van Essen, F.W. & Anthony, D.W.: Susceptibility of nontarget organisms to Nosema algerae (Microsporida: Nosematidae), a parasite of mosquitoes. J. Invertebr. Pathol. 28, 77-85 (1976).

28. Weiser, J.: The intermediary host for the fungus Coelomomyces chironomi. J. Invertebr. Pathol., 28, 273-274 (1976).

29. Whisler, H.C., Zebold, S.L., Shemanchuk, J.A.: Alternate host for mosquito parasite, *Coelomomyces*. Nature, *251*, 5477, 715-716 (1974).

30. Whisler, H.C., Zebold, S.L., Shemanchuk, J.A.: Life history of *Coelomomyces psorophorae*. Proc. Nat. Acad. Sci., *72*, 2, 693-696 (1975).

31. WHO. Ecology and Control of Vectors in Public Health. Wld. Hlth. Org., Geneva, Tech. Rpt. Ser. 561, 35 pp., (1975).

Microbial Ecology of Plants

# The Rhizosphere

## The Ultrastructure of the Rhizosphere of *Trifolium subterraneum* L.

R.C. FOSTER and A.D. ROVIRA

### A. Introduction

Early investigations of root fine structure using transmission electron microscopy (TEM) of whole mounted roots showed that the surface was coated with an amorphous granular mucilage (8) in which bacteria may be embedded. More recently the rhizosplane has been examined by scanning electron microscopy (SEM) (6) but neither of these techniques allows bacteria which have similar external morphology to be distinguished. Morphologically similar bacteria can be distinguished, however, in ultrathin sections of rhizospheres examined in the TEM (10, 11, 12, 18).

This investigation is part of a general study of the ultrastructure of the rhizospheres of agricultural or forest plants aimed at widening our understanding of rhizosphere ecology with particular reference to root pathogens. In this paper we describe the rhizosphere of Trifolium subterraneum L. and compare it with that of plants grown in the same soil (12).

### B. Materials and Methods

Undisturbed cores each containing one plant of Trifolium subterraneum L. cv Bacchus Marsh were obtained as described previously (22).

### C. Results

Figures 1 and 2 are generalised low-power views of rhizospheres of lateral roots of clover plants at the flowering stage. To the left in both figures are the outer cortical cells of the root (R), perhaps overlain by collapsed root cap remnants. The cells have been colonised by soil microorganisms and the cell walls have been lysed leaving the electron-dense lignified cell wall corners.

The rhizosphere organisms are embedded in a fibrous matrix which occupies most of the space between the root surface and the clay and organic soil particles (Figures 1 and 2). Many kinds of microorganism can be distinguished by their morphology (Figs. 3-6) and internal ultrastructure as well as by the structure of their capsule material (Figs. 7-9). The bacteria are closely packed, especially when close to the root surface (see Table 1), and many occur in distinctive small colonies about 5 to 10 µm in diameter so that only 4 or 5 cells appear in an ultrathin section. Some bacteria are surrounded by an extended capsule which can be

---

Fig. 1. General view of the rhizosphere. All the plates are transmission electron micrographs of transverse ultrathin sections of fine lateral roots of Trifolium subterraneum cv Bacchus Marsh. The colonies are labelled a-o in order of appearance on the plates. The bar indicates 1 µm unless otherwise stated on the figure. The following abbreviations are used: C, capsule material; M, soil minerals, mainly clay particles; PCWM, plant cell wall material; O, soil organic matter; R, clover root; V, virus ▶

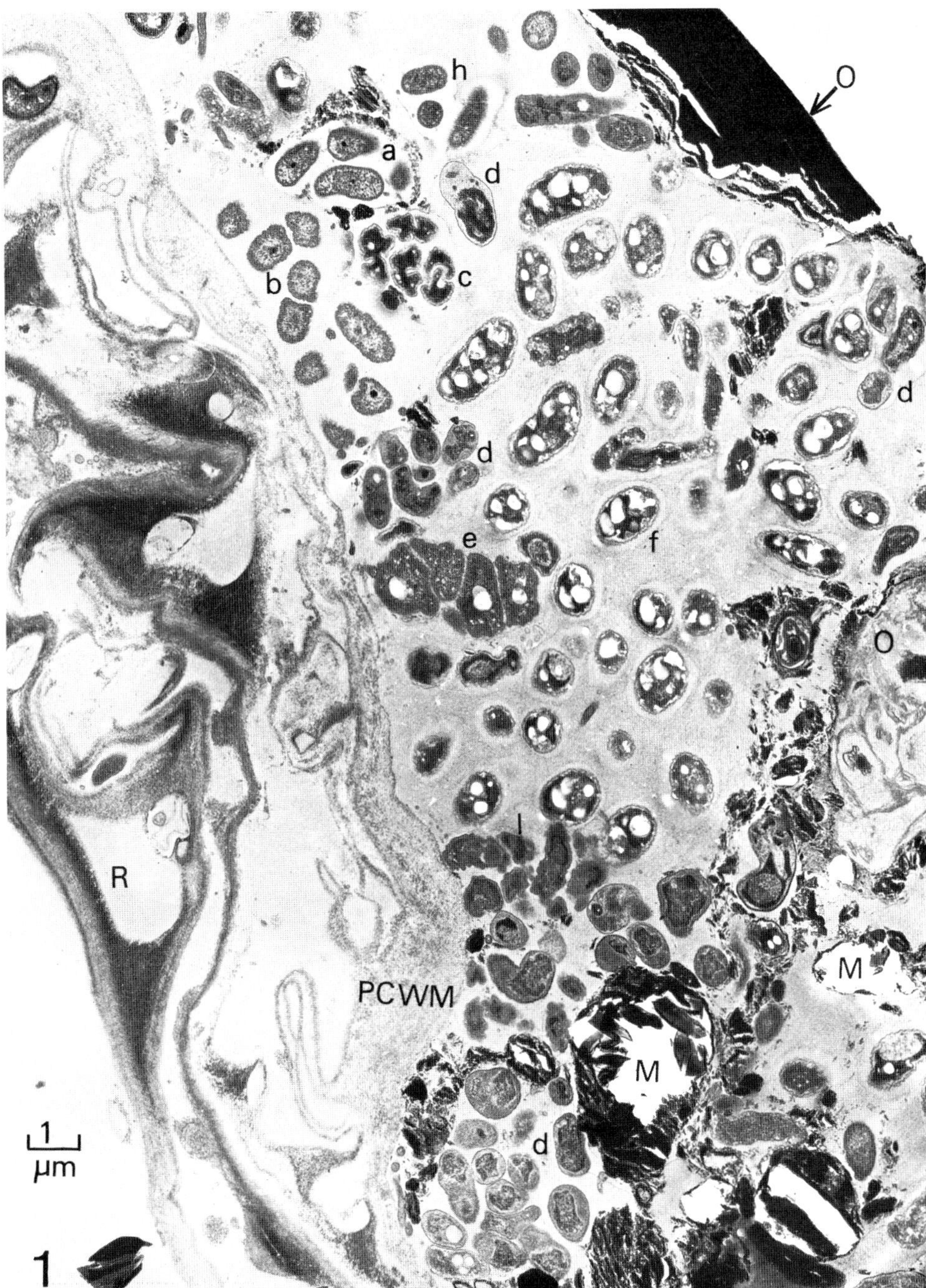


Fig. 1

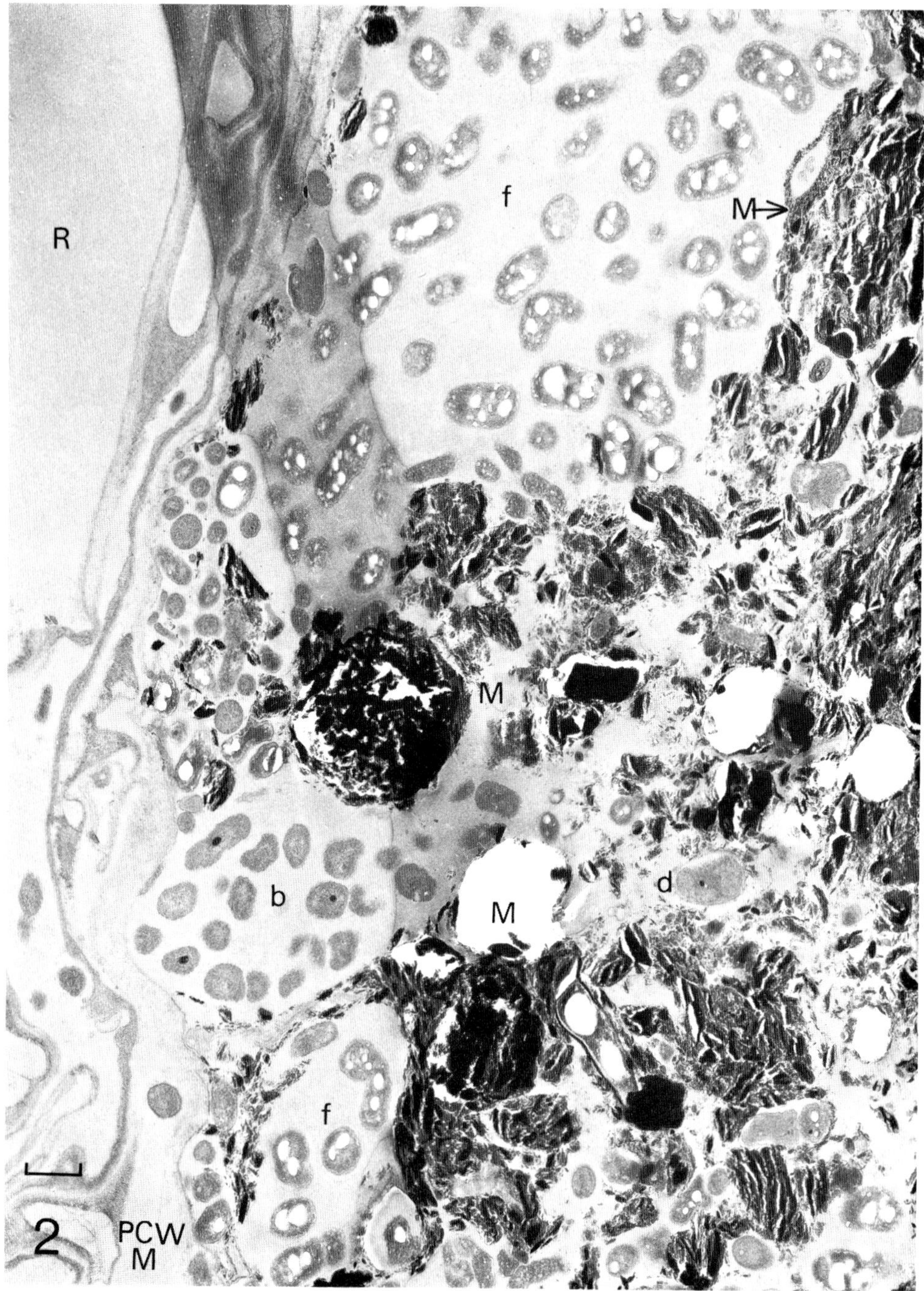


Fig. 2. General view of the rhizosphere (see legend to Fig. 1)

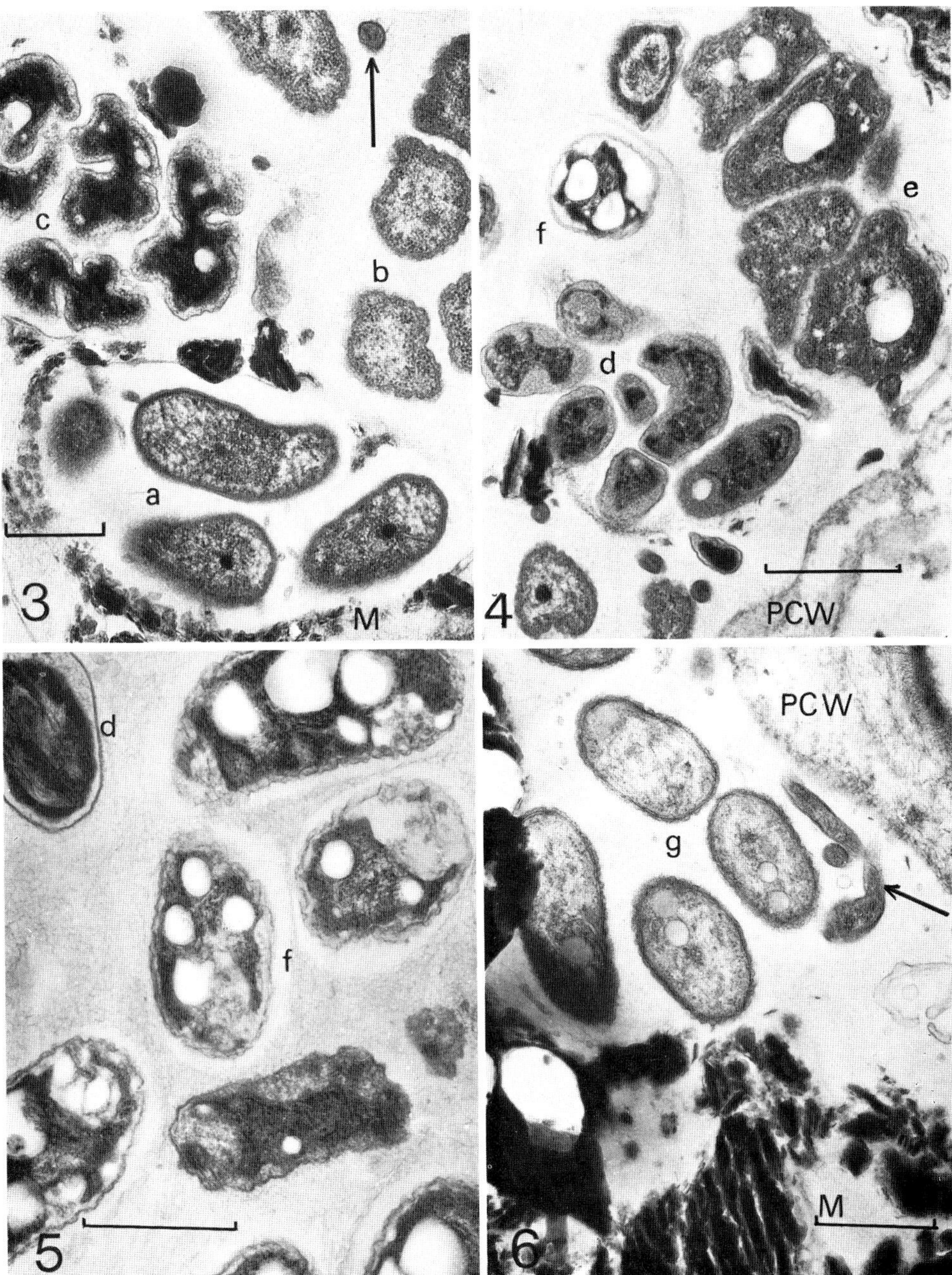


Figs. 3-6. Examples of rhizosphere organisms (see legend to Fig. 1)

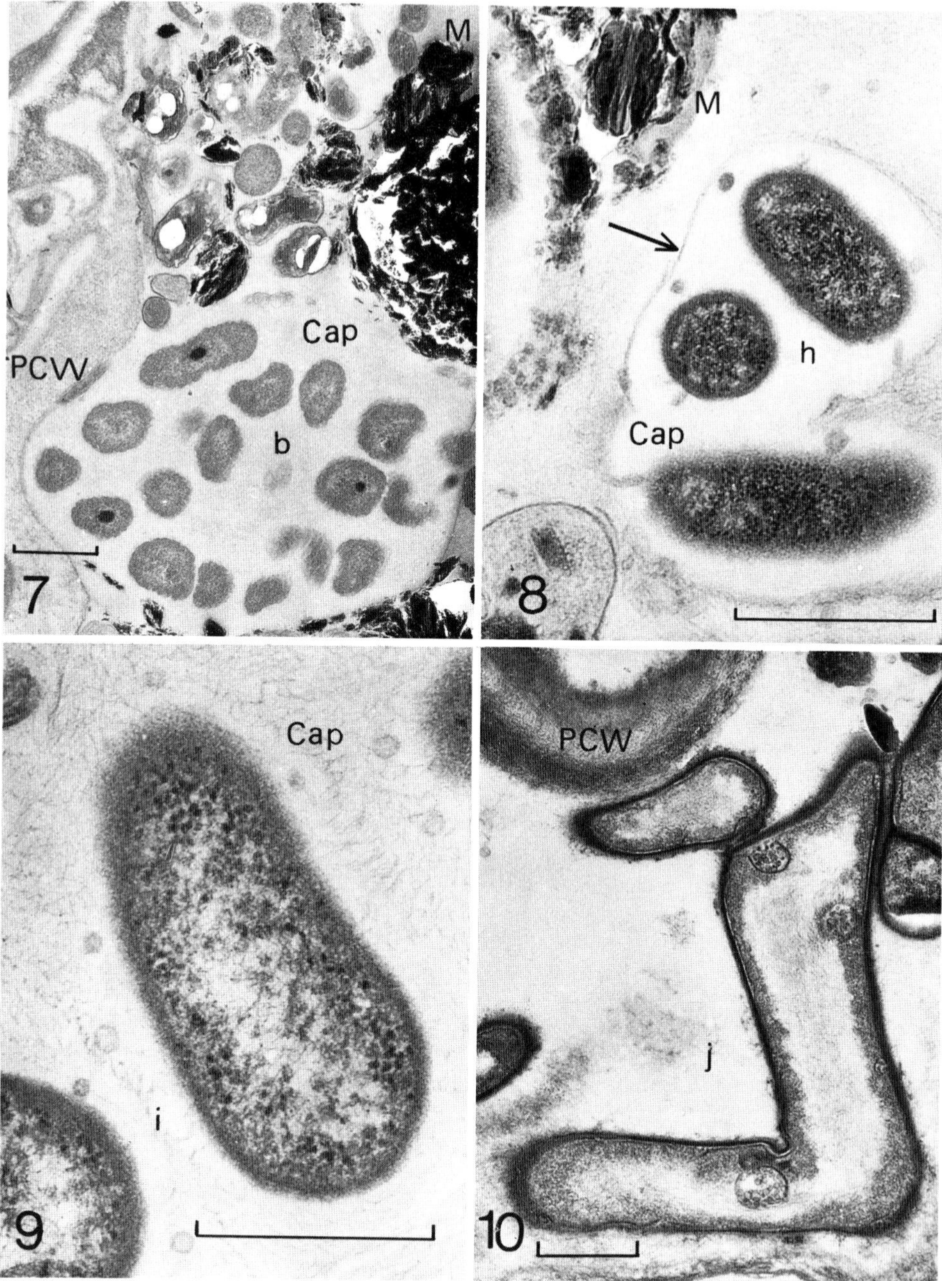


Figs. 7-10. Examples of rhizosphere organisms (see legend to Fig. 1)

readily distinguished from the general rhizosphere matrix by its physical organisation and staining properties (Figs. 3-14).

After close examination of several low power electron micrographs we consider that there is evidence of zonation within the rhizosphere. Most of the distinctive small colonies appear to occur within 5 μm of the rhizoplane (Table 2) and the cells are more densely packed in this region (Table 1). Measurements of several hundred microorganisms from the rhizosphere indicated that the mean diameter was about 1 μm and we here define the rhizoplane as that extending from the root surface to 1 μm into the soil. It is convenient for purposes of comparison to define an inner rhizosphere 0 to 10 μm including the rhizoplane and an outer rhizosphere 10 - 20 μm from the root surface.

Some of the organisms encountered in clover rhizospheres are illustrated in detail in Figures 3-14 and they have been labelled alphabetically to correspond to the types in Table 2.

The most common organism profiles, are types 'd' and 'f'. Type 'f' has similar densities in the innermost zones but it comprises 70% of all the organisms in the outer rhizosphere compared with only 19% at the rhizoplane.

Table 1. Numbers of microorganisms in the rhizosphere

| Zone | Distance from root in μm | Number of kinds of organism which can be distinguished cytologically | Estimated frequency cells per cc. |
|---|---|---|---|
| RHIZOPLANE | 0-1 | 11 | $120 \times 10^9$ |
| INNER RHIZOSPHERE | 0-5 | 12 | $96 \times 10^9$ |
| | 5-10 | 5 | $41 \times 10^9$ |
| OUTER RHIZOSPHERE | 10-15 | 2 | $34 \times 10^9$ |
| | 15-20 | 2 | $13 \times 10^9$ |

Table 2. Populations of microorganisms in various zones of the rhizosphere

| Zone | Distance from the root surface in μm | Numbers of individuals per cc of soil ($\times 10^{-9}$) | | | | | | | | | | |
|---|---|---|---|---|---|---|---|---|---|---|---|---|
| | | a* | b | c | d | e | f | g | h | l | m | ind** |
| RHIZOPLANE | 0 - 1 | - | 13 | - | 27 | 3 | 23 | 10 | - | 13 | 3 | 3 |
| INNER RHIZOSPHERE | 0 - 5 | 2 | 5 | 2 | 21 | 3 | 24 | 2 | - | 8 | 2 | 7 |
| | 5 - 10 | 1 | - | 2 | 10 | - | 25 | - | 4 | - | - | 1 |
| OUTER RHIZOSPHERE | 10 - 15 | - | - | - | 9 | - | 23 | - | - | - | - | 2 |
| | 15 - 20 | - | - | - | 2 | - | 4 | - | - | - | - | 2 |

* type of organism profile, see figures and text

** type indistinguishable cytologically

In Figure 2 the outer regions of the rhizosphere have been included in the section and they contain clay minerals in higher concentration than in the inner rhizosphere, but bacteria can be found amongst the grains over 20 μm from the rhizoplane. In some cases the bacteria appear to be closely enclosed by the clay grains and it is possible that these are inside specialised microsites where bacterial growth may be inhibited by lack of space or nutrients. They may, however, be sites in which bacteria can survive changes in soil pH due to the buffering action of ions on the mineral surfaces and changes in soil moisture since water may be protained in the fine capillaries within the clay particles. They may also be protected from predators (21). Some microorganisms have bizarre shapes (Figure 3, colony 'c', Figures 11 to 18) but even where they have more usual shapes the organisms can often be distinguished by their characteristic ultrastructure.

Some Trifolium rhizospheres contain actinomycete-like organisms often very close to the root ('j' Figure 10). These organisms have branching, filamentous cells with septae and prominent mesosomes. They have a Gram positive type of wall which encloses a granular, densely packed peripheral cytoplasm surrounding a fibrous nucleoid region. There is no extended capsule and the cells lie in the general rhizosphere matrix.

Figures 11-14 show some of the more uncommon lobed forms which, in our experience, are usually found only within the root or very near the rhizoplane. Similar forms have been seen by us in wheat and Pinus radiata rhizospheres, again always within a few microns of the host cell wall or in the crevices between host epidermal cells and they seem to be restricted to this specialized environment. Associated with these cells is an unusual helically lobed organism 'n' about 0.25 μm in diameter and 0.5 μm long, similar to an organism recently described by Old et al.(19) in rhizospheres of grasses from dune sands. Occasionally large lobed (protozoan ?) organisms are encountered, 4 or 5 μm long. Protozoa are common in soils and may be expected in areas with high bacterial populations such as the rhizosphere. Such unusual types of microorganisms are seldom isolated in this form from soils by the usual plating techniques, and it is possible that they are too firmly enclosed by the rhizosphere mucigel to be removed by washing, or that they have stringent nutrient requirements which cannot be met in the usual culture media.

## I. POSSIBLE PARASITIC RHIZOSPHERE ORGANISMS

Close examination of Figure 1 reveals that throughout the rhizosphere tubular organisms which are much smaller than the majority of bacteria occasionally occur (Figs. 15-19); they are circular in transverse section and are about 0.2 to 0.4 μm in diameter but their length cannot be determined from ultrathin sections. They have dense cytoplasm which shows some differentiation into a central fibrous nucleoid region (Fig. 15) and a peripheral granular region. They are bounded by a double membrane (Figs. 15 & 16) of the Gram negative type. They occurred close to several quite distinct types of organism as shown in Figures 15 to 18. It does not seem possible to us therefore that these tubular profiles can be parts of the larger bacteria, and we believe that they are independent soil organisms. Similar elongated forms have been illustrated in whole-mount, negatively-stained preparations (14) and in ultrathin sections (1) and identified as Bdellovibrio. Bdellovibrio are known to occur in Australian agricultural soils, and to attack Rhizobium and other soil bacteria (20).

Also embedded in the rhizosphere matrix there are clusters of very small particles, usually associated with 'f'-type bacteria, (Figs. 20 & 22). The particles lie in an amorphous zone surrounded by matrix material (Figs. 20 & 21) and do not appear to have any particular orientation with respect to the adjacent bacterial

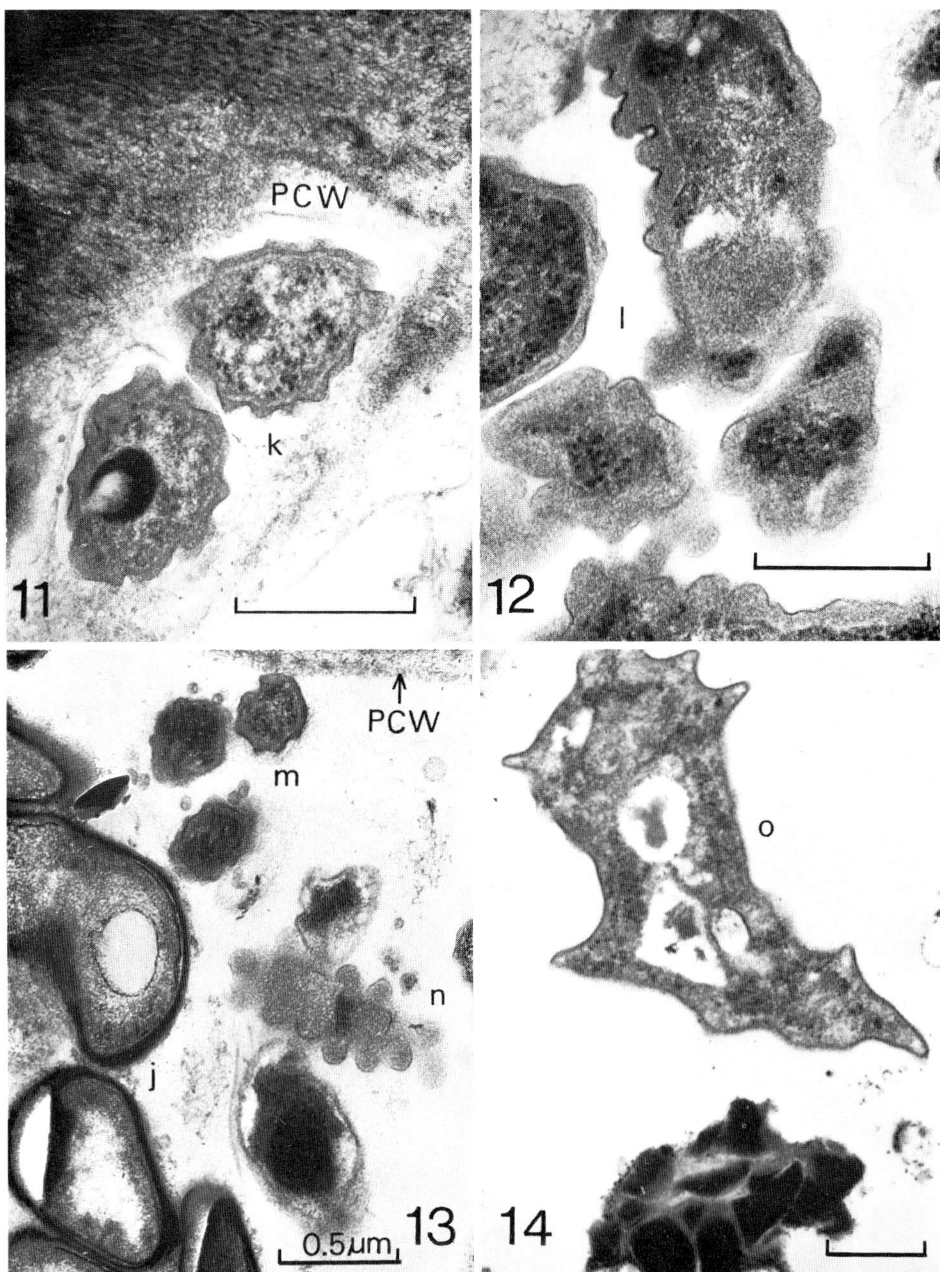


Figs. 11-14. Lobed rhizoplane organisms (see legend to Fig. 1)

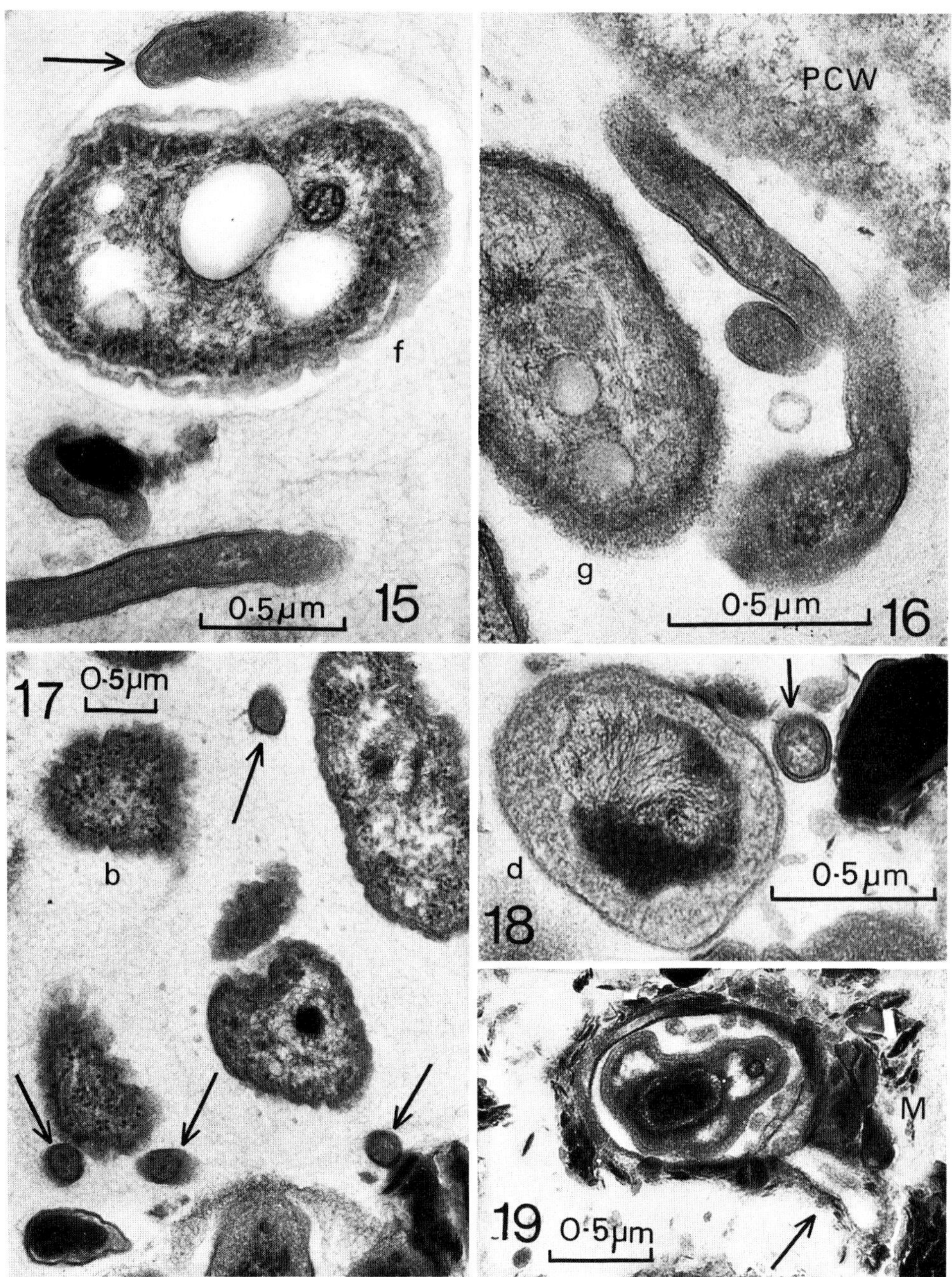


Figs. 15-19 (see legend to Fig. 1)

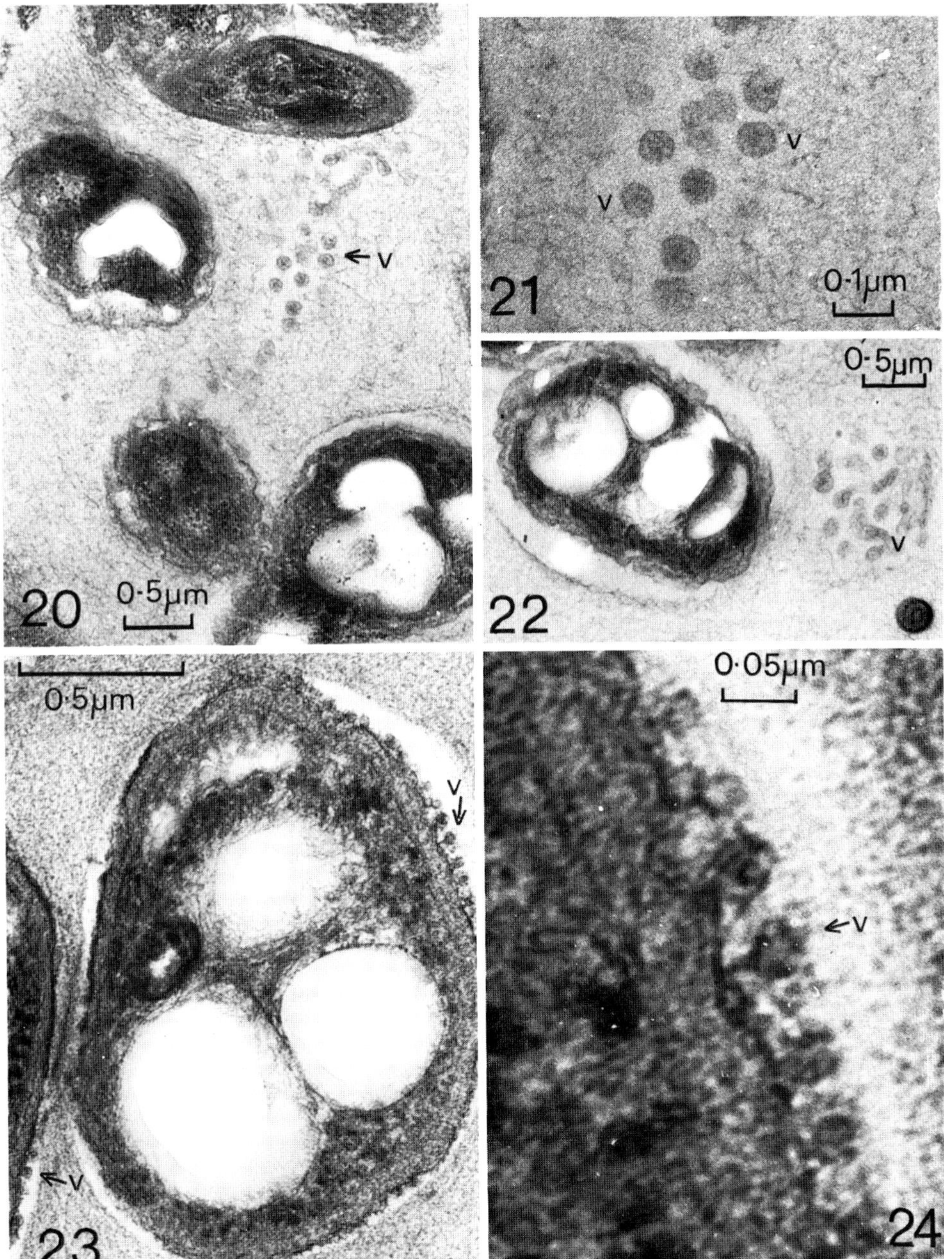


Figs. 20-24. Virus-like particles (see legend to Fig. 1)

cells. We believe that these particles are bacteriophage as they are about 0.05 - 0.01 μm diameter, which is of the right order for bacteriophage (2) and they appear to be angular in cross section with an electron-dense core (Figs. 21-24).

## D. Discussion

This study has shown that the whole of the rhizosphere of lateral roots of clover is permeated by a matrix of fine fibrils in which individual bacteria and colonies are embedded. The matrix extends to at least 20 μm from the root surface in some places, penetrating between the larger soil mineral grains and pushing the smaller clay particles to one side so that they form electron dense boundaries to the smaller bacterial colonies. Sometimes there is a distinct space between the matrix and the colonies of bacteria and between the matrix and individual bacteria suggesting that the matrix is different from the bacterial capsule materials.

It has been known for many years that roots secrete a wide range of organic materials into the rhizosphere (4, 17, 22). Barber and Martin (5) have recently shown that non-sterile roots grown in either simulated or natural soils secrete many times as much carbohydrate into the rhizosphere as sterile roots. Some of these secretions form a distinct mucilagenous layer on the surface of roots (6, 7, 16) and can be seen on roots grown under sterile conditions (7) as well as on roots grown in field soils (12, 13). In other TEM studies we have shown that the mucigel may be very extensive indeed forming a layer 10 or 20 μm thick over the surface of the root and it may have a fibrous appearance.

It is also well known however, that many species of bacteria produce extensive external deposits of fibrous carbohydrate both in pure culture and in mixed microbial associations in natural environments (9, 15). Bacteria obtained by centrifugation of soil suspensions sometimes have fibrous coats (3). Jones _et al_ (15) showed that there is sometimes a transparent amorphous 'void' between the bacteria and the fibrous layer similar to that seen in Figures 6 and 15. These facts suggest that the fibrous matrix in the rhizosphere of the lateral roots of older clover consists mainly of capsule material produced by the bacterium type 'f' and not mucigel secreted by the host.

The density of microorganisms in the _Trifolium_ rhizosphere ($5 \times 10^{10} cm^{-3}$) is about three orders of magnitude greater than that obtained by conventional plating techniques. Moreover the volume of soil sampled in Figure 1 contains at least eight different kinds of organism, since this number can be distinguished cytologically. This must mean that the root has contacted at least eight morphologically distinct founder microorganisms, which is many more than would be expected in so small a volume of soil, and it would appear that in order to account for the variety of organisms seen in Figure 1 there must have been migration of organisms from the surrounding soil into the rhizosphere.

In view of the richness of the rhizosphere in bacterial types and numbers it would be surprising if predatory organisms had not exploited this environment. Both _Bdellovibrio_ and bacteriophage are usually obtained from soils and we believe that our interpretation of our electron micrographs is not inconsistent with the data presented in the literature (14) and that both these organisms are present in the clover rhizospheres. Close examination of wheat rhizospheres from the same soil shows that these contain _Bdellovibrio_-like organisms also,(12) and they may play an important role in the control of rhizosphere organisms. If our interpretation is correct, this is to our knowledge the first demonstration by direct observation of _Bdellovibrio_ and bacteriophage in rhizospheres of plants grown under field conditions.

## E. Acknowledgements

We are pleased to acknowledge the considerable skill and patience shown by Mrs P. Udompongsanon in the preparation of this material for electron microscopy and we thank Mr. T. Cock for preparing the plates for publication.

## F. References

1. Abram, D. & B.K. Davis: Structural properties and features of Bdellovibrio bacteriovorus. J. Bact. 104: 948-965 (1970).
2. Ackerman, H.W.: The Morphology of Bacteriophages. In Laskin A.I. & H.A. Lechatilier, Handbook of Microbiology, Vol. 1. pp. 573-607 (1972). CRC Press, Cleveland, Ohio, USA.
3. Bae, H.C., E.H. Cota-Robles & L.E. Casida: Microflora of soil as viewed by transmission electron microscopy. Appl. Microbiol. 23: 637-641 (1972).
4. Barber, D.A. & K.B. Gunn: The effect of mechanical forces on the exudation or organic substances by the roots of cereal plants grown under sterile conditons. New Phytol. 73: 39-45 (1974).
5. Barber, D.A. & J.K. Martin: The release of organic substances by cereal roots into soil. New Phytol. 76: 69-80 (1976).
6. Campbell, R. & A.D. Rovira: The study of the rhizosphere by scanning electron microscopy. Soil Biol. Biochem. 6: 747-752 (1973).
7. Dart, P.J. & F.V. Mercer: The legume rhizosphere. Archiv. f. Mikrobiol. 47: 344-378 (1964).
8. Dawes, C.J. & F. Bowler: Light and electron microscope studies on the cell wall structure of root hairs of Raphinus sativus. L. Amer. J. Bot. 46: 561-565 (1959).
9. Fletcher, M. & G.D. Floodgate: An electron microscopic demonstration of acid polysaccharide involved in the adhesion of a marine bacterium to solid surfaces. J. Gen. Microbiol. 74: 325-334 (1973).
10. Foster, R.C. & G.C. Marks: Observations on the mycorrhizas of forest trees. The rhizosphere of Pinus radiata D. Don. Aust. J. Biol. Sci. 20: 915-926 (1967).
11. Foster, R.C. & A.D. Rovira: The rhizosphere of wheat roots studied by electron microsccpy. Bull. Ecol. Res. Comm. (Stockholm) 17: 93-102 (1973).
12. Foster, R.C. & A.D. Rovira: Ultrastructure of the wheat rhizosphere. New Phytol. 76: 343-352 (1976).
13. Greaves, M.P. & J.F. Darbyshire: The ultrastructure of the mucilaginous layer on plant roots. Soil Biol. Biochem. 4: 443-449 (1972).
14. Horowitz, A.T., M. Kessel & M. Shilo: Growth cycle of predaceous Bdellovibrios in host free extract system and some properties of the host extract. J. Bacteriol. 117: 270-282 (1974).
15. Jones, H.C., I.C. Roth & W.M. Saunders: Electron microscopy study of a slime layer. J. Bacteriol. 99: 316-325 (1969).
16. Jenny, H. & K. Grossenbacher: Root-soil boundary zones as seen in the electron microscope. Soil Sci. Soc. Amer. Proc. 27: 273-277 (1963).
17. Martin, J.K.: $^{14}C$-labelled material leached from the rhizosphere of plants supplied continuously with $^{14}CO_2$. Soil Biol. Biochem. 7: 395-399 (1975).
18. Old, K.M. & T.H. Nicolson: Electron microscopical studies of the microflora of sand dune grasses. New Phytol. 74: 51-88 (1975).
19. Old, K.M., S. Hallam & J.H. Nicholson: Helicallylobed bacteria in plant roots. Soil Biol. Biochem. 7: 73-75 (1975).

20. Parker, C.A. & P.L. Grove: Bdellovibrio bacteriovorus parasitizing Rhizobium in Western Australia. J. Appl. Bacteriol. 33: 253-255 (1970).
21. Roper, M.M. & K.C. Marshall: Modification of the interaction between E. coli and bacteriophage in saline sediments. Microbial Ecol. 1: 1-13 (1974).
22. Rovira, A.D.: Plant root exudates. Bot. Rev. 35: 35-57 (1969).

# The Root Cortex as Part of a Microbial Continuum

K. M. OLD and T. H. NICOLSON

## A. Introduction

Those who have studied symbiotic root associations such as nodulation and endomycorrhizas have long been conscious of cortical penetration. It is only recently, however, mainly by the use of electron microscopy, that it has been realised that the general rhizosphere microflora extends into cortical tissues. The general concept which is presented here is that the root cortex should be included as part of the soil/root microbial environment. While this should be regarded as a microbial continuum it is convenient to separate the interrelated components of:
(a) The rhizosphere soil; (b) The endophytic hyphasphere since most plants are mycorrhizal; (c) The root surface or rhizoplane, and (d) the root epidermis and cortex.

Fine structure aspects of (a) are being dealt with in this symposium (6). This article covers some aspects of the other three microbial 'niches'.

## B. Materials and Methods

The plants utilised in this study were 4-week old seedlings of barley (c.v. Imber) grown in 12 cm pots of light sandy soil with 20 plants per pot. Portions of the root systems were dissected as quickly as possible after removal from the soil and fixed in 3% gluteraldehyde in phosphate buffer (pH 7) for 24 h. Portions of the roots, approximately 5 mm long taken from various parts of the root axis, were critical point-dried and coated with gold for scanning electron microscopy (SEM) observation. Other root segments were prepared for transmission electron microscopy (TEM) by post fixation with 1% $OsO_4$ in phosphate buffer for 2 h. After dehydration in ethanol they were embedded in Araldite. Ultra-thin sections were prepared and stained for 2 min in a saturated solution of uranyl acetate in 50% ethanol and approximately 7 min in lead acetate.

## C. Results and Discussion

The microbiology of young root surfaces is greatly complicated by the presence of mucilaginous materials generally termed 'mucigel' and presumably mostly of plant origin. The outer surfaces of the epidermal cells near the root tip may be coated with this material (Fig. 1). It is hoped that current studies (3) will shed more light on the structure and function of the mucigel. On older surfaces it seems to be largely removed (Fig. 2) presumably by microorganisms, though similar material of microbial origin may remain (12). On such surfaces complex bacterial colonies may appear (Fig. 3).

In establishing a cortical microflora, penetration takes place by two chief avenues. First, bacteria may penetrate between epidermal and cortical cells (Fig. 7), concentrating in the grooves between epidermal cells (Fig. 2), which was first noted by light microscopy (15), and is presumably a response to a greater secretion of nutrient at that juncture. Second, a more striking manner of penetration is directly through cell walls. In many cases individual or groups of bacteria erode through and form distinct apertures in the walls (5, 12) as shown in Fig. 6. Single apertures can enlarge and join forming gullies leading to considerable rupturing and fraying of the surface layers (Fig. 4). The surfaces of Endogonaceous hyphae are also similarly eroded and holes may appear (Fig. 5). While perforation and erosion of the walls of these cortical and fungal cells appears to be due chiefly to bacterial action, it is possible that giant amoebae may also be involved. Recently it has been found (10, 11) that such organisms are responsible for the perforation of a variety of fungal spores and cells of other organisms in natural soils. These delicate naked amoebae would be extremely difficult to reveal by electron microscopy of root tissue.

The majority of bacteria involved in cortical penetration are 'normal-appearing cells without unique characteristics' (1, 5) although some more exotic types have been encountered. Examples are shown in Figs. 8 and 9. In Fig. 8 a cell of unknown origin is occupied by a group of smaller bacterial cells, and is similar to examples published by Mishustin and Nikitina (9) and identified as '<u>Bdellovibrio</u> from soil in cells of <u>Tuberoidobacter nulans</u>'. A frequent occupant of cortical cells either as single bacteria or colonies is a 'helically-lobed' organism (Fig. 9) which is similar to frequent colonisers of lysing fungal cells (13).

The root has been described as 'the interface between terrestrial life and the mineral substrate supplying all other essential elements' (4). It is becoming increasingly evident that this interface is three-dimensional extending deeply into the root cortex and out into the mineral soil. The existence of mucigel as a layer of pectic gel-like substance that could facilitate transfer of nutrient ions from the soil solution to the cells of the plant was proposed by Jenny and Grossenbacher (7). Leppard (8) showed that wheat roots bore an extensive web of fine strands consisting primarily of polygalacturonic acid that extended deeply into the surrounding soil. The various niches afforded by the cortex, epidermis, mucigel and associated structures, and the adjacent mineral soil matrix are occupied by a microflora that differs quantitatively and qualitatively from the soil at large. Microbial components, such as endophytic hyphae, may also similarly provide ecophysiological substrates (Fig. 5). All substances entering the root must pass through this complex continuum of animate and inanimate structures. Recently, Pauli and Deelman (14) have suggested that integration of these is so complete that they behave, and should be regarded, as a single unit. Apart from special cases

---

Fig. 1. Surface of epidermal cells adjacent to the root tip covered with mucigel ▶

Fig. 2. Epidermal cells intact but lacking mucigel. Bacteria (arrows) are most numerous in the groove between cells

Fig. 3. Many bacteria on the epidermal cell surface. Some are partially embedded in mucilage (arrow)

Fig. 4. Ruptured and perforated epidermal cell wall

Fig. 5. Bacteria adhering to a fungal hypha. The hyphal wall shows signs of erosion (arrows)

Fig. 6. Bacteria penetrating directly through an epidermal cell wall

Fig. 7. Bacteria (arrows) which have penetrated between cortical cells

Fig. 8. Very small bacteria apparently within a prokaryotic cell

Fig. 9. Helically-lobed bacterium in a lysed cortical cell

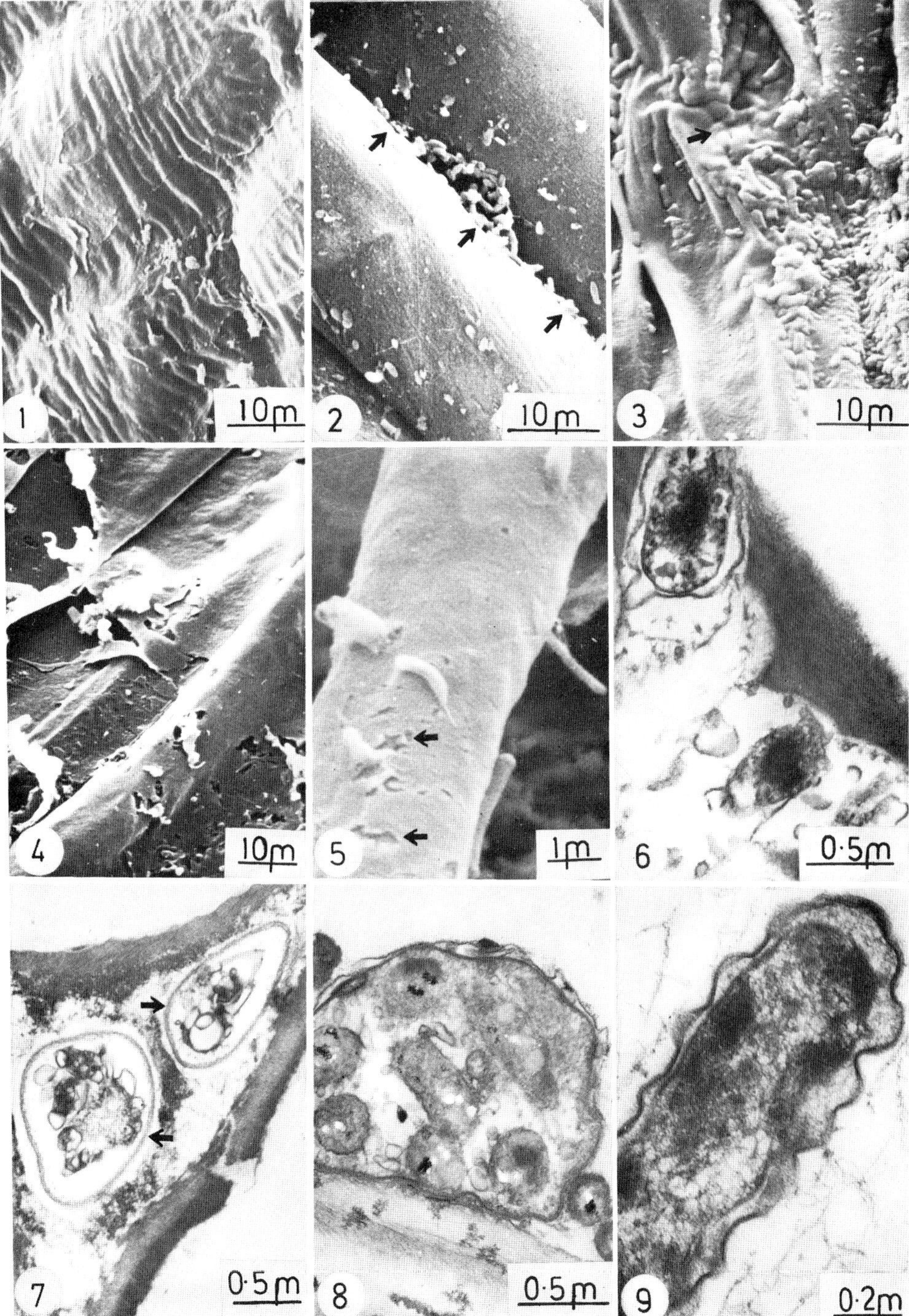


Figs. 1 - 9

however, (2) little is known of the extent to which the microflora mediates nutrient uptake. The contribution of electron microscopy has been to demonstrate the spatial relationships which exist between the microflora and the cells of roots grown in natural soils (5, 12). There is now a need for a more thorough investigation of the effects of this microflora on root function and biodegradation.

## D. References


1. Bae, H.C., Cota-Robles, E.H., Casida, L.E.: Microflora of soil as viewed by transmission electron microscopy. Appl. Microbiol. 23, 637-648 (1972).
2. Barber, D.A., Bowen, G.D., Rovira, A.D.: Effects of microorganisms on absorption and distribution of phosphate in barley. Aust. J. Plant Physiol. 3, 801-808 (1976).
3. Bowen, G.D., Rovira, A.D.: Microbial colonization of plant roots. Ann. rev. Phytopath. 14, 121-144 (1976).
4. Epstein, E.: A blind spot in biology. Science, 176, 235 (1972).
5. Foster, R.C., Rovira, A.D.: Ultrastructure of wheat rhizosphere. New Phytol. 76, 343-353 (1976).
6. Foster, R.C., Rovira, A.D.: The ultrastructure of the rhizosphere. Proc. in Life Science Series Microbial Ecology. Springer Verlag (1978).
7. Jenny, H., Grossenbacher, K.: Root soil boundary zones as seen in the electron microscope. Soil Sci. Soc. Proc. 27, 273-277 (1963).
8. Leppard, G.G.: Rhizoplane fibrils in wheat; demonstration and derivation. Science, 185, 1066-1067 (1974).
9. Mishustin, E.M., Nikitin, E.C.: Parasitic soil microorganisms. Bull. Ecol. Res. Comm. (Stockholm) 17, 37-44 (1973).
10. Old, K.M.: Giant amoebae cause perforation of conidia of Cochliobolus sativus. Trans. Br. mycol. Soc. 68, 277-281 (1977).
11. Old, K.M., Darbyshire, J.F.: Soil fungi as food for giant amoebae. Soil Biol. Biochem. (in press) (1978).
12. Old, K.M., Nicolson, T.H.: Electron microscopical studies of the microflora of sand dune grasses. New Phytol. 74, 51-58 (1975).
13. Old, K.M., Wong, J.N.F.: Helically-lobed soil bacteria from fungal spores. Soil Biol. Biochem. 4, 39-41 (1972).
14. Pauli, F.W, Deelman, J.C.: Soil-plant interface in the root-hair zone as a unity of opposites. Perspectives in Biol. and Med. 19, 493-499 (1976).
15. Rovira, A.D.: A study of the development of the root surface microflora during initial stages of plant growth. J. Appl. Bact. 19, 72-79 (1956).

# Studies of Microbial Colonization of Wheat Roots and the Manipulation of the Rhizosphere Microflora

B. SCHIPPERS and J.W.L. VAN VUURDE

## A. Introduction

Plant growth is directly influenced by the saprophytic root microflora. Rhizosphere organisms may supply, or compete for nutrients, may produce growth substances, toxic metabolites and plant cell wall degrading enzymes. The root saprophytes may also be of indirect influence by interacting with root pathogens and thus affecting the incidence of root diseases.

Some studies indicate that the saprophytic root microflora composition can be manipulated in a way which is advantageous to plant growth and detrimental to root pathogens e.g. by treatment of leaves with chemicals (4, 6) or by inoculation of seeds or roots with selected saprophytic micro-organisms (1). Results, however, often are controversial and unpredictable, probably due to the interference of many factors with microbial colonization of the roots e.g. the type and nutrient level of the soil and the age and physiological state of the plant. More basic knowledge regarding microbial colonization of roots is needed. Research can then be directed toward the manipulation of the rhizosphere microflora to the benefit of the plant.

## B. Methods and Experiments

A plant-soil model system was used to study the colonization of roots by bacteria and actinomycetes and the effect of urea leaf treatments on the rhizosphere composition. Wheat seedlings were grown in sloping perspex boxes in controlled environment growth cabinets. The boxes were filled with sandy loam with controlled water matrix potential. From emergence until sampling, control seedlings were sprayed daily with 2 ml water. Urea leaf treatment was carried out by spraying 2 ml 1.5% urea in water (instead of water) from day 4 till day 7 after seedling emergence.

After approximately two week's growth, each 11 day old seminal root was dissected into 11 portions, the portions being formed during the 11 successive days. The portions are called here 'day-segments' I (root tip) to XI (root base). Any lateral roots were removed from the day-segments and adhering soil particles were dislodged by gently shaking the roots in sterile water. Portions 10 mm in length were taken at random from each day-segment, shaken with glass beads in water and the suspension plated on a special peptone agar. Another random 10 mm portion of each day-segment was treated with fluorochromes. A direct assessment of the proportion of the root surface covered by micro-organisms was made using fluorescence microscopy. Successive staining of the roots with coriphosphine as the main fluorochrome and with congo red and acridine orange overcomes the intensive background fluorescence of cortical cells which has so far hampered the use of fluorescence microscopy for direct observation of the root surface microflora.

The percentage of root surface covered with bacteria and actinomycetes was determined by taking six microscope fields at random per 10 mm root portion per day-segment.

Micro-organisms developing from the plated suspension on agar were tested for their antagonistic properties towards Gaeumannomyces graminis using the triple agar layer test of Herr (2).

## C. Results

The graph of Fig. 1 represents the mean percentages of root surface covered by bacteria of replicate day-segments of three seminal roots. A rise in root surface colonization always occurs on the day-segments III, IV or V of individual roots. A second rise on day-segments VII, VIII or IX is usually followed by a decrease in numbers of bacteria but not of actinomycetes.

Sources of nutrients released by the roots and supposed to be responsible for the rises in coverage of the root surface by bacteria are schematically presented (Fig. 1).

The first colonization peak at day segments III, IV or V corresponds to the first major zone of nutrient release, the ± 0.5 cm zone of cell elongation at the root tip. Most of this nutrient is released in insoluble form and comprises sloughed-off root cap cells and polysaccharide compounds which remain in the immediate vicinity of the roots. Soluble simple carbon compounds that diffuse into the surrounding soil are also released (5). The gradual slope of the root colonization graph on day-segments III to V (Fig. 1) is due to the fact that the day-segment that is most completely colonized differs among the individual roots. This may be due to differences between roots in the daily release of nutrients from the

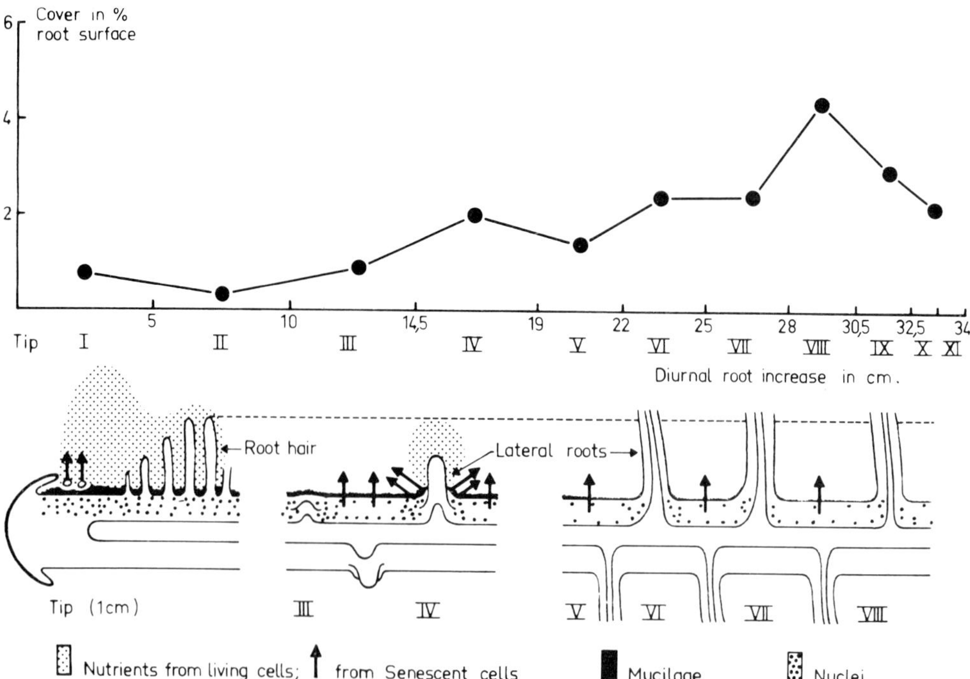


Fig. 1. Graph of the mean percentages of root surface covered by bacteria of replicate day-segments of three seminal roots

root tip and from dying cells and in efficiency of nutrient consumption by early root colonizers.

The second colonization peak at day-segments VII to IX appears about four days after the first lateral root breaks through the cortex and corresponds to the second major zone of nutrient release. In this zone there are also increasing numbers of senescent epidermal cells, root hairs, and cortical cells, recognized by the lack of stainable nuclei (3). Nutrients diffusing from these senescent cells may lead to an increase in the population of root surface micro-organisms.

The decrease in cover by bacteria of day-segments X and XI can be explained by a decrease in easily consumable nutrients in these older root segments. Alternatively this decrease of bacteria may be due to the noticed increase in actinomycetes that utilize the more complex compounds present at these sites.

The dilution plate counts gave root colonization patterns similar to that estimated by fluorescence microscope analysis.

The proportion of the root surface covered by micro-organisms was increased after urea leaf treatments, on segments older than II days. This resulted in a more gradual increase in coverage of the surface of individual roots, but at a higher population density compared to the controls. Similar results were obtained with dilution plate counts (7).

Urea leaf treatment also significantly increased numbers of micro-organisms that are antagonistic to G. graminis in vitro. This increase, however, only took place on day-segments IV to VII, that were formed four days after the four successive urea leaf treatments, respectively.

## D. Conclusions

Our experiments indicate that information on the underlying principles of microbial colonization of roots can be extended by direct observation using fluorescence microscopy of roots grown in model systems. It provides information needed for the direction of research towards manipulation of the root microflora in a manner beneficial to plant growth.

Leaf treatments appear to be a good tool for studying root colonization in relation to the major sites of nutrient release. Basic knowledge of root microflora colonization will help to formulate criteria for the isolation and selection of 'candidate antagonists' to be used against pathogens and the selection of microorganisms advantageous to plant growth. Such micro-organisms are likely to be fast growing organisms that easily settle and migrate on the root surface.

## E. References


1. Brown, Margaret E.: Seed and root bacterization. A. Rev. Phytopath. 12, 181-197 (1974).
2. Herr, L. J. : A method of assaying soils for numbers of actinomycetes antagonistic to fungal pathogens. Phytopathology 49, 270-273 (1959).
3. Holden, J.: Use of nuclear staining to assess rates of cell death in cortices of cereal roots. Soil Biol. Biochem. 7, 333-334 (1975).
4. Khalifa, O.: Effect of urea on pea wilt. Rep. Rothamsted Experimental Station. Report for 1963, 116-117 (1964).
5. Rovira, A. D., Davey, C. B.: Biology of the rhizosphere. In: The Plant Root and Its Environment. Carson, E. W. (ed.). Charlottesville: Univ. Press of Virginia pp. 153-204 (1974).

6. Vraný, J.: The effect of foliar application of urea on the root fungi of wheat growing in soil artificially contaminated with *Fusarium* spp. Folia microbiol., Praha 17, 500-504 (1972).
7. Van Vuurde, J.W.L. and Schippers, B.: Effect of foliar application of urea on rhizosphere and rhizoplane microflora of wheat. EPPO Bull. 5, 395-405 (1975).

# The Variation with Plant Age of Root Carbon Available to Soil Microflora

J.K. MARTIN

## A. Introduction

The amount and chemical nature of the organic carbon released into soil from actively growing wheat roots has been established with growth cabinet experiments in which $^{14}CO_2$ was supplied to plant shoots (1, 10, 11). This paper presents data for the nature of the water-soluble $^{14}C$-labelled material released into soil after the wheat had ripened and the soil was air-dried under summer conditions. These results are combined with those from the experiments with growing plants to give a generalized scheme for the different classes of plant-derived organic material which become available to the rhizosphere microflora at different stages of plant development.

## B. Materials and Methods

Pots containing $^{14}C$-labelled wheat plants (8) were kept in a glasshouse at air temperatures of 15-30$^{o}$C for 7 months. The desiccated plant tops were clipped 0.5 cm from the soil surface and the soil blocks, containing the undisturbed roots, were cooled to 4$^{o}$C. The soil (1 kg/pot) was saturated with cold distilled water and leached with cold distilled water 1, 2 and 5 days after wetting, giving 100 ml leachate at each collection. The pots were brought to room temperature and further leachings (each 100 ml) with distilled water were collected 8, 15, 22, 29 and 36 days after the initial wetting. The soil blocks were air-dried and stored in a glasshouse for 16 months after which the leaching sequence described above was repeated.

$^{14}C$-labelled material in the water leachates was partially fractionated on columns of anion-exchange resin (carbonate form) and by ultra-filtration through a membrane with a nominal cut-off mol. wt. - $10^3$ (9).

## C. Results

The $^{14}C$ activity in the successive leachates is shown in Figure 1. The initial water-soluble activity, which was 1.8 times higher than the maximum value obtained for leachates from pots containing growing plants (8), must have been released during air-drying since the pots were leached for several weeks after the wheat had ripened (8). This view was confirmed when air-drying restored the $^{14}C$ activity in each leachate of the second sequence to near the corresponding value obtained in the first sequence (Fig. 1).

In both sequences, there were clear cut differences between the first and final leachates for the proportion of the $^{14}C$ activity behaving as low molecular weight "neutral" material (resin fraction (i) in Table 1). There were also large differences in the amounts of anthrone reactive material, with values representing ca. 40% of the organic carbon in the first leachates but < 3% of the carbon in the final leachates. Thin-layer chromatography and electrophoresis followed by radio-

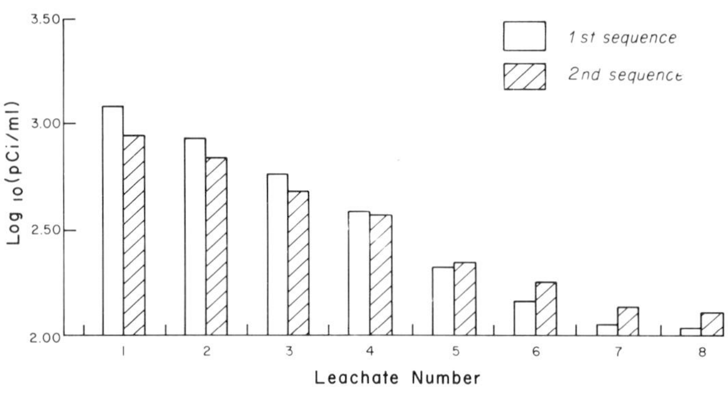


Fig. 1. $^{14}C$ activity in water leachates from soil containing dead wheat roots (mean of 5 replicates)

autography showed that much of the $^{14}C$ activity in leachates 1+2 had mobilities comparable to simple hexoses. Gas chromatography of alditol acetates prepared directly without acid hydrolysis confirmed the presence of free sugars in the first leachates. A value of 378 μg free glucose/mg C was obtained, together with smaller quantities of other free monosaccharides (values expressed as a percentage of total free sugar were - glucose 64.7, galactose 6.0, mannose 18.6, xylose 2.7, arabinose 4.6, ribose 2.6, fucose 1.5, rhamnose 1.3). Comparable treatment of combined leachates 7+8 showed only traces of glucose and mannose.

Table 1. Partial characterization of the water-soluble organic matter released from dead wheat roots in undisturbed soil

| | 1st leaching sequence | | 2nd leaching sequence | |
|---|---|---|---|---|
| Combined leachates | 1+2 | 7+8 | 1+2 | 7+8 |
| Resin fraction[a] | | | | |
| (i) Not retained | 36 | 4 | 32 | 4 |
| (ii) Exchangeable | 26 | 13 | 26 | 17 |
| (iii) Not recovered | 38 | 83 | 42 | 79 |
| Mol. wt $<10^3$ [a] | 52 | 25 | 60 | 38 |
| Hexose carbon[b] | 960 | 77 | 948 | 102 |

[a] Defined in Martin (9). Values are % of water-soluble $^{14}C$ activity recovered in the fraction.

[b] Anthrone reactive material (6) expressed as μg glucose equivalent/mg organic C in the combined leachates.

## D. Discussion

The rewetting of an air-dry soil results in a rapid evolution of $CO_2$, followed by a slower steady rate of $CO_2$ evolution. The effect has been explained by Birch (2) as the result of air-drying increasing the surface area of colloidal organic matter exposed to microbial attack and by Jenkinson (7) as the result of air-drying causing a partial sterilization of the soil flora, followed by decomposition by surviving organisms of the material released from the dead cells. It seems likely that both effects operate (13). Regardless of the releasing mechanism, the present results provide a simple explanation for the flush of $CO_2$. Two fractions of organic material would be available to the surviving microorganisms - a finite amount of carbohydrate-rich material and a much larger amount of negatively charged material which could be maintained in soil solution at a steady low concentration. These results, in conjunction with those from my earlier studies, provide a basis for the proposal that there is a succession of different classes of plant-derived organic material available to rhizosphere microflora with increasing plant age.

The first material available to the soil microflora would be root exudates and root mucilage. Very large populations of microorganisms have been shown to occur within grooves between epidermal cells, on the root surface, or within the mucilage layer surrounding roots (3, 5). These organisms could have very specific growth requirements, limiting their growth on conventional solid media. Some of the disparity between the numbers of soil microorganisms estimated from plate counts and from direct microscopic measurement could then be explained if many of the microbial cells found in soil were survivors of populations which had grown in intimate association with root surfaces.

The next major source of bacterial nutrients arises from the autolysis of degenerate epidermal and cortical tissue and sloughed root cap material (10). Cell lysis appears to be part of normal root development but may be accelerated by some components of the soil microflora (10) and particularly by localized water stress (12). Some 20% of the $^{14}C$ present in root-free soil from the growth cabinet experiments consisted of low mol. wt. material with the chemical properties of simple sugars and amino acids (11). Root lysates, diffusing into the soil from the roots, are most probably responsible for the stimulation of fluorescent pseudomonads in the rhizosphere of wheat plants from early tillering to flowering (9).

Subsequent to cell lysis, there will be extensive colonization and degradation of the structural polysaccharides of the cell wall by bacteria, fungi and actinomycetes (4).

Finally, there is the release of low mol. wt. material, rich in simple sugars, which follows plant maturation and/or drying of the topsoil. This material will probably persist in the dry soil but will be rapidly decomposed when the soil is rewet and accounts for the results presented in this paper.

Although the classes of organic material defined above have been considered separately, they will undoubtedly become available to soil microflora at the same point of time but at different sites on the root system. One possible interaction is the production of antibiotics by fungi or actinomycetes degrading cell walls of senescent tissue adjacent to active root tissue. Organisms, such as the fluorescent pseudomonads, which are resistant to antibiotics, would have a selective advantage in utilizing the sugars, amino acids, etc present in root exudates and root lysates.

## E. Acknowledgement

The author is grateful to Dr. M. Oades, Waite Agricultural Research Institute, for the G.L.C. analysis of free sugars.

## F. References


1. Barber, D.A., Martin, J.K.: The release of organic substances by cereal roots into soil. New Phytol. 76, 69-80 (1976).
2. Birch, H.F.: Further observations on humus decomposition and nitrification. Pl. Soil 11, 262-286 (1959).
3. Bowen, G.D., Foster, R.C.: Dynamics of microbial colonization of plant roots. In: Somiplan Symposium Soil Microbiology and Plant Nutrition. Broughton, W.J., John, C.K. (eds.). (In press).
4. Foster, R.C., Rovira, A.D.: Ultrastructure of wheat rhizosphere. New Phytol. 76, 343-352 (1976).
5. Foster, R.C., Rovira, A.D.: The ultrastructure of the rhizosphere of *Trifolium subterraneum*. L. Proc. in Life Sciences Microbial Ecology. Springer Verlag (1978).
6. Herbert, D., Phipps, P.J., Strange, R.E.: Chemical analysis of microbial cells. In: Methods in Microbiology. Norris, J.R., Ribbons, D.W. (eds.). London: Academic Press, Vol. 5B pp. 209-344 (1971).
7. Jenkinson, D.S.: Studies on the decomposition of plant material in soil - II Partial sterilization of soil and the soil biomass. J. Soil Sci. 17, 280-302 (1966).
8. Martin, J.K.: $^{14}C$-labelled material leached from the rhizosphere of plants supplied with $^{14}CO_2$. Aust. J. Biol. Sci. 24, 1131-1142 (1971).
9. Martin, J.K.: $^{14}C$-labelled material leached from the rhizosphere of plants supplied continuously with $^{14}CO_2$. Soil Biol. Biochem. 7, 395-399 (1975).
10. Martin, J.K.: Factors influencing the loss of organic carbon from wheat roots. Soil Biol. Biochem. 9, 1-7 (1977).
11. Martin, J.K.: The chemical nature of the carbon-14-labelled organic matter released into soil from growing wheat roots - effects of soil microorganisms. In: Soil Organic Matter Studies. Vienna: IAEA, Vol. 1, pp. 197-203 (1977).
12. Martin, J.K.: Effect of soil moisture on the release of organic carbon from wheat roots. Soil Biol. Biochem. 9, 303-304 (1977).
13. Tuckwell, S.B., Jenkinson, D.S.: Effects of air-drying on the release of organically-bound nutrients by soils. Rep. Rothamsted Exp. Stat. Part 1, p. 197 (1974).

# Tracing Populations of Plant Stimulatory *Bacillus* spp. in Alien Rhizosphere Environments

P. FAHY

## A. Introduction

Renewed interest in the use of phytostimulatory Bacillus in Australia has followed the work of Broadbent et al (2) and a number of field trials have shown promising increases in growth and yield of a range of nursery, vegetable, cereal and field crops (1, 5, 6, 7). Use, effectiveness and modes of action of stimulatory bacteria have recently been reviewed (3).

Responses to seed and soil inoculations are not always consistent and this paper reports on some preliminary observations on the tracing of Bacillus subtilis populations in effective and non-effective situations. The usefulness and potential of various methods of tracing bacteria on root surfaces are discussed.

## B. Materials and Methods

I. Pot trials. Wheat (var. Festival) and subterranean clover (vars. Seaton Park and Woogenellup) were grown in a temperature controlled glasshouse ($24^{o}$C, 14 hr day, $18^{o}$C night) in peat/sand potting mix supplemented with complete nutrients and treated with aerated steam ($60^{o}$C/30 min). Clover was also grown in untreated basaltic loam and a granite clay loam. Clover tops were harvested at flowering and wheat seed at maturity.

II. Bacillus inoculations. Inoculum was grown in nutrient broth (48 hr, $25^{o}$C) and injected into finely ground peat ($\gamma$ irradiated 5 x $10^{6}$ rad) to a moisture content of 50% and further incubated for 1 week ($25^{o}$C). Three methods of inoculation were used: (a) incorporation of peat mix into the soil (approximately $10^{5}$ cells $g^{-1}$), (b) lime pelleting with methyl cellulose stuck on to the seed, (c) dipping 6 day old seedlings into washed cells from a nutrient broth culture (48 hr, $25^{o}$C).

III. Bacterial counts. Populations of Bacillus DD32 (2) selected for resistance to streptomycin sulphate (100 $\mu$g $ml^{-1}$) were counted by replicated spread plate dilutions on congo red agar (2) supplemented with streptomycin sulphate (75 $\mu$g $ml^{-1}$ and cycloheximide 50 $\mu$g $ml^{-1}$.)

IV. Fluorescence microscopy. The indirect FITC labelling technique was used with narrow band blue incident light fluorescence (Leitz, Ortholux II, 100 w quartz halide). Antisera to vegetative cells, isolates DD32 and A13, grown on 1% glucose nutrient agar to inhibit sporulation, were prepared by eight intravenous injections of rabbits (3-4 day intervals). Counterstaining with carbol fuschin and normal horse serum was used to reduce non-specific staining of methanol fixed tissue preparations.

V. Scanning electron microscopy. Specimens were fixed in 2½% glutaraldehyde in 0.1 M cacodylate buffer (pH 7.4) with 0.1% ruthenium red for 1 hr, dehydrated in a graded alcohol series, washed twice in acetone, critical point dried with liquid $CO_2$ and gold coated. A Cambridge Stereoscan II at 10 Kv was used. Latex particle (0.5 $\mu$m) - antibody labelling of fixed specimens was attempted (4).

## C. Results and Discussion

I. Pot trials. 1. Clover. Mode of inoculation is important in development of stimulation. In a repeated series of trials seed coated inoculum of *Bacillus* DD32 produced no growth response but soil inoculum yielded significant increases in total dry matter of between 23-50%. Seedling dip inoculum produced an intermediate growth response of 19%. A streptomycin resistant strain produced a similar trend in results: soil inoculum 16-22%, seedling dip 16%, seed coating 0%. Atomic absorption spectrophotometric analysis of 11 major and minor elements (R. Weir, Biological and Chemical Research Institute, Rydalmere, New South Wales) revealed no specific alteration in nutrient uptake in any treatments.

2. Wheat. Seed coating significantly increased grain harvest from between 13-23%. Much larger increases in yield are obtained with soil inoculations (P. Broadbent, unpublished data).

II. Tracing streptomycin resistant *Bacillus* DD32. 1. Seedling dipping of clover. DD32 populations declined to undetectable levels within 24 days on root tips in potting mix and untreated soils. On mature roots levels of 100 to 1,000 cells $mm^{-1}$ root were retained till harvest in all three soils. Soil levels of DD32 rose to approximately 2,000 cells $g^{-1}$ at harvest. Pots were replanted with uninoculated seed after drying (10 days) but DD32 failed to recolonize germinating plants in the three soils tested.

2. Comparison of soil and seed inoculation for clover and wheat. Figures 1-3 trace populations of DD32 over 14 days of growth on seed coats, growing tips and mature roots in potting mix. DD32 failed to multiply on germinating seed or establish on roots of coated clover but continual recolonization with soil inoculum was evident. The bacterium on coated wheat grew on germinating seed and multiplied to the level of the soil inoculum on mature roots.

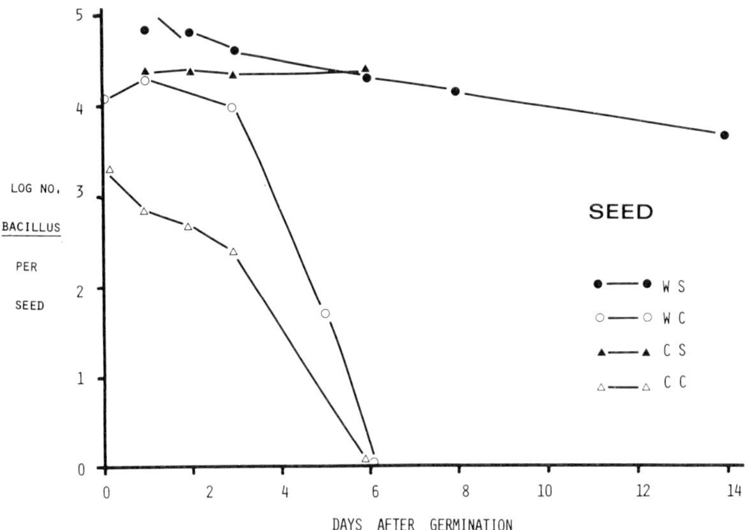


Fig. 1. Growth of streptomycin resistant *Bacillus* DD32 in clover and wheat. WS, Wheat soil inoculum, WC, wheat coated seed, CS, clover soil inoculum, CC, clover coated seed

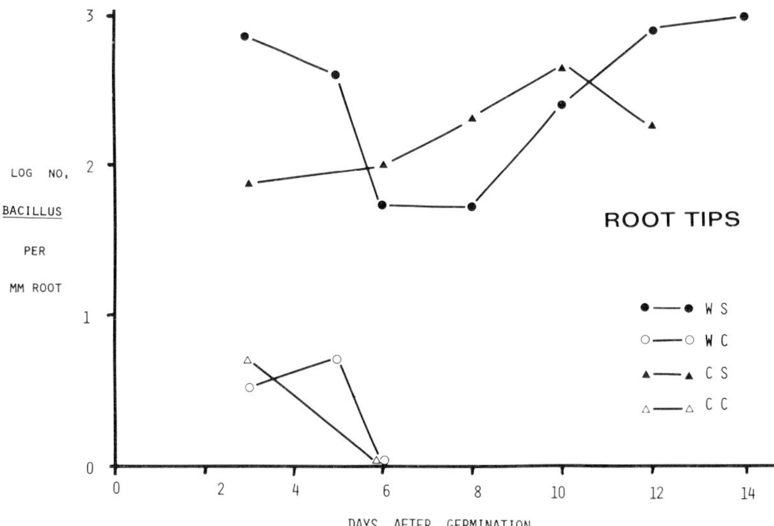


Fig. 2 (legend see Fig. 1)

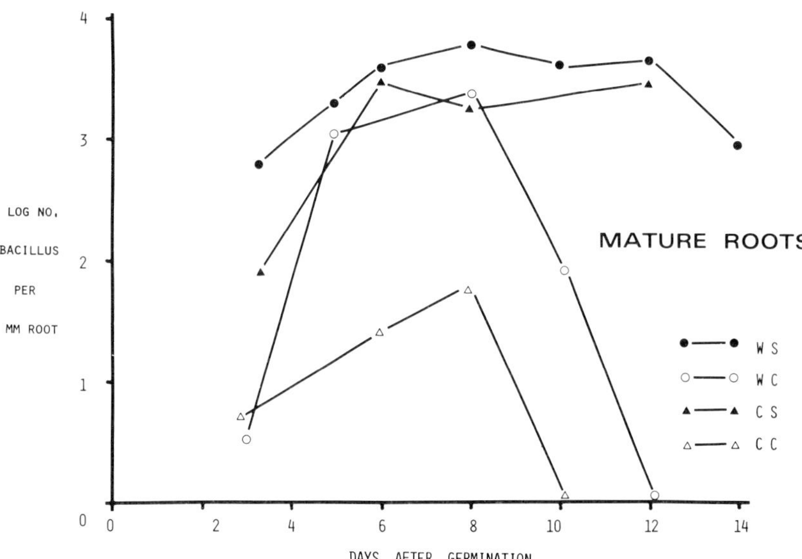


Fig. 3 (legend see Fig. 1)

These differences in establishment on clover and wheat may be sufficient to account for responses observed in pot trials.

3. Double antibiotic marker tracing. High populations of streptomycin resistant bacteria grew on germinating clover and wheat and inhibited growth of *Bacillus* on isolation plates, particularly when the *Bacillus* were in low numbers. Nalidixic acid (75 $\mu g\ ml^{-1}$) inhibited these isolates. A range of nalidixic acid resistant

mutants were selected (at 100 μg $ml^{-1}$) from an effective streptomycin resistant isolate of DD32 and screened for effectiveness on clover and wheat. Growth studies with an effective isolate are in progress.

4. Fluorescent antibody techniques. Bacillus could be detected on the surface of freeze microtome-sectioned roots of 2-3 day old inoculated clover and wheat seedlings but were difficult to observe on whole root preparations due to topography. Most Bacillus were loosely attached to the rhizosplane. Bacillus were easily washed from roots dipped in aqueous solutions, 90% being lost within two minutes with gentle agitation reducing the effectiveness of any quantitative method based on direct observation. Indirect techniques could prove more useful. Antisera to Bacillus subtilis tend to be species specific (8) but absorptions with a range of Bacillus isolates and estimations of fluorescence intensity using a series of neutral grey filters allowed specific detection of Bacillus DD32 and A13.

5. Scanning E.M. studies. Bacillus on germinating seeds can be readily observed by SEM. The addition of ruthenium red to the glutaraldehyde fixative appears to aid in retention of cells on root surfaces. On coated wheat seed cells readily colonized the growing point and developed root hairs during the first two days of root growth.

Antibody coated latex particles were used in an attempt to specifically label Bacillus DD32 on wheat roots. Specific labelling was attained in pure culture but some non-specific binding was apparent on root surfaces. Further development of this technique appears promising.

## D. References


1. Broadbent, P., Baker, K.F., Franks, N.: Effect of Bacillus spp. on increased growth of seedlings in steamed and in nontreated soil. Phytopathology - in press.
2. Broadbent, P., Baker, K.F., Waterworth, Y.: Bacteria and actinomycetes antagonistic to fungal root pathogens in Australian soils. Austral. J. Biol. Sci. 24, 925-944 (1971).
3. Brown, M.E.: Seed and root bacterization. Ann. Rev. Phytopathology 12, 181-192 (1974).
4. Lo Buglio, A.F, Rinehart, J.J., Balcerzak, S.P.: A new immunologic marker for scanning electron microscopy. In: Proc. 5th S.E.M. Symposia, Part II. IIT Res.Inst. Chicago, pp. 313-320 (1972).
5. Merriman, P.R., Birkenhead, W.E.: Effects of Bacillus subtilis and Streptomyces griseus on growth of vegetables. Austral. Plant Pathol. Newsletter 6, 24 (1977).
6. Merriman, P.R., Price, R.D., Baker, K.F.: The effect of inoculation of seed with antagonists of Rhizoctonia solani on the growth of wheat. Austral. J. Agric. Res. 25, 213-218 (1974).
7. Merriman, P.R., Price, R.D., Kollmorgen, J.F., Piggott, T., Ridge, E.H.: Effect of seed inoculation with Bacillus subtilis and Streptomyces griseus on the growth of cereals and carrots. Austral. J. Agric. Res. 25, 219-226 (1974).
8. Norris, J.R., Wolf, J.: A study of antigens of the aerobic sporeforming bacteria. J. Appl. Bacteriol. 24, 42-56 (1961).

# Association of Free-Living Nitrogen Fixing Bacteria with Plantroots in Temperate Region

K. VLASSAK and L. REYNDERS

## A. Introduction

High nitrogenase activities have been recorded in the rhizosphere of several composite plant species (4, 7) and on the roots of tropical graminaceous plants (1, 3, 8). The possibility exists that the association between plantroots and nitrogen-fixing organisms tends towards a semi-symbiotic relationship showing a relatively high energy efficiency. This gives rise to great speculation on the agricultural value of this phenomenon.

This paper deals with the presence of nitrogenase activity on the roots of some plant species from a temperate region.

## B. Materials and Methods

I. Plant cultures. Plants of Zea mays L., Cichorium intybus L. and Helianthus annuus L. were sown in a locally sampled soil containing small numbers of Spirillum lipoferum, while Helianthus tuberosus L. was planted as small root-stocks. The plants were grown under greenhouse conditions, inoculation taking place two and (or) three weeks after sowing by topical application of adequately diluted bacterial cultures ($10^6$ a $10^7$/g soil). The bacteria used as inoculum were obtained as follows - Spirillum lipoferum; str 7 from ATCC 29145; strains M2 and T2 isolated respectively from the roots of Zea mays cv. Anjou 210 and Triticum aestivum cv. Clement; strains S11, S19 and S631 isolated from soil samples under grass, mixed deciduous forest and maize cover; G and R are yet unidentified organisms associated with the T2 isolate. After a growing period ranging from 6 to 13 weeks, plant roots were harvested early in the afternoon and assayed for nitrogenase activity using the acetylene reduction technique (8).

II. Bacterial cultures. For the isolation of Spirillum lipoferum semi-solid, N-deficient malate medium was used (5); yeast extract 0.01% was added when large volumes of cultures were required. Species purity was checked on nutrient agar, while stock cultures of the isolated strains were kept on nutrient agar slants at room temperature.

## C. Results and Discussion

In a previous study (6) high nitrogenase activity was demonstrated on the roots of Zea mays, Triticum aestivum and Cichorium intybus, a plant belonging to the composites. To ascertain whether this activity is found on other inulin containing plants, nitrogenase activities of Cichorium intybus, Helianthus annuus and Helianthus tuberosus, were compared. As can be seen in Table 1 very high activities were recorded, especially with the root-stock forming Jerusalem artichoke, Helianthus tuberosus. On the non-inoculated roots of all three plant-species very high numbers of nitrogen-fixing Spirillum sp. were detected using enrichment cultures of semi-solid, N-deficient malate medium. A very spectacular time-dependent

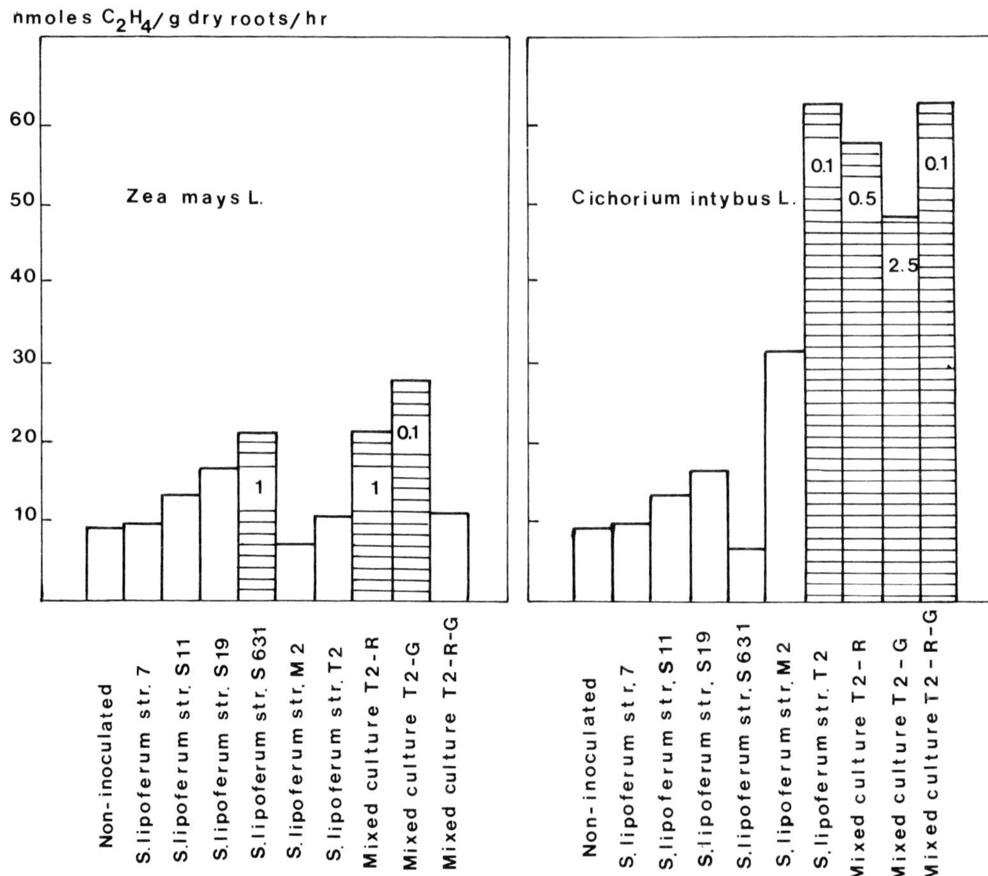


Fig. 1. Nitrogenase activity of Zea mays and Cichorium intybus inoculated with selected bacterial cultures. Shaded treatments differ significantly from non-inoculated control; level indicated percentagewise (LSD, n = 6)

mechanism could be observed, with plants showing significant nitrogenase activity only after 80 to 90 days of growth. The same phenomenon was observed when working on intact plant-soil systems of two composite species (7). This new observation clearly shows the prime importance of the soil-free root in the nitrogenase activity of the plant-soil system.

Because of the variable effect of inoculation on the nitrogenase activity of different plants (6) and because of the wide distribution of nitrogen-fixing Spirillum sp. in Belgian soils (5), an experiment was conducted to test the efficiency or plant specificity of a number of isolated bacteria. As shown in Fig. 1 very significant differences were found between the inocula tested. The low nitrogenase activities of the uninoculated control plants and those inoculated with Spirillum lipoferum strains 7, S11 and S19, amounting to the same level in both plant species, supports the idea that each plant species has its preference for specific strains. The non interacting organisms fixed only small amounts of $N_2$, probably as simple asymbiotic nitrogen-fixers. The factors influencing this apparently delicate association are under investigation.

Table 1. Nitrogenase activity (nmoles $C_2H_4$ plant$^{-1}$ $h^{-1}$) on the roots[a] of some composite plant species after 80 - 90 days of growth

| Plant species | Treatment | Nitrogenase activity range | Mean activity |
|---|---|---|---|
| Cichorium intybus | Non-inoculated | 6.4 - 12.3 | 9.9 |
| | Inoculated [b] | 24.4 - 55.8 | 42.3 |
| Helianthus annuus | Non-inoculated | 12.7 - 118.8 | 56.9 |
| | Inoculated | 12.0 - 299.8 | 89.5 |
| Helianthus tuberosus | Non-inoculated | 207.0 - 631.9 | 380.8 |
| | Inoculated | 155.8 - 797.6 | 474.7 |

[a] Average root dry weight ± 1 gram.

[b] Inoculated with *Spirillum lipoferum* str. Sll.

## D. Acknowledgements

The authors are grateful to N.F.W.O. (Nationaal Fonds voor Wetenschappelijk Onderzoek) and I.W.O.N.L. (Instituut tot Aanmoediging van het Wetenschappelijk Onderzoek in Nijverheid en Landbouw) for financial support.

## E. References

1. Balandreau, J., Rinaudo, G., Ibtissam Fares-Hamad, Dommergues, Y.: Nitrogen fixation in the rhizosphere of rice plants. In: Nitrogen fixation by free-living microorganisms. Stewart W.D.P. (ed.) I.B.P. 6 Cambridge: University Press, 57-70 (1975).
2. Beijerinck, M.W.: Ueber ein *Spirillum* welches freien Stickstoff binden kann? Zentralbl. Bakteriol. Parasitenkd. Infektionskr. Hyg. Abt. 2, 63, 353-359 (1925)
3. Dobereiner, J., Day, J.M.: Nitrogen fixation in the rhizosphere of tropical grasses. In: Nitrogen fixation by free-living microorganisms. Stewart W.D.P. (ed.) I.B.P. 6 Cambridge: University Press, 39-56 (1975).
4. Jain, M.K., Vlassak, K.: Assay of nitrogenase activity in intact plant systems. Ann. Microbiol. (Inst. Pasteur), 126 B, 405-408 (1975).
5. Reynders, L., Vlassak, K.: Nitrogen fixing *Spirillum* species in Belgian soils. Agricultura 24, 329-336 (1976).
6. Reynders, L., Vlassak, K.: Nitrogen fixation by *Spirillum* - plantroot associations. Third Intern. Symp. on Environmental Biogeochemistry. Wolfenbuttel, Germany. (March 27/April, 1977).
7. Vlassak, K., Jain, M.K.: Biological nitrogen fixation studies in the rhizosphere of *Cichorium intybus* and *Taraxacum officinale*. Rev. Ecol. Biol. Sol 13, 411-418 (1976).
8. von Bulow, J.F.W., Dobereiner, J.: Potential for nitrogen fixation in maize genotypes in Brasil. Proc. Nat. Acad. Sci. U.S.A. 72, 2389-2393 (1975).

# Endomycorrhizal Fungi Stimulate P Fertiliser Uptake from Soil by Ryegrass and Clover

C. Ll. POWELL

## A. Introduction

Many New Zealand hill country soils still have low to moderate fertility and pasture growth is often highly responsive to superphosphate. All hill country soils so far examined contain endomycorrhizal fungi, most of which greatly stimulate growth and phosphate uptake by clover and ryegrass (5, 6, 8, 10). It has been shown (3, 4, 7) that highly efficient mycorrhizal fungi can be introduced into unsterilized field soils and can stimulate more legume growth than the indigenous mycorrhizal fungi (IMF). It was also shown (3, 4) that efficient mycorrhizal fungi utilised rock phosphate more effectively than the indigenous mycorrhizal fungi did in unsterilised acid soils. Results presented in this paper show that the mycorrhizal fungus *Glomus tenuis* (Greenall) Hall was more efficient than the indigenous mycorrhizal fungi in three New Zealand soils at tapping rock phosphate and stimulating the growth of ryegrass and white clover.

## B. Materials and Methods

I. Soils. Three hill country soils with low pH and low phosphate status were used (Table 1). Soils were collected at 0-10 cm depth from established pasture, passed through a 1 cm diam sieve and packed into 10 cm diam pots holding 310 ml soil. The pots sat on and absorbed water from moist sand trays in an unheated glasshouse.

Table 1. Properties of test soils

| Soil | pH | $NaHCO_3$ extractable P µg P/ml |
|---|---|---|
| Dunmore silt loam | 5.0 | 9 |
| Mahoenui silt loam | 5.4 | 11 |
| New Plymouth brown loam | 5.8 | 4 |

II. Fertilisers. Six phosphate fertilisers were mixed with each test soil at the rate of 129 mg P/litre soil, equivalent to 62 kg P/ha on a superficial basis. The fertilisers used were rock phosphates (90% of particles passing through 150 µm diam sieve) from Nauru Island, Queensland, Chatham rise (NZ), Peru and Christmas Island 'C' grade, and superphosphate made from Nauru Island rock phosphate. Calcined Christmas Island rock was used in Dunmore soil and uncalcined rock in the other two soils. There was also an unfertilised treatment in each soil. Nutrient solution containing all other essential elements except phosphate (1) was added weekly to all pots.

III. Plants and inoculum. Ryegrass (Lolium perenne L cv Grasslands Ruanui) and white clover (Trifolicum repens L cv Grasslands Huia) were grown from seed in dishes of sterilised Dunmore soil highly infested with Glomus tenuis or with the IMF from each test soil. Each seedling dish received filtered soil washings bulked from all seedling dishes to ensure that the contaminating bacteria were the same for all seedlings. Within three or four weeks of seed germination the seedlings had become mycorrhizal and were planted out with ryegrass and 1 clover seedling per pot. In each soil, plants inoculated with the IMF or G. tenuis were grown at each phosphate regime. There were eight replicate pots of the fourteen treatments in each soil (seven phosphate regimes x 2 mycorrhizal fungi).

IV. Measurements. After 64-75 days growth, shoot material was cut at 2 cm above level, dried at 50$^{o}$C overnight and weighed.

## C. Results

In Figures 1-3, the phosphate fertilisers are ranked left to right in order of increasing ability to stimulate clover and ryegrass growth. Figure 1 shows that clover and ryegrass growth in Dunmore soil were highly responsive to mycorrhizal infection and rock phosphate fertiliser. In all phosphate treatments, plants inoculated with G. tenuis were larger than those inoculated with the IMF. This shows that G. tenuis had successfully infected the seedling roots despite competition from the large population of IMF already in the soil and was more efficient at stimulating shoot growth. Efficient mycorrhizal infection invariably stimulates better phosphate uptake by plants (2) especially in soils of low P status in which all other elements have been supplied.

The largest clover growth responses to G. tenuis infection were in unfertilised soil and in soil given Christmas Island or Chatham rise rock phosphate. Similarly, the largest ryegrass growth responses to G. tenuis infection were in unfertilised soil and soil given Chatham rise and Peruvian rock phosphates.

Clover was very responsive to rock phosphates in New Plymouth brown loam (Figure 2). Glomus tenuis again stimualted much more clover growth than the IMF did in unfertilised soil and in soils given Chatham rise and Peruvian rock phosphates. Glomus tenuis was more efficient than the IMF at stimulating ryegrass growth in all phosphate treatments (Figure 2), especially in unfertilised soil and in soil given Christmas Island and Peruvian rock phosphates.

Ryegrass and clover responded very differently to mycorrhizal infection and phosphate fertilisers in Mahoenui silt loam (Figure 3). Glomus tenuis was much more efficient than the IMF at stimulating clover growth, especially in soil fertilised with Chatham rise rock and superphosphate. Clover plants (especially those infected with the IMF) showed very small responses to phosphate fertiliser. There were large growth responses in ryegrass to phosphate fertiliser (Figure 3) but only in soil given Chatham rise and Peruvian rock phosphates was G. tenuis superior to the IMF at stimulating growth.

## D. Discussion

These results confirm Mosse's findings (3) that endomycorrhizal fungi can be introduced into unsterilised soils having high populations of IMF, and can stimulate much more plant growth than the IMF. In general, plants inoculated with G. tenuis grew much better than plants infected with the IMF in unfertilised soils and in soils given Chatham rise and Peruvian rock phosphate. The growth benefit from G.

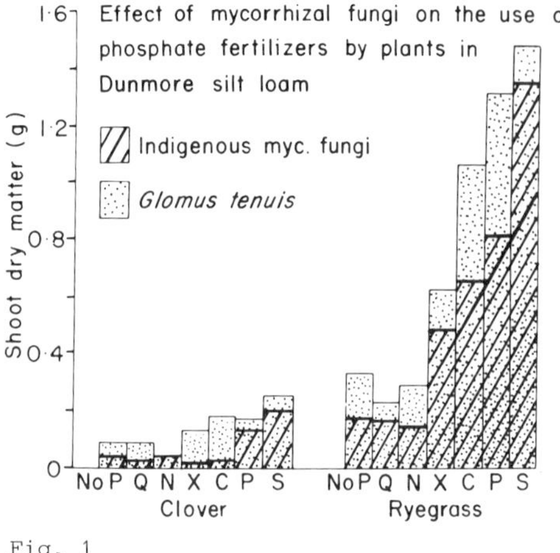


Fig. 1

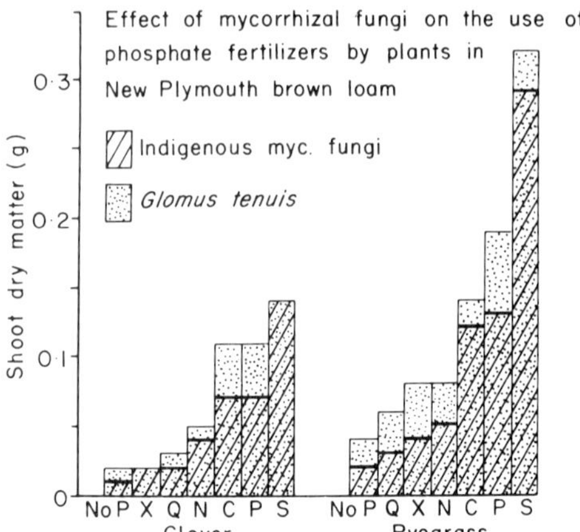


Fig. 2

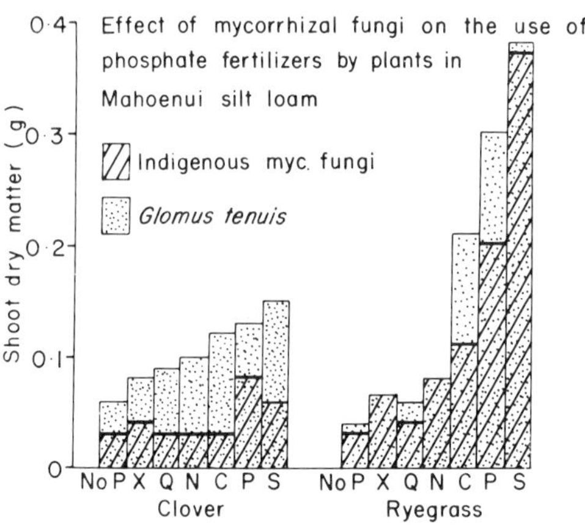


Fig. 3

Figs. 1-3. Shoot dry matter production of clover and ryegrass inoculated with the indigenous mycorrhizal fungi (IMF) or Glomus tenuis and grown in 3 unsterilized soils with several phosphate fertilisers. No P, unfertilised soil; Q, Queensland rock; N, Nauru Island rock; X, Christmas Island 'C' grade rock; C, Chatham rise rock; P, Peruvian rock; S, superphosphate. Fertilisers added at rate equivalent to 62 kg P/ha

tenuis infection will probably increase as soils become depleted of phosphate with repeated cropping (8, 9).

These results do show however that even with G. tenuis - an efficient endophyte in a wide range of soils (Powell, paper in preparation) - the size of the mycorrhizal growth response depends on which host species, soil type and fertilisers are used. In New Plymouth loam for example, G. tenuis was more efficient than the IMF at stimulating ryegrass but not clover growth in soil given Christmas Island rock. Conversely G. tenuis stimulated much more clover but only little more ryegrass growth in soil given Chatham rise rock (Figure 2).

The consistently efficient use of Chatham rise and Peruvian rock phosphates by G. tenuis in three soils suggests that the external hyphae are well suited to spreading through the soil and absorbing phosphate as soon as it dissociates from particles of these two rock phosphates.  If current attempts to introduce efficient mycorrhizal fungi into the field are successful, suitable rock phosphates could be good maintenance fertilisers for well developed pasture.

## E. Acknowledgements

My thanks are due to M.W. Brown for supplying and grading the rock phosphates and manufacturing the superphosphate and to J.S. Park and Mrs J. Daniel for maintaining the experiments.

## F. References

1. Middleton, K.R., Toxopeus, M.R.J.: Diagnosis and measurement of multiple soil deficiencies by a subtractive technique. Pl & Soil 38, 219-226 (1973).
2. Mosse, B.: Advances in the study of vesicular-arbuscular mycorrhiza. Ann. Rev. Phytopath. 11, 171-196 (1973).
3. Mosse, B.: Plant growth responses to vesicular-arbuscular mycorrhiza X. Responses of *Stylosanthes* & maize to inoculation in unsterile soils. New Phytol. 78, 277-288 (1977).
4. Mosse, B., Powell, C.Ll., Hayman, D.S.: Plant growth responses to vesicular-arbuscular mycorrhiza. IX. Interactions between VA Mycorrhiza, rock phosphate and symbiotic nitrogen fixation. New Phytol 76, 331-342 (1976).
5. Powell, C.Ll.: Mycorrhizal fungi stimulate clover growth in New Zealand hill country soils. Nature 264, 436-438 (1976).
6. Powell, C.Ll.: Mycorrhizas in hill country soils. II Effect of several mycorrhizal fungi on clover growth in sterilised soils. New Zealand J. Agric. Res. 20, 59-62 (1977a).
7. Powell, C.Ll.: Mycorrhizas in hill country soils. III Effect of inoculation on clover growth in unsterile soils. N.Z. J. Agric. Res 20, 343-348 (1977b).
8. Powell, C.Ll.: Mycorrhizas in hill country soils. V Growth responses in ryegrass. New Zealand J. Agric. Res. 21 in press (1978)
9. Powell, C.Ll., Daniel, J.: Mycorrhizal fungi stimulate uptake of soluble and insoluble phosphate fertiliser from a phosphate-deficient soil. New Phytol. 80, in press (1978)
10. Powell, C.Ll., Sithamparanathan, J.: Mycorrhizas in hill country soils. IV Infection rate in grass and legume species by indigenous mycorrhizal fungi under field conditions. New Zealand J. Agric. Res 21, in press (1978).

# The Effect of Mycorrhizas on the Phosphate Content, Nitrogen Fixation and Growth of *Medicago sativa*

S.E. SMITH and M.J. DAFT

## A. Introduction

Symbiosis of legumes with micro-organisms frequently involves both Rhizobium spp. which form $N_2$-fixing root nodules, and fungi, forming vesicular-arbuscular mycorrhizas. As in other hosts the development of mycorrhizas results in increased uptake of mineral nutrients (particularly phosphate) and in increased growth (5). Nodulation and nitrogenase activity may be similarly affected (6).

Physiological comparisons between mycorrhizal and non-mycorrhizal plants of different sizes are complicated by the fact that nutrient absorption is influenced by the size of the root system, as well as by the presence of fungal hyphae which increase the volume of soil exploited. Also, nodulation and nitrogen fixation are affected by botn nutrient supply and by availability of photosynthate; the latter is in turn influenced by the size and activity of the shoot. Tinker and co-workers (9) have used measurements of root length and hyphal weight to calculate phosphate inflows into plants via the roots and hyphae. It may also be helpful to compare plants matched in dry weight when assessing the effects of mycorrhizal infection.

In this paper we consider data from experiments on Medicago sativa (described in greater detail elsewhere, (6)) which allow such comparisons of matched plants in terms of phosphate content, nodulation, nitrogen fixation and total plant nitrogen. Changes in these parameters taking place during the growth of plants are also considered.

## B. Materials and Methods

Medicago sativa cv Europe seedlings were raised on sterile John Innes compost for 2 weeks and then transplanted (40 per tray) to trays containing 5 kg of either calcareous soil or acid washed sand. Both soil and sand were air-dried before autoclaving and were used either unamended or with the addition of $Na_2HPO_4$ at the rates of 1.6 and 1.2 m equivalents phosphate per kilogram respectively.

At transplanting all seedlings were inoculated with Rhizobium meliloti. Plants in half the trays also received mycorrhizal inoculum as a suspension of chlamydospores of Glomus mosseae (strain L1 (1)) and infected maize root segments. The remaining plants received filtered washings from infected roots. Plants were grown in a glasshouse with supplementary light from mercury vapour lamps on a 14 h day.

At each harvest thirty plants were removed from each tray and divided into bunches of 10. On each bunch measurements were made of nitrogenase activity (acetylene reduction), nodulation (on an arbitrary scale 0-9), dry weight of roots and shoots, percentage of root length infected by mycorrhizal fungi (following clearing and staining with lactophenol/cotton-blue), phosphate content (by a molybdenum blue method) and nitrogen content (using a Hewlett Packard 185B CHN Analyser). Details of these methods are given elsewhere (6).

## C. Results

Plants grown in soil were harvested 12 weeks after transplanting. Total plant dry weight and nodulation were increased both by phosphate fertilization and by mycorrhizal infection (Table 1). Mycorrhizal plants without added P were approximately the same weight as non-mycorrhizal plants with added P. However, the latter plants had fewer nodules (Table 1, treatments 2 and 3). All plants grown with added phosphate had higher phosphate contents than unfertilised plants (Fig. 1 A and C). Within a fertilizer treatment mycorrhizal plants contained more P than equivalent non-mycorrhizal plants. Nitrogenase activity reflected the degree of nodulation (Fig. 1 B and D).

The effect of both mycorrhizal infection and increased phosphate supply in reducing the proportion of dry weight in the root system can be seen from Table 2. Comparison of treatments 2 and 3 indicates that these plants had similar root:shoot dry weight ratios as well as similar dry weights (Table 1). However, mycorrhizal plants had a higher proportion of their total P in the roots.

Plants grown in sand were harvested at 7 and 12 weeks. Nodulation at both harvests was less affected by the different treatments than in the first experiment. No mycorrhizally induced growth increases had occurred by 7 weeks and, in the presence of added P, mycorrhizal plants were actually smaller than controls (Table 3). However, growth increases in the presence of both phosphate and mycorrhizas were apparent by 12 weeks. Figures 2 and 3 show that while P content and acetylene reduction on a whole plant basis were unaffected by mycorrhizal infection at 7 weeks

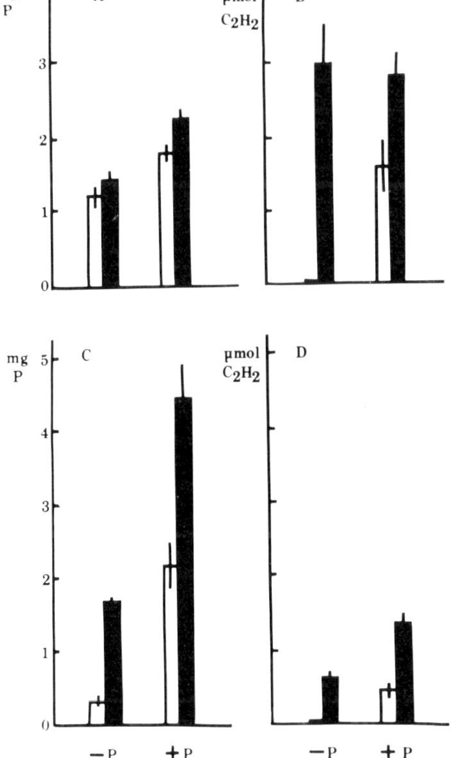


Fig. 1 A-D. *Medicago sativa* plants grown in soil with (+P) and without (-P) the addition of 1.6 m equivalents phosphate/Kg supplied as $Na_2HPO_4$. Phosphate content expressed as mg P/g dry weight (A) and mg P/plant (C). Acetylene reduction expressed as µmol $C_2H_2$ reduced/h/g dry weight root (B) and µmol $C_2H_2$ reduced/h/10 plants (D). (± standard error of mean of 6 bunches)

Table 1. Percentage of roots infected by mycorrhizal fungi (% mycorrhiza) nodulation (assessed on an arbitrary scale 0-9) and dry weight (g) of *Medicago sativa* plants grown on calcareous soil for 12 weeks, in the presence and absence of mycorrhizal fungi (*Glomus mosseae*) and $Na_2HPO_4$ supplied at 1.6 m equiv/kg. (Mean of 6 bunches of 10 plants ± standard error of mean)

| Treatment | Phosphate | | Mycorrhiza (%) | Nodules (0-9) | Dry Weight (g/10 plants) |
|---|---|---|---|---|---|
| 1 | -P | non-mycorrhizal | 0 | <1 | 0.266 ± 0.023 |
| 2 | -P | mycorrhizal | 57 | 5.8 | 1.163 ± 0.093 |
| 3 | +P | non-mycorrhizal | 0 | 2.5 | 1.184 ± 0.147 |
| 4 | +P | mycorrhizal | 52 | 4.5 | 2.202 ± 0.124 |

Table 2. Root: Shoot ratios of dry weight, total P (P in root : P in shoot) and P concentration (mg/P/g Dry weight root : mgP/g Dry weight shoot) in *Medicago sativa* grown in calcareous soil for 12 weeks in the presence and absence of mycorrhizal fungi (*Glomus mosseae*) and $Na_2HPO_4$ , supplied at 1.6 m equiv/kg. (Mean of 6 bunches of 10 plants)

| Treatment | Phosphate | | Root : Shoot Ratios Dry weight | Total P | mg/P/g DW |
|---|---|---|---|---|---|
| 1 | -P | non-mycorrhizal | 0.70 | 0.70 | 1.10 |
| 2 | -P | mycorrhizal | 0.45 | 0.35 | 1.10 |
| 3 | +P | non-mycorrhizal | 0.43 | 0.28 | 0.54 |
| 4 | +P | mycorrhizal | 0.32 | 0.33 | 1.04 |

(C), both were increased on a dry weight basis (A). By 12 weeks the larger mycorrhizal plants had higher total P and nitrogenase activity (D), but there was no difference in phosphate content between treatments on a dry weight basis. Phosphate fertilization, but not mycorrhizal infection, increased nitrogenase activity on the same basis (B).

Nitrogen content of plants is shown in Table 4. Mycorrhizal plants had higher % and total N than non-mycorrhizal plants in the same fertilizer treatment. Plants in treatments 2 and 3, which in this experiment had somewhat different dry weights, had very similar nitrogen contents as well as similar values for mgP/g dry weight.

## D. Discussion

The demonstration that, at any particular soil P level, mycorrhizal infection increased total plant P, growth, nodulation and nitrogen fixation confirms the results of other workers on legumes (3).

Reduction in the root:shoot dry weight ratio and change in the distribution of phosphate in mycorrhizal plants has been observed in other species (e.g. onions (4)).

Table 3. Percentage of roots infected by mycorrhizal fungi (% mycorrhiza) nodulation (on an arbitrary scale 0-9) and dry weight (g) of *Medicago sativa* plants grown in acid washed sand in the presence and absence of mycorrhizal fungi (*Glomus mosseae*) and $Na_2HPO_4$ supplied at 1.2 m equiv/kg, and harvested at 7 and 12 weeks. (Mean of 3 bunches of 10 plants ± standard error of mean)

| Treatment | Phosphate | | 7 weeks Mycorrhiza (%) | 7 weeks Nodules (0-9) | 7 weeks Dry wt. (g/10 plants) | 12 weeks Mycorrhiza (%) | 12 weeks Nodules (0-9) | 12 weeks Dry wt. (g/10 plants) |
|---|---|---|---|---|---|---|---|---|
| 1 | -P | non-mycorrhizal | 0 | 1.0 | 0.26 ± 0.05 | 0 | 3.0 | 0.45 ± 0.12 |
| 2 | -P | mycorrhizal | 18 | 1.3 | 0.26 ± 0.02 | 32 | 3.7 | 0.58 ± 0.04 |
| 3 | +P | non-mycorrhizal | 0 | 2.3 | 0.49 ± 0.09 | 0 | 4.0 | 0.76 ± 0.05 |
| 4 | +P | mycorrhizal | 25 | 3.2 | 0.28 ± 0.02 | 45 | 3.0 | 1.29 ± 0.06 |

Table 4. Nitrogen content of dried tissue of *Medicago sativa*, grown for 12 weeks on acid washed sand in the presence and absence of mycorrhizal fungi (*Glomus mosseae*) and $Na_2HPO_4$ supplied at 1.2 m equiv. per kg

| Treatment | Phosphate | | N content Root (%) | N content Shoot (%) | N content Total (mg/10 plants) |
|---|---|---|---|---|---|
| 1 | -P | non-mycorrhizal | 1.9 | 2.3 | 9 |
| 2 | -P | mycorrhizal | 2.1 | 2.9 | 20.4 |
| 3 | +P | non-mycorrhizal | 2.7 | 2.9 | 21.0 |
| 4 | +P | mycorrhizal | 4.7 | 4.4 | 57.0 |

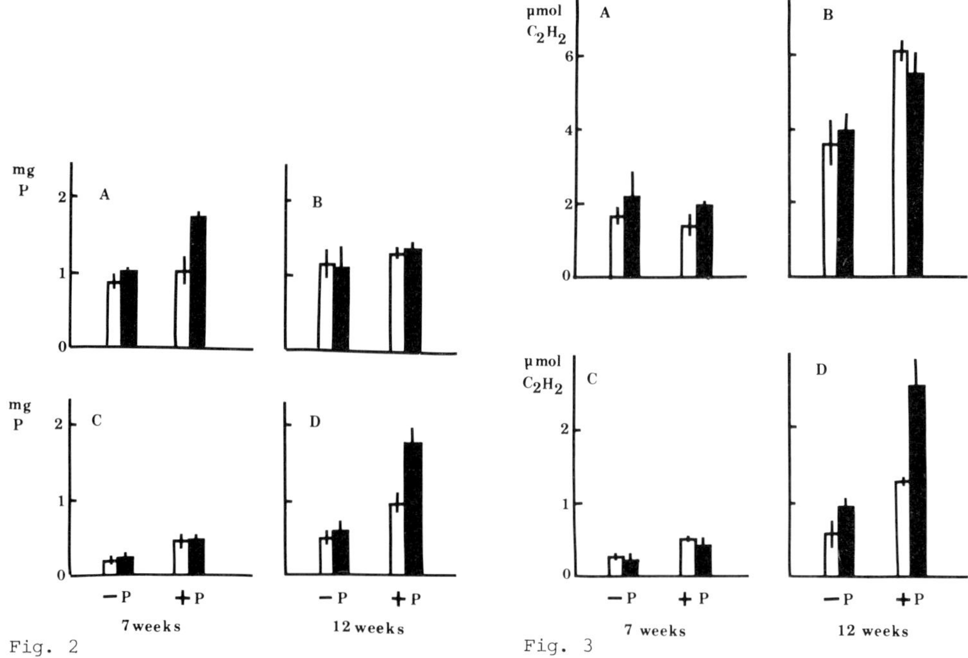


<u>Fig. 2 A-D.</u> <u>Medicago sativa</u> grown on acid washed sand with (+P) and without (-P) the addition of 1.2 m equivalents phosphate/kg supplied as $Na_2HPO_4$, and harvested at 7 weeks and 12 weeks. Phosphate content expressed as mg/g dry weight (A and B) and mg/10 plants (C and D). (± standard error of mean of 3 bunches)

<u>Fig. 3 A-D.</u> <u>Medicago sativa</u> grown on acid washed sand with (+P) and without (-P) the addition of 1.2 m equivalents phosphate/kg supplied as $Na_2HPO_4$ and harvested at 7 weeks and 12 weeks. Acetylene reduction expressed as μmol $C_2H_2$ reduced/h/g dry weight root (A and B) and μmol $C_2H_2$ reduced/h/10 plants (C and D). (± standard error of mean of 3 bunches

Our results, showing similar root:shoot dry weight ratios in mycorrhizal and non-mycorrhizal plants of the same total dry weight, are compatible with the suggestion that the reduction in root:shoot ratio is due to improved nutrition (4). High root P in mycorrhizal plants has been attributed to storage of P in fungus and host cells. In our experiments it was associated with improved nodulation and very much higher nitrogenase activity. It may be that the changes in P distribution favour P accumulation in the nodules, which have been shown to have high P content compared with adjacent root tissue (3). High nodule P may enhance nitrogen fixation which is dependent upon high levels of this nutrient (7).

Plants grown in sand and harvested at 7 weeks show a growth reduction in mycorrhizal plants. This can be attributed to the use of photosynthate by the fungus, rather than to phosphorus toxicity (2) as the plants had lower P contents than others grown at higher levels of phosphate fertilization in the same experiment (6) and subsequently grew better than non-mycorrhizal controls.

Under the nutrient limiting conditions of these experiments, high values of dry weight and nitrogen content at the final harvest are associated with high P contents (mg/g DW) at 7 weeks. This illustrates that differences in P content between treatments must be at a maximum in young plants and be progressively reduced as growth takes place, resulting in the observed insignificant differences in P (mg/g DW) at the final harvest. Measurements of growth and mineral content and nitrogenase activity of plants throughout the course of an experiment (8) can greatly assist the assessment of mycorrhizal effects on plant growth.

## E. Acknowledgements

We would like to thank Mr. D.G. Bowen for many useful discussions and Drs. J.A. Raven and F.A. Smith for reading the manuscript.

## F. References

1. El Giahmi, A.A., Nicolson, T.H., Daft, M.J.: Endomycorrhizas in some Libyan crop plants. Trans. Br. mycol. Soc. 67, 278-282 (1976).
2. Mosse, B.: Plant growth responses to vesicular arbuscular mycorrhiza. IV. In soil given additional phosphate. New Phytol. 72, 127-136 (1973).
3. Mosse, B., Powell, C.L., Hayman, D.S.: Plant growth responses to vesicular-arbuscular mycorrhiza. IX. Interactions between V.A. mycorrhiza, rock phosphate and symbiotic nitrogen fixation. New Phytol. 76, 331-342 (1976).
4. Sanders, F.E.: The effect of foliar-applied phosphate on the mycorrhizal infection of onion roots. In: Endomycorrhizas. Sanders, F.E., Mosse, B., Tinker, P.B. (eds.). Academic Press pp. 261-276 (1975).
5. Smith, S.E.: Mycorrhizal fungi. C.R.C. Critical Reviews in Microbiol. 3, 275-313 (1974).
6. Smith, S.E., Daft, M.J.: Interactions between growth, phosphate content and $N_2$ fixation in mycorrhizal and non-mycorrhizal *Medicago sativa*. Aust. J. Plant Physiol. 4, 403-413 (1977).
7. Stewart, W.D.P., Alexander, G.: Phosphorus availability and nitrogenase activity in aquatic blue green algae. Freshwat. Biol. 1, 389-404 (1971).
8. Stribley, D.P., Read, D.J., Hunt, R.: The biology of mycorrhiza in the Ericaceae V. The effects of mycorrhizal infection, soil type and partial soil sterilisation (by gamma irradiation) on the growth of Cranberry (*Vaccinium macrocarpon* Ait.) New Phytol. 75, 119-130 (1975).
9. Tinker, P.B.: Effects of vesicular-arbuscular mycorrhiza on higher plants. Symp. soc. Exp. Biol. 29, 325-349 (1975).

# The Use of Mycorrhizal Fungi for Improving Establishment and Growth of *Pinus* Species Used for High-Altitude Revegetation

B.J. WILLS and A.L.J. COLE

## A. Introduction

Problems with the revegetation of high altitude terrain are posed by uncompromising climatic, edaphic and biotic conditions in these areas, particularly where reafforestation is proposed. Soil movement on slopes, frost heave, the short growing season and animal predation all result in high tree seedling mortality rates. For revegetation to be successful, early establishment and rapid growth of the seedling is essential. The development of a good mycorrhizal association is of great importance in this respect to fulfil these requirements.

This investigation is concerned with the problems of establishing effective mycorrhizal associations between *Pinus mugo* and *Suillus* (*Boletus*) *luteus* (L. ex Fries). Mountain pine is an exotic species found to grow successfully under N.Z. conditions up to 1600 m. It is resistant to wind and snow-pack damage, growing well on dry slopes. Its low spreading canopy makes it suitable for erosion control. *S. luteus* is believed to be the main mycorrhizal fungus associated with *Pinus* species at higher altitudes in N.Z. Because of its known association with *P. mugo* it was used in this study.

Pre-inoculation of seeds with the myco-symbiont would offer many advantages. The most effective association can be established between host and available symbionts, especially where the latter are absent from the soil. Viable ecto-mycorrhizas may thus be introduced to the revegetation sites with the seeds ensuring early establishment due to increased growth rates in the first season. An effective mycorrhizal inoculum needs to be found and suitable methods of inoculation devised. Problems such as non-viability under field conditions, deterioration under extended storage times and insufficient supplies of the inoculum source must, however, be overcome.

Previous work (3, 11, 12, 13) has shown pure cultured mycelial inocula to be successful. If growing conditions are stable or can be controlled, as in the nursery or growth room, this method is effective (1, 4, 5, 14, 15, 20). Success of such inoculations under severe conditions has not been evaluated, but deterioration of inoculum in the field and the vast quantities needed limit its use.

Little work has been done to develop a basidiospore inoculum and information is lacking on basidiospore germination and viability. Spore germination levels are characteristically low with many of the mycorrhizal basidiomycetes. Levels of spores must often exceed $10^6$ or $10^7$ per seed before mycorrhizal associations are established (9, 10, 18, 19). Where spores are stored or freeze dried, these levels must be increased 10 or 100 fold to be effective.

## B. Materials and Methods

Seeds of *P. mugo* were supplied by the N.Z. Forest and Range Experimental Station Rangiora, from collection sites at Broken River (Canterbury). They were surface sterilized in 30% hydrogen peroxide for 1 hr and washed in sterile distilled water.

S. luteus sporophores are produced from February until April. Collections were made throughout this season and spores were dropped on to large glass sheets by placing the inverted pileus on these overnight. The spores (representing a mixed sample of 20-30 sporophores) were then scraped into Bijou bottles and stored at 4°C or freeze dried.

Soil was obtained from Broken River and sterilised for 4 hrs at 85°C. It is a High Country Yellow Brown Earth (pH 6.0), with high total phosphorous and high phosphorous retention.

## C. Results and Discussion

Many attempts have been made to induce germination of various basidiospores *in vitro*. Germination of *Agaricus bisporus* spores was obtained by stimulation with living mycelium and with isovaleric acid (7, 8). Difficulties in obtaining germination of several *Boletus* species was overcome by using an activator, *Rhodotorula*, and the spores germinated on a malt extract medium (2). *S. luteus* spores have been germinated in this laboratory when incubated at 25°C on a modified medium (M40) (17); initial germination of a few spores overcame the dormancy of adjacent spores. The period of incubation necessary varied from 2-4 weeks and only fresh spores, dropped from sporophore tissue on to the agar surface, germinated. Spores stored for any length of time have not been germinated successfully. Average germination percentage figures *in vitro* are as low as 0.1% and a yeast activator had no effect. Spores of *Laccaria laccata* could not be germinated *in vitro* (16), but a 46% inoculation level of Douglas fir seedlings was achieved in pot trials. A 30% germination level for spores of *Pholiota aurivella* was obtained after 5 days on malt-extract agar, and the spores maintained their viability for 6 months (6).

Several possible germination stimulants were tested by adding sterile solutions to the modified M40 agar. Citric acid, $CaHPO_4$ and malt extract had some effect and inositol and nicotinic acid stimulated germination sufficiently to warrant further investigation (Table 1). So far germination of stored spores has not been achieved using these substances. During these tests it was noted that a spore concentration effect was evident. Where spore drops from hymenial tissue exceeded 1000 per sq. mm, no spores germinated. Germination was most prevalent at concentrations of 500-600 per sq. mm. The presence of a water soluble germination inhibitor in P. tinctorius has been suggested (9) and a similar effect is possibly being observed here.

Although *in vitro* studies to assess spore germination are difficult, pot trials have shown that the spores are viable although viability is difficult to assess. Use of acridine orange and fluorescent microscopy (16) showed that 25% of *L. laccata* spores were viable and could produce mycorrhiza on Douglas fir seedlings. In our experiments spores of *S. luteus* were mixed with a known volume (50 cc) of the carriers, sand, vermiculite, soil and distilled water, which were then added to the top of soil filled pots (10 cm diameter). Seedlings grown in these were assessed for mycorrhizal short root development at monthly intervals (Figure 1). The spore levels, 0.1 g and 0.01 g, correspond to $3.6 \times 10^8$ and $3.6 \times 10^7$ (± 9%) total numbers of spores respectively. It can be seen that a lag period occurs before the mycorrhizal association is formed, and is more pronounced at the 0.01 g spore level. The lag may be due to the time taken for production of lateral roots, on which the mycorrhizal short roots are borne. Seedlings take 2-3 weeks to emerge and another 3-5 weeks before the secondary laterals become well established. Additional delays may be caused by slow spore germination and the slowness of the mycelium to contact

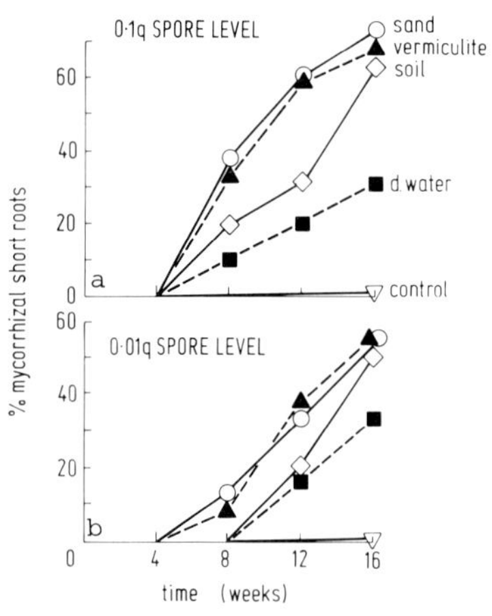


Fig. 1. Effect of spore carriers on S. luteus spore germination as measured by the ability to form mycorrhizal associations with P. mugo

Table 1. Effect of Chemical Stimulants on Basidiospore Germination

| Treatment | Nicotinic Acid | Inositol | Citric Acid | Malt Extract | $CaH_2PO4$ | Control (M40 Agar) |
|---|---|---|---|---|---|---|
| % Germination | 0.92% | 0.82% | 0.18% | 0.15% | 0.06% | 0% |

Table 2. Effect of Spore Concentration on Mycorrhizal Development

| Treatment (spore concentration) | $10^4$ | $10^5$ | $10^6$ | $10^7$ | $10^8$ | Control |
|---|---|---|---|---|---|---|
| * Soil Trial | 0% | 8% | 21% | 31% | 46% | 0% |
| Coated Seeds | 2.5% (5 x $10^7$ spores/seed) | | | | | |

*Figures expressed as % mycorrhizal short roots/plant.
Spores applied in distilled water suspension.

the root. The sand and vermiculite were better inoculum carriers than soil and distilled water was least effective.

The ability of the spore inoculum to form mycorrhizal associations was studied at different spore concentrations. A dilution series of spores in distilled water was used to inoculate seeds which were then planted. After 3 months the seedlings were assessed for mycorrhizal associations (Table 2). The results show that numbers of mycorrhizal short roots increase steadily up to the maximum concentration of spores ($10^8$ spores) used. Some seeds were coated with spores in the same trial but few mycorrhizal associations resulted. This may have been due to hypocotyl extension and subsequent raising of the testa out of the soil taking many spores with it; the few remaining may be insufficient to allow mycorrhizal development.

The results can be summarised as follows. The use of basidiospores of Suillus luteus to provide an effective mycorrhizal inoculum source for application to Pinus mugo seeds has been investigated. In vitro germination of basidiospores was very low but additions of inositol and nicotinic acid stimulated germination. Mycorrhizal development in seedlings was enhanced when sand and vermiculite were used as basidiospore carriers. A concentration of $10^8$ spores in the carrier medium was most effective in mycorrhizal short root formation.

## D. References


1. Bryan, W.C. & Zak, B.: Synthetic Culture of Mycorrhizae of Southern Pines. For. Sci. 7 (2), 123-, (1961).
2. Fries, N.: Chemical Factors in the Germination of Spores of Basidiomycetes. In: The Fungus Spore. Madelin, M.F. (ed.) p. 189- Butterworths (1966).
3. Gobl, F.: An approach to the cultivation of mycorrhizal inoculum. Z. Ges Forstwesen 92 (4), 2- (1975). Translation in Environment, Canada OOENV TR-1113.
4. Hacskaylo, E.: Pure Culture Synthesis of Pine Mycorrhizae in Terralite. Mycologia 45, 971- (1953).
5. Lamb, R.J. & Richards, B.N.: Effect of Mycorrhizal Fungi on the Growth and Nutrient Status of Slash and Radiata Pine Seedlings. Aust. For. 35 (1), 1- (1971).
6. Lavallee, A. & Lortie, M.: Quelques Observations sur la Germination et la Viabilite des Basidiospores du Pholiota aurivella. Phytoprotection 52, 112-, (1971).
7. Losel, D.M.: The stimulation of Spore Germination in Agaricus bisporus by living mycelium. Ann. Bot. (N.S.) 28, 541- (1964).
8. Losel, D.M.: The stimulation of Spore Germination in Agaricus bisporus by organic acids. Ann. Bot. (N.S.) 31, 417-, (1967).
9. Marx, D.H.: Synthesis of Ectomycorrhizae on Loblolly Pine Seedlings with basidiospores of Pisolithus tinctorius For. Sci. 22 (1), 13-, (1976).
10. Marx, D.H. & Ross, E.W.: Aseptic Synthesis of Ectomycorrhizae on Pinus taeda by Basidiospores of Thelephora terrestris. Can. J. Bot. 48, 197-, (1970).
11. Moser, M.: Artificial Mycorrhiza Inoculation of Forest Plants, I. Forstw. Cbl. 77, 32-, (1958a).
12. Moser, M.: II. Peat Culture. Ibid 77, 273-, (1958b).
13. Moser, M.: III. Inoculation Methods. Ibid 78, 193-, (1959).
14. Riffle, J.W.: Pure Culture Synthesis of Ectomycorrhizae on Pinus ponderosa with Species of Amanita, Suillus and Lactarius. For. Sci. 19 (4), 242-, (1973).
15. Shoulders, E.: Mycorrhizal Inoculation Influences Survival Growth and Chemical Composition of Slash Pine Seedlings. U.S. Dept. Agric. - For. Res. Paper SO-74 (1972).
16. Stack, R.W., Sinclair, W.A. & Larsen, A.O.: Preservation of Basidiospores of Laccaria laccata for use as Mycorrhizal Inoculum. Mycologia 67, 167-, (1975)
17. Stevens, R.B. (ed.): Mycological Guidebook 700 pp. Washington (1974).
18. Theodorou, C.: Introduction of Mycorrhizal Fungi into Soil by Spore Inoculation of Seed. Aust. For. 35, 23-, (1971).
19. Theodorou, C., & Bowen, G.D.: Inoculation of Seeds and Soil with Basidiospores of Mycorrhizal Fungi. Soil Biol. Biochem. 5, 765-, (1973).
20. Trappe, J.M.: Pure Culture Synthesis of Douglas-Fir Mycorrhizae with Species of Hebeloma Suillus, Rhizopogon and Astraeus. For. Sci. 13, 120-, (1967).

# The Biology of Mycorrhiza in Heathland Ecosystems with Special Reference to the Nitrogen Nutrition of the Ericaceae

D.J. READ

## A. Introduction

Heathland ecosystems dominated by ericaceous and epacridaceous plants occur widely in both the northern and southern hemispheres. Such heaths are found over a wide range of climatic conditions. At one extreme are the sub-arctic *Cassiope* and *Vaccinium* heaths of the northern hemisphere, at the other the *Erica* heaths of southern Africa and epacridaceous heaths of Australia which occur in areas of high summer temperature and low annual rainfall. Despite the wide geographical and climatic ranges of these communities it appears that the dominant plants are characterised by the possession of mycorrhizas which are comparable in structure and probably have a similar fungal endophyte (1, 3, 4, 5). Recent work suggests that the endophyte is probably an Ascomycete of the genus *Pezizella* (6). While it is known that the structure of ericoid mycorrhizas is comparable in different heathlands, studies of their function have so far been largely restricted to mor-humus soils of the northern hemisphere. In this situation it has been shown that mycorrhizal infection leads to enhanced nitrogen content (7, 8, 9) and it has been suggested that the success of ericaceous plants in mor-humus soils, all of which have inherently low available nitrogen status, is attributable to an increased efficiency of nitrogen absorption.

Heathland soils of warmer climates differ markedly from the organic mor-humus soils of colder regions. They normally have much lower organic content and mineralisation of nitrogen occurs much more rapidly. For such soils in New Zealand it has been suggested that the primary influence of ericoid mycorrhizas is upon phosphorus nutrition (2). It is therefore of interest to compare the function of ericoid mycorrhizas in the two apparently distinct soil situations. Experiments have been designed to compare the growth of mycorrhizal and non-mycorrhizal plants of a characteristic South African species, *Erica bauera*, under conditions of nitrogen stress. For the major experiments the *Erica* plants were inoculated with an indigenous South African endophyte obtained from the same host plant, but cross inoculation trials with *Pezizella* were also carried out in order to compare the effects produced by microbial associates of northern and southern heaths.

## B. Methods

Seed of *Erica bauera* was sown in sterile potting compost. This species was selected because it germinates more uniformly and rapidly than do most other southern Africa species. After germination, seedlings were transferred to tubes containing leached B horizon soil from a sandy heathland profile developed over Table Mountain Sandstone (TMS) in Cape Province, South Africa. The soil was first sterilised by autoclaving. No problems of heat induced toxicity were experienced with this soil probably because of its low organic content (<0.8%). Half the tubes were inoculated with a mycelial macerate of an indigenous South Africa endophyte obtained from roots of *Erica bauera* growing in natural heathland soil. The remainder were uninoculated. Both sets of seedlings were grown in a glass house for eight weeks after which time the inoculated plants were examined for mycorrhizal infection and a sample of 6 plants of each category harvested. Each plant was divided into root and shoot; small samples of the roots were stained and examined for mycorrhizal infection while the remainder and the shoots were oven dried and weighed. Total nitrogen content of the shoots was determined following Kjeldahl digestion.

Twenty-four plants of the mycorrhizal and of the non-mycorrhizal category were left in soil, while the same number of each category were transferred either to pure silica sand containing a dilute nutrient solution lacking nitrogen or to liquid culture containing the same nutrient solution again without nitrogen. Samples of 6 mycorrhizal and 6 non-mycorrhizal seedlings were taken from each of the three treatments, via: soil, sand-N and liquid culture-N at weekly intervals for 4 weeks. At each harvest plants were analysed for root and shoot weight, mycorrhizal infection and total N content as described previously.

In a subsequent experiment seedlings of *E. bauera* were inoculated with *Pezizella ericae* in sterilised TMS soil and in sterilised British heathland soil, B horizon Fox House soil (8). The plants were later analysed for levels of infection and total nitrogen content in order to determine the extent to which the southern hemisphere species was comparable in its response to an endophyte of different origin.

## C. Results

The results of the first experiment are summarised in Table 1. The responses of the plants to culture in the different media can conveniently be considered separately.

I. Growth and nitrogen nutrition in soil. All inoculated plants were heavily infected after 8 week's growth in TMS soil. Mycorrhizal plants had significantly higher N contents and yields compared with their non-mycorrhizal counterparts while their root-shoot ratios were decreased.

This pattern was maintained for the subsequent four weeks of growth on soil. By the final harvest non-mycorrhizal plants were pale green in colour, indicating nitrogen deficiency. Shoots of the mycorrhizal plants were healthy in appearance although there was some evidence of reduced infection intensity in their roots.

II. Growth and nitrogen nutrition in pure silica sand without nitrogen. Analysis of the mycorrhizal status of plants at the end of the four week growth period showed that infection had decreased considerably and that though the cortical cells had not collapsed only traces of intracellular infection could be observed. The nitrogen status of the shoots however was maintained at a level comparable with that found in the plants at the time of transfer from soil. In contrast, the shoot N levels of non-mycorrhizal plants declined and were significantly lower than those of the mycorrhizal plants by the end of the four week period of nitrogen stress.

III. Growth and nitrogen nutrition in liquid culture without nitrogen. This treatment produced a completely aberrant response in the mycorrhizal plants and the results are included largely to demonstrate the sensitive nature of the balance occurring between host and endophyte. Transfer of mycorrhizal plants from soil to liquid medium led to an 'explosive' outburst of fungal growth, such that in the first week after transfer fungal mycelium was visible in the medium surrounding all infected roots. By the third week the whole root system of each plant was enclosed in a ball of fungal mycelium and shoots were beginning to show discolouration. Shoot N content of mycorrhizal plants dropped significantly in this treatment. The responses of non-mycorrhizal plants to this treatment were similar to those found in non-mycorrhizal plants grown in sand.

Table 1. The relationship between mycorrhizal infection, growth and nitrogen nutrition in *Erica bauera* grown in heathland soil for 12 weeks or transferred after 8 weeks in soil to silica sand minus N, or liquid culture minus N. Each figure represents the mean of 6 plants. Asterisks denote significant difference between mycorrhizal and non-mycorrhizal plants at 5% level

| Harvest | Mycorrhizal Plants | | | | | | Non-Mycorrhizal Plants | | | | |
|---|---|---|---|---|---|---|---|---|---|---|---|
| | Root Weight MGS | Shoot Weight MGS | Total Plant Weight MGS | Root Shoot Ratio | % N in Shoots | Mycorrhizal Infection % | Root Weight MGS | Shoot Weight MGS | Total Plant Weight MGS | Root Shoot Ratio | % N in Shoots |
| First Harvest after 8 weeks culture in heath soil | 24 ± 1 | 30 ± 5 | 54* | 0.80 | 1.4* | 84 | 26 ± 6 | 23 ± 4 | 49 | 1.1 | 0.9 |
| Final Harvest after further 4 weeks culture on heath soil | 25 ± 4 | 34 ± 4 | 59* | 0.73 | 1.2* | 78 | 28 ± 9 | 25 ± 3 | 53 | 1.1 | 0.8 |
| Final Harvest after 4 weeks culture in sand - N | 21 ± 4 | 35 ± 6 | 56* | 0.60 | 1.3* | 26 | 28 ± 7 | 24 ± 3 | 52 | 1.2 | 0.5 |
| Final Harvest after 4 weeks culture in liquid medium - N | 24 ± 1 | 24 ± 3 | 48 | 1.0 | 0.5 | - | 26 ± 7 | 25 ± 7 | 51 | 1.0 | 0.6 |

IV. Inoculation with P. ericae. Growth of Erica bauera in TMS and Fox House soils was comparable. Infection levels on both soils were the same whether they were inoculated with indigenous endophyte or P. ericae. A similar enhancement of shoot N status occurred in both soils when inoculated with Pezizella (Table 2).

Table 2. Responses of E. bauera seedlings to inoculation with Pezizella ericae, originally isolated from Calluna vulgaris, and with the normal host endophyte, in sterilised British (Fox House) and African (Table Mountain Sandstone) soils

| Inoculum | Fox House Soil | | | TMS | | |
|---|---|---|---|---|---|---|
| | Yield after 8 weeks in mg | % N in shoot | % Mycorrhizal infection | Yield after 8 weeks in mg | % N in shoot | % Mycorrhizal infection |
| Pezizella | 65 | 1.4 | 88 | 53 | 1.4 | 76 |
| South African endophyte | 68 | 1.6 | 87 | 54 | 1.4 | 84 |

## D. Discussion

The response to mycorrhizal infection shown by Erica bauera is clearly comparable with that found in species of northern heathlands. The major role of infection appears to be enhancement of nitrogen uptake. The mechanism whereby this enhancement of uptake is achieved, however, may be different in view of the different patterns of nitrogen turnover found in the two climatic regimes. In warm semi-arid heathland soils showing a more rapid natural turnover of organic matter, mineralisation of N may occur in flushes coinciding with or following periods of rainfall. Mycorrhizal infection in this circumstance may facilitate optimal absorption of recently mineralised N so that losses due to leaching or uptake by competitors is reduced. Field studies of root systems of South African Erica spp. suggest that active root growth and maximum infection occur during and shortly following periods of rainfall. It is only during such periods that their root systems are healthy in appearance and show infection levels comparable with those of their northern counterparts. During the dry season most of the fine roots are moribund, the cortical cells having collapsed to leave only the vascular strand and some pericycle tissue. Despite this degeneration of the root system, the shoots retain a healthy appearance; many of the Erica species of southern heaths flower and set seed during this period. The results of the sand culture experiment suggests a possible explanation for this phenomenon. It is likely that $NH_4$ or organic N compounds absorbed by the endophyte during growth in soil were locked up in intra-cellular hyphal complexes and then mobilised during the subsequent period of nitrogen stress to support shoot growth and maintain its N status. This mobilisation presumably arises from digestion of the intracellular hyphal material, and would lead to the diminution of infection intensity found during the period of sand culture. Such a hypothesis would also explain the decline of infection intensity found in the field during periods of drought which once again represent times of nitrogen stress. Seen in this light mycorrhizal infection has a dual role. It not only facilitates increased uptake of N, but the endophyte acts as a residual pool of nitrogenous material in the root which is subsequently made available for controlled release by digestion during periods of stress. Progressive reduction of infection intensity would not only release N to support shoot growth but would reduce the drain upon host carbohydrate supplies during the critical period of flowering and seed production.

These experiments together with field observations therefore suggest that the role of mycorrhizal infection in both northern and southern hemisphere ericaceous species is similar. Ericoid mycorrhizas are seen as structures facilitating the uptake and storage of N in both types of habitat. In this connection it is interesting to observe that in the heathlands of both northern and southern hemispheres, insectivorous plants particularly of the genus Drosera are common associates of the ericaceous and epacridaceous dominants. Insectivory has long been recognised as a special adaptation to enhance N supply. It now seems likely that mycotrophy in ericaceous plants is a parallel adaptation yielding a similar enhancement of N supply. Mycorrhizas may play a similar role in epacridaceous heath but detailed studies of this group remain to be carried out.

## E. Acknowledgements

I would like to acknowledge the support of the Royal Society and the Natural Environmental Research Council.

## F. References

1. Bain, H.F.: Production of synthetic mycorrhiza in the cultivated cranberry. J. Agric. Res. 55, 811-35 (1937).
2. Brook, P.J.: Mycorrhiza of Pernettya macrostigma, New Phytol. 51, 388-396 (1952).
3. McLennon, E.: Non-symbiotic development of seedlings of Epacris impressa. New Phytol., 34, 55-63 (1935).
4. McNabb, R.F.R.: Mycorrhiza in the New Zealand Ericales. Aust. J. Bot., 9, 57-61 (1961).
5. Pearson, V., & Read, D.J.: The biology of mycorrhiza in the Ericaceae. I. The isolation of the endophyte and the synthesis of mycorrhizas in aseptic culture. New Phytol., 72, 371-383 (1973).
6. Read, D.J.: Pezizella ericae sp. nov., the perfect state of a typical mycorrhizal endophyte of Ericaceae. Trans. Brit. Mycol. Soc., 63, 381-419 (1974).
7. Read, D.J. & Stribley, D.P.: The effect of mycorrhizal infection on nitrogen and phosphorus nutrition of ericaceous plants. Nature, Lond. 244, 81-82 (1973).
8. Stribley, D.P., Read, D.J., & Hunt, R.: The biology of mycorrhiza in the Ericaceae. V. The effects of mycorrhizal infection, soil type and partial soil sterilisation on growth of cranberry (Vaccinium macrocarpon Ait.) New Phytol. 75, 119-130 (1975).
9. Stribley, D.P. & Read, D.J.: The biology of mycorrhiza in the Ericaceae. VI. The effects of mycorrhizal infection and concentration of ammonium nitrogen on growth of cranberry in sand culture. New Phytol., 77, 63-72 (1976).

# Plant Diseases

## Soil Ethylene Production Specifically Triggered by Ferrous Iron

A.M. SMITH, P.J. MILHAM, and W.L. MORRISON

### A. Introduction

Ethylene is commonly produced in soil and has considerable importance in soil biology (6, 9). It is well established that this ethylene is produced only in anaerobic microsites which form in soil as a consequence of aerobic microbial activity (6). Not only must the microsites be oxygen-free but their redox potentials must fall sufficiently before ethylene is produced (6). The actual source of this ethylene, however, is in dispute although the fungus, Mucor hiemalis (2) and anaerobic bacteria (6, 8) have been implicated. Many other microorganisms isolated from soil produce $C_2H_4$ in pure culture especially when amended with methionine (1, 5), but their relevance to $C_2H_4$ production in soil remains to be demonstrated.

In this paper, ferrous iron ($Fe^{2+}$) is identified as a specific trigger for ethylene production in soil. Furthermore, the data show that ethylene production in soil is non-biological in origin although intense aerobic microbial activity is essential for establishment of the anaerobic microsites in which $Fe^{2+}$ is mobilized. The potential importance of $Fe^{2+}$ in ethylene production was highlighted in a series of preliminary experiments on the effects of amending soils with various reducing agents.

### B. Materials and Methods

I. $C_2H_4$ production in soil amended with $Fe^{2+}$. Four replicated 5 g samples of dried and sieved Mt. Wilson rainforest soil (7) were placed in 14 ml glass vials (8), wetted to about -0.5 bars (3 ml $H_2O$ per 5 g soil) with distilled water amended with $Fe^{2+}$ ($FeSO_4.7H_2O$) at 0, 5 x $10^2$, $10^3$, $10^4$, 5 x $10^4$ and $10^5$ ppm, sealed immediately with rubber septa and incubated at 25°C. $C_2H_4$ production was monitored by gas chromatography (7) at various times from 15 min to 3 days after the start of the experiment.

The experiment was repeated using ferrous gluconate as an amendment at 0 and $10^4$ ppm $Fe^{2+}$ to ensure that any effects on $C_2H_4$ production were attributable to $Fe^{2+}$ and not to $SO_4^{2-}$.

Additionally, in the vicinity of 100 soils of widely different origins and histories from throughout Australia were wetted to about field capacity with $FeSO_4.7H_2O$ solutions at 0 and $10^4$ ppm $Fe^{2+}$ to confirm that any effects on $C_2H_4$ production were not confined to a particular soil or group of soils.

To establish whether $Fe^{2+}$ stimulated $C_2H_4$ release from soil by reacting with some biological or non-biological system, four vials of Mt. Wilson rainforest soil were wetted with 1:1 chloroform-methanol (3ml per 5 g soil) containing either 0 or $10^4$ ppm ferrous perchlorate. $C_2H_4$ production was recorded after 15 min and 3 hr at 25°C.

II. Differentiation between effects of $Fe^{2+}$ and its Eh on $C_2H_4$ production. Lots (100 g) of Mt Wilson rainforest soil were wetted with 60 ml of distilled water, 0.1, 0.2 and 0.3N $H_2SO_4$ or 0.1, 0.2 and 0.3N NaOH, left for 3 hr, then placed in thin layers on trays and dried overnight at 37°C in a forced draught oven.

Four replicate 5 g samples of dried, sieved soil from each treatment were wetted with solutions of $10^4$ ppm $Fe^{2+}$ as already described and $C_2H_4$ production recorded after 15 min and 3 hr. pH measurements were made on duplicate samples from each treatment after addition of $Fe^{2+}$ solutions (1:2 in 10 mM $CaCl_2$).

III. Effect of various reducing agents on $C_2H_4$ production from soil and relationship to $Fe^{2+}$ mobilization. Mt. Wilson soil was treated in the standard way with solutions of a range of reducing agents and their ability to liberate $C_2H_4$ from soil was compared with $Fe^{2+}$ at 0, $10^3$, 5 x $10^3$ and $10^4$ ppm. The reducing agents used were ascorbate, cysteine hydrochloride, thioglycollic acid, sodium dithionite, sodium bisulphite, sodium sulphite, sodium thiosulphate, hydrazine hydrochloride and hydrogen sulphide. Quantities of reducing agents added to soil were standardized on reducing equivalents $\left(\frac{M.wt}{e-}\right)$ using $10^4$ ppm $Fe^{2+}$ as the standard. The only exception was the gas $H_2S$ which was added to evacuate vials of dry soil to completely replace the air atmosphere. The soil in these vials was then wetted with distilled water syringed through the septa.

$C_2H_4$ production was read from four replicates of each treatment after incubation at 25°C for 15 min and 3 hr.

$Fe^{2+}$ concentrations in the soil were determined on an identical series of treatments. Determinations were made 15 min after addition of the reducing agents because relatively high concentrations of $C_2H_4$ are released within this period. The vials were evacuated twice and filled with oxygen-free $N_2$. Seven ml degassed 10mM $CaCl_2$ were syringed into each vial which was shaken vigorously, centrifuged at 750 x g for 10 min. Great care was taken to prevent oxidation of $Fe^{2+}$ because the resulting $Fe^{3+}$ would not have been extracted by this method. Two ml clear supernatant were syringed from each vial and added to 5 ml 1N HCl in a test tube. The concentrations of Fe were determined in the atomic absorption spectrophotometer against standard solutions made up in 1N HCl. A simple calculation enabled the $Fe^{2+}$ concentration to be determined in the soil solutions when soil was wetted to -0.5 bar for the standard $C_2H_4$ test.

All reducing agents (except $H_2S$) at the concentrations used in soil were added to standard solutions of 5 x $10^2$ ppm $Fe^{2+}$ in 1N HCl to ascertain whether any chemical reactions occurred that could interfere with Fe determination. These solutions were read on the atomic absorption spectrophotometer 3 hr after the reducing agents and $Fe^{2+}$ were mixed.

IV. Relationship between naturally produced soil $C_2H_4$ and $Fe^{2+}$ liberated soil $C_2H_4$. Vials of Mt Wilson rainforest soil and Mt Tamborine avocado soil (Soil a (7)) were wetted to about -0.5 bars with distilled water, sealed with rubber septa, and incubated at 25°C. After 1, 2, 3, 5, 7, 10 and 14 days, four vials of each soil were measured for $C_2H_4$ production and were then destructively sampled to measure $Fe^{2+}$ concentration by the method already described. After 14 days' incubation, an additional four vials of each soil were dried at 37°C for 24 hr, rewetted with excess $Fe^{2+}$ solution (5 x $10^4$ ppm), sealed with rubber septa, incubated for 3 hr and read for $C_2H_4$ production. For comparison, four vials of each soil which had not been incubated, were treated with $Fe^{2+}$ and $C_2H_4$ production was recorded after 3 hrs.

## C. Results

I. $C_2H_4$ production in soil amended with $Fe^{2+}$. All amendments of Mt Wilson rainforest soil with $Fe^{2+}$ caused a marked increase in $C_2H_4$ production (Fig. 1). Furthermore, $C_2H_4$ production was readily detected within 15 min of addition of the $Fe^{2+}$ solution with the maximum concentration being recorded after 3 hr. Highest

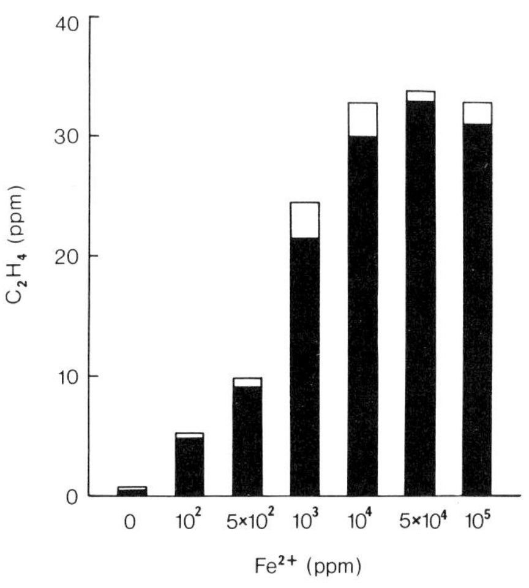


Fig. 1. $C_2H_4$ yield from Mt Wilson soil 3 h after treatment with various concentrations of $Fe^{2+}$. Unshaded areas represent standard error

production of $C_2H_4$ was obtained with amendments of $10^4$ ppm or more of $Fe^{2+}$ (Fig. 1). Ferrous gluconate gave a similar effect to $FeSO_4.7H_2O$ indicating that the $Fe^{2+}$ is involved and not the accompanying anion.

Also, all the various soils tested, including forest, grassland, plantation and intensively cultivated, have produced $C_2H_4$ on addition of $10^4$ ppm $Fe^{2+}$ confirming that this effect is not confined to particular soils. Big differences were recorded in $C_2H_4$ yields, from <1 ppm to >$10^2$ ppm, on amendment of soils with $Fe^{2+}$ in our vial tests. $C_2H_4$ yields from these soils following amendment with $Fe^{2+}$ were always directly related to their ability to produce ethylene under the standard soil incubation test (8).

Mt Wilson rainforest soil produced the same amount of $C_2H_4$ when the soil was wetted with ferrous perchlorate in 1:1 chloroform-methanol as it did when treated with aqueous $Fe^{2+}$. This is strong evidence that this $C_2H_4$ is not biological in origin.

II. Differentiation between $Fe^{2+}$ and its Eh on $C_2H_4$ production. Altering the pH 4.1 to 5.61 causes a shift in the Eh of the $Fe^{3+} \rightleftharpoons Fe^{2+}$ redox couple of -272mV

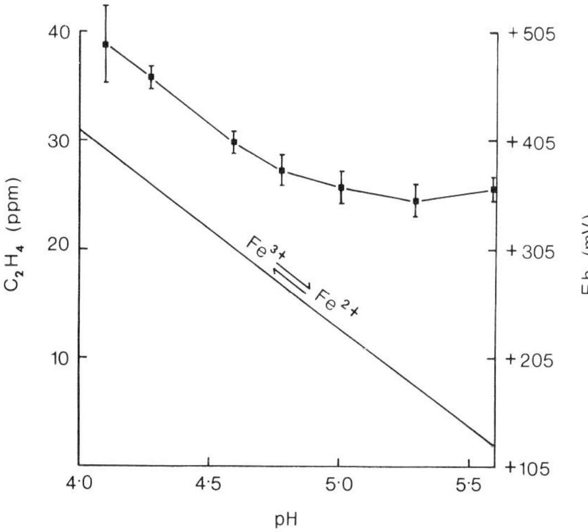


Fig. 2. $C_2H_4$ yield from Mt Wilson soil 3 h after treatment with $10^4$ ppm $Fe^{2+}$ with soil adjusted to a range of pH. Standard errors indicated by bars. Theoretical Eh of $Fe^{3+} \rightleftharpoons Fe^{2+}$ redox couple at the different pH included for comparison

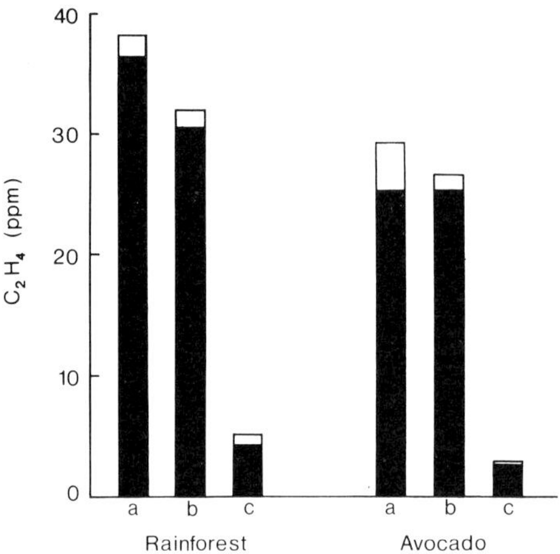


Fig. 3. $C_2H_4$ yield from Mt Wilson rainforest and Mt Tamborine avocado soils when treated a $10^4$ ppm $Fe^{2+}$, b incubated moist at 25°C for 14 days, c treated with $10^4$ ppm $Fe^{2+}$ after incubation for 14 days. Unshaded areas indicate standard errors

(4) as indicated in Fig 2. However, this marked shift in pH and Eh had only a minor influence on the stimulatory effect of $Fe^{2+}$ on $C_2H_4$ production (Fig. 2). This result indicates that it is $Fe^{2+}$ and not its Eh that triggers $C_2H_4$ production. The slight increase in $C_2H_4$ production at more acid pH is consistent with the slower rate of immobilization of $Fe^{2+}$ by oxidation in soil under acid conditions.

III. Effect of various reducing agents on $C_2H_4$ production from soil and relationship to $Fe^{2+}$ mobilization. All reducing agents including the three concentrations of $Fe^{2+}$ but with the exception of sodium thiosulphate, sodium bisulphite and sodium sulphite, triggered the production of $C_2H_4$ from Mt.Wilson soil (Table 1). High concentrations of $C_2H_4$ were recorded 15 mins after addition of reducing agents with maximum production recorded after 3 hrs. All reducing agents which triggered $C_2H_4$ production from soil released high concentrations of $Fe^{2+}$ into the soil solution (Table 1). $H_2S$ released less $Fe^{2+}$ but this probably results from the formation of FeS (4). The relatively large loss of $Fe^{2+}$ from the soil solution with addition of the three $Fe^{2+}$ solutions may result from oxidation of the $Fe^{2+}$ and formation of complexes as well as exchange reactions in soil (4). None of the reducing agents tested formed direct complexes with $Fe^{2+}$ that interfered with atomic absorption determination (Table 1).

In an attempt to explain why certain reducing agents did not trigger $C_2H_4$ release or liberate $Fe^{2+}$, vials of Mt Wilson soil were wetted with solutions containing the calculated concentrations of sodium sulphite and sodium bisulphite, but with $10^4$ ppm $Fe^{2+}$ also added. After 3 hrs, $C_2H_4$ production reached almost 30 ppm in both treatments which is identical to the $Fe^{2+}$ treatment alone (c.f. Table 1). Also, 1604 and 2147 ppm $Fe^{2+}$, respectively, were recovered from the soil solutions. This experiment indicates that these reducing agents do not make $Fe^{2+}$ unavailable in some way nor does their presence in soil specifically inhibit $C_2H_4$ production.

IV. Relationship between naturally produced soil $C_2H_4$ and $Fe^{2+}$ liberated soil $C_2H_4$. The source of $C_2H_4$ released by incubating soils or by treating with $Fe^{2+}$ appears to be identical (Fig. 3). With the Mt Wilson rainforest soil, 37 ppm $C_2H_4$ were liberated by $Fe^{2+}$ treatment whereas after 14 days' incubation (and no exogenous $Fe^{2+}$ treatment) this soil liberated 31 ppm $C_2H_4$. When this incubated soil was dried and then treated with $Fe^{2+}$ only 5 ppm $C_2H_4$ were released. Thus incubation for 14 days followed by an $Fe^{2+}$ treatment liberates the same total quantity of $C_2H_4$ as an $Fe^{2+}$ treatment without prior incubation. A similar result occurred with the avocado soil (Fig. 3).

Table 1. $C_2H_4$ production and $Fe^{2+}$ concentration in Mt Wilson soil after treatment with various reducing agents

| Reducing agent | $C_2H_4$ ppm after 3 hrs | $Fe^{2+}$ ppm after 15 min | Interaction between $Fe^{2+}$ and reducing agents, $Fe^{2+}$ ppm* |
|---|---|---|---|
| $Fe^{2+}$, $10^3$ ppm | 17 | 8.8 | - |
| $Fe^{2+}$, $5 \times 10^3$ ppm | 28 | 238 | - |
| $Fe^{2+}$, $10^4$ ppm | 31 | 2465 | - |
| Ascorbate | 21 | 264 | 359 |
| Thioglycollic acid | 22 | 6431 | 355 |
| Cysteine | 15 | 689 | 343 |
| Hydrazine | 16 | 150 | 352 |
| Dithionite | 20 | 188 | 362 |
| Thiosulphate | 2 | 7.6 | 367 |
| Bisulphite | 2 | 1.5 | 344 |
| Sulphite | 1 | 2.0 | 355 |
| Hydrogen sulphide | 18 | 21.3 | - |
| Water control | 0.4 | 0 | - |

* Less $5 \times 10^2$ ppm recovered because of dilution factor.

Additional evidence that the $C_2H_4$ released by both treatments is identical comes from work with autoclaved soils. Autoclaved soils produce no $C_2H_4$ even if recontaminated with unsterilized soil and incubated aerobically or anaerobically. Such soils do not produce $C_2H_4$ either when treated with $Fe^{2+}$.

The results in Fig. 4 are consistent with the specific triggering role of $Fe^{2+}$ in $C_2H_4$ production in soil. $Fe^{2+}$ is detectable in the soil solution by the time $C_2H_4$ production starts. The concentration of $Fe^{2+}$ is relatively low until day 7, but this could be expected because some of the initial $Fe^{2+}$ released would undergo complexation and exchange reactions in these high organic matter soils.

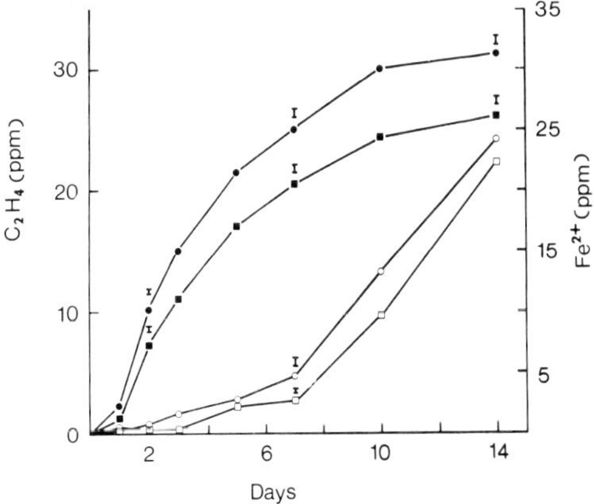


Fig. 4. $C_2H_4$ (●, ■) yields and $Fe^{2+}$ (○, □) mobilization from Mt Wilson rainforest (●, ○) and Mt Tamborine avocado (■, □) soils when incubated moist over 14 days at 25°C

D. Discussion

Amendment of soils with $Fe^{2+}$ causes an instantaneous release of $C_2H_4$. The speed of this reaction makes it highly unlikely that any biological system is involved. Confirmation that this $Fe^{2+}$ stimulated release of $C_2H_4$ from soil originates from a non-biological system was obtained from the experiment using ferrous perchlorate in 1 : 1 chloroform-methanol. This treatment liberated $C_2H_4$ from soil as effectively as aqueous $Fe^{2+}$.

When $Fe^{2+}$ triggers $C_2H_4$ production in soil, it is difficult to determine whether the $Fe^{2+}$ has some direct effect on a precursor or else simply establishes favourable redox conditions in soil for $C_2H_4$ production. By altering the pH of soil over the range pH 4.1 - 5.6, however, the redox potential of the $Fe^{3+} \rightleftharpoons Fe^{2+}$ couple shifts through -272 mV (4). This major shift in Eh had virtually no effect on the amount or speed of $C_2H_4$ production. The effect of $Fe^{2+}$ on $C_2H_4$ production in soil, therefore, appears to result from some direct action.

The results obtained from amending soil with an array of reducing agents also strongly implicates $Fe^{2+}$ as a specific trigger for $C_2H_4$ release. Any reducing agent that mobilized $Fe^{2+}$ in soil, stimulated $C_2H_4$ production. Conversely, those that did not mobilize $Fe^{2+}$ in soil also failed to stimulate $C_2H_4$ production. Additional evidence for the specific nature of the $Fe^{2+}$ stimulus of $C_2H_4$ production came from amendment of soil with other metals including $Co^{2+}$, $Ni^{2+}$, $Mn^{2+}$, and $Cu^{2+}$. None of these metals increased $C_2H_4$ production from soil (unpublished data).

The results obtained when soils were amended with $Fe^{2+}$ before and after incubation for 14 days, indicate strongly that $Fe^{2+}$ released $C_2H_4$ and the $C_2H_4$ produced normallyin soil, are from an identical source. Thus, the $C_2H_4$ released from soil by $Fe^{2+}$ treatment does not appear to be a laboratory artifact and so does have relevance to $C_2H_4$ production in field soils.

In this study, the soils required wetting with high concentrations of $Fe^{2+}$ ($10^4$ ppm $Fe^{2+}$ or more) to ensure maximum release of $C_2H_4$. This raises questions about the relevance of $Fe^{2+}$ for $C_2H_4$ release in field soil especially since in our incubation study (Fig. 4), $Fe^{2+}$ was present by the time $C_2H_4$ was produced but its concentration in soil was low. When $Fe^{2+}$ is added exogenously to dry soil, however, much of it will be oxidized to immobile $Fe^{3+}$ forms, some will be complexed with organic compounds in soil, and considerable amounts will undergo exchange reactions in soil. Thus, most of the $Fe^{2+}$ added becomes unavailable for triggering of $C_2H_4$ release in soil. Conditions are completely different when endogenous soil $Fe^{2+}$ is mobilized during incubation of soil. Reductive processes begin presumably in the centre of soil crumbs where $Fe^{2+}$ will be mobilized only after all the oxygen is consumed. Endogenous $Fe^{2+}$, then, is much less prone to oxidation than exogenus $Fe^{2+}$. Also, other cations and especially $Mn^{2+}$ will be released prior to $Fe^{2+}$ reduction (4). This means that there will be competition for exchange sites and a higher proportion of the released $Fe^{2+}$ should remain in the soil solution. Thus, much lower concentrations of endogenous $Fe^{2+}$ should be effective in triggering $C_2H_4$ production in soil.

The key role of $Fe^{2+}$ in triggering $C_2H_4$ production in soil explains clearly why oxygen-free and relatively highly reduced conditions are essential in microsites in soil before $C_2H_4$ forms. $Fe^{2+}$ is mobilized only under such conditions (3, 4, 10). This requirement for $Fe^{2+}$ also explains the potency of nitrate nitrogen ($NO_3^-$-N) as an inhibitor of $C_2H_4$ production in soil (6, 8). $NO_3^-$-N acts as an efficient terminal electron acceptor for a range of facultative anaerobic bacteria in the absence of oxygen. In microbiologically active soils adequately supplied with $NO_3^-$-N, hydrogen donors, and suitable species of bacteria, the redox potential will be poised sufficiently high to stop electrochemical reactions, including $Fe^{3+}$ reduction, that occur at lower redox potentials (6). The $Mn^{4+} \rightleftharpoons Mn^{2+}$ redox couple also occurs at a higher redox potential in soil than the $Fe^{3+} \rightleftharpoons Fe^{2+}$ (3, 4, 10). Manganese, however, will have only a minor role in inhibiting $C_2H_4$ production in soil as compared with $NO_3^-$-N. $NO_3^-$-N is in a high state of oxidation, is soluble in water and, therefore, highly mobile in soil, is used as an electron acceptor by many microorganisms, and occurs in appreciable concentrations at least in agricultural soils. Manganese, on the other hand, is insoluble in the oxidized form, and is thus immobile and is a minor constituent of most soils (6).

This work clearly establishes that $C_2H_4$ in soil is not produced directly by any particular microorganism or group of microorganisms. However, intense aerobic microbial activity is still essential in soil before $C_2H_4$ forms because oxygen-free and highly reduced microsites must develop before conditions become favourable for the reaction leading to $C_2H_4$ production.

These results mean that our postulated oxygen-ethylene cycle in soil (6, 8) which appears to have such important implication for turnover rates of organic matter, recycling of plant nutrients, and incidence and control of soil-borne plant pathogens, requires the following modification:

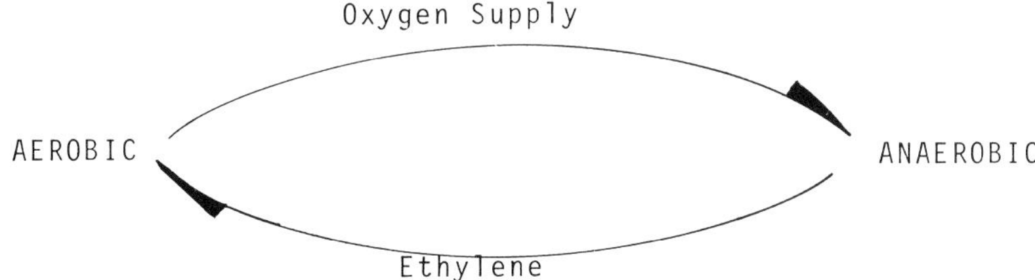


When sufficient organic nutrients become available in soil from freshly incorporated organic residues or exudates from plant roots, aerobic microorganisms proliferate and deplete the oxygen supply at least at microsites in soil. Anaerobic microsites result and provided conditions become sufficiently reduced, $Fe^{2+}$ is mobilized and triggers a chemical reaction in the microsite resulting in the release of $C_2H_4$. $C_2H_4$ then diffuses out through the soil and inactivates sensitive aerobic microorganisms. This results in a reduction in the demand for oxygen which diffuses back into the microsites either limiting or stopping $C_2H_4$ production. Aerobes then increase in activity until the cycle is repeated.

## E. References


1. Dasilva, E. J., Henriksson, E., Henriksson, L. E.: Ethylene production by fungi. Plant Science Lets. 2, 63-66 (1974).
2. Lynch, J. M.: Ethylene in soil. Nature 256, 576-577 (1975).
3. Patrick, W. H. Jr., Mikkelsen, D. S. : Plant nutrient behaviour in flooded soil. In: Fertilizer Technology and Use. 2nd Ed. Olsen, R. A. (ed.). Madison, Wis. : Am. Soc. Soil Sci. pp. 187-215 (1971).
4. Ponnamperuma, F. N. : The chemistry of submerged soils. Adv. Agron. 24, 29-96 (1972).
5. Primrose, S. B. : Ethylene-forming bacteria from soil and water. J. gen. Microbiol. 97, 343-346 (1976).
6. Smith, A. M. : Ethylene in soil biology. Ann. Rev. Phytopathol. 14, 53-73 (1976).

7. Smith, A. M. : Ethylene production by bacteria in reduced microsites in soil and some implications to agriculture. Soil Biol. Biochem. 8, 293-298 (1976).
8. Smith, A. M., Cook, R. J. : Implications of ethylene production by bacteria for biological balance of soil. Nature 252, 703-705 (1974).
9. Smith, K. A., Jackson, M. B. : Ethylene, waterlogging and plant growth. Agric. Res. Counc. Letcombe Lab. Ann. Rept. 1973, 60-75 (1974).
10. Takai, Y., Kamura, T. : The mechanism of reduction in waterlogged paddy soil. Folia Microbiol. Prague 11, 304-313(1966).

# Microbiological Problems in Seedling Establishment

J.M. LYNCH

## A. Introduction

When seeds are drilled so that they come into close contact with straw from a previous crop, poor plant establishment can result (3). This problem may be severe in the autumn, when the straw has not decomposed, particularly if heavy rainfall results in the development of anaerobic conditions around the seed and straw. Initial studies (3) showed that water soluble substances which impaired the development of seedlings were present in the decomposing straw and subsequently it was found that when straw was degraded by soil microorganisms under anaerobic conditions in a solution of mineral salts, acetic acid was the only phytotoxin identified at concentrations which could affect the growth of seedlings (7). Although there are probably also other toxins produced from straw in the field (6, 11), the purpose of this paper is to report some factors which govern its accumulation in soil and to discuss other ways microorganisms affect seed germination.

## B. Methods

Soil (200 g) and distilled water (600 ml), with or without chopped straw (20 g), were added to a 1 litre conical flask, sealed with a rubber bung. Alternatively, soil (300 g), ground wheat straw (50 g) and distilled water (3.2 l) were incubated in a chemostat vessel, similar to that described for the cultivation of moulds (13).

Organic acids were determined by gas chromatography and their effects on the germination and root extension of barley (Hordeum vulgare) investigated in sand culture (7).

The direct effects of microorganisms were studied by soaking seeds in washed suspensions. The quantity of soluble organic material produced by surface sterilized seed (1) was estimated by measuring the total carbon (5) released into solution.

## C. Results and Discussion

I. Microbial production of phytotoxins. In the presence of wheat (Triticum aestivum) straw, acetic, propionic and butyric acids in the ratio 40:2:1 accumulated in solutions from peat, loam and clay soils. Acetic acid was the only substance detected in phytotoxic concentrations. There was no clear correlation between the concentration of acid present and the inhibitory effects on the root extension of young barley seedlings, probably because the plant growth was influenced by substances other than acetic acid which are produced, but not detected.

Plant residues from barley, wheat, oat (Avena sativa), rape (Brassica napus) and couch grass (Agropyron repens) added to the loam soil gave rise to phytotoxic concentrations of acetic acid, which affected the growth of barley, rape, wheat, clover (Trifolium repens) and maize (Zea mays). In the absence of plant residues, insufficient acetic acid was produced to inhibit growth.

Fig. 1. Colonization of barley seed by Gliocladium roseum

Acetic acid can be lost from soil as a result of microbial respiration and under anaerobic conditions it provides a substrate for methanogenic bacteria. Its degradation was least rapid in flooded soil but even then it proceeded to a rate of 13 m moles $100g^{-1}$ soil $d^{-1}$. The rate of accumulation of the acid in flooded soil never exceeded 1 m mole $100g^{-1}$ soil $d^{-1}$ but clearly the true rate of production must have been in excess of the rate of degradation. The balance between production and degradation appears to be governed by redox potential (12, 14) and investigations on the breakdown of straw in the chemostat showed that the accumulation of the acid was associated with reductions in dissolved oxygen tension, redox potential and pH; at the same time there were increases in the rate of accumulation of soluble carbon and in the solubilization of iron and manganese (9).

The movement of acetic acid through the soil from the straw is restricted because it can be degraded biologically and immobilized abiotically. In a non-calcareous silty loam soil, it was found that 1.5 cm from the straw, the concentration was one half of that present on the surface of the straw. Hence seeds are most likely to be affected by straw when they are in close proximity to it.

II. Interactions between microbes and seeds. The fungus Gliocladium roseum, which commonly occurs on decaying straw in soil (2), colonized barley seeds at the tip which is closest to the embryo (10) and if the inoculum concentration was high enough it could prevent germination (Fig. 1). By coating the seed at various points with silicone grease, it was demonstrated that this tip, where there is a fibrous plug formed as a continuation of the inner layer of the husk, is the pathway for

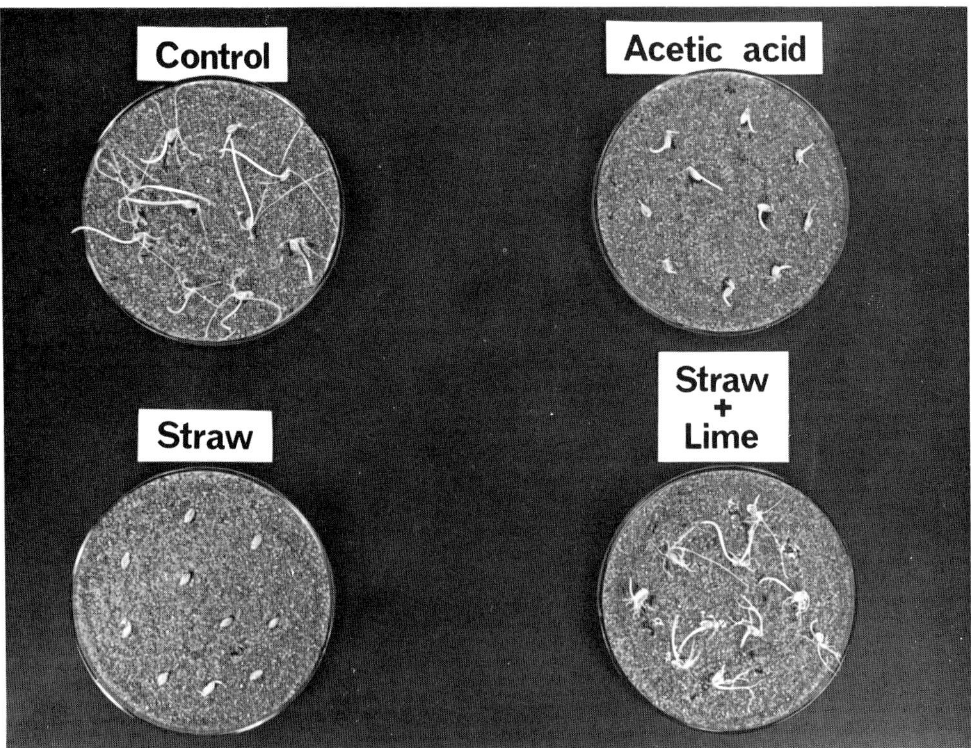


Fig. 2. Effect of acetic acid (15 mM) and a solution from fermented wheat straw containing a similar concentration of acetic acid. Barley seeds were placed on sand moistened either with distilled water (control) or the suspected toxin and kept at 20°C for 5 days. Mitigation of the effect by coating the seed with lime is also shown

the entry of oxygen and water. It is also the site of exudation from the seed of organic materials, which promote the growth of microorganisms. Studies of germination in a respirometer and under defined oxygen concentration indicated that competition between the fungus and the seed for oxygen was a likely cause of the inhibition of germination. The exudation of soluble carbon compounds by seeds almost doubled when the concentration of oxygen in the solution was reduced from 20.8% to zero. Consequently if the supply of oxygen is restricted, as can happen in wet soils, particularly when energy substrates such as straw are added to the soil, seeds become especially vuenerable to microbial attack since the enhanced exudation promotes microbial growth and increased competition for the limited amount of available oxygen.

Another common soil fungus, Mucor hiemalis, produced similar effects to G. roseum.

III. Mitigation of the problems. Where possible burning of the straw appears to be the most effective remedy to the toxicity problem (4), but if this is not possible, other solutions must be considered. It was shown earlier (7) that acetic acid was most toxic at low pH and it has now been demonstrated using $C^{14}$- acetate that the uptake at pH 3.2 is four times greater than that at pH 7.0. This accords with the view that the undissociated state of the acid ($pK_a$ = 4.75) penetrates the lipid components of the seed most readily. Consequently it seemed that if the pH of the immediate environment of the seed could be increased, the effect of the acid might be mitigated. This was brought about by coating the seeds with powdered chalk (Fig. 2).

The inhibitory effect of fungi on germination, which is not restricted to the presence of straw residues, can be largely overcome by coating seeds with a suitable bacterial inoculum (8), but as this would be difficult to achieve under field conditions, treatment with chemicals would probably be more practicable.

## D. References


1. Barber, D.A.: The effect of microorganisms on the absorption of inorganic nutrients by intact barley plants. I. Apparatus and Culture Technique. J. exp. Bot., 18, 163-169 (1967).
2. Domsch, K.G., Gams, W.: Fungi in agricultural soils. London: Longman, pp. 82-85 (1972).
3. Ellis, F.B., Barber, D.A., Graham, J.P.: Seedling development in the presence of decaying straw residues. ARC Letcombe Lab. Ann. Rept. 1974, 39-40 (1975)
4. Ellis, F.B., Lynch, J.M.: Why burn straw? ARC Res. Revs. 2, 29-33 (1977).
5. Gunn, K.B.: An automated method for the determination of total organic carbon in root exudates. ARC Letcombe Lab. Ann. Rept. 1974, 76-77 (1975).
6. Lynch, J.M.: Products of soil microorganisms in relation to plant growth. CRC Crit. Rev. Microbiol. 5, 67-107 (1976).
7. Lynch, J.M.: Phytotoxicity of acetic acid produced in the anaerobic decomposition of wheat straw. J. appl. Bact, 42, 81-87 (1977).
8. Lynch, J.M.: Microbial interactions around imbibed barley seeds. Ann. appl. Biol, in press.
9. Lynch, J.M., Gunn, K.B.: Chemical aspects of the production and movement of acetic acid in soil. ARC Letcombe Lab. Ann. Rept. 1976, 56-58 (1977).
10. Lynch, J.M. Pryn, S.J.: Interaction between a soil fungus and barley seed. J. gen. Microbiol., 103, 193-196 (1977).
11. McCalla, T.M., Norstadt, F.A.: Toxicity problems in mulch tillage. Agric. Environ., 1, 153-174 (1974).
12. Ponnamperuma, F.N.: The chemistry of submerged soils. Adv. Agron., 24, 29-96 (1972).
13. Rowley, B.I., Bull, A.T.: Chemostat for the cultivation of moulds. Lab. Practice, 22, 286-289 (1973).
14. Takai, Y., Kamura, T.: The mechanism of reduction in waterlogged paddy soil. Folia Microbial, 11, 304-313, (1966).

# Indole-3-Ylacetic Acid Metabolism of *Corynebacterium fascians*

D.R. KEMP

## A. Introduction

Margaret Lacey (4, 5, 6, 7, 8) pioneered the work on certain fasciation diseases of plants. Such diseases are characterised by a stunting of the main stem coupled with a stimulation of lateral bud development, a mass of lateral shoots often developing that are not separated by the normal length of stem internode. She proved that this disease is caused by infection with a bacterium, now known as Corynebacterium fascians, that the organism could produce the disease without wounding, and that culture extracts from the parasite would simulate the symptoms when applied to the buds of a susceptible host.

The disease symptoms were shown to be due, at least in part, to the destruction of indole-3-ylacetic acid (IAA) in the host and various strains of Corynebacterium fascians isolated by Lacey were found to degrade IAA in vitro.

Synthesis of IAA from tryptophan by both micro-organisms and higher plants is well established. It has been demonstrated (13) however that many reports of such synthesis in higher plants may be attributed to the presence of bacteria. Increased auxin synthesis from tryptophan is characteristic of tobacco plants infected with the vascular pathogen, Pseudomonas solanacearum, and it has been shown that this organism produces indole-3-ylacetaldehyde as an intermediate in cell free systems (10). Similarly, Agrobacterium tumefasciens forms indole-3-ylacetaldehyde during IAA synthesis from tryptophan with 3-indole-lactic acid and tryptophol being formed as well (3).

In higher plants synthetic pathways of IAA from tryptophan vary. In tobacco, tryptamine is an intermediate (10), while in Lentil (Lens culinaris) indole-3-ylacetaldehyde and tryptophol are formed (2). The main pathway probably involves deamination or transamination of tryptophan to indole-pyruvic acid, which is decarboxylated to indole-3-ylacetaldehyde using lipoic acid as a cofactor. The indole-3-ylacetaldehyde is converted to indole-acetyl Co A, and finally to IAA by loss of Co A coupled with hydrolysis.

## B. Materials and Methods

Metabolism of IAA by Corynebacterium fascians was studied by growing the organism in an unbuffered nutrient salt solution, pH 6.9, and the breakdown of added IAA followed colorimetrically by the Salkowski reaction (1), sampling every one or two hours. To follow whether the enzymes involved could be induced by previous growth in the presence of IAA, the inoculum in some cases was grown in culture medium containing 100 $\mu g\ ml^{-1}$ IAA. It was assumed that the organism would use IAA as either a carbon or a nitrogen source and the induction medium lacked either the carbon source (mannitol) or the nitrogen source (ammonium chloride). Non-induced inocula were grown on full nutrient medium lacking IAA. The inoculum was grown for three to six days with shaking in 100 ml of medium in a 500 ml conical flask. The cells were centrifuged and washed in sterile distilled water once, and resuspended in 5 ml of sterile distilled water, and added to 100 ml of medium lacking a carbon or nitrogen source but containing IAA at a concentration of 50 μg $ml^{-1}$. The concentration of bacterial protein was monitored (9).

## C. Results

Degradation of IAA occurred in both nitrogen and carbon deficient media, but the rate varied with the induction medium. When cells were induced in a nitrogen deficient medium, the rate of degradation was considerably more rapid if the cells were transferred into a similar medium than if transferred to a carbon deficient medium. However, where cells were induced in a carbon deficient medium the rate was little different on transfer to either a nitrogen or a carbon deficient medium. The rate was, furthermore, considerably less than the fastest rate obtained (Table 1 and Figure 1). Cells that had not been induced by prior growth in a nitrogen or carbon deficient medium containing IAA showed no degradation of the auxin during the eight hour experimental period.

Table 1. Rate of indole-3-ylacetic acid degradation during eight hours by *C. fascians* in induced cultures

| | μmoles indole-3-ylacetic acid degraded per minute per mg bacterial protein | | | |
|---|---|---|---|---|
| Induction medium | Nitrogen deficient medium | | Carbon deficient medium | |
| | Flask 1* | Flask 2* | Flask 1* | Flask 2* |
| Nitrogen deficient | 0.048 | 0.050 | 0.014 | 0.016 |
| Carbon deficient | 0.030 | 0.029 | 0.028 | 0.029 |

* Flasks 1 and 2 are duplicates

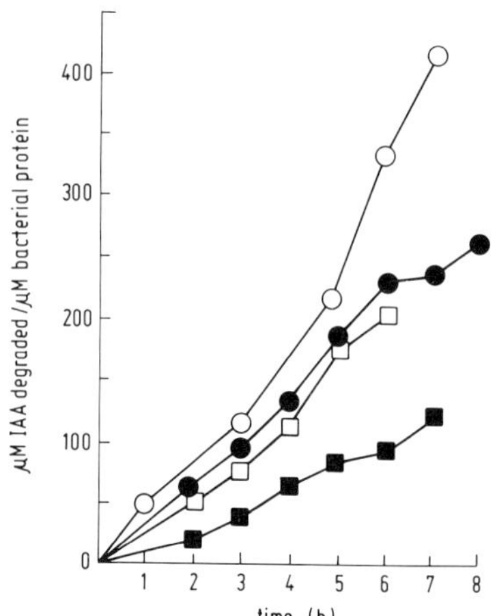


Fig. 1. μM IAA degraded by *C. fascians* per mg bacterial protein plotted against time in hours. The cultures were induced in a medium containing IAA but deficient in nitrogen (-N) or carbon (-M). ○—○—○ Induced in -N medium and incubated in a similar medium; ●—●—● induced in -M medium and incubated in -N medium; □—□—□ induced in -M medium and incubated in a similar medium; ■—■—■ induced in -N medium and incubated in -M medium

These results appear to indicate that under the conditions stated there are at least two pathways of IAA degradation depending on which nutrients are lacking from the growth medium.

In other experiments degradation of IAA in complete medium was studied. Following the finding that in one strain C. fascians degradation of IAA was so rapid during lag or early log phases of growth that products could not be recognised but, if older cultures were used, where the cells were in late log phase and the culture had become slightly acid, products could be distinguished, IAA was added to three day old cultures in which the pH of the medium had fallen to approximately 3.0. Initially samples were removed every 3 h and subsequently at 38 h, 69 h, 7, 10 and 20 days for assay of IAA.

Figure 2 shows a typical result for one of the nine strains studied. At first the absorbance increased, reaching a maximum at 17 h, (it occurred later for some strains) but eventually fell, during which time the colour produced in the reaction changed from pink, characteristic of IAA, to orange and finally to yellow, characteristic of tryptophol. Only one strain, the growth of which was unpigmented instead of the usual orange, was different, and showed a slow fall in absorbance without the initial rise. Growth, measured by turbidity, increased slightly during the three days after the addition of IAA, but was not considered to be an important factor.

The absorption spectra of the yellow solutions produced by samples in the Salkowski assay matched that produced by tryptophol, and by reference to a standard curve the concentration of tryptophol in the samples was calculated. In one typical strain this was found to be 33 $\mu g\ ml^{-1}$ corresponding to a 66% conversion of IAA into tryptophol.

Paper chromatography was used to analyse samples taken at 38 h and 7 days. The results obtained are shown in Table 2. From a comparison of Rf values and colour reactions of unknowns with the standards the main products at both times of sampling were anthranilic acid, indole-3-ylacetaldehyde and tryptophol. Small quantities of indole were also identified.

In order to add support to the above results, degradation products of IAA formed by C. fascians were extracted in bulk from 400 ml of medium. After concentration, the products were used in chromatography with ten different solvents and the presence of tryptophol and indole-3-ylacetaldehyde confirmed.

## D. Discussion

It seems evident that the main degradative pathway of IAA by C. fascians under acid conditions involves two reductive steps. The first is a reversal of the final step in the synthesis of IAA from tryptophan, resulting in the production of indole-3-ylacetaldehyde, while the second produces tryptophol. In cucumber plants high levels of IAA lead to the production of tryptophol (12). The tryptophol may

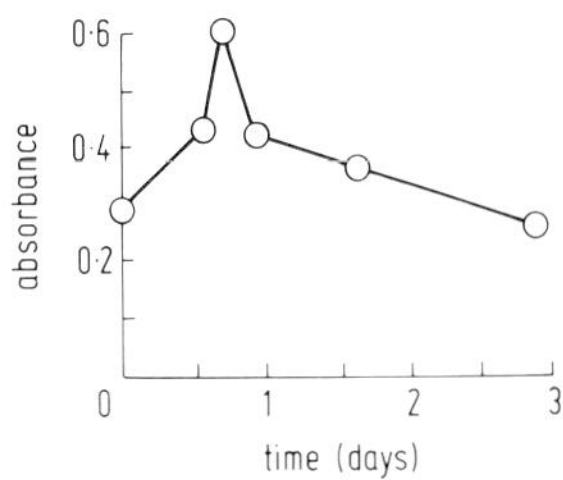


Fig. 2. Absorbance in the Salkowski reaction to estimate IAA degradation by an active strain of C. fascians plotted against time in days. The rise in absorbance during the first day after addition of IAA to three day old cultures in complete medium was typical of eight of the nine strains tested

Table 2. Colour Reactions and Rf values of standards and products of degradation of indol-3-ylacetic acid by *C. fascians*

| Standard | Rf value | Test | | | Probable nature of product |
|---|---|---|---|---|---|
| | | u.v. light | Ehrlich's Reagent | DMCA Reagent | |
| Indole-3-ylacetic acid | 0.31 | Purple | Mauve | Dark mauve | |
| Anthranilic acid | 0.33 | Bright purple | Bright yellow | Pink - soon fading | |
| Indole-3-ylcarboxylic acid | 0.23 | Purple | Mauve | Slowly blue | |
| Tryptophol | 0.85 | Violet | Purple → grey | Blue, later with a yellow centre, finally grey | |
| Indole-3-ylacetaldehyde | 0.79 | Faint blue | Faint violet → yellow | Faint purple | |
| Indole-3-ylaldehyde | 0.85 | Purple | Yellow | Faint purple | |
| Indole | 0.88 | No fluorescence | Pink | Green | |
| Degradation Product* (38 hour samples) | | | | | |
| 1 | 0.31 | Bright purple | Yellow → mauve | Dark mauve | Anthranilic acid and indole-3-ylacetic acid |
| 2 | 0.83 | No fluorescence | Yellow | Pink → faint purple | Indole-3-ylacetaldehyde |
| 3 | 0.86 | Violet | Yellow → purple | Deep blue | Tryptophol |
| 4a | 0.71 | Grey-blue | No colour | No colour | Not known |
| (7 day samples) 1 | 0.32 | Violet | Mauve | Dark mauve | Anthranilic acid and indol ylacetic acid |
| 2 | 0.82 | Pale blue | Yellow | Pink, soon fading | Indole-3-ylacetaldehyde |
| 3 | 0.86 | Violet | Yellow → purple | Deep blue | Tryptophol |
| 4b | 0.87 | No fluorescence | Pink | No colour | Indole |

*Numbers refer to spots on the chromatogram developed in isopropanol/0.88 ammonia/water (10/1/1 v/v). Spots 1, 2, and 3 are equivalent at both times of sampling. Of the nine strains examined, spot 1 was formed by all strains at both times; spot 2 was formed by only four strains at 38 hours and all strains at 7 days; spot 3 was formed by all strains except the unpigmented one at both times, while spot 4a was formed by all strains and spot 4b by only three strains.

serve as a reserve indole derivative, that under conditions of low auxin levels in the plant can be drawn upon to form more IAA. The effect of the parasite on such a regulatory system in the host of C. fascians must be profound, since IAA would be degraded and converted into tryptophol, so that high levels of the latter would be paralleled by low levels of the former, a situation the reverse of the normal. The auxin activity of tryptophol varies according to the plant tested, although a number of species show no response (12). It thus appears that in plants infected with C. fascians the level of active auxin would fall with resultant effect on the growth of the host and the manifestations typical of the fasciation disease.

## E. References


1. Dean, C.: Study of certain metabolic activities of C. fascians, which bear on the Pathogenicity and classification of that Organism. D.I.C. Dissertation: University of London. (1961).
2. Hofinger, M., & Gasper, T.: Biosynthese d'auxine in situ a Partir de l-tryptophane. Bull. Soc. r. Bot. Belg. 102, 197 (1969).
3. Kaper, J.M., & Veldstra, H.: On the Metabolism of Tryptophan by Agrobacterium tumefaciens. Biochim. Biophys. Acta, 30, 401, (1958).
4. Lacey, Margaret S.: Studies in Bacteriosis XXII I. The Isolation of a bacterium associated with 'fasciation' of sweet pease, 'cauliflower' of strawberry plants and 'leafy gall' of various plants. Ann. appl. Biol. 23, 302, (1936a).
5. Lacey, Margaret S.: Further studies on a bacterium causing fasciation of sweet peas. Ann. appl. Biol. 23, 743, (1936b).
6. Lacey, Margaret S.: Studies in Bacteriosis XXV. Studies on a bacterium associated with Leafy-galls, fasciations and 'cauliflower' disease or various plants. Part IV. The Inoculation of strawberry plants with Bacterium fascians (Tilford). Ann. Appl. Biol. 29, 11, (1942).
7. Lacey, Margaret S.: The Cytology and relationships of Corynebacterium fascians. Trans. Brit. mycol. Soc. 38, Part 1, 49, (1955).
8. Lacey, Margaret S.: The development of filter-passing organisms in Corynebacterium fascians cultures. Ann. Appl. Biol. 49, 634, (1961).
9. Lowry, O.H., Rosenbrough, N.J., Farr, A.L. & Randall, R.H.: Protein measurement with the Folin phenol reagent. The Biochemical Journal, 193, 265, (1951).
10. Phelps, R.H., & Sequeira, L.: Auxin biosynthesis in a host-parasite complex. In: Biochemistry and Physiology of plant growth substances. Wightman, F. and Setterfield, G. (ed.). The Runge Press Ltd., Ottawa, pp. 197-212 (1968).
11. Rayle, D.L. & Purves, W.K.: Isolation and identification of indole-3-ethanol (tryptophol) from cucumber seedlings. Plant Physiology. Wash. 42, 520 (1967a).
12. Rayle, D.L. & Purves, W.K.: In: Biochemistry and Physiology of Plant Growth Substances. Wightman, F. & Setterfield, G. (ed.). The Runge Press Ltd., Ottawa. pp. 243-257 (1968).
13. Wichner, S. & Libbert, E.: Interaction between plants and epiphytic bacteria regarding their auxin metabolism. Physiologia Plantarum, 21, 227 (1968).

# The Source of Infections by Basidiomycete Fungi Causing a Decline and Replant Disease in Central Otago, New Zealand

J.B. TAYLOR

## A. Introduction

Central Otago is an arid region in New Zealand where fruit trees are grown under irrigation on river terraces adjacent to the Clutha River. The original vegetation was a xerophytic scrub community dominated by matagouri (*Discaria toumoutou* Raoul), with the following woody plants being locally important - *Chordispartium stevensonii* Cheesm.; *Hebe pimeleoides* (Hook. f.) Ckn. & Allan var. *rupestris* Ckn. & Allan; *Hymenanthera alpina* (Kirk) W.R.B. Oliver; *Leptospermum ericoides* A. Rich.; *Muehlenbeckia australis* (Forst. f.) Meissn.; *Olearia virgata* Hook. f. var. *lineata* Kirk and *Pimelea pseudo-lyallii* Allan.

Most of the original vegetation was cut for fuel during the 'Gold Rush' 1868-1880, and replaced, on suitable terrace soils, by cherry and apricot orchards. Other land on which matagouri was cut was colonized by briar-rose (*Rosa rubiginosa* L.).

A serious decline and associated replant problem in Central Otago orchards has affected cherry trees, and to a lesser degree, apricot trees. The presence of characteristic cankers and lesions on roots suggested that basidiomycete fungi were the cause of the problem. Isolation of the pathogen *Peniophora sacrata* G.H. Cunn. from diseased roots confirmed this view (2). An investigation was made into the possibility that *P. sacrata* or other basidiomycetes could infect stone fruits either directly from matagouri or indirectly from briar-rose which had replaced the matagouri.

## B. Materials and Methods

I. Source of isolates. The root systems of plants on three sites near Alexandra were dug up and studied for the presence of basidiomycete cankers. The sites were (a) a mature matagouri community, (b) briar-roses and (c) cherry trees. Plants in the first two showed no decline symptoms above ground whereas the cherry trees showed advanced decline, and there was a pronounced replant problem on this site.

After washing,roots showing lesions were incubated for 3-7 days at 100% R.H. and any white mycelium growing from the lesions plated on to potato dextrose agar incorporating selective agents (3). Two or three days later phycomycetous fungi not eliminated by the selective agents were removed. Basidiomycetes were recognized by the presence of clamp-connections, or other characteristic mycelial structures. The 112 cultures retained were studied together with reference isolates (3, 4) making a total of 156.

II. Biochemical coding. In previous work, basidiomycete reactions in a series of 'instant' and 'incubated' biochemical tests were used (3, 4) to study isolates. Isolates were placed in a species when their reactions could be matched with an isolate obtained from an identified fruiting body. Isolates with similar reactions, but which could not be matched against an identified strain remained unidentified and were placed in a 'Group' pending formal identification. A similar procedure was used to identify the isolates in this work.

Biochemical tests were carried out as follows: Each isolate was grown on B3 agar and 5 mm diameter plugs removed. Plugs were placed on individual Petri dishes of B3 agar incorporating either lecithin (1) (10% v:v), milk (4) or rutin (3). The plates were incubated for 3-14 days. B3 agar was prepared as follows to avoid fluctuation in pH caused by heating the malt. 'Oxoid' potato dextrose agar 4.4 g, Davis agar 13.3 g, water 1 l, pH 5.4 was sterilized and decanted into 90 ml aliquots. Malt was added to this agar as 10 ml 'Oxoid' malt solution pH 5.4 previously sterilized at 105 kPa (15 lb/in$^2$) for 15 min. in a pressure cooker. Selective agents were added (3).

## C. Results

Two-thirds of the isolates could be placed in recognised groups or species and the remaining isolates were all different except for five pairs of similar isolates (Table 1). Several fungi found on roots of declining fruit trees were also found on roots of matagouri, or *Olearia virgata* var. *lineata*. Representatives of five species or groups, *P. sacrata*, *Pholiota adiposa*, Groups 4, 6 and 7 were found on roots of stone fruit and either on briar-rose or plants of the matagouri community. *C. drucei* and Group 11 were found on roots of both briar-rose and matagouri. *Armillaria sp*. and representatives of Groups 10 and 12 were found on roots of briar-

Table 1. Numbers and kinds of basidiomycete fungi found on roots of cherry, briar-rose and plants of a matagouri community in Central Otago

| Basidiomycete fungus | Cherry | Briar-rose | Matagouri community+ | Pathogenicity* | No. of records in N.Z. |
|---|---|---|---|---|---|
| *Armillaria* sp. | 0 | 0 | 2 | pathogenic | many |
| *Collybia drucei* (Stevenson) Horak | 1 | 1 | 4 | pathogenic (5) | 25 |
| *Peniophora sacrata* | 2 | 0 | 2 | pathogenic (2) | many |
| *Pholiota adiposa* Fr. | 2 | 11 | 0 | ? | 21 |
| Group 4 | 1 | 0 | 3 | ? | 8 |
| Group 6 | 4 | 0 | 2 | ? | 11 |
| Group 7 | 2 | 0 | 6 | pathogenic (5) | 24 |
| Group 8 | 5 | 0 | 0 | ? | 10 |
| Group 10 | 0 | 5 | 0 | pathogenic (5) | 4 |
| Group 11 | 5 | 1 | 3 | ? | 25 |
| Group 12 | 0 | 2 | 2 | ? | 6 |
| Number identified | 22 | 20 | 24 | | |
| Number unidentified | 1 | 17 | 22 | | |
| Total | 23 | 37 | 46 | | |

+Isolates were all from matagouri except those of Group 4 which were from *O. virgata* va. *lineata*.

*Pathogenicity of these fungi was only established on other hosts e.g. strawberry, apple, or pine.

| | | | | | | |
|---|---|---|---|---|---|---|
| 10 | 12<br>12 | | | | | |
| | | | | | | |
| | | | | | | |
| | | | | | | |
| P.a. | | 10 | 11 | | | |
| P.a. | P.a.<br>P.a.<br>P.a. | P.a. | 10<br>10 | | | P.a.<br>10 |
| | | | | | P.a.<br>P.a.<br>C.d. | P.a. |
| | | | | | P.a. | |

Fig. 1. Distribution of fungi in 5 m grids on the briar-rose site. An empty grid indicates that no recognized species or groups of basidiomycetes were isolated from the briar-rose roots. C.d., *Collybia drucei* (Group 3), P.a., *Pholiota adiposa* (Group 2), 10, Group 10, 11, Group 11, 12, Group 12

rose or matagouri, but not on roots of fruit trees, although *Armillaria* sp. and Group 12 have been found elsewhere on roots of stone fruit trees. The distribution of recognised fungi in the briar-rose site (Fig.1) showed that Group 11 and 12 and *C. drucei* were rare but *P. adiposa* and Group 10 were abundant.

## D. Discussion

Eight basidiomycete fungi suspected of being pathogens or known to be pathogenic to strawberry (5) were isolated from roots of cherry trees showing severe decline symptoms. Of these seven were also found on roots of matagouri and/or briar-rose. Failure to isolate representatives of all eleven fungal types from roots in each of the three sites was due to sporadic distribution of these fungi. That distribution was sporadic is shown in Fig. 1 and it is evident that basidiomycetes were absent from many parts of this site.

The strains of *C. drucei*, *P. adiposa* and *P. sacrata*, found on the roots of cherry trees, briar-rose and/or matagouri are indigenous, and it is possible that strains in Groups 4, 6, 7, 8, 10, 11, and 12 are also indigenous. The presence of the same fungal types on cherry trees, on matagouri and/or briar-rose and their absence on cherry trees produced by nurseries suggests that the cherry roots become infected from diseased roots on plants of the matagouri community or the briar-rose, left in the soil after land clearing. This would account for the widespread occurrence of decline and replant disease of stone fruit in Central Otago.

With this background infection on roots of matagouri and briar-rose, any stone fruit orchards replacing these communities are likely to become infected with basidiomycete fungi. Ideally, these infections on roots of matagouri, briar-rose or existing stone fruit trees should be eradicated before fruit trees are planted.

The long delay between the first observations of decline and the identification of the indigenous fungi responsible was due in part to the expectation that a single cosmopolitan pathogen produced the decline and replant problem. It is possible that decline and replant problems elsewhere are caused by indigenous basidiomycetes.

## E. Acknowledgements

I wish to acknowledge technical help from J.E. Hawkins, J.D. McLaren and M.A. Prentice and the co-operation of the Annon Bros., J. & P. Taylor and A.L. Weir, fruitgrowers in Central Otago.

## F. References

1. Lelliot, R.A., Billing, E., Hayward, A.C.: A determinative scheme for the fluorescent plant pathogenic pseudomonads. J. appl.Bact. 29, 470-489 (1968).
2. Taylor, J.B.,: Root-canker disease of apples caused by Peniophora sacrata. N.Z.J. Bot. 7, 262-279 (1969).
3. Taylor, J.B.: Biochemical tests for the identification of mycelial cultures of basidiomycetes. Ann. appl. Biol. 78, 118-123 (1974).
4. Taylor, J.B.: Biochemical characterization of some additional cultures of basidiomycetes. Ann. appl. Biol. 85, 181-193 (1977a).
5. Taylor, J.B.: A pathogenicity test for basidiomycete fungi isolated from roots. N.Z.J. Exp. Agric., in press (1977b).

# Systems Analysis as a Strategy for Agroecosystem Management: The Barley Leaf Rust Epidemic

P.S. TENG, M.J. BLACKIE, and R.C. CLOSE

## A. Introduction

A plant disease epidemic is a complex pathogen-host system where growth of pathogen and host, and their interactions are affected by innumerable environmental factors. These features render the system ideal for analysis and simulation (5). The environment in which a disease epidemic develops can be considered an agroecosystem, or artificial man-made ecosystem (2). Systems analysis represents the tool by which an agroecosystem can be dissected, understood, and its behavioural patterns defined as a prerequisite to managing the agroecosystem. The plant disease manager (usually the farmer) endeavours to manipulate the agroecosystem to maximise economic yield, by means of rapid acquisition and analysis of data concerning the status of the agroecosystem.

Barley leaf rust caused by the fungus *Puccinia hordei* Otth presented an opportunity to use systems analysis to formulate disease management schemes. The disease had been shown to be of importance in New Zealand (1) and to be capable of causing considerable yield loss (6).

## B. Analysis of the Barley Leaf Rust Agroecosystem and Model Development

The barley leaf rust agroecosystem comprises two biological subsystems - the fungus and the host plant, with the interaction of these over time constituting the disease epidemic that is observed in the field. In analysing the *P. hordei* epidemic, both barley and fungus growth must be considered. The rust epidemic that occurs in the field is a polycyclic epidemic which has resulted from a series of overlapping single infection cycles called monocycles (5). Each monocycle in turn consists of distinctive activities or components of the fungus when it exists within and outside the leaf. The activities of the barley leaf rust monocycle which have been defined for modelling (Fig. 1) are spore production during an infectious period, spore liberation, spore survival, spore deposition, spore germination, penetration by germ tube, and latent period (7). Systems analysis enables all available quantitative data to be assembled as a simulation model of the agroecosystem, and makes possible experimentation with the agroecosystem. Three phases were evident during development of the agroecosystem model described in this paper with each phase being represented by its own model version:

Phase 1 - an initial system model which simulated the urediniospore stage of leaf rusts.

Phase 2 - a detailed system model which simulated a leaf rust epidemic occurring on 'Zephyr' barley under specific field conditions.

Phase 3 - a modified phase 2 model for prediction. The structure of the initial system model (phase 1) was based solely on published knowledge about rusts, and included no empirical relationships. This initial system model was used for experimentation and the response patterns generated used for locating control points in the agroecosystem (5). The biological phase of the monocycle, consisting

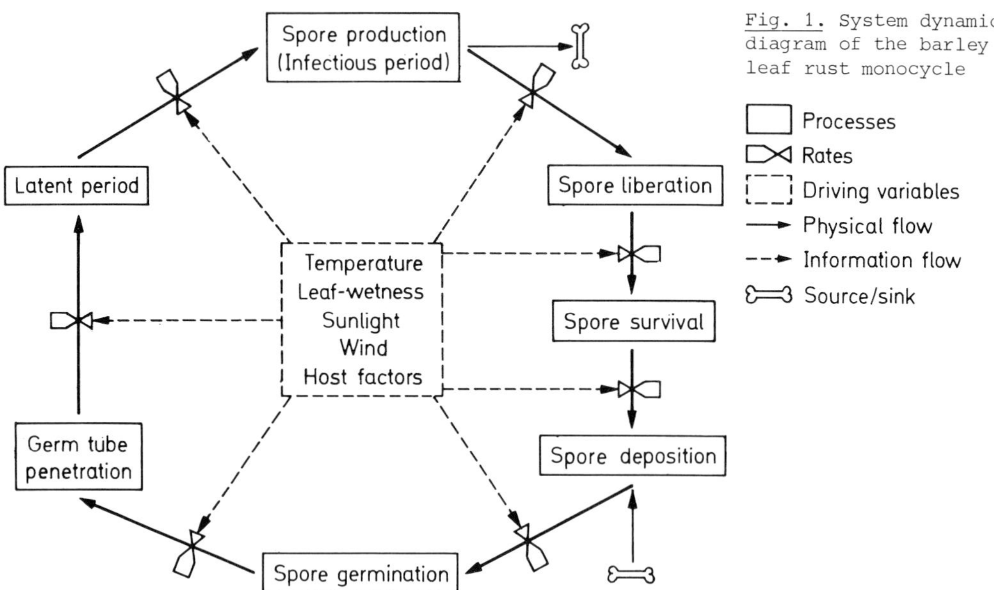


Fig. 1. System dynamic diagram of the barley leaf rust monocycle

of the fungal activities of spore production, penetration and latent period, was found to be a critical determinant of epidemic rate, and consequently due attention was paid to the design of experiments to quantify these activites for the phase 2 and 3 models. For example, it was necessary to construct special spore collectors for accurate measurement of spore production and length of infectious period (8). The phase 3 model is currently being designed and will be tested during the cropping season of 1977/78.

## C. Data Collection and Analysis for Model Development

From a series of laboratory and field experiments at Lincoln College, between mid-1975 and mid-1977, data was obtained for the phase 2 and 3 models. These experiments determined the effects of the following variables on each component of the rust monocycle.

1. Spore production (infectious period) - effect of temperature, pustule age, pustule density of leaf, rust race/host cultivar combination.
2. Spore liberation - effect of wind, leaf wetness.
3. Spore survival - effect of temperature, light intensity.
4. Spore deposition - effect of wind.
5. Spore germination - effect of temperature, light, leaf wetness, spore density on leaf.
6. Germ tube penetration - effect of race/cultivar combination.
7. Latent period - effect of temperature, pustule density, race/cultivar.

The effect of the rust polycycle on barley grain yield was determined by five field trials, where different levels of disease were generated by varying sowing date and number of sprays of systemic fungicides. All the above data were subjected to least squares regression analyses to determine the response surface of an activity to the relevant variables (9). The functions were accepted for use in the model if they satisfied five statistical criteria - high $R^2$ value, significant F-test in analysis of variance, low standard error of estimate of dependent variable, significant multiple correlation coefficient, and significant t-tests for partial regression coefficients. Where a function was used for predictive purposes, as in the disease/loss relationship, two further criteria - no autocorrelation and homoscedasticity - were imposed (4).

## D. Managing the Barley Leaf Rust Agroecosystem

The case for rationalising fungicide use based on knowledge of the yield loss caused by an epidemic and the extra revenue generated by each fungicide spray has been put (3) on economic and pollution grounds. This approach requires a departure from scheduled calendar spraying, necessitates a rapid means of digesting information derived from surveillance of crop and fungus and an accurate method of predicting the course of an epidemic. The design of the phase 3 model will enable prediction of the progress of a rust epidemic and the yield reduction to be expected from this epidemic. The model also will simulate the situation when a fungicide spray is applied, and predict the expected yield with control. Management schemes to minimise fungicide use and to rationalise the use of each spray on a cost-benefit basis arise logically from an awareness that agricultural practices must be viewed, not only from the perspective of a single farm, but in the light of their total effects on the environment. Most modern fungicides are petroleum products and require substantial amounts of energy to manufacture and to apply. We believe that systems theory and computer technology provide a means of achieving savings in energy through the rational use of fungicides.

## E. References

1. Arnst, B.J., Fenwick, J.E.: A survey of barley diseases in New Zealand, Proc. 26th N.Z. Weed & Pest Control Conf. 157-160 (1973).
2. Harper, J.L.: Agricultural ecosystems. Agroecosystems 1, 1-6 (1974).
3. James, W.C.: Assessment of plant diseases and losses. Ann. Rev. Phytopath. 12, 27-48 (1974).
4. Koutsoyiannis, A.: Theory of econometrics - an introductory exposition of econometric methods, London: Macmillan Press Ltd., pp. 175-222 (1973).
5. Teng, P.S., Blackie, M.J., Close, R.C.: A simulation analysis of crop yield loss due to rust disease. Agric. Systems 2, 189-198 (1977).
6. Teng, P.S., Close, R.C.: A preliminary comparison of benodanil and MEB 6447 for control of leaf rust of barley. APPS Newsletter 5, In press (1977a).
7. Teng, P.S., Close, R.C.: The effect of temperature and uredinium density on urediniospore production, latent and infectious periods of *Puccinia hordei* Otth. N.Z.J. Agric. Res. In press (1977b).
8. Teng, P.S., Close, R.C.: Mass efficiency of two urediniospore collectors. N.Z.J. Exptl. Agric. 5, 197-199 (1977c).
9. Throsby, C.D.: Fitting production functions to experimental data. Rev. Mktg. Ag. Econs. 29, 112-147 (1961).

# *Phytophthora cinnamomi* in Native Forests of Australia and New Zealand: Indigenous or Introduced?

F.J. NEWHOOK

## A. Introduction

Phytophthora cinnamomi Rands is found in native forests of Australia and New Zealand in three degrees of association: 1. causing death of eucalypts and woody herbaceous species in several genera on a large scale, extending annually: notably in jarrah (Eucalyptus marginata Sm.) forests of Western Australia (16, 18) and in the Brisbane Ranges, Victoria (28) and Wilson's Promontory National Park, Victoria (23, 25): 2. periodic, often extensive mortality following abnormally heavy rainfall in summer and autumn, e.g. east Gippsland, Victoria (11, 26) and near Auckland New Zealand in regenerating kauri (Agathis australis Salisb.) (17): 3. readily detectable, but apparently tolerated by a flora otherwise comparable in composition to those in Western Australia and Victoria, e.g. New South Wales (5, 19, 20), parts of Queensland (19) and New Zealand (17).

Various hypotheses have been advanced to explain the different degrees of disease expression. It has been suggested that the environmental contribution to the 'disease triangle' is significant (11, 14, 23), e.g. high soil moisture in warm periods, presence or absence of water stress after periods of root death. Evidence is accumulating too, which suggests that differences in microbial saprophyte populations as well as other soil factors may greatly influence pathogenic activity by P. cinnamomi (4, 6, 24).

While there seems no doubt that in Western Australia, the Brisbane Ranges and Wilson's Promontory National Park Phytophthora cinnamomi must be a new introduction encountering a flora that has not had an opportunity to develop resistance, a small group of workers believes that P. cinnamomi is indigenous in the east where, they claim, the flora has developed genetic resistance during a long period of coevolution with the pathogen. They argue that there is no justification for local quarantines and management practices aimed at preventing spread of contaminated soil: that the fungus is ubiquitous and that disease expression results from an upset to the natural balance by milling, road making and other causes of disturbance to a forest. Newhook and Podger (14) considered the evidence and logic for and against the indigenous theory and decided strongly in favour of P. cinnamomi being a pathogen introduced into Australia in fairly recent times and into New Zealand perhaps during Maori colonisation. We warned of the dangers of missing out on opportunities that still remain for preventing spread. The warning has since been highlighted by the continuing occurrence of new disease outbreaks similar to those in the Brisbane Ranges and Wilson's Promontory, with P. cinnamomi associated with dieback in previously healthy areas of native forest in the Grampian Ranges of Western Victoria (7), in two major reserves in South Australia (9), in 'Wallum' heathland, Queensland (15) and in rainforest of North Queensland (2).

I continue to believe that the basic assessment which Podger and I made in 1972 still holds and that our judgement has been reinforced by much of the work that has been published since. I will summarise the main post-1972 contributions to both sides of the controversy and offer my personal comments. While the debate of necessity deals with purely circumstantial evidence there are nevertheless substantial practical issues at stake. Important amongst these are forest management decisions which involve large budgets. They also have an important bearing on the success of economic forestry in large areas of certain states and also on the aesthetic and recreational future of some important national parks and reserves.

## B. The Indigenous Hypothesis

Belief that *P. cinnamomi* is indigenous seems to be based mainly on the following statements of Pratt and Heather (19): 1. The fungus is a common component of the soil microflora and has been present for an exceedingly long time in many hardwood forests in coastal and semicoastal Australia along the 2000 mile (3200km) sampling zone from Cairns to southern Tasmania. 2. The pathogen occurs in isolated areas of forest where spread by man is unlikely. 3. The fungus may have been partly instrumental in eastern Australia in determining the present distribution of eucalypts and other sensitive plant species, i.e. resistant in wetter and susceptible in drier sites. 4. Long association of a pathogen with vegetation would lead to development of disease resistance in the habitat occupied by the pathogen.

Additional support, at least for the northern half of eastern Australia, has been published (1, 4) on the basis of a study of microbial antagonists of *P. cinnamomi* and of the bacteria which stimulate sporangial formation.

The first of the above statements is untenable at least in Victoria where considerable data (7, 23, 25, 27, 28, 29) has been amassed from observations and inoculations and from many hundreds of isolations from affected and healthy forest in the Brisbane Ranges, Wilson's Promontory and the Grampians. In these areas the pattern of *P. cinnamomi* distribution (present in diseased areas and absent beyond the margins), and the fact that most of the flora is highly susceptible leaving behind virtually only grasses and sedges, precludes the fungus from being anything but an invader of plant communities, as in Western Australia, that have had no chance of evolutionary selection for tolerance. Two more similar situations have recently been reported from South Australia (9).

Pratt *et al*. (20) have been questioned about (22) and in turn have defended (21), their belief that an area in New South Wales to which they continually refer was unaffected by man's activities, the basis of their second assumption. In a sense the doubt is immaterial in a mainland situation where, independent of man, there are opportunities for native and introduced animals to act as vectors for infested soil. *P. cinnamomi* has been isolated (8) from feet of feral pigs on Hawaii where the pathogen is rapidly extending its range in native forest on lava fields where passive movement of surface soil is otherwise hard to visualise. Investigation (2) of a recently-discovered dieback syndrome in rainforest of northern Queensland, showed frequent association of feral pig wallows with *P. cinnamomi* and dieback. In fact it may not take an obvious major animal to transport *P. cinnamomi* inoculum effectively. I have reported (13) the frequent and widespread occurrence of the fungus in soil in virgin kauri forest on Little Barrier Island, a sanctuary with rugged terrain 14km from the nearest land and lacking obvious animal vectors. This situation, incidentally, would qualify for indigenous status better than any so far reported from Australia.

Regarding the hypothesis that *P. cinnamomi* could have influenced the distribution of susceptible and tolerant plants and have been a selection factor in evolution of the flora of eastern Australia, Weste and Marks (26) point out that most of the forests of Victoria are dominated by root rot sensitive *Eucalyptus* spp. and that these and many other susceptible species can be found in gullies as well as on drier sites. In fact it is in the relatively flat coastal forests of eastern Victoria that the highly susceptible communities have been hit four times in the last 19 years following abnormally heavy summer rainfall (10). Such a sequence of disease occurrence is not compatible with any theory of pathogen-dictated distribution or evolutionary influence over 'an exceedingly long time'. In east Gippsland in 1970 I visited an area where Eucalyptus sieberi had died in extensive fashion in the late 1950s in one of the fairly recent epidemic outbreaks. Saplings of E. sieberi often 5m tall were regenerating in the previously devastated area. Although they were outwardly healthy they could be pushed over readily, the roots were confined to an inextensive surface plaque and most were dead (Figs. 1 and 2): hardly an example of genetic resistance to root rot! No wonder that the local forester was disturbed by a not-infrequent condition that seemed to him to be 'sudden death' but which was in fact a case of water depletion as a result of shoot/root imbalance. Other plants of the community including acacias and epacrids showed the same root disease syndrome.

If it should happen that a convincing case for geographic or topographic selection for genetic resistance to P. cinnamomi in eastern communities should be put forward, heed must be taken of recent work (3) which demonstrated that there is parallel susceptibility and resistance amongst a wide range of Eucalyptus spp. to P. cinnamomi and several other Phytophthora and Pythium species. The selective agent would not need to have been P. cinnamomi. As Baker and Cook (1) have said, '...many plant species have resistance to parasites to which they have never been exposed'.

In considering the evidence offered by the investigation of soils that suppress sporulation and disease incitement by P. cinnamomi (4) I readily accept the phenomenon but remain puzzled about the aspect of support for an indigenous status for the

Fig. 1. Sapling of Eucalyptus sieberi readily pushed over, showing how roots are restricted to a shallow plaque of roots, restricted in radial spread

Fig. 2. Same tree as Figure 1. Note healthy crown despite almost non-existent root system

pathogen. The claim that the only known soils truly suppressive to P. cinnamomi are in south Queensland is surely premature. The authors list only one truly suppressive soil amongst 52 tested and that was from an intensively managed avocado orchard and not from native forest. Recent reports (6, 12) suggest that there are also forest soils in southern New South Wales and Western Australia which are suppressive to P. cinnamomi. While there undoubtedly are cases (1) where long uninterrupted association between a pathogen and local microorganisms has led to development of a suppressive microflora this does not necessarily apply in the Australian examples of forests on moist but well-drained, organic-rich soils. In fact a more attractive theory for control in these cases comes from the same source (1): '... the greater the variety and number of microorganisms supported by organic amendments, the greater the probability of suppressing the pathogen'.

The proponents for an indigenous status for *P. cinnamomi* have many times suggested that the local forest communities have reached a state of balanced co-existence with a widespread pathogen and that disturbance by man upsets the balance and leads to disease expression. This should be readily demonstrable in controlled trials if it were the case. Meantime it is hard to reconcile the theory with past examples of man disturbing forest communities through logging and road making yet failing to induce spreading disease. Where man has recently done the latter there is good evidence that he unwittingly introduced the pathogen. Why should the disease continue to spread beyond the disturbance, even uphill, if such a supposed state of balance had developed? Would not historic and prehistoric fires have opened up communities that were supposedly in balance and created as much or more of a change in soil temperature and water relations as has man's quite recent disturbance? Yet there are no indications that such events have induced the type of disease that parts of eastern and western Australia are now experiencing: epidemics that are eliminating a high proportion of the flora and changing forests to open woodland, heathland, or grassland. Others have already posed the same questions.

We must consider the time scale required for selective factors such as a soil pathogen to have influenced evolution of a flora. With long-lived, often coppicing, dominants such as eucalypts, we have no alternative in Australia but to accept that such a selective process would need millenia to be effective. If the assumptions associated with indigenous status for *P. cinnamomi* in Australia are valid are they compatible with what might seem to be equally valid claims for indigenous status for *P. cinnamomi* in New Zealand, North America or Europe? Do we really have in either of our countries a generalised genetic resistance to *P. cinnamomi* root rot, or, rather, degrees of tolerance influenced by local or regional differences in climate or in microbial and abiotic factors in the soil?

I believe that work published in the five years since we previously discussed the origin of *P. cinnamomi* in Australia and New Zealand (14) has strengthened rather than diminished the view that the pathogen is introduced rather than indigenous in our areas. I believe, too, the correct, responsible advice to the forest administrators is that they do still have opportunities to prevent spread of *P. cinnamomi* and that the appropriate action is urgent.

## C. References


1. Baker, K.F., Cook, R.J.: Biological control of plant pathogens. San Franciscô: Freeman, pp. 68-9 (1974).
2. Brown, B.N.: *Phytophthora cinnamomi* associated with patch death in tropical rain forests in Queensland. Aust. Pl. Path. Soc. Newsletter 5, 1-4 (1976).
3. Brown, B.N.: Aspects of interaction between environmental factors, *Eucalyptus* spp., and Pythiaceous fungi, especially *Phytophthora cinnamomi*. Ph.D. thesis, University of Auckland, (1977).
4. Broadbent, P., Baker, K.F.: Behaviour of *Phytophthora cinnamomi* in soils suppressive and conducive to root rot. Aust. J. Agric. Res. 25, 121-137 (1974).
5. Fraser, L.R.: unpublished
6. Halsall, D.M.: Examination of a forest soil suppressive to *Phytophthora cinnamomi*. Proc. Int. Symp. Microbial Ecology, Dunedin, N.Z. (Abs. D36) (1977).

7. Kennedy, J., Weste, G.: Phytophthora cinnamomi in the Grampians, Victoria. Aust. Pl. Path. Soc. Newsletter 6, 23-4 (1977).

8. Kliejunas, J.T., Ko, W.H.: Dispersal of Phytophthora cinnamomi on the island of Hawaii. Phytopathology 66, 457-460 (1976).

9. Lee, T.C., Wicks, T.J.: Phytophthora cinnamomi in native vegetation in South Australia. Aust. Pl. Path. Soc. Newsletter 6, 22-23 (1977).

10. Marks, G.C., Kassaby, F.Y., Reynolds, S.T.: Dieback in the mixed hardwood forests of eastern Victoria: a preliminary report. Aust. J. Bot. 20, 141-154 (1972).

11. Marks, G.C., Kassaby, F.Y., Fagg, P.C.: Dieback tolerance in eucalypt species in relation to fertilization and soil populations of Phytophthora cinnamomi. Aust. J. Bot. 21, 53-65 (1973).

12. Nesbitt, H.J., Malajczuk, N., Glenn, A.R.: Bacterial colonization of Phytophthora cinnamomi. Rands. Proc. Int. Symp. Microbial Ecology, Dunedin, N.Z. (Abs. D39) (1977).

13. Newhook, F.J.: Climate and soil type in relation to Phytophthora attack on pine trees. Proc. N.Z. Ecol. Soc. 7, 14-15 (1960).

14. Newhook, F.J., Podger, F.D.: The role of Phytophthora cinnamomi in Australian and New Zealand forests. Ann. Rev. Phytopathology 10, 299-326 (1972).

15. Pegg, K.G., Alcorn, J.L.: Phytophthora cinnamomi in indigenous flora in southern Queensland. Search 3, 257 (1972).

16. Podger, F.D.: Phytophthora cinnamomi, a cause of lethal disease in indigenous plant communities in Western Australia. Phytopathology 62, 972-981 (1972).

17. Podger, F.D., Newhook, F.J.: Phytophthora cinnamomi in indigenous plant communities in New Zealand. N.Z. J. Bot. 9, 625-638 (1971).

18. Podger, F.D., Doepel, R.F., Zentmyer, G.A.: Association of Phytophthora cinnamomi with a disease of Eucalyptus marginata forest in Western Australia. Plant Dis. Reptr. 49, 943-7 (1965).

19. Pratt, B.H., Heather, W.A.: The origin and distribution of Phytophthora cinnamomi Rands in Australian native plant communities and the significance of its association with particular plant species. Aust. J. Biol. Sci. 26, 559-73 (1973).

20. Pratt, B.H., Heather, W.A., Shepherd, C.J.: Recovery of Phytophthora cinnamomi from native vegetation in a remote area of New South Wales. Trans. Br. Mycol. Soc. 60, 197-204 (1973).

21. Pratt, B.H., Heather, W.A., Shepherd, C.J.: Comment on the article by R. and V. Routley. Trans. Br. Mycol. Soc. 63, 419 (1974).

22. Routley, R., Routley, V.: Note on the recovery of Phytophthora cinnamomi from the Budawang Ranges. Trans. Br. Mycol. Soc. 63, 413-419 (1974).

23. Weste, G.: Phytophthora cinnamomi: The cause of severe disease in certain native communities in Victoria. Aust. J. Bot. 22, 1-8 (1974).

24. Weste, G.: The effect of Phytophthora cinnamomi on microbial populations associated with the roots of forest trees. Proc. Int. Symp. Microbial Ecology, Dunedin, N.Z. (Abs. D38) (1977).

25. Weste, G., Law, C.: The invasion of native forest by Phytophthora cinnamomi. III: Threat to the National Park, Wilson's Promontory, Victoria. Aust. J. Bot. 21, 31-51 (1973).

26. Weste, G., Marks, G.C.: The distribution of Phytophthora cinnamomi in Victoria. Trans. Br. Mycol. Soc. 63, 559-572 (1974).

27. Weste, G., Ruppin, P.: Factors affecting the population density of Phytophthora cinnamomi in native forests of the Brisbane Ranges, Victoria. Aust. J. Bot. 23, 77-85 (1975).

28. Weste, G., Taylor, P.: The invasion of native forest by Phytophthora cinnamomi. I: Brisbane Ranges, Victoria. Aust. J. Bot. 19, 281-294 (1971).

29. Weste, G., Ruppin, P., Vithanage, K.: Phytophthora cinnamomi in the Brisbane Ranges: patterns of disease extension. Aust. J. Bot. 24, 201-208 (1976).

# Examination of a Forest Soil Suppressive to *Phytophthora cinnamomi*

D.M. HALSALL

## A. Introduction

Soils in the Tamborine Mt. area of southeastern Queensland were first demonstrated (1, 2) to be suppressive to Phytophthora cinnamomi. The soil discussed here was obtained from Tallaganda State Forest in southern New South Wales. Characteristics of the Tallaganda and Tamborine Mt. soils are compared in Table 1. Tallaganda Forest is a wet sclerophyll forest where Eucalyptus viminalis, E. fastigata and E. radiata are the dominant species with an understorey of Banksia marginata, Acacia melanoxylon and Leucopogon lanceolatus.

Table 1. Comparison of some properties of suppressive soils from (a) Tallaganda State Forest (b) Tamborine Mt.

| | Tallaganda State Forest | Tamborine Mt. |
|---|---|---|
| Description | grey brown solodic soil with yellow clay subsoil | red basaltic soil |
| Drainage | slow draining | well drained |
| Total Nitrogen | low 0.12% | high 0.617% |
| Organic Carbon | low 3.63% | high 8.14% |
| Exchangeable Ca | low 7 me% | high 20-25 me% |
| pH | 5.8 | 5.5 - 7.0 |

In Tallaganda forest, areas of dieback are coincident with wet gully sites and the presence of Phytophthora cryptogea which is not as serious a pathogen of Eucalyptus sp. as is P. cinnamomi but it may prevent regeneration in affected areas (5). Neither P. cinnamomi nor P. cryptogea could be isolated from apparently healthy gullies adjacent to the dieback areas. When P. cinnamomi mycelium and chlamydospores were added to Tallaganda soil in the laboratory, the fungus could not be recovered after 2-4 weeks. It was decided to examine the soil to see if it possessed characteristics similar to that described for soils of the Mt. Tamborine area (1).

## B. Materials and Methods

I. Sporangial production in soil leachates. Soil leachates (1:5, soil:water) were used to stimulate sporangial production on mycelial mats (4). When required, leachates were sterilised by Millipore filtration (0.45 μm) prior to the addition of sterile nutrients. Sterility was confirmed by testing on V8 agar.

II. Sporangial production in soil. Sporangia were produced on mycelial mats covered with soil moistened to field capacity and incubated 2-4 days at 25°C (1).

III. Pathogenicity tests. Susceptible plants (E. sieberi) were grown in soil, untreated or oven-heated to 100°C. Inoculation was by a zoospore suspension added to the soil surface or by mycelium and chlamydospores mixed through the soil.

## C. Results and Discussion

I. Criteria of a suppressive soil. When E. sieberi seedlings were grown in Tallaganda or in conducive soils inoculated with zoospore suspensions, deaths due to P. cinnamomi infection in the Tallaganda soil were considerably reduced compared with deaths in the conducive soil (Table 2). Deaths due to P. cryptogea did not differ significantly between the two soils.

The numbers of sporangia produced on mycelial mats of P. cinnamomi buried in Tallaganda soil were reduced compared with those formed in conducive soil. There was no difference between the numbers of P. cryptogea sporangia formed in the two soils. In a Tallaganda soil leachate, sporangial production by P. cinnamomi was suppressed (2-3 zoosporangia/field) compared with P. cryptogea (80-100 zoosporangia /field). Thus, using the three criteria of Broadbent and Baker (1), the Tallaganda soil can be said to be suppressive to P. cinnamomi but conducive to P. cryptogea.

II. Studies on a suppressive soil leachate. Dilution of the soil leachate did not reverse its suppressive action on P. cinnamomi but did decrease its conducive action on P. cryptogea.

The effects of adding different nutrients to the leachate are shown in Fig. 1; in all cases the pH of the leachate was maintained at 5.8. The increase in sporangial production by P. cinnamomi in the presence of $NH_4NO_3$ was further stimulated by the addition of phosphate or calcium. However $NH_4NO_3$ effectively inhibited sporangial production in P. cryptogea.

The effects of different nitrogen sources on sporulation of P. cinnamomi are shown in Fig. 2, $NH_4{}^+$ or $NH_3$ reduced the suppressive action of the leachate but this effect was reversed under sterile conditions. With P. cryptogea, the reverse occurred, $NH_4{}^+$ inhibiting sporulation in both sterile and non-sterile preparations.

For conducive soils, sterilised leachates usually do not stimulate sporulation of P. cinnamomi (3). When suppressive soil leachates were sterilised, sporangial numbers were not affected but the stimulatory effect of $NH_4NO_3$ was lost. The importance of the microbial component was demonstrated by adding a suppressive

Table 2. Infection of Eucalyptus sieberi seedlings

| | | Seedlings surviving | |
|---|---|---|---|
| | Control | + P. cinnamomi | + P. cryptogea |
| **Suppressive Soil** | | | |
| 7 days | 50 | 36 | 24 |
| 14 days | 50 | 35 | 23 |
| 28 days | 50 | 29 | 16 |
| **Conducive Soil** | | | |
| 7 days | 50 | 17 | 29 |
| 14 days | 50 | 6 | 26 |
| 28 days | 50 | 0 | 19 |

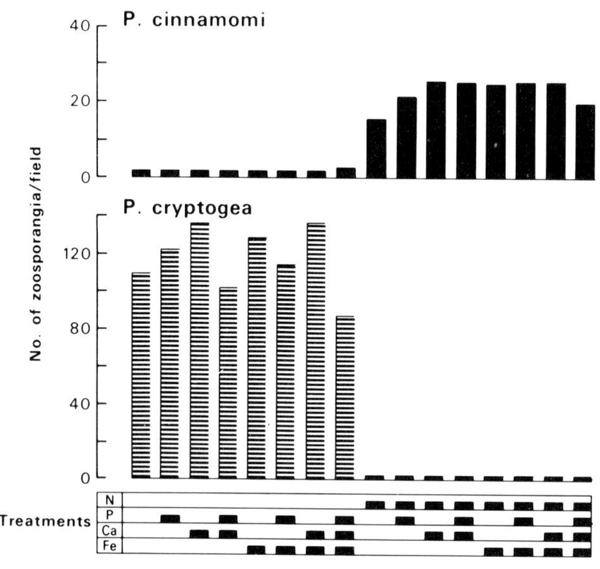


Fig. 1. Effects of the addition of nitrogen, phosphate, calcium and iron to Tallaganda soil leachates on sporangium formation in P. cinnamomi and P. cryptogea. Nitrogen as ammonium nitrate (2 mM); phosphate as disodium hydrogen phosphate (1 mM); calcium as calcium chloride (1 mM); iron as ethylenediaminetetra-acetic acid ferric sodium salt (FeNaEDTA) (0.1 mM). pH 5.8

microbial residue to a sterile filtrate of a conducive soil and a conducive residue to a sterile suppressive filtrate (Table 3). The response of the resulting leachates was largely determined by the microbial component.

An attempt, by Broadbent and Baker (1), to identify the microbial association in suppressive soils involved steam-air treatment of soil. "Antagonists" associated with the suppression of P. cinnamomi survived treatment at 60°C for 30 min but did not survive 100°C for 30 min thus a suppressive soil so treated became conducive.

In the present work, because of the lack of steam-air facilities, heat treatment of Tallaganda soil leachates was used. Heating the leachate to 62°C for 30 min or 100°C for 10 min did not reduce the suppression of P. cinnamomi but these treatments prevented the release of suppression by $NH_4NO_3$.

III. Survival of P. cinnamomi mycelium and chlamydospores in suppressive soil. In pots of fallow soil, P. cinnamomi could not be recovered either by baiting or by direct plating after one month. When this soil was heated to 100°C prior to

Table 3. Effects of removing and replacing the microbial component of soil leachates

| | Number of zoosporangia/field | | |
|---|---|---|---|
| | No additions | +suppressive residue | +conducive residue |
| Sterile Filtrate | | | |
| Conducive | 0.5 | 0.8 | 14.6 |
| Suppressive | 0.7 | 0.8 | 7.8 |
| Non-sterile Leachate | | | |
| Conducive | 16.1 | | |
| Suppressive | 0.9 | | |

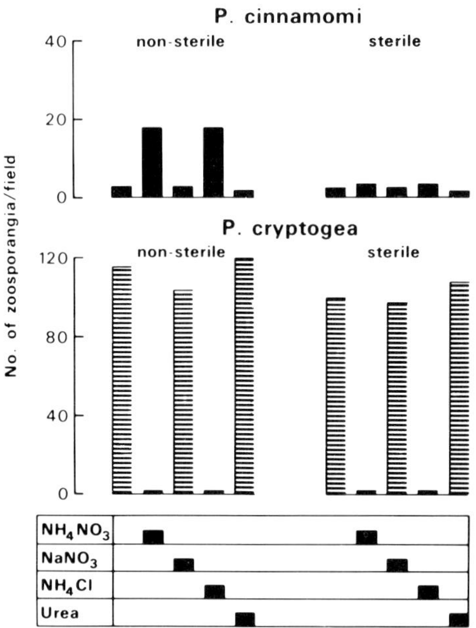


Fig. 2. Effects of different nitrogen sources on the sporulation of P. cinnamomi and P. cryptogea in Tallaganda soil leachate. All nitrogen sources added to give a final concentration of 2 mM. pH 5.8

inoculation with chlamydospores, P. cinnamomi was recovered at a low frequency (10%) after one month but not after three months.

When inoculated soil planted with non-host plants was examined after three months, the fungus could not be recovered. When inoculated soil planted with host plants was examined, however, P. cinnamomi was recovered at a low frequency (10-40%).

## D. Conclusion

Suppression of P. cinnamomi by Tallaganda soil differs in several important aspects from that described for the Tamborine Mt. soils. It is specific to P. cinnamomi whilst being conducive to P. cryptogea. Suppression is not reduced by heat treatment of the leachate, but it is possible that dry heat treatment of the soil modifies suppression. It remains possible that suppression is mediated by a microorganism(s), a microbial product(s) or by a virus.

## E. References


1. Broadbent, P., & Baker, K.F.: Behaviour of Phytophthora cinnamomi in soils suppressive and conducive to root rot. Aust.J.Agr.Res.25, 121-138 (1974).
2. Broadbent, P. & Baker, K.F.: Soils suppressive to Phytophthora root rot in Eastern Australia. In: Biology and Control of Soil-borne Plant Pathogens. Bruehl, G.W. ed. St. Paul, Minn. American Phytopathological Society pp. 152-157, (1975).
3. Chee, K.H., & Newhook, F.J.: Relationship of Micro-organisms to sporulation of Phytophthora cinnamomi. Rands. N.Z. J. Agric. Res. 9, 32-43 (1966).
4. Halsall, D.M., & Forrester, R.I.: Effects of certain cations on the formation and infectivity of Phytophthora zoospores. 1. Effects of calcium, magnesium, potassium and iron ions. Can. J. Microbiol. 23, 994-1001 (1977).
5. Shepherd, C.J., & Pook, E.W.: The association of P. cryptogea with decline of Australian native plant species. Aust. J. Bot. In press.

# The Response of Eucalypt Roots to Infection by *Phytophthora cinnamomi*

J.T. TIPPETT

## A. Introduction

The changes induced in eucalypt roots infected with Phytophthora cinnamomi Rands have been compared with changes induced by water-logging and with several host responses observed on infection of the resistant Acacia pulchella. Comparison of the ultrastructure of infected roots with roots subjected to waterlogging was made in an attempt to recognize those changes associated with autolysis or altered host metabolism.

Histological and ultrastructural study of fungal establishment in roots of the tolerant blue gum E. stjohnii did not reveal any specific ultrastructural changes considered to reflect activation of possible defence mechanisms. For example, very few wall lesions were observed and death of cells invaded by intracellular hyphae was slow (3, 4). In contrast to these observations, post-penetration events in the resistant A. pulchella included formation of wall lesions and rapid death of invaded host cells.

## B. Materials and Methods

Plant and fungal culture methods and procedures used in processing tissues for light and electron microscopy have been described (4). A. pulchella seedlings were two months old when roots were inoculated with zoospores. Preliminary investigation of ultrastructural changes caused by waterlogging or anoxia was made in roots of E. sieberi, sampled from pots two days after they were submerged in buckets of water and the seedlings showing signs of wilting.

The presence of aniline blue positive material (callose) in thin sections of infected roots was detected after removal of the epoxy resin (2) and staining (1) with the fluorochrome aniline blue (0.01%, aqueous).

## C. Results and Discussion

In eucalypt roots infected for up to 24 h, root cell protoplasts often appeared collapsed or 'plasmolyzed' (Fig. 1); extensive wall hydrolysis was also evident in cells of the disorganized root tip (Fig. 3). These changes may have been initiated by the production of fungal phytotoxins or fungal extracellular enzymes (4). It was suggested however, that some changes in root ultrastructure, especially once necrosis of root cells was advanced, may have been autolytic in nature and not

---

Figs. 1-4. Light micrographs of cortical cells (cut in longitudinal section) close ▶ to the apex of E. sieberi roots. H, hypha; P, collapsed protoplast; W, host cell wall. (1) Infected cortex, 24 h after inoculation. (2) Cortex of root subjected to waterlogging for 48 h. (3) Electron micrograph of invaded cortical cells. These cells were in close proximity to those shown in (1). (4) Electron micrograph of cells from waterlogged root, similar to those shown in (2)

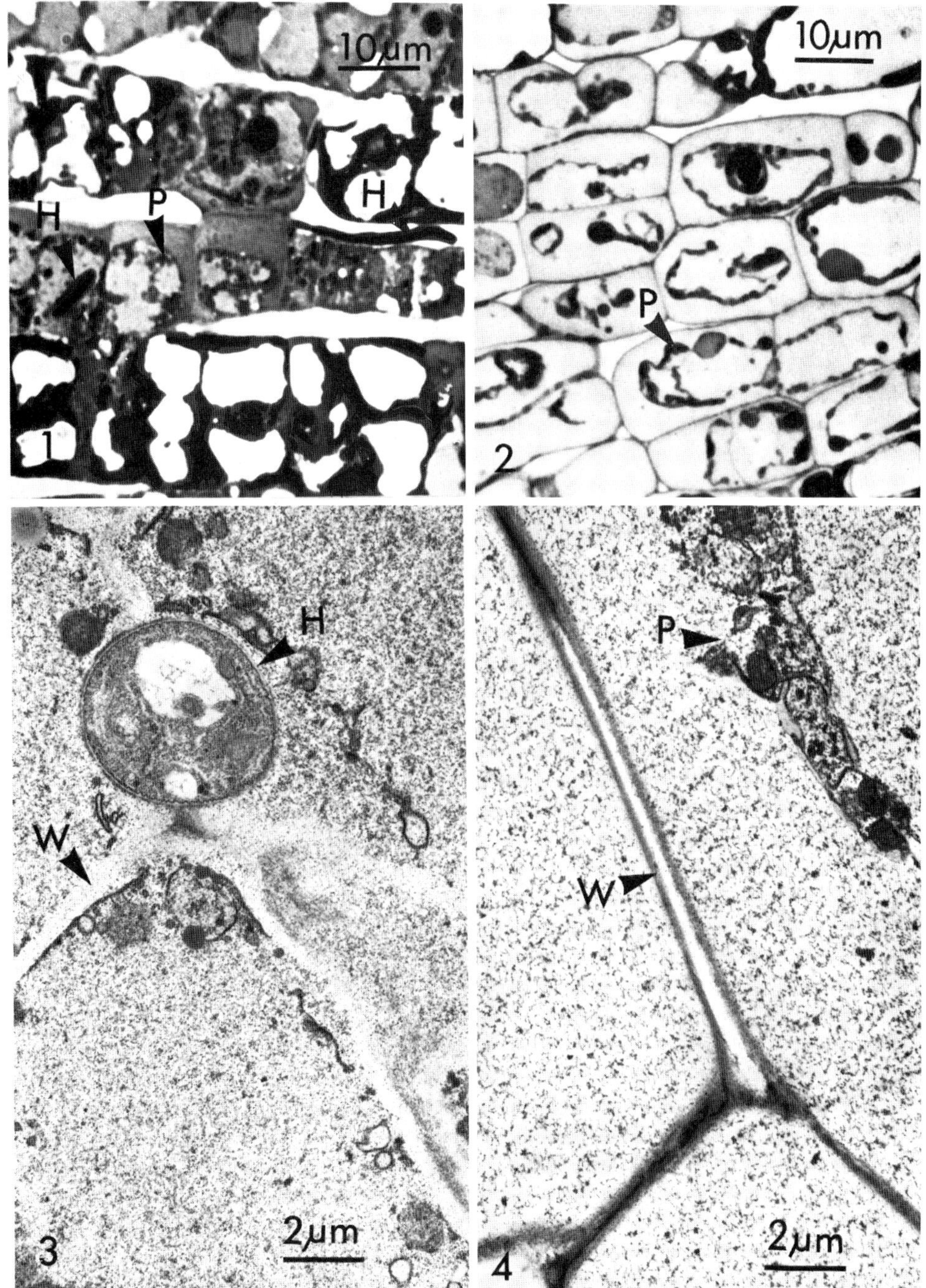


Figs. 1-4

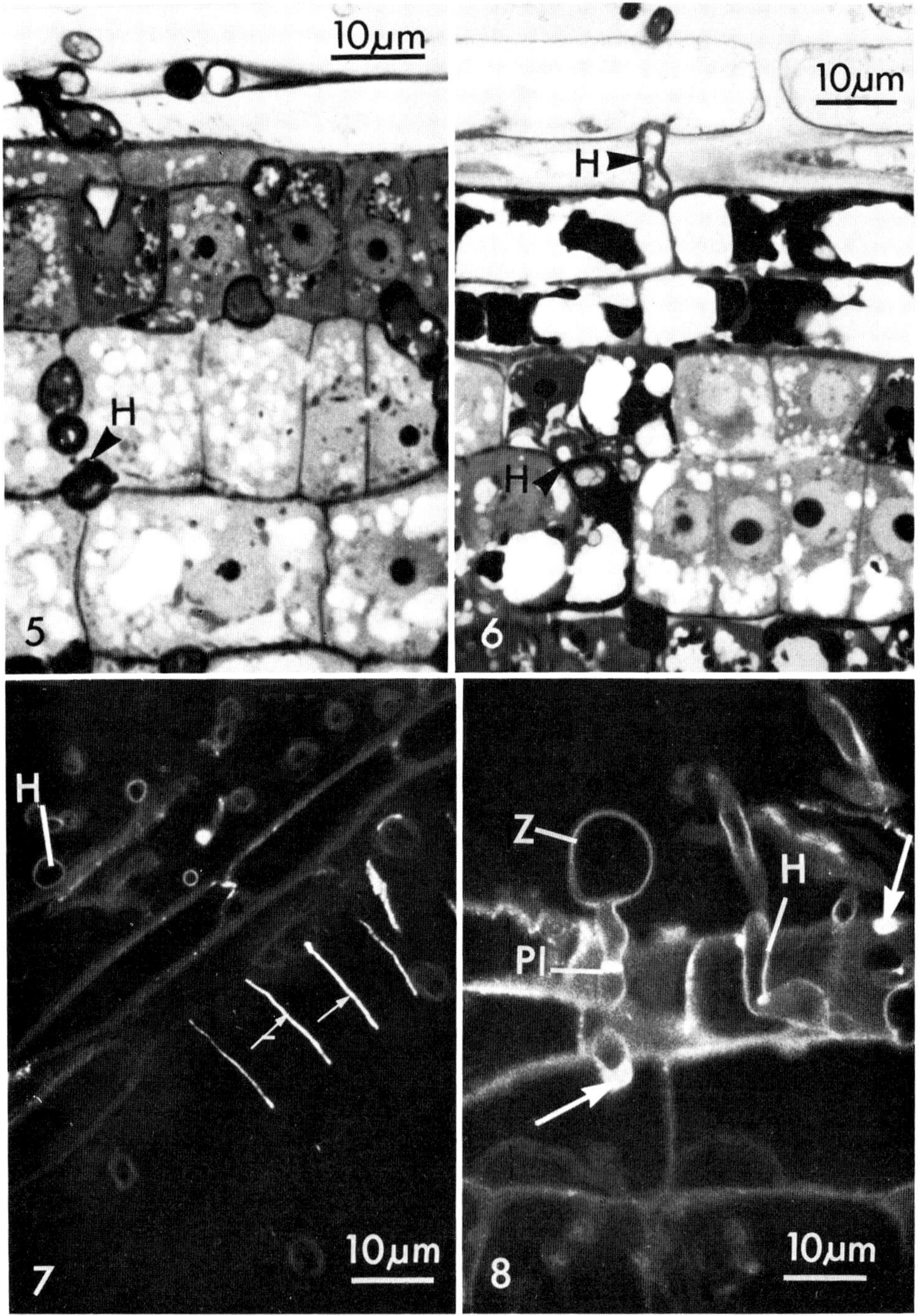


Figs. 5-8

specifically associated with the interaction under study. The light micrograph Fig. 2 shows the appearance of necrotic cortical cells close to the tip of an E. sieberi root, sampled 48 h after waterlogging. For comparison Fig. 1 shows infected cortical cells. Protoplast collapse, observed in roots which had been subjected to waterlogging, was not dissimilar to that observed in infected roots, although polyphenol accumulation was apparently more rapid in inoculated roots (Fig. 1). The noticeable difference in the electron micrographs is that walls are initially little affected in the 'waterlogged' root (Fig. 4) whereas wall hydrolysis is rapid in the invaded root cortex (Fig. 3).

The sequence of events associated with penetration and postpenetration of P. cinnamomi germ tubes in A. pulchella has been studied by light and electron microscopy (Malajczuk and Tippett, in preparation). Comparison of the light micrographs, (Fig. 5, E. obliqua; susceptible, with Fig. 6, A. pulchella; resistant), illustrated that death of cells was more rapid in A. pulchella and was associated with polyphenol accumulation. Wall lesions were common in A. pulchella in close proximity to invading hyphae. The lesions fluoresced bright yellow when stained with the fluorochrome aniline blue and viewed under UV excitation, thus suggesting the presence of callose (Fig. 8). Wall lesions were rarely observed in the infected eucalypts (Fig. 7).

## D. Conclusions

Some of the changes associated with necrosis of eucalypt root tips during the establishment of P. cinnamomi were similar to the changes induced by waterlogging alone and hence they are not considered to directly reflect fungal activity. Although it is impossible to separate the ultrastructural changes observed in infected tissue into two categories of pathogen induced and autolytic, it does seem that rapid and extensive wall hydrolysis is initiated by the fungal infection. In comparing changes occurring in the eucalypts with those occurring in the resistant acacia it was evident that only the acacia roots exhibited responses often considered to be associated with resistance. As no qualitative differences have been observed in the response of different Eucalypt spp. to infection the reason for the differing susceptibilities of eucalypts is still unknown.

## E. References

1. Eschrich, W., Currier, H.B.: Identification of callose by its diachrome and fluorochrome reactions. Stain Techn. 39, 303-307 (1964).

2. Lane, B.P., Europa, D.L.: Differential staining of ultrathin sections of epon-embedded tissues for light microscopy. J. Histochem. Cytochem. 13, 579-582 (1965).

---

◀ Figs. 5-8. (5) Light micrograph of E. obliqua root 4 h after inoculation. (Longitudinal section in zone of elongation.) H, hypha. (6) A. pulchella root 4 h after inoculation. Invaded cells have accumulated polyphenols. (7) Fluorescence micrographs of sections treated with aniline blue for the detection of callose. E. obliqua 4 h after inoculation, only intense fluorescence was associated with cell plates (small arrows). (8) A. pulchella 6 h after inoculation. Empty zoospore cyst (Z) seen at root surface. The cyst is sealed off from the germ tube by a plug (Pl) containing callose. Wall lesions or papilla are arrowed. H, hypha

3. Tippett, Joanna T., Holland, A.A., Marks, G.C., O'Brien, T..P.: Penetration of *Phytophthora cinnamomi* into disease tolerant and susceptible eucalypts. Arch. Microbiol. 108, 231-242 (1976).

4. Tippett, Joanna T., O'Brien, T.P., Holland, A.A.: Ultrastructural changes in eucalypt roots caused by *Phytophthora cinnamomi*. Physiol. Plant Path. 11, (1977).

# Environmental Factors Controlling Severity of Disease Due to *Phytophthora cinnamomi* in Victoria

G. WESTE

In Victorian native forests, disease due to Phytophthora cinnamomi Rands was observed to vary with the season and symptoms usually appeared following wet periods in spring and summer. Preliminary investigations revealed a parallel seasonal variation in pathogen population density.

Apart from seasonal variation, disease due to P. cinnamomi in different sclerophyll forests varied from severe, in which all susceptible plants died and extension was rapid, to moderate, in which only the most susceptible died and others showed a range of dieback symptoms. The interaction between pathogen population density and soil characteristics such as temperature, water potential and microbial populations was investigated for three different ecosystems, each containing native forest which had never been cultivated or fertilized and had not been fired for approximately 30 years.

For all sites, pathogen populations were small in winter (P.D.1. 0-4) as long as maximum soil temperatures remained less than 10$^{o}$C. When maximum soil temperatures rose above 10$^{o}$C, for example briefly in July at Narbethong, pathogen populations increased rapidly. There was no reduction in pathogen populations with high soil temperatures such as 33$^{o}$C at Wilson's Promontory provided the soil water potential was higher than -9 bar.

Pathogen populations declined in late summer-early autumn for sites in the Brisbane Ranges and Narbethong when soil water potential declined to values lower than -9 bar, although temperatures remained optimum. Immediately soil water potentials increased, pathogen populations also increased.

Correlation coefficients for pathogen population density and soil temperature were highly significant ($p < 0.001$) for the three sites individually and for the whole area in the Brisbane Ranges from 27.4.74 to 18.12.74. During this period, soil moisture remained relatively high and relatively constant. From 18.12.74 to 7.5.75, there was a highly significant correlation ($p < 0.001$) between soil moisture and pathogen population density for site 2 and a significant correlation ($p < 0.05$) for site 1. Results for Wilson's Promontory and Narbethong showed the same trends, but were not significant, probably because of the shorter period involved.

A survey of the microbial populations of the three forest soils demonstrated that Wilson's Promontory and Narbethong soils supported three to ten times the microbial populations found in Brisbane Ranges soil. These differences were highly significant for actinomycetes, aerobic and anaerobic bacteria and significant for aerobic sporing bacteria and fungi. In the Brisbane Ranges, severe disease resulted in a further reduction of soil microflora to 1/5 or 1/6 of its low normal population. In spring, following severe disease, soil fungi, actinomycetes and aerobic bacteria were reduced by factors of 10, 6 and 5 respectively. The reduction was probably associated with the destruction of 60-70% of the understorey, thereby reducing both microbial substrate and colonizing surface as evidenced by the large increase from 30-90% of bare ground. These results were not observed in communities at Wilson's Promontory and Narbethong where disease was less severe and microbial populations remained relatively unaffected.

Microbial populations varied with the season. Maximum counts of total micro-organisms associated with roots were obtained during the autumn-winter period and minimum numbers from spring to summer. This seasonal variation was characteristic for test samples from all sites, whether from rhizosphere or rhizoplane of susceptible or tolerant host roots, from diseased or unaffected areas. Actinomycetes, fungi, aerobic and aerobic-sporing bacteria were always maximal during autumn-winter and minimal in spring-summer.

Conversely, populations of *P. cinnamomi* were lowest during winter and approached zero from May to August; then reached a maximum in the spring-summer period, from the end of September to the end of January. Numbers decreased during dry periods in autumn when soil water potential fell to lower than -9 bar. This difference in phase is a function of the Victorian climate and may be an important factor enabling the invading pathogen to upset the ecological balance of the forest community by avoiding biological control factors such as microbial competition, fungistasis and antagonism.

Five changes were observed in the microflora associated with disease due to *P. cinnamomi*. These were:- (a) decreased numbers of actinomycetes from diseased rhizospheres sampled compared with rhizospheres from unaffected forests; (b) increased numbers of anaerobes from the rhizospheres of both susceptible and tolerant roots sampled from all diseased sites. Soil moisture levels increased, as plant transpiration decreased; (c) increased microbial numbers, particularly of aerobic bacteria, on the rhizoplanes of all susceptible and tolerant hosts examined; (d) the large increase in numbers of micro-organisms, particularly of aerobic bacteria in the rhizospheres sampled from both tolerant and susceptible plants from severely diseased forests of the Brisbane Ranges and associated with increased substrate in the form of dead roots; and (e) the final decrease in soil microbial numbers accompanying increased bare ground and associated with reduced substrate. This must eventually result in reduced fertility accompanying severe disease.

Investigations have demonstrated that certain environmental factors are associated with severity of disease in native forests. These factors may be listed:-

1. Shallow soils with impeded drainage.
2. Water saturation when soil temperatures are greater than 10$^{o}$C allowing zoospore production, dispersal and infection.
3. A plentiful supply of host roots, susceptible both to the pathogen and to water saturation.
4. Subsequent drought period during which host deaths occur.
5. Low microbial populations characteristic of soils poor in substrate and therefore in general fungistasis.

All these characteristics occur in Brisbane Ranges forests. Victorian dry sclerophyll forests represent a severe hazard because (a) approximately 75% of the plants are susceptible, including the structural dominants, and (b) the climate fluctuates from warm and wet to dry periods during spring and summer when *P. cinnamomi* populations are maximal, but aerobic bacteria and actinomycetes are minimal, thereby reducing the probability of biological control.

# Bacterial Colonization of *Phytophthora cinnamomi* Rands

H.J. NESBITT, N. MALAJCZUK, and A.R. GLENN

## A. Introduction

A number of workers (2, 6, 7, 8) have implicated bacteria in the lysis of fungi in the soil. Extensive bacterial colonization has been observed on hyphae and sporangia of P. cinnamomi and it has been suggested that this interaction may be important in the biological control of this pathogen (5). This paper describes further work to elucidate the role of bacteria colonizing P. cinnamomi in natural soil leachates.

## B. Materials and Methods

I. Cultural conditions. Cellophane (1.5 $cm^2$) covered by 1 $cm^2$ of actively growing hyphae (5) was incubated at 27°C in soil leachates derived from an organic rich soil taken from Dwellingup, Western Australia. The controls consisted of glass fibre of similar diameter to the hyphae and were incubated on cellophane in the same leachate. At daily intervals, hyphae and controls (in triplicate) were sampled and washed (x 5) in 20 ml of sterile water to remove any loosely adsorbed, or free floating bacteria.

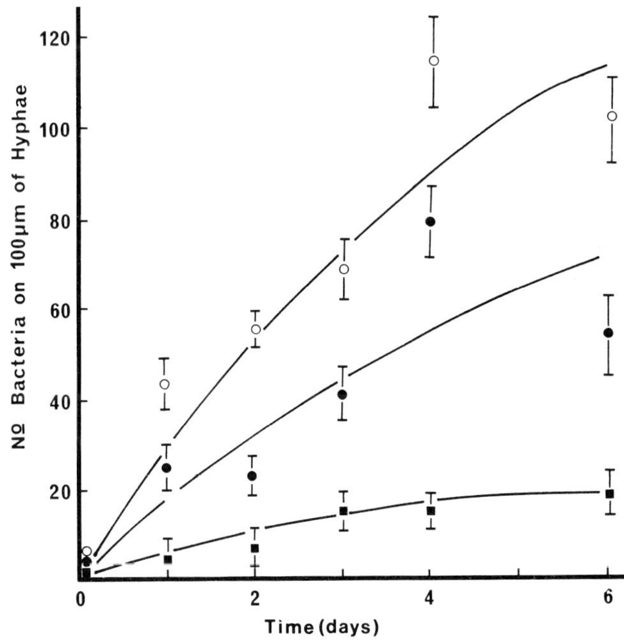


Fig. 1. Number of bacteria colonizing live (—●—) and lysed (—○—) hyphae and glass fibre controls (—■—) with time. Vertical bars represent standard errors

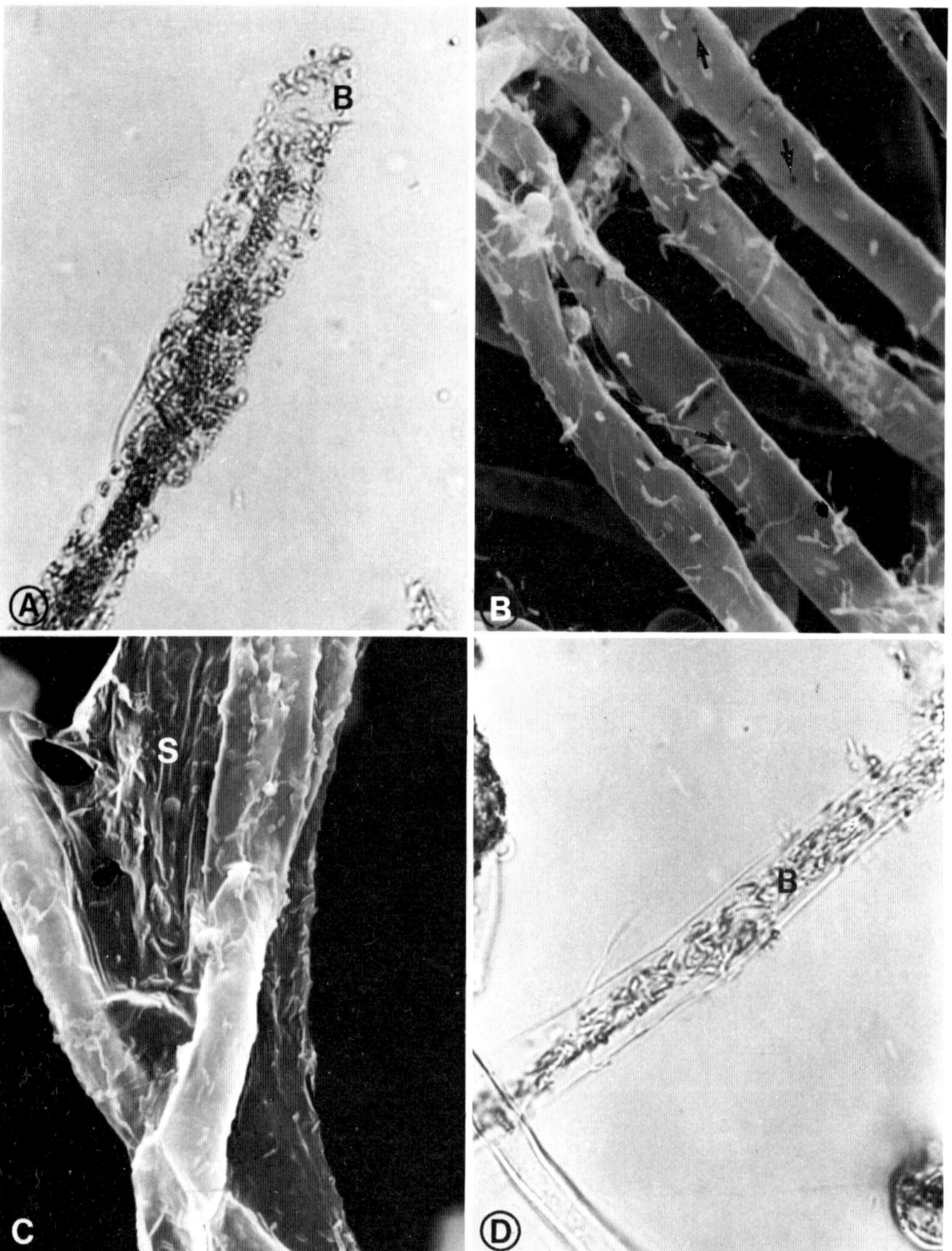


Fig. 2. (A) Bacteria B enveloping P. cinnamomi hypha (X 1600). (B) Scanning electron micrograph showing P. cinnamomi hyphae with different morphological types of bacteria on the surface. Arrows indicate holes in the hyphae (X 1900). (C) Scanning electron micrograph of P. cinnamomi hyphae enveloped with slime S layer (X 1900). (D) Bacterial cells B within P. cinnamomi hyphae (X 1600)

II. Assessment of hyphal lysis and direct bacterial counts. Hyphal lysis was estimated by the presence or absence of cytoplasm after staining with an aniline-blue acetic acid solution (4). Direct bacterial counts were taken off hyphae and glass fibre controls using a x 100 objective. Hyphal lysis and numbers of bacteria were estimated using a Zeiss III photomicroscope containing an eyepiece graticule attachment. Electron microscopy was carried out as described previously (5).

## C. Results and Discussion

The number of bacteria firmly adhering to the hyphae increased with time and there were two to five times more bacteria on live hyphae than on the controls. On lysed hyphae the bacterial numbers were even greater (Fig. 1). Light and scanning electron microscopy of hyphal mats and glass fibres showed differences in the distribution of the colonizing bacteria. On glass fibres there was a fairly even distribution of bacteria one cell thick. Bacteria on the hyphae were often in clusters several cells deep (Fig. 2A) many of which were joined by slime strands (Fig. 2A) which eventually developed into slime layers enveloping the hyphae (Fig. 2C). Slime production was not observed in the sterile controls. It is not known at present if the slime is of fungal or bacterial origin.

A range of morphologically diverse organisms became associated with the hyphae (Fig. 2B). The numbers of viable, bacteria isolated on one tenth strength tryptic soy agar (Difco) after maceration of the hyphal mats was consistently less than 50% of the direct microscopic counts. This may have been due to the fact that several bacterial cells become bound together in the slime material and formed only one colony on an agar plate. In addition not all bacterial types were isolated; e.g. actinomycetes were occasionally observed on the hyphae but these were not isolated on the tryptic soy medium. Preliminary evidence from these experiments shows a predominance of Gram negative bacteria (B. Reynolds, unpublished data). On occasion bacteria were also observed within the hyphae but this did not seem to be

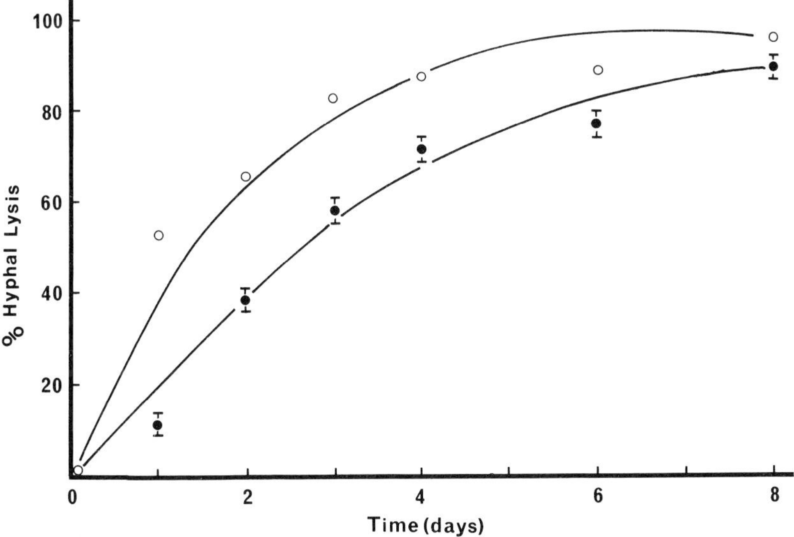


Fig. 3. Lysis of randomly selected hyphae (—●—) and 50 μm of hyphal tip (—○—) with time. Vertical bars represent standard errors

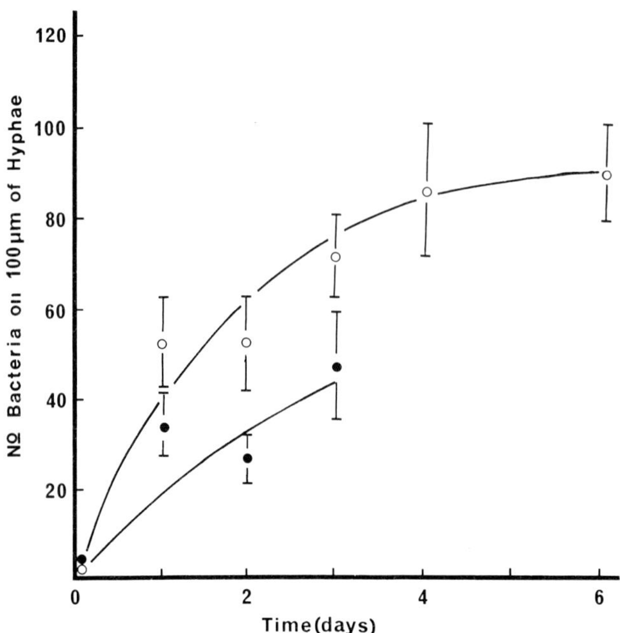


Fig. 4. Number of bacteria colonizing live (—●—) and lysed (—○—) hyphal tips. Vertical bars represent standard errors

related to increasing bacterial numbers. A number of holes were observed in the hyphal cell wall, similar to that described by other workers using different fungi (2, 6, 7, 8) (Fig. 2B). The occasional penetration of bacteria into a hypha may be due to damage other than that caused by bacterial activity (Fig. 2D).

With increasing bacterial numbers there was also an increase in the numbers of ciliate and flagellate protozoa. At the present time we do not know if these organisms exert any controlling effect on the bacterial population (3).

Correlated with the increase in bacterial numbers there was also an increase in hyphal lysis (Fig. 3). Although it is tempting to suggest that increasing bacterial numbers caused the lysis it can be argued that the adhering bacteria are utilizing the products of hyphal lysis. It should be noted that in control experiments there was little lysis of hyphae in sterile leachates. It is also possible that many of the bacteria residing in the "hyphasphere" may exude extracellular metabolites which cause breakdown of hyphal cytoplasm.

During the assessment of hyphal lysis it was noted that hyphal tips (terminal 50 µm of the hyphae) lysed more rapidly than the rest of the hyphae (Fig. 3) and pseudosepta formed separating live from lysed hyphae. Hyphal lysis was so rapid that insufficient live hyphal tips were available to determine the numbers of bacteria colonizing during the latter stages of this experiment (Fig. 4). There were, however, the same number of bacteria on the tips as on the rest of the hyphae. It is possible that specific hyphalytic micro-organisms may be attracted to hyphae by fungal exudates (1) representing early colonizers. These bacteria may cause lysis and the fungal metabolites then released would attract a different bacterial population.

Bacterial colonization of fungal hyphae reported in this paper represents another aspect of microbial ecology with implications in the biological control of fungal pathogens.

## D. References

1. Chet, I., Vogel, S. & Mitchell, R.: Chemical detection of microbial prey by bacterial predators. J. Bact. 106, 863-867 (1971).
2. Clough, K.S. & Patrick, Z.A.: Characteristics of the perforating agent of chlamydospores of Thielaviopsis basicola (Berk. & Br) Ferraris. Soil Biol. Biochem. 8, 473-478 (1976).
3. Habte, M., & Alexander, M.: Further evidence for the regulation of bacterial populations in soil by protozoa. Arch. Microbiol. 113, 181-183 (1977).
4. Jones, P.C.T. & Mollison, J.E.: A technique for the quantitative estimation of soil micro-organisms. J. Gen. Microbiol. 35, 54-60 (1948).
5. Malajczuk, N., Nesbitt, H.J. & Glenn, A.R.: A light and electron microscope study of the interaction of soil bacteria with Phytophthora cinnamomi Rands. Can. J. Microbiol. In press, 1977.
6. Old, K.M. & Patrick,Z.A.: Perforation and lysis of spores of Cochliobolus sativus and Thievaliopsis basicola in natural soils. Can. J. Bot 54, 2798-2809 (1976).
7. Rovira, A.D. & Campbell, R.: A scanning electron microscope study of interactions between microorganisms and Gaeumannomyces graminis (Syn.Ophiobolus graminis) on wheat roots. Microbiol. Ecology 2, 177-185 (1975).
8. Wong, J.F., & Old, K.M.: Electron microscopical studies of the colonization of conidia of Cochliobolus sativus by soil microorganisms. Soil Biol. Biochem. 6, 89-96 (1974).

# Rhizobium –Legume Symbiosis

## The Intracellular Environment of *Rhizobium japonicum* in Soybean Root Nodules

F.J. BERGERSEN

### A. Introduction

The study of the ecology of microorganisms is often hampered by the difficulty of measuring various parameters within the microenvironments which constitute the microbial habitat. This difficulty is often cited as the main cause for failure to account for observations of microbial activity in terms of parameters of the macroenvironment. In response, the microbiologist often retreats to the laboratory and defines his problem in terms of pure cultures and controlled conditions, only to meet the criticism that "it is not like that in nature".

The $N_2$-fixing root nodules of legumes provide a unique opportunity to study a natural microbial environment in quantitative terms. The microstructure is observable by optical and electron microscopy, the microbial population is uniform and is monospecific. It is possible to measure or calculate concentrations and fluxes of important metabolites. Furthermore, it has been possible to conduct laboratory experiments with dissected components of the nodule system and with laboratory cultures, and to relate the results to the in vivo situation. There are still some aspects of uncertainty, some of them crucial to complete understanding, but enough is known to begin an integrated description of the intracellular microenvironment in soybean root nodules. These have been the most studied biochemically and physiologically and quantitative microscopical data are available. In some instances it has been necessary to use data from nodules of other legumes. Some of the quantitative data used in this account come from the references cited in the text. Other data have been obtained from unpublished experiments and from measurements of electron micrographs and direct microscopic examination of material used by Bergersen and Goodchild (11,12).

### B. The Host Cell

The description given here refers to 30-day old nodules on Lincoln soybeans inoculated with R. japonicum strain CB1809 and grown in a glasshouse in Canberra. Some of the parameters change with time as the nodules increase in size, others are remarkably constant. The age of 30 days was chosen as being typical of the 20-day period during which $N_2$ was fixed at the maximum rate (12) and resembling mature nodules from field-grown plants.

I. Structure of nodule tissue. The central, $N_2$-fixing tissue consists of a roughly spherical matrix of enlarged parenchyma cells packed with bacteroids, uninfected interstitial cells and gas-filled intercellular spaces (11). The average 30-days old nodule contains $2.3 \times 10^5$ (st. dev. $0.25 \times 10^5$) (28) infected cells, which occupy 78% of the central tissue volume (12). The central tissue is not uniform. The outer part consists of radially elongated, bacteroid-filled cells: near the centre, these cells are more symmetrical and there is a greater proportion of interstitial cells interspersed among them. Figure 1 presents the basic features of the relationships between components of the central tissue. The average

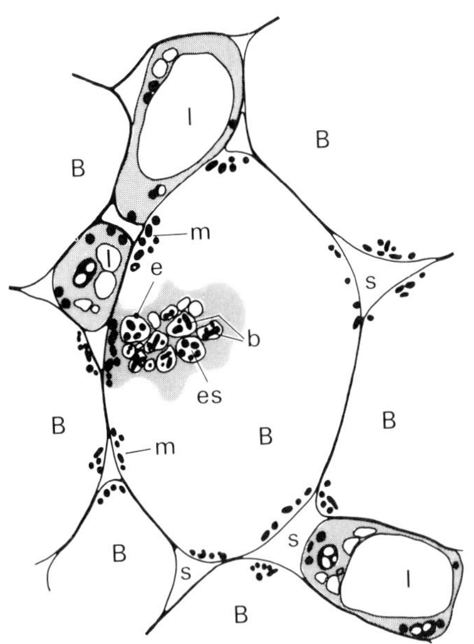


Fig. 1. Features of the central tissue of soybean nodules during maximum rates of nitrogen fixation. A, Tracing from a representative electron micrograph; B, bacteroid-filled host cells: volume 10-12 x $10^{-5}$ $mm^3$; e, membrane envelopes: 3-10 x $10^3$ per cell, diam. 3 μm; b, bacteroids: 3-4 x $10^4$ per cell; I, interstitial cell: present in ratio of 1-1.6 per bacteroid filled cell; S, gas-filled intercellular space: cross section 1-5 $\mu m^2$; 5-7 per bacteroid-filled cell; e,s, envelope space: 50% of host cell volume

bacteroid-filled cell has a volume of 10-12 x $10^{-5}$ $mm^3$ and its major axis a length of 38-62 μm. It has interfaces with other bacteroid-filled cells, (occupying 20-60% of the surface) with interstitial cells (20-46%) and with 5-7 gas-filled intercellular spaces (20-35% of the surface) each of whose cross section area is 1-5 $\mu m^2$ (10,11). The interfaces with interstitial cells may be concerned with the exchange of solutes while most of the $O_2$, $CO_2$ and $N_2$ probably passes across the intercellular space interfaces, near which the mitochondria of the bacteroid-filled cells are concentrated. Of the total volume of bacteroid-filled cells, only about 15% is cytoplasm, while 83% is occupied by 3-10 x $10^3$ membrane-bounded envelopes (each about 3 μm in diameter) containing a total of 3-4 x $10^4$ bacteroids; about 50% of the volume of the bacteroid-filled cells is space between the membrane envelopes and the bacteroid surfaces (12).

II. Leghaemoglobin. Soybean nodules aged 30 days contain 12 nmol of leghaemoglobin (12), which is confined to the bacteroid-containing cells (27). This unique plant haemoglobin (2) is intimately concerned with the supply of $O_2$ to the bacteroids and is therefore an important feature of their environment. The intracellular location of leghaemoglobin has been a controversial matter. It could be distributed in the cytoplasm, and the membrane envelope space, or be confined in either of these locations. Bergersen and Goodchild (12) presented evidence that most, or all of it was in the envelope space, in contact with the bacteroids. Work with other legumes also supports this location (18, 21, 31). However, leghaemoglobin is synthesized by the host cells and recently-reported experiments failed to detect it in the envelope space by immuno-electron microscope methods (32). It has also been impossible to prepare intact membrane envelopes containing leghaemoglobin from nodule homogenates. For the purpose of this account, we will assume that the weight of evidence shows that the membrane envelope space is the location of the leghaemoglobin, where its concentration is about 1.5 mM (12). In doing this, we recognize the need for unequivocal resolution of the location of this important haemoprotein.

Leghaemoglobin in vivo is in the reduced form and spectrophotometry of intact, small soybean nodules showed that the haemoprotein was 25% oxygenated (1). It has not been possible to examine the larger, 30-day old nodules in this way, but the same degree of oxygenation will be assumed.

The role of leghaemoglobin in the supply of $O_2$ to the bacteroid surfaces has been developed mainly from in vitro experiments (3, 13, 14, 17, 33) which will be considered later in this paper. Leghaemoglobin also plays an important part in the distribution of $O_2$ within the host cell. The in vivo state of oxygenation indicates that the average concentration of free $O_2$ in the cell is about $1 \times 10^{-8}$ M. At this low concentration only low diffusive fluxes are possible. Stokes and Bergersen (28) used the cytological dimensions of nodule cell structures and respiratory rates of bacteroids, to model the distribution of $O_2$ within the cell. They concluded that leghaemoglobin facilitated distribution, and that for a given flux, $O_2$ concentration gradients would need to be about 10 times greater in the absence of leghaemoglobin. A significant feature of the model was the relatively large effect of the boundary layers associated with the frequently-encountered membrane envelopes.

III. Enzymes of host cytoplasm. The intracellular environment is also characterised by the close proximity of large concentrations of host enzymes, which modify substrates supplied to and metabolic products from the bacteroids.

Soybean nodules contain about 3.6 μmol $NH_3$/g f.w. nodules (6). It is assumed that about ½ of this is present in the nodule cells as newly-fixed N. It now seems clear that host enzymes are responsible for the rapid assimilation of $NH_3$ produced by bacteroid nitrogenase (20, 24). Cytosol glutamine synthetase accounts for 90% of total activity in soybean nodules and the enzyme is present to the extent of 2% ot total cytosol protein (22). Scott et al. (26) have proposed for lupin nodules, that the primary step in assimilation of $NH_3$ is via glutamine synthetase followed by reactions involving glutamate synthase, glutamate-aspartate transaminase and asparagine synthetase. The products of these reactions are translocated via the xylem and used in plant growth.

The fuel for the nodule reactions is new photosynthetic carbon (4) and concentrations of sucrose and (+)-pinitol were found to be related to nitrogenase activity (29). However, the carbon compounds which are actually metabolized by the bacteroids have yet to be identified. It seems likely that they are products of mitochondrial or cytosol enzymes of the host cell. For the purposes of this paper they have been treated as units of hexose. In assessing the flux of carbon substrates for reactions close to the bacteroids, the scheme of Scott et al. (26) has been used as the model for assimilation of $NH_3$. In this model, one molecule of asparagine is produced from one molecule of oxaloacetate, one of NADH and 3 molecules of ATP, thus assimilating 2 molecules of $NH_3$. Thus, 0.33 mol hexose/mol $NH_3$ is required as oxaloacetate, plus 0.04 mol hexose/mol $NH_3$ for ATP production (assuming P/O = 3) and 0.08 mol hexose/mol $NH_3$ for NADH, making a total of 0.45 mol hexose/mol of $NH_3$.

## C. The Bacteroids

The quantitative activities of bacteroids have been determined from experiments with suspensions prepared anaerobically from nodule homogenates and used in no-gas-phase assays of nitrogenase and respiratory activity in the presence of partially oxygenated leghaemoglobin (13, 14).

I. Nitrogenase activity. Bacteroid suspensions have optimum nitrogenase activities of 4-5 nmol $C_2H_4$/min/mg d.w. when receiving $O_2$ from 25% oxygenated leghaemoglobin (Bergersen and Turner, unpublished) with gluconate as substrate. This

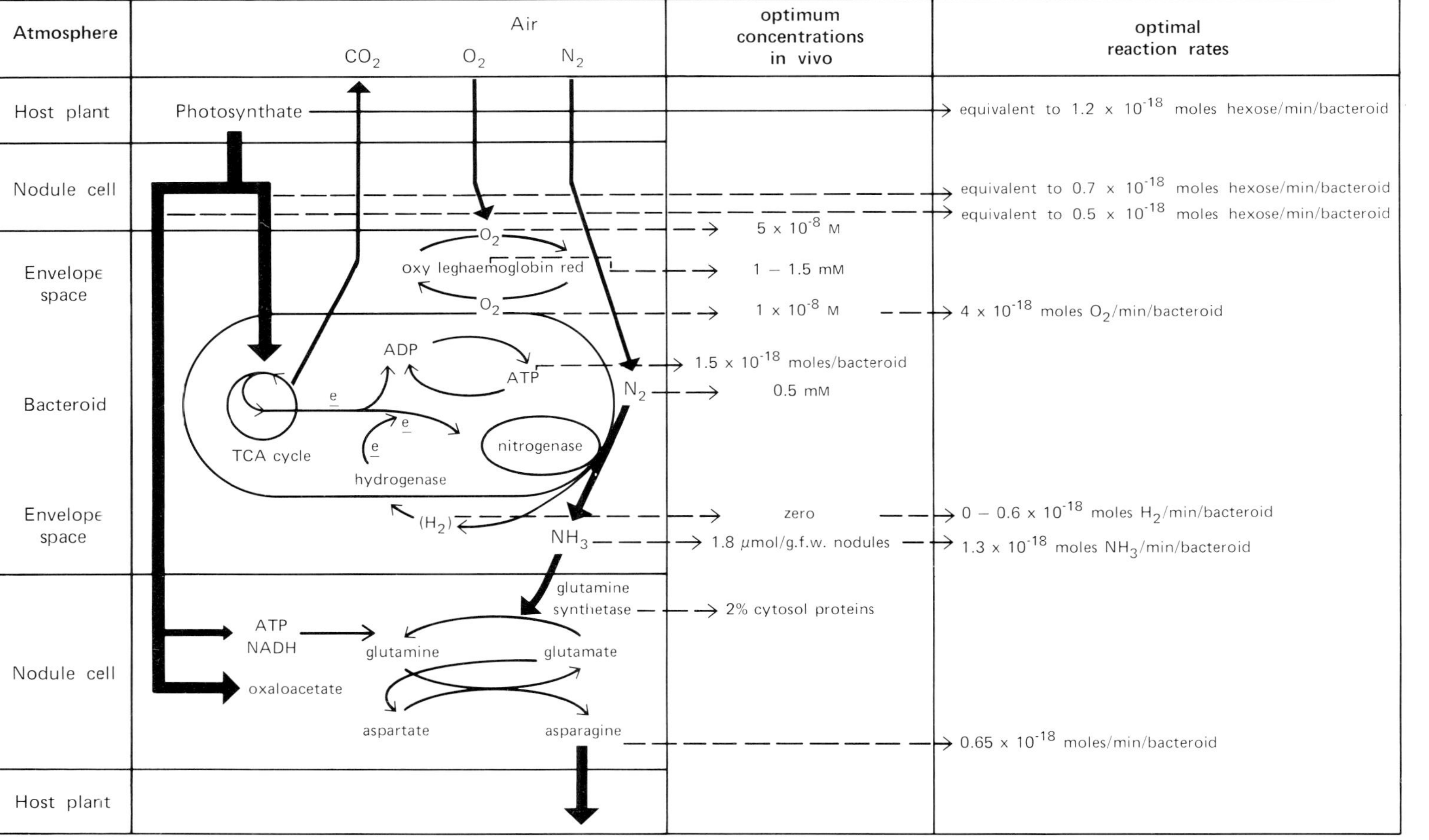


Fig. 2. Quantitative parameters of reactions occurring near or in the bacteroids of soybean nodules. For details and sources see text

corresponds to about 2.6 n mol $NH_3$/min/mg d.w. There are 2 x $10^9$ bacteroids in 1 mg d.w. (28). Therefore each bacteroid fixes about 1.3 x $10^{-18}$mol $NH_3$/min.

Nodulated soybean roots have peak nitrogenase activities of about 30 µmol $C_2H_4$/h/g f.w. and using the $C_2H_4$ : $NH_3$ ratio of 3.3 established with $^{15}N_2$ for that material (7), this corresponds to 9.1 µmol $NH_3$/h/g f.w. The average 30-day nodule weighs 40 mg f.w. (12) and contains about 2 x $10^5$ cells, each containing about 3 x $10^4$ bacteroids. Therefore bacteroids in vivo fix

$$\frac{9.1 \times 10^{-6} \times 0.04}{2 \times 10^5 \times 3 \times 10^4 \times 60} = 1 \times 10^{-18} \text{ mol } NH_3\text{/min/bacteroid}$$

which agrees well with results from the experiments with bacteroid suspensions.

II. Respiration. In the assays in which nitrogenase activity of bacteroid suspensions was measured with gluconate as substrate while receiving $O_2$ from 25% oxygenated leghaemoglobin, (above), the respiration rate was about 8 nmol $O_2$/min/mg d.w. or 4 x $10^{-18}$ mol $O_2$/min/bacteroid (calculated as above).

Again, the rates of respiration of intact soybean nodules can be compared with this value. Bergersen (5) reported 3 µl $O_2$/h/mg d.w. (6.3 µl/min/ g.f.w.) and Tjepkema (30) reported 5-7 µl $O_2$/min/ g.f.w. These values convert to about 13 nmol $O_2$/min/nodule. On the basis of numbers of bacteroids per nodule this is equivalent to 2 x $10^{-18}$ mol $O_2$/min/bacteroid. This does not correspond very well to the value for bacteroid suspensions, since nodule respiration must contain additional components due to repiration of the plant cells. Nevertheless, the values are in fair agreement, considering the cumulative effects of errors in the relatively crude estimates used. In addition, different strains of R. japonicum were used. The estimates of $O_2$ flux for bacteroid respiration in vivo are therefore reasonable.

During deoxygenation of leghaemoglobin, bacteroid oxidases produce ATP very efficiently and hence nitrogenase activities are high. When oxygenation is about 25%, as considered here, ATP concentration in the bacteroids was 3 nmol/mg (14) which converts to 1.5 x $10^{-18}$ mol/bacteroid. While considering ATP used for nitrogenase, it should be mentioned that with some strains of R. japonicum, there can be considerable wastage due to evolution of $H_2$ from nitrogenase (25). It can account for half of the ATP used by nitrogenase (referred to in Fig. 2). The R. japonicum strain CB1809 considered here evolves little $H_2$ from its nodules and this is believed to be due to the action of a hydrogenase which conserves reducing power (cf. 16).

The relatively low oxygenation of leghaemoglobin in vivo (1) and the response of nodule respiration to increased $O_2$ concentration (5) indicate that the bacteria in the nodule cells are $O_2$-limited. Recent results with $N_2$-fixing chemostat cultures of rhizobia indicate that nitrogenase synthesis is greatest and least susceptible to repression by $NH_3$, in $O_2$-limited conditions (9, 15, 16).

III. Carbon substrate supply. In estimating the flux of carbon substrates to bacteroids we have considered that one molecule of hexose equivalent is completely combusted to $CO_2$ and $H_2O$, consuming 6 molecules of $O_2$ and that the R.Q. is 1.1 (8). The additional carbon flux for accumulation of poly-β-hydroxybutyrate in the bacteroids has been ignored because it is likely to be comparatively small. Thus a respiratory rate of 4 x $10^{-8}$ mol $O_2$/min/bacteroid requires 0.67 x $10^{-18}$ mol hexose/min/bacteroid.

## D. Conclusions

The structure and quantitative data described in this paper allow the formulation of a preliminary account of concentrations and fluxes of important metabolites involved in the support of $N_2$ fixation by the bacteroids of soybean nodules (Fig. 2). When values obtained from experiments with bacteroid suspensions were used together with nodule structure data, the calculated fluxes of N, $CO_2$ and $O_2$ agreed fairly well with values observed with intact nodules. Further, the partitioning of carbon flux almost equally between that used for bacteroid respiration and that utilized in assimilation of fixed N is similar to the results of Minchin and Pate (23) who studied carbon balance in intact nodulated plants of *Pisum sativum*. Therefore, despite the lack of precision in some of the measurements made, it is concluded that this compilation indicates that the results of recent laboratory studies may be used to describe some important features of the intracellular environment.

## E. Acknowledgements

The participation of G.L. Turner in many of the experiments, and the stimulation of many discussions with colleagues in the CSIRO Division of Plant Industry, are gratefully acknowledged. A.N. Stokes is thanked also for permission to quote results from model studies in advance of their publication.

## F. References

1. Appleby, C.A.: Properties of leghaemoglobin *in vivo* and its isolation as ferrous oxyleghaemoglobin. Biochim. Biophys. Acta 188, 222-229 (1969).
2. Appleby, C.A.: Leghaemoglobin. In: The Biology of Nitrogen Fixation. Quispel A. (ed.). Amsterdam: North Holland, pp. 521-554 (1974).
3. Appleby, C.A., Turner, G.L., Macnicol, P.K.: Involvement of oxyleghaemoglobin and cytochrome P-450 in an efficient oxidative phosphorylation pathway which supports nitrogen fixation in *Rhizobium*. Biochim. Biophys. Acta 387, 461-474 (1975).
4. Bach, M.K., Magee, W.E., Burris, R.H.: Translocation of photosynthetic products to soybean nodules and their role in nitrogen fixation. Pl. Physiol. 33, 118-124 (1958).
5. Bergersen, F.J.: The effects of partial pressure of oxygen upon respiration and nitrogen fixation by soybean root nodules. J. gen. Microbiol. 29, 113-125 (1962).
6. Bergersen, F.J.: Ammonia - on early stable product of nitrogen fixation by soybean root nodules. Aust. J. biol. Sci. 18, 1-9 (1965).
7. Bergersen, F.J.: The quantitative relationship between nitrogen fixation and the acetylene-reduction assay. Aust. J. biol. Sci. 23, 1015-1025 (1970).
8. Bergersen, F.J.: Biochemistry of symbiotic nitrogen fixation by legumes. Ann. Rev. Pl. Physiol. 22, 121-140 (1971).
9. Bergersen, F.J.: Nitrogenase in chemostat cultures of rhizobia. In: Proceedings II Int. Symp. $N_2$ fixation. Postgate, J.R. and Rodriguez-Barrueco, C., (eds.). New York, Academic Press (in press).
10. Bergersen, F.J.: Legume symbiosis. In: Interactions between Soil and Microorganisms and Plants. Dommergues, Y. and Krupa, S.V. (eds.). Amsterdam. Elsevier (in press).
11. Bergersen, F.J., Goodchild, D.J.: Aeration pathways in soybean root nodules. Aust. J. biol. Sci. 26, 729-740 (1973a).
12. Bergersen, F.J., Goodchild, D.J.: Cellular location and concentration of leghaemoglobin in soybean root nodules. Aust. J. biol. Sci. 26, 741-756 (1973b).
13. Bergersen, F.J., Turner, G.L.: Leghaemoglobin and the supply of $O_2$ to nitrogen-fixing root nodule bacteroids: studies of an experimental system with no gas phase. J. gen. Microbiol. 89, 31-47, (1975a).

14. Bergersen, F.J., Turner, G.L.: Leghaemoglobin and the supply of $0_2$ to nitrogen-fixing root nodule bacteroids: presence of two oxidase systems and ATP production at low free $0_2$ concentration. J. gen. Microbiol. 91, 345-354 (1975b).

15. Bergersen, F.J., Turner, G.L.: The role of $0_2$-limitation in control of nitrogenase in continuous cultures of Rhizobium sp. Biochem. Biophys. Res. Comm. 73, 524-531 (1976).

16. Bergersen, F.J., Turner, G.L.: Activity of nitrogenase and glutamine synthetase in relation to availability of oxygen in continuous cultures of a strain of cowpea Rhizobium sp. supplied with excess ammonium. Biochim. Biophys. Acta (in press).

17. Bergersen, F.J., Turner, G.L., Appleby, C.A.: Studies of the physiological role of leghaemoglobin in soybean root nodules. Biochim. Biophys. Acta 292, 271-282 (1973).

18. Dilworth, M.J., Kidby, D.K.: Localization of iron and leghaemoglobin in the legume root nodule by electron microscope autoradiography. Exptl. Cell. Res. 49, 148-159 (1968).

19. Dixon, R.O.D.: Hydrogenase in legume root nodule bacteroids: occurrence and properties. Arch. Mikrobiol. 85, 193-201 (1972).

20. Dunn, S.D., Klucas, R.V.: Studies on possible routes of ammonium assimilation in soybean root nodule bacteroids. Canad. J. Microbiol. 19, 1493-1499 (1973).

21. Gourret, J-P., Fernandez-Arias, H.: Etude ultrastructurale et cytochimique de la différenciation des bactéroides de Rhizobium trifolii Dang. dans les nodules de Trifolium repens L. Canada. J. Microbiol. 20, 1169-1181 (1974).

22. McParland, R.H., Guevara, J.G., Becker, R.R., Evans, H.J.: The purification and properties of the glutamine synthetase from the cytosol of soya-bean root nodules. Biochem. J. 153, 597-606 (1976).

23. Minchin, F.R., Pate, J.S.: The carbon balance of a legume and the functional economy of its nodules. J. exptl. Bot. 24, 259-271 (1973).

24. Robertson, J.G., Farnden, K.J.F., Warburton, M.P., Banks, J.M.: Induction of glutamine synthetase during nodule development in lupin. Aust. J. Pl. Physiol. 2, 265-272 (1975).

25. Schubert, K.R., Evans, H.J.: Hydrogen evolution: a major factor affecting the efficiency of nitrogen fixation in nodulated symbionts. Proc. Nat. Acad. Sci. USA 73, 1207-1211 (1976).

26. Scott, D.B., Farnden, K.J.F., Robertson, J.G.: Ammonia assimilation in lupin nodules. Nature 263, 703-705 (1976).

27. Smith, J.D.: The concentration and distribution of haemoglobin in the root nodules of leguminous plants. Biochem. J. 44, 585-591.

28. Stokes, A.N., Bergersen, F.J.: Oxygen distribution in host cells in nitrogen-fixing root nodules: a model using data for soybean nodules. Aust. J. Pl. Physiol. (In press).

29. Streeter, J.G., Bosler, M.E.: Carbohydrates in soybean nodules: identification of compounds and possible relationships to nitrogen fixation. Pl. Sci. Lett. 7, 321-329 (1976).

30. Tjepkema, J.D.: Oxygen transport in the soybean nodule and the function of leghemoglobin. Ph.D. Thesis. University of Michigan (1971).

31. Truchet, G.: Mise en évidence de l'activité peroxydasique dans les différentes zones des nodules radiculaires de pois (Pisum sativum L.). Localisation de la leghémoglobine. C.r. hebd. Seanc. Acad. Sci., Paris, 274, 1290-1293 (1972).

32. Verma, D.P., Bal, A.K.: Intracellular site of synthesis and localization of leghemoglobin in root nodules. Proc. Nat. Acad. Sci. U.S.A. 73, 3843-3847 (1976).

33. Wittenberg, J.B., Bergersen, F.J., Appleby, C.A., Turner, G.L.: Facilitated oxygen diffusion: the role of leghemoglobin in nitrogen fixation by bacteroids isolated from soybean root nodules. J. biol. Chem. 249, 4057-4066 (1974).

# Root Exudates of Pea Seedlings and Their Effect Upon *Rhizobium leguminosarum*

A.W.S.M. VAN EGERAAT

## A. Introduction

In the rhizosphere, micro-organisms are stimulated by organic compounds exuded by roots. Leguminous plants strongly stimulate bacteria of the genus *Rhizobium*. In the rhizosphere of non-legumes rhizobial stimulation is much less or absent. The cause of this differential stimulation is not known but the composition of the root exudate is likely to be of the utmost importance.

In this paper the behaviour of *Rhizobium leguminosarum* on the root surface and in the rhizosphere of pea seedlings is described in connection with the release of ninhydrin-positive compounds by the roots. The role of some of these amino compounds on growth of *Rh. leguminosarum* is discussed.

This study is part of a more extensive work on the influence of root exudates (and of the chemical composition of the root itself) on bacterial growth, in particular on growth of *Rhizobium*.

## B. Materials and Methods

To study the growth of *Rhizobium leguminosarum* on the root surface of young pea plants, roots of sterile seedlings (root length 2-2.5 cm, no lateral roots present) were inoculated with *Rhizobium*. Inoculation was carried out by dipping the roots for 2-3 sec in a suspension of *Rhizobium* cells (3). Bacterial counts were made in the period between 0 and 72 h after inoculation.

For studying the promotion of growth of rhizobia in the rhizosphere, agar (carbon source present), inoculated with *Rhizobium leguminosarum* was used as a medium for root growth (3).

Root extracts, root-tip exudates and root exudates were obtained as described by van Egeraat (2). The ninhydrin-positive compounds in the samples were analysed with the aid of an amino-acid analyser.

Experiments to obtain information on the influence of homoserine upon growth of rhizobia belonging to different cross-inoculation groups were performed in culture tubes containing 5 ml of nutrient solution. The medium was enriched with glutamic acid, homoserine and mannitol, added separately or in combination (4).

The inhibition of *Rhizobium leguminosarum* by isoxazolin-5-one was studied on yeast-extract mannitol agar. Prior to solidification, the agar was mixed with a suspension of the bacteria. The compound (1 mg in water) was added to discs, placed on top of the agar.

## C. Results and Discussion

I. Growth of *Rhizobium leguminosarum* on the surface of pea-seedling roots. Estimations of the bacteria present on the root surface of pea seedlings are given in Table 1. From the moment of inoculation till 24 hr later a sharp decrease in the number of *Rhizobium* cells on the root surface had occurred. Forty-eight h after inoculation a strong increase was observed and continued to 72 h. The increase of the Rhizobium cells on the root surface coincided with the formation of lateral roots, a process which causes severe damage of the main root (Fig. 1) and the release of considerable amounts of organic compounds from the wounds. Growth of the bacteria due to the release of these compounds is very likely. Multiplication of the bacteria probably only takes place on that part of the root where lateral roots are present. To confirm this, main roots were cut into three pieces, 72 h after inoculation. The rhizobia present on each of the root segments were counted and the results are given in Table 2. It is obvious that multiplication of Rhizobium cells occurred primarily on that segment of the root where all the lateral roots were present. By comparison few bacteria were found on the root segments without lateral roots and it appears that growth conditions on the surface of intact roots of pea seedlings do not favour multiplication of Rhizobium.

Table 1. Growth of Rhizobium leguminosarum, strain PRE on the root surface of pea seedlings

| Time in hours after inoculation | 0 | 6 | 24 | 48 | 72 |
|---|---|---|---|---|---|
| Number* of bacteria found per root | 2580 | 490 | 295 | 12,920 | 277,000 |

* Mean values of 3-5 replicates.

Table 2. Numbers or Rhizobium leguminosarum, strain PRE, found on the root surface of a pea seedling, 72 hours after inoculation. Root divided into three segments of equal length

| Segment | Lateral roots | Numbers** of bacteria found |
|---|---|---|
| 1* | + | 325,000 |
| 2 | - | 2,080 |
| 3 | - | 1,090 |

* Upper part of the root.

** Values of a typical experiment.

II. The absence of rhizobial stimulation in the rhizosphere of pea seedlings. As organic compounds are exuded by growing roots, it was expected that bacteria in close proximity to roots would be stimulated. Furthermore it was suggested that pea roots would cause a stronger stimulation than roots derived from smaller seeds. The results obtained showed that strains of Rhizobium leguminosarum were stimulated by root-tip exudates of a number of leguminous plants, but contrary to expectation, pea-seedling roots were the only ones causing no stimulation. Only where lateral roots emerged from the main root was a stimulation of Rhizobium observed after some time. Experiments on the absence of rhizobial stimulation in the rhizosphere of pea seedlings are described elsewhere (3).

Fig. 1. Damage to main root resulting from the emergence of lateral roots

III. Ninhydrin-positive compounds present in root extracts of pea seedlings. Figure 2 gives a chromatogram of acid and neutral ninydrin-positive compounds (n.p. c.) present in roots of 6- days old pea seedlings. As the basic amino compounds were only present in very small concentrations, the chromatogram showing these compounds is omitted. Compounds X and Y, initially unknown, were identified as γ-L-glutamyl-D-alanine and 2-alanyl-isoxazolin-5-one (5). As can be seen from the chromatogram, homoserine is quantitatively the most important amino compound present in root extracts of pea seedlings.

IV. Ninhydrin-positive compounds present in root-tip exudates. Figure 3 shows a chromatogram representing the acid and neutral n.p.c. found in exudates of tips of pea-seedling roots. Similar patterns are obtained from exudates of whole roots, before the formation of lateral roots. The n.p.c. exuded in greatest amount is 2-alanyl-isoxazolin-5-one, with γ-L-glutamyl-D-alanine next. Homoserine is present in root-tip exudates in very small amounts.

V. Ninhydrin-positive compounds in exudates of roots with lateral roots. The influence of lateral root formation on the composition of the n.p.c. in the root exudates is shown in Figure 4. Lateral root formation obviously causes exudation of homoserine and this amino acid is most likely liberated from the main root by the wounds originating from the formation of lateral roots. In absence of these wounds, homoserine is exuded by the seedling root only in minute quantities in spite of the high concentration of this amino acid in the roots (Figure 2).

VI. Homoserine and growth of *Rhizobium leguminosarum*. *In vitro* experiments showed that homoserine favours the growth of those *Rhizobium* strains which are capable of producing nodules with peas, e.g. strains belonging to the cross-inoculation group *Rhizobium leguminosarum*. These bacteria can use homoserine as both carbon and nitrogen source. Detailed studies as to the influence of homoserine on rhizobial growth have been described (4).

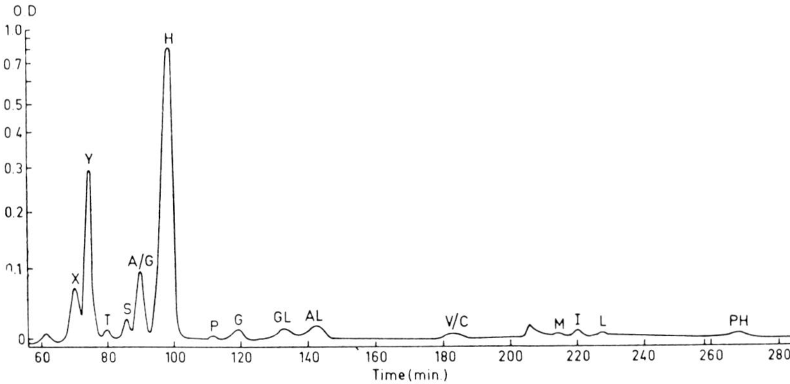


Fig. 2. Acid and neutral ninhydrin-positive compounds (n.p.c.) present in roots of 6-day old pea seedlings. X, γ-L-glutamyl-D-alanine; Y, 2-alanyl-isoxazolin-5-one; A, aspartic acid; T, threonine; S, serine; A/G, asparagine + glutamine; H, homoserine; p, proline; G, glutamic acid; GL, glycine; AL, alanine; v, valine; C, cystine; M, methionine; I, isoleucine; L, leucine; TY, tyrosine; Ph, phenylalanine

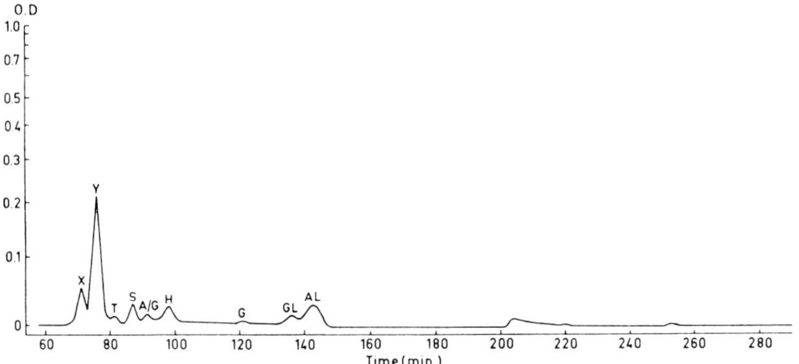


Fig. 3. Chromatogram showing ninhydrin-positive compounds exuded by tips of pea-seedling roots. For explanation of symbols see legend of Figure 2

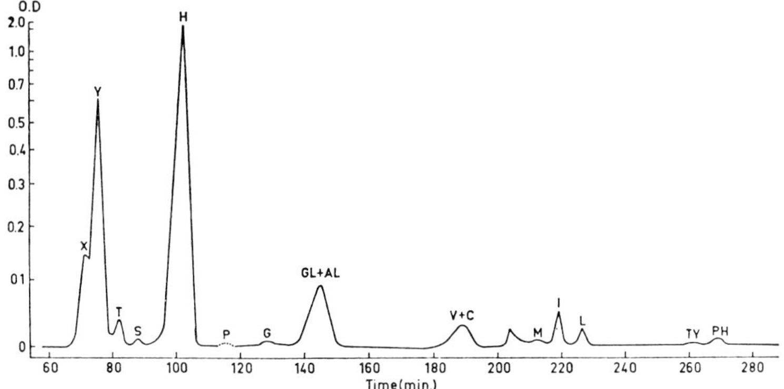


Fig. 4. Acid and neutral ninhydrin-positive compounds present in root exudates of pea seedlings with lateral roots. For explanation of symbols see legend of Figure 2

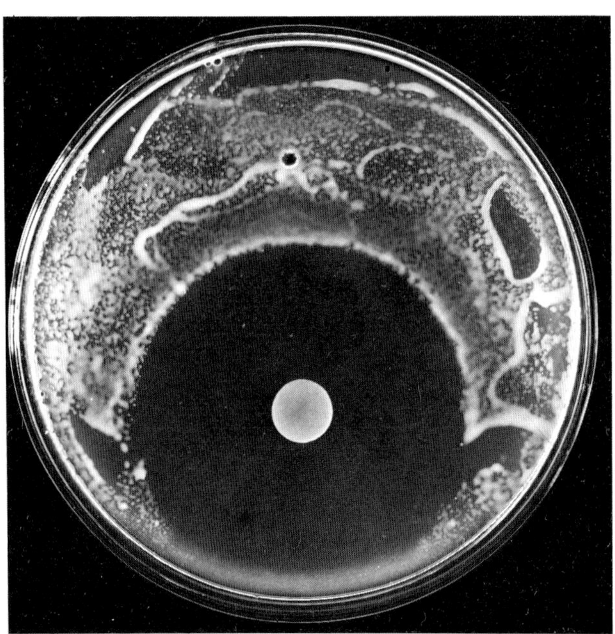

Fig. 5. Growth inhibition of Rhizobium leguminosarum by isoxazolin-5-one

VII. The possible role of γ-L-glutamyl-D-alanine and 2-alanyl-isoxazolin-5-one on growth of Rhizobium leguminosarum. As γ-L-glutamyl-D-alanine and 2-alanyl-isoxazolin-5-one are not commercially available: both compounds must be isolated in sufficient amounts or must be synthesized before their effect on the growth of Rhizobium can be assessed. This work is in progress.

If Rhizobium is able to split the dipeptide, L-glutamic acid and D-alanine would be liberated. In vitro experiments with Rhizobium leguminosarum showed that using equal amounts of L-glutamic acid and D-alanine, growth was completely inhibited at a D-alanine concentration of 0.4 $gl^{-1}$. The same was also true when using L-glutamic acid concentrations of 1.0 $gl^{-1}$. Consequently γ-L-glutamyl-D-alanine might be responsible for the growth inhibition of Rhizobium leguminosarum on the root surface and in the rhizosphere of pea seedlings. The isoxazolinone derivative may also be responsible for the bacterial growth inhibition. Isoxazolinones are known to inhibit certain decarboxylating reactions in mammals and bacteria (7). Furthermore, the antibiotic cycloserine is an isoxazolinone (4-amino-isoxazoline-3-one). Recently, two isoxazole amino acids with antibiotic activity (6) have been isolated from Streptomyces sviceus.

We have succeeded in synthesizing isoxazolin-5-one. The synthesis of this compound was first described by De Sarlo (1). As is shown in Figure 5, the compound inhibits rhizobial growth and it is most likely that the alanine derivative does too.

VIII. Putting things together. Root-tip exudates of pea seedlings inhibit growth of Rhizobium leguminosarum (and other bacteria). The inhibiting action is most likely caused by 2-alanyl-isoxazolin-5-one. When lateral roots emerge from the main root a relatively large quantity of homoserine is liberated. This amino acid has proved to be an excellent nutrient for Rhizobium leguminosarum. Rhizobia belonging to other cross-inoculation groups are not or are little stimulated by this amino acid. It appears that Rh. leguminosarum after being initially inhibited by root-tip exudates is then selectively stimulated by homoserine on those portions of the root or in the rhizosphere where lateral roots emerge. The inhibiting action caused by the isoxazolinone amino acid is then eliminated. This may occur as follows. Apart from homoserine other n.p.c. are released during the formation of lateral roots. Recent experiments have shown that one of these compounds is uracil-alanine. As this compound is very similar to 2-alanyl-isoxazolin-5-one, it is possible that uracil-alanine serves as a competitor against the isoxazolinone amino acid, thus diminishing the inhibiting action of the latter.

In a later stage of root growth the exudation of 2-alanyl-isoxazolin-5-one as well as the release of homoserine sharply decreases. As a consequence the bacterial inhibition becomes less distinct and the specific stimulation of Rh. leguminosarum disappears.

## D. References


1. De Sarlo, F., Dini, G., Lacrimini, P.: Isoxazolin-5-one. J. Chem. Soc. (C), 86-89 (1971).
2. Egeraat, A.W.S.M. van: Exudation of ninhydrin-positive compounds by pea-seedling roots: a study of the sites of exudation and of the composition of the exudate. Plant and Soil 42, 37-47 (1975).
3. Egeraat, A.W.S.M. van: The growth of Rhizobium leguminosarum on the root surface and in the rhizosphere of pea seedlings in relation to root exudates. Plant and Soil 42, 367-379 (1975).
4. Egeraat, A.W.S.M. van: The possible role of homoserine in the development of Rhizobium leguminosarum in the rhizosphere of pea seedlings, Plant and Soil 42, 381-386 (1975).
5. Egeraat, A.W.S.M. van: Enzymatic studies on ninhydrin-positive compounds exuded by root tips of pea seedlings. Plant and Soil (in press).
6. Lambein, F., Kuo, Y.H., Van Parijs, R.: Isoxazolin-5-ones. Chemistry and biology of a new class of plant products. Heterocycles 4, 567-593 (1976).
7. Sourkes, T.L., D'Iorio, A.: Inhibitors of catechol amine metabolism. In: Metabolic Inhibitors Vol. II. Hochster, R.M. & Quastel, I.H. (eds.) Academic Press, New York - London, pp. 79-99 (1963).

# An Appraisal of Methods for Distinguishing Between Strains of *Rhizobium* spp. and Their Application to the Identification of Isolates from Field Environments

J. BROCKWELL, A. DIATLOFF, and E.A. SCHWINGHAMER

## A. Introduction

Important criteria for assessing the efficacy of legume seed inoculation include success of the inoculum (Rhizobium spp.) in colonizing the host plant rhizosphere, in forming root nodules, and in establishing itself as a permanent member of the soil microflora (3, 31). The study of this area of bacterial ecology leans heavily upon the techniques used to identify the strains of rhizobia reisolated from rhizosphere, nodule, and soil. In turn, the validity of these techniques depends on the stability of the markers used to distinguish one strain from another. In this paper, we consider six different types of markers that have been used in our laboratories for distinguishing between Rhizobium strains and discuss their relative merits in terms of stability, ease of recognition, and convenience of handling.

## B. Methods

The six types of markers are listed in Table 1. The parallel use of two different markers on the one Rhizobium strain provided a means of confirming the actual stability of each marker independently. Seven strains were used as inocula for seed of different legumes sown into the field under a range of conditions. Subsequently, nodules were selected, isolation of rhizobia made in most cases, and the reisolates identified using both types of marker. Instances of disagreement in recognition between the two markers represent a change in one or other of them of the characteristic that permits strain identification. In some experiments, three markers were used simultaneously.

## C. Results

I. Serology and symbiotic characteristics. Table 2 shows the results of an experiment in which serotyping and symbiotic behaviour were used in parallel to identify strains of R. trifolii. The group of ten field isolates (i) which were identified serologically as the inoculant strain NA30 could be differentiated by their symbiotic characteristics from groups (ii) and (iii) but not from group (iv) (5 isolates). This represents 81% agreement between the two techniques for identification of NA30 18 years after its introduction into the environment. This, incidentally, is the longest recorded period of persistence of an inoculant strain in the field.

II. Serology and symbiotic characteristics and colony colour. Table 3 presents data showing 100% agreement between serotyping and symbiotic behaviour as means of identifying rhizobia appropriate to Lotononis bainesii. The distinctive pink colony colour was a slightly less reliable marker (95%) for Lotononis rhizobia since there were four instances of possible confusion with Protaminobacter rubra.

III. Serology and antibiotic resistance. The results presented in Table 4 show that the correspondence between serological and antibiotic resistance charact-

Table 1. Details of markers used for identification of strains of *Rhizobium* spp., and some relevant literature references

| Marker type | Specific markers | References to methods | Presentation of data | References to other relevant work |
|---|---|---|---|---|
| (i) Serology a. Gel immune diffusion | Antigens of, and specific anti-serum against, 1. *R. trifolii* NA30; 2. CC2480*a*; 3. CC276; *Rhizobium* sp. 4. CB360; 5. CB376; 6. CB1923 | (13, 30) | 1. Table 2; 2. Table 4; 3. Table 6; 4,5. Table 3; 6. Table 5 | (14,21,33,42) |
| b. Immuno-fluorescence | not used | - | - | (1,34,46) |
| c. Agglutination | not used | - | - | (6,8,18,26,43,47, 48,49) |
| (ii) Colony characteristics | Pink colour of *Rhizobium* sp. CB360, CB376 | (27) | Table 3 | (11,24) |
| (iii) Nodule characteristics | | | | |
| a. Colour | Black nodules of *C. pubescens* x CB1923 | (7) | Table 5 | (12) |
| b. Morphology | not used | - | - | (11,19,20) |
| (iv) Symbiotic characteristics | 1. Ineffectiveness of *T. subterraneum* cv. Howard x NA30; 2. Effectiveness of *L. bainesii* x CB360 and x CB376 | 1. (17) 2. (28) | 1. Table 2; 2. Table 3 | (2,4) |
| (v) Genetic markers a. Antibiotic resistance | 1. Streptomycin resistant CC2480*a*; 2. streptomycin resistant CC276 and CB782; 3. Streptomycin resistant CB1923, spectinomycin resistant CB1923 | 1,2,3. (29) 3. (39) | 1. Table 4; 2. Table 6; 3. Table 5 | (5) |
| b. Auxotrophy | not used | - | - | (22) |
| (vi) Antagonism | | | | |
| a. Bacteriocin production and sensitivity | CB782 (producer strain) *v.* CC276 (sensitive strain) | (37) | Table 6 | (25,32,38,40,41,44,45) |
| b. Phage production and sensitivity | not used | - | - | (25,32,38,40,41,44,45) |

Table 2. Serological and symbiotic characteristics of 27 *Rhizobium trifolii* strains isolated from *Trifolium subterraneum* inoculated with NA30 and sown at Murrumbateman, New South Wales, 18 years previously

| No. of field isolates | Serological identity with NA30 | Effectiveness with *T. subterraneum* cv. Howard | cv. Bacchus Marsh |
|---|---|---|---|
| 10 (i) | + | – | + |
| 10 (ii) | – | + | + |
| 2 (iii) | – | – | – |
| 5 (iv) | – | – | + |
| Control = NA30 | + | – | + |

Table 3. Serological and symbiotic characteristics of pink bacteria isolated from *Lotononis bainesii* inoculated with CB360 or CB376 and sown some years before at Coolum, Queensland. (14 isolates from each site.) (After Diatloff (10))

| Site | Age of pasture (years) | Pink colonies | Serological identity with: CB360 | CB376 | neither | Effective with *L. bainesii* |
|---|---|---|---|---|---|---|
| (i) | 10 | 14 | 6 | 8 | 0 | 14 |
| (ii) | 13 | 14 | 10 | 4 | 0 | 14 |
| (iii) | (a) | 14 | 1 | 13 | 0 | 14 |
| (iv) | 14 | 14 | 1 | 11 | 2 (c) | 12 |
| (v) | 12 (b) | 14 | 1 | 11 | 2 (c) | 12 |
| (vi) | (a) | 14 | 4 | 10 | 0 | 14 |
| Control | CB360 | 7 | 7 | 0 | 0 | 7 |
| Control | CB376 | 7 | 0 | 7 | 0 | 7 |

(a): Volunteer stand of uncertain age;
(b): Oversown 6 years after original sowing;
(c): Subsequently identified as *Protaminobacter rubra*.

ers as markers of identity of *R. trifolii* strain CC2480*a* was never less than 95% at any of five harvests over a period of 41 months.

IV. Serology and antibiotic resistance and nodule colour. The results of an experiment using *Centrosema pubescens* are presented in Table 5. Agreement between serological and antibiotic resistance markers was 100%. The black nodule characteristic of *C. pubescens* inoculated with CB1923 was a fairly reliable marker for identification of CB1923 isolated from black nodules, the agreement between antibiotic resistance and nodule colour markers being 89% (or 97%, if the inconsistent result, probably due to cross-contamination following heavy rain, from the fourth sampling of CB1923str$^{r}$ plots is ignored). However, agreement between these two markers was only 24% for identification of CB1923 isolated from non-black nodules. These were usually small and presumably most of them were immature CB1923 nodules where the black pigment had not yet developed.

Table 4. Agreement between serotyping and antibiotic resistance for identification of Rhizobium trifolii isolated at intervals from subterranean clover inoculated with streptomycin-resistant CC2480a and sown at Murrumbateman, New South Wales. (After Brockwell, Schwinghamer, and Gault (5))

| Time from inoculation | No. of isolates | Positive agreement (a) | Negative agreement (b) | Total agreement (a) + (b) | Disagreement (c) |
|---|---|---|---|---|---|
| 1 | 92 | 84.8 | 11.9 | 96.7 | 3.3 |
| 6 | 64 | 82.8 | 14.1 | 96.9 | 3.1 |
| 23 | 150 | 38.0 | 57.3 | 95.3 | 4.7 |
| 36 | 158 | 35.4 | 63.3 | 98.7 | 1.3 |
| 41 | 158 | 41.8 | 56.3 | 98.1 | 1.9 |
| Total | 622 | 49.9 | 47.4 | 97.3 | 2.7 |

(a): Isolate identified as CC2480a by serotyping and by antibiotic resistance (inoculum strain);
(b): Isolate not identifiable with either marker (naturally-occurring strain);
(c): Isolate identified as CC2480a with one marker, but not with the other.

V. Symbiotic characteristics and bacteriocin production. Forty eight R. trifolii isolates from T. semipilosum were identified as the inoculum strain CB782-$str^r$ by streptomycin resistance. All were effective on T. semipilosum and 46/48 produced a bacteriocin antagonistic to sensitive strain CC276, a 96% agreement (Table 6). All 17 naturally-occurring strains tested were streptomycin sensitive, ineffective on T. semipilosum and did not produce an anti-CC276 bacteriocin.

VI. Serology and bacteriocin sensitivity. Twenty four isolates from T. subterraneum were identified by the streptomycin resistance of the inoculum strain CC276$str^r$. Twenty two of the isolates were serologically identical with the inoculum strain and all 24 were sensitive to the bacteriocin produced by CB782. However, 11/17 of the naturally-occurring strains tested were also sensitive to the CB782 bacteriocin (Table 6).

## D. Discussion

The stability of immunological and antibiotic resistance markers and their reliability for identification of strains of Rhizobium spp. were demonstrated by the high level of agreement between the two techniques in distinguishing inoculum strains reisolated from the field. Bacteriocin production was also a reliable marker but bacteriocin sensitivity was not, many naturally-occurring strains being sensitive to a bacteriocin produced by CB782. Colony colour was a completely stable, reliable marker although, with the highly specific combinations examined (L. bainesii x CB360 or x CB376), the mere formation of nodules was unequivocal evidence of infection of the host by the inoculum. With the nodule colour marker, unpigmented nodules frequently contained the inoculum strain, due perhaps to slow development of pigmentation; in contrast, the inoculum strain was almost invariably found in black nodules. A highly specific symbiotic character (T. semipilosum x

Table 5. Serological (a) and antibiotic resistance (b) characteristics of *Rhizobium* isolates from black nodules on *Centrosema pubescens* inoculated with antibiotic resistant CB1923 and sown at Brisbane, Queensland

| Time from inoculation (months) | Nodule colour | Plots inoculated with CB1923 $str^r$ (b) | | | Plots inoculated with CB1923 $spc^r$ (b) | | |
|---|---|---|---|---|---|---|---|
| | | No. of isolates | Streptomycin resistant | Streptomycin sensitive | No. of isolates | Spectinomycin resistant | Spectinomycin sensitive |
| 3 | Black | 29 | 28 | 1 | 37 | 37 | 0 |
| | Not black | 3 | 3 | 0 | 2 | 1 | 1 |
| 4 | Black | 28 | 28 | 0 | 29 | 29 | 0 |
| | Not black | 6 | 6 | 0 | 2 | 2 | 0 |
| 6 | Black | 32 | 32 | 0 | 27 | 25 | 2 |
| | Not black | 8 | 7 | 1 | 7 | 3 | 4 |
| 9 | Black | 40 | 18 | 22* | 39 | 36 | 3 |
| | Not black | 0 | 0 | 0 | 1 | 0 | 1 |
| Total | Black | 129 | 106 | 23 | 132 | 127 | 5 |
| | Not black | 17 | 16 | 1 | 12 | 6 | 6 |

(a): whenever serotyping was done there was 100% agreement between serological and antibiotic resistance markers;

(b): CB1923 $str^r$ (streptomycin resistant), CB1923 $spc^r$ (spectinomycin resistant).

* Subjected to flooding, cross contamination due to water flow.

Table 6. Observations on the use of antibiotic resistance, symbiotic character, serological, and antagonism markers to identify isolates from Trifolium semipilosum and T. subterraneum inoculated with streptomycin resistant strains of R. trifolii and grown at Canberra, A.C.T.

| Inoculum strain | No. of isolates examined | Strepto-mycin resistant | Effective on T. semipilosum | Serologic-al identity with inoculum | Production of bacteriocin (anti-CC276) | Sensitive to bacteriocin (produced by CB782) |
|---|---|---|---|---|---|---|
| CB782 | 48 | 48 | 48 | (a) | 46 | (a) |
| CC276 | 24 | 24 | (a) | 22 | (a) | 24 |
| Naturally-occurring strains | 17 | 0 | 0 | 1 (CC276) | 0 | 11 |

(a): not attempted or inappropriate observations.

CB782, L. bainesii x CB360 or x CB376) was a reliable marker, but when the character was less specific (T. subterraneum cv. Howard x NA30) identification was often inaccurate.

Of the markers examined, only bacteriocin sensitivity was so unreliable as to be useless for recognition of Rhizobium strains. The value of the companion marker, bacteriocin production, is limited by the small number of producer strains available. It was quickly found that symbiotic markers, even highly specific ones, were tedious to use; less specific ones, requiring the use of more than one host species were very time-consuming indeed. In addition to C. pubescens x CB1923, Rhizobium strains producing black nodules are also available for Lablab purpureus, Centrosema mucunoides, Glycine wightii, Indigofera hirsuta, and Macroptilium atropurpureum (7, 9, 12), thus providing a range of material suitable for ecological studies of tropical Rhizobium spp. Colony characteristics are likely to be of limited value in field work but have been useful in closed systems such as tube culture (24). Of possible genetic markers for identifying rhizobia, only streptomycin-, spectinomycin-, and kanomycin-resistant mutants have been used to any extent (5, 29, 39). This is partly because loss or modification of effectiveness is often associated with mutation to antibiotic resistance (35, 36). However, the range of antibiotics suitable for the production of mutants seems to be widening (23). In addition, the use of mutant strains simultaneously resistant to more than one antibiotic may be feasible (29).

The work described in this paper confirmed the long-term experience in our laboratories that identification of rhizobia is best done by means of immunology. Personal preference often seems to determine the choice of which immunological technique is used. Excellent work over a long period has been done using agglutination procedures, which are particularly apt for simple systems. In more complex microbiological situations, fluorescent antibody methods or gel immune diffusion serology are required. The gel immune diffusion technique that we have employed appears to be ideal for identification of rhizobia, providing unambiguous results with a minimum of effort. With large nodules, the act of crushing usually yields sufficient antigen for immune diffusion. Dealing with smaller nodules takes a little more time since rhizobial isolates need culturing to provide enough antigen. Equipment is now available to streamline the procedures (15, 16).

## E. References


1. Bohlool, B.B., Schmidt, E.L.: Soil Sci. 110, 229-236 (1970).
2. Brockwell, J.: Aust. CSIRO Div. Plant Ind. Fld Stn Rec. 10, 51-58 (1971).
3. Brockwell, J.: In: A Treatise on Dinitrogen Fixation. Section IV. Agronomy and Ecology. Hardy, R.W.F., Gibson, A.H. (ed.). New York: John Wiley and Sons, pp. 277-309 (1977).
4. Brockwell, J., Hely, F.W.: Aust. J. Agric. Res. 17, 885-899 (1966).
5. Brockwell, J., Schwinghamer, E.A., Gault, R.R.: Soil Biol. Biochem. 9, 19-24 (1977).
6. Bushnell, O.A., Sarles, W.B.: J. Bacteriol. 38, 401-410 (1939).
7. Cloonan, M.J.: Aust. J. Sci. 26, 121 (1963).
8. Date, R.A., Decker, A.M.: Can. J. Microbiol. 11, 1-8 (1965).
9. Diatloff, A.: Aust. Plant Path. Soc. Newsl. 1, 28-29 (1972).
10. Diatloff, A.: Soil Biol. Biochem. 9, 85-88 (1977).
11. Diatloff, A., Langford, S.: Qld J. Agric. Anim. Sci. 32, 95-100 (1975).
12. Dobereiner, J.: In: As Leguminosas na Agricultura Tropical. Dobereiner, J., de Eira, P.A., Franco, A.A., Campelo, A.B. (ed.). Brazil: Instituto de Pesquisa Agropecuaria do Centro-Sul, pp. 181-192 (1971).
13. Dudman, W.F.: J. Bacteriol. 88, 782-794 (1964).
14. Dudman, W.F., Brockwell, J.: Aust. J. Agric. Res. 19, 739-747 (1968).
15. Gault, R.R., Byrne, P.T., Brockwell, J.: Lab. Pract. 4, 292-293 (1973).
16. Gault, R.R., Koci, J., Brockwell, J.: Lab. Pract. 5, 11-13 (1974).
17. Gibson, A.H.: Aust. J. Agric. Res. 15, 37-49 (1964).
18. Gibson, A.H., Dudman, W.F., Weaver, R.W., Horton, J.C., Anderson, I.C.: Plant Soil special vol., pp. 33-37 (1971).
19. Hely, F.W.: Aust. J. Agric. Res. 16, 575-589 (1965).
20. Hely, F.W., Costin, A.B., Wimbush, D.J.: Aust. CSIRO Div. Plant Ind. Fld Stn Rec. 3 (1), 55-62 (1964).
21. Holland, A.A.: Antonie van Leeuwenhoek J. Microbiol. Serol. 32, 410-418 (1966).
22. Johnston, A.W.B., Beringer, J.E.: J. Gen. Microbiol. 87, 343-350 (1975).
23. Levin, R.A., Montgomery, M.P.: Plant Soil 41, 669-676 (1974).
24. Marques Pinto, C., Yao, P.Y., Vincent, J.M.: Aust. J. Agric. Res. 25, 317-329 (1974).
25. Marshall, K.C.: J. Aust. Inst. Agric. Sci. 15, 273-281 (1956).
26. Means, U.M., Johnson, H.W., Date, R.A.: J. Bacteriol. 87, 547-553 (1964).
27. Norris, D.O.: Aust. J. Agric. Res. 9, 629-632 (1958).
28. Norris, D.O., t'Mannetje, L.: E. Afr. Agric. For. J. 29, 214-235 (1964).
29. Obaton, M.: C.R. Hebd Seances Acad. Sci. Ser. D. Sci. Natur. (Paris) 272, 2630-2633 (1971).
30. Ouchterlony, O.: Prog. Allergy 5, 1-78 (1958).
31. Parker, C.A., Trinick, M.J., Chatel, D.L.: In: A Treatise on Dinitrogen Fixation. Section IV. Agronomy and Ecology. Hardy, R.W.F., Gibson, A.H. (ed.). New York: John Wiley and Sons, pp. 311-352 (1977).
32. Roslycky, E.B.: Can. J. Microbiol. 13, 431-442 (1967).
33. Scheffler, J.G., Louw, H.A.: S. Afr. Med. J. 40, 701 (1966).
34. Schmidt, E.L., Bankole, R.A., Bohlool, B.B.: J. Bacteriol. 95, 1987-1992 (1968).
35. Schwinghamer, E.A.: Can. J. Microbiol. 10, 221-233 (1964).
36. Schwinghamer, E.A.: Antonie van Leeuwenhoek J. Microbiol. Serol. 33, 121-136 (1967).
37. Schwinghamer, E.A.: Soil Biol. Biochem. 3, 355-363 (1971).

38. Schwinghamer, E.A., Belkengren, R.P.: Arch. Mikrobiol. 64, 130-145 (1968).
39. Schwinghamer, E.A., Dudman, W.F.: J. Appl. Bacteriol. 36, 263-272 (1973).
40. Schwinghamer, E.A., Pankhurst, C.E., Whitfeld, P.R.: Can. J. Microbiol. 19, 359-368 (1973).
41. Schwinghamer, E.A., Reinhardt, D.J.: Aust. J. Biol. Sci. 16, 597-605 (1963).
42. Škrdleta, V.: Soil Biol. Biochem. 2, 167-171 (1970).
43. Stevens, J.W.: J. Infect. Dis. 33, 557-566 (1923).
44. Szende, K., Ördögh, F.: Naturwissenschaften 17, 404-405 (1960).
45. Takahashi, I., Quadling, C.: Can. J. Microbiol. 7, 455-465 (1961).
46. Trinick, M.J.: J. Appl. Bacteriol. 32, 181-186 (1969).
47. Vincent, J.M.: Proc. Linn. Soc. N.S.W. 66, 145-154 (1941).
48. Vincent, J.M.: Proc. Linn. Soc. N.S.W. 87, 8-38 (1962).
49. Wright, W.H., Sarles, W.B., Holst, H.G.: J. Bacteriol. 19, 39 (1930).

# *Rhizobium* Strains Isolated from Wild and Cultivated Legumes: Suppression of Nodulation by a Non-Nodulating *Rhizobium* Strain

T.A. LIE, R. WINARNO, and P.C.J.M. TIMMERMANS

## A. Introduction

Wild and primitive pea genotypes often fail to form root nodules with Rhizobium strains (European strains) derived from cultivated peas, growing in temperate regions (1, 3, 4). Recently, a specific group of bacteria was isolated, which form effective nodules in the primitive pea cv. Afghanistan (5, 6) that was formerly assumed to be a non-nodulating plant (1, 3, 4). During a survey, it was established that these specific Rhizobium strains (Middle-East strains) are widespread in Turkey and Israel but absent in Europe, North Africa and many tropical countries (6). These results showed that in the "Centre of Origin" of leguminous plants (2), the Rhizobium strains present differ from those isolated from cultivated plants.

So far, experiments reported have been conducted under controlled conditions, using plants growing with the roots under aseptic conditions (3, 4, 5, 6). In some experiments, using non-sterile soils in Wageningen, pea cv. Afghanistan failed to nodulate with a Turkish Rhizobium strain, which formed nodules in the same plant under aseptic conditions. This paper reports that non-nodulating Rhizobium strains, although not capable of forming nodules in a particular plant, can compete with a nodulating strain and prevent the latter forming nodules. These findings might be important for inoculation of soils containing a native Rhizobium population.

## B. Materials and Methods

Pea plants cv. Afghanistan, non-nodulating with the European Rhizobium strains PRE, $PF_2$, $310^a$ and 313 but nodulating with the Turkish strain Tom, were grown, with the roots under aseptic conditions in a nitrogen-free nutrient solution, in a climate room at 20$^{o}$C. The plants were inoculated either with a single culture or with a mixture of Rhizobium strains at a rate of 5000 bacteria per ml nutrient solution, ten to fourteen days after germination of the seeds. The process of nodule formation was checked daily and the nodules counted and identified two or three weeks after inoculation.

Besides the above-mentioned Rhizobium strains PRE, $PF_2$, $310^a$, 313 and Tom, the mutant strains $PF_2^{++}$ and $Tom^{++}$, resistant to streptomycin and acriflavin, were used. The Rhizobium strains in the nodules were identified by culturing on yeast-extract-mannitol agar (YEM) and YEM containing antibiotics (9).

Identification involved taking a whole root system instead of detaching the nodules and treating each nodule individually. The roots were washed thoroughly with water containing a drop of detergent (Teepol), surface-sterilized with ethanol 70% (1 minute) followed by 6% $H_2O_2$, containing a drop of Teepol (5 minutes). Afterwards they were wrapped in sterile paper towels to remove adherent water and the roots were spread to locate the position of the nodule. Each nodule was pricked with a sterile preparation needle (or wooden tooth pick) and the inoculum was transferred to YEM and to YEM containing 25 µg streptomycin and 2.5 µg acriflavin per ml agar. When there were too many nodules on the roots, one out of two or three nodules was identified. Using this method large numbers of nodules can be handled using a minimum of glassware (5).

The number of bacteria was determined using the spot plate method, using a repeating dispenser delivering 0.02 ml drops. The number of colonies was counted 48-72 hours after incubation at 25$^{o}$C, using a microscope at low magnification.

## C. Results

I. Plants inoculated simultaneously with mixtures of Rhizobium strains. The plants were inoculated with a mixture of a non-nodulating and a nodulating Rhizobium strain at the same time. For comparison, plants inoculated with a single culture of Rhizobium were included. The results in Table 1 show, in agreement with our previous findings (3, 4) that the European Rhizobium strains PRE, $PF_2$, $310^a$ and 313 do not form nodules on pea cv. Afghanistan and the plants turned yellow as a result of nitrogen deficiency. The roots, however, were covered with small swellings (6). A large number of red nodules were formed by strain $Tom^{++}$ and the plants were dark green.

When $Tom^{++}$ was applied together with strains $PF_2$ or 313, no nodules at all were obtained, whereas with strain $310^a$ the number of nodules was reduced and nodulation was delayed. On the other hand, no such suppressing effect was found when strain PRE was applied together with strain $Tom^{++}$. All the nodules, within the limit of experimental errors, were identified as being derived from Rhizobium strain $Tom^{++}$.

Table 1. Root-nodule formation in pea cv. Afghanistan, inoculated either with a single culture or with a mixture of a nodulating ($Tom^{++}$) and non-nodulating Rhizobium strains (PRE, $PF_2$, $310^a$ and 313)

| Rhizobium strains | Number of Nodules per plant | Nodulation time[1)] (days) |
|---|---|---|
| $PF_2$ | 0 | - |
| 313 | 0 | - |
| $310^a$ | 0 | - |
| PRE | 0 | - |
| $Tom^{++}$ | 90 | 8 |
| $Tom^{++}$ + $PF_2$ | 0 | - |
| $Tom^{++}$ + 313 | 0 | - |
| $Tom^{++}$ + $310^a$ | 31 | 13 |
| $Tom^{++}$ + PRE | 101 | 8 |

1) Time to obtain 100% nodulated plants.

II. Plants inoculated consecutively with different Rhizobium strains. In the following experiment the second inoculum was applied three days after the first inoculum. As shown in Table 2 both Rhizobium strains $PF_2$ and its mutant $PF_2^{++}$ did not form nodules either as a single culture or as in mixtures. Rhizobium strain Tom and its mutant $Tom^{++}$ produced equal numbers of nodules and, when applied together, nearly all the nodules resulted from the first-applied inoculum. Rhizobium strain $PF_2$ (and $PF_2^{++}$) suppressed nodulation, when applied 3 days earlier than Tom (or $Tom^{++}$)

to the plant. However, when Tom (or $Tom^{++}$) was applied three days earlier than $PF_2$ (or $PF_2^{++}$) many nodules were formed. All the nodules formed could be identified as being derived from Rhizobium strains Tom or its mutant $Tom^{++}$.

Table 2. Competition between nodulating (Tom and $Tom^{++}$) and non-nodulating ($PF_2$ and $PF_2^{++}$) Rhizobium strains in nodulation of pea cv. Afghanistan

| 1st Inoculant Day 0 | | 2nd Inoculant Day 3 | Number of Nodules per plant | Nodules derived from 1st inoculant (%) |
|---|---|---|---|---|
| $PF_2$ | | - | 0 | - |
| $PF_2^{++}$ | | - | 0 | - |
| $PF_2$ | + | $PF_2^{++}$ | 0 | - |
| $PF_2^{++}$ | + | $PF_2$ | 0 | - |
| Tom | | - | 29 | 100 |
| $Tom^{++}$ | | - | 28 | 90 |
| Tom | + | $Tom^{++}$ | 26 | 91 |
| $Tom^{++}$ | + | Tom | 30 | 97 |
| $PF_2$ | + | $Tom^{++}$ | 0 | - |
| $PF_2^{++}$ | + | Tom | 0 | - |
| Tom | + | $PF_2^{++}$ | 20 | 100 |
| $Tom^{++}$ | + | $PF_2$ | 23 | 95 |

III. Multiplication of Rhizobium in the rhizosphere. In some experiments the number of bacteria was counted at the end of the experiment in the nutrient solution, surrounding the roots. In both cases when $PF_2$ or Tom was applied as a single culture, about the same number of bacteria was found at harvest time. The number at inoculation was 5000 per ml nutrient solution and it was increased to 2-4 x $10^6$ $ml^{-1}$. Unfortunately, rather erratic results were obtained when the bacteria were counted in the mixed cultures, and no definite conclusions can be shown. Both cultures of Rhizobium strain however, multiply in the nutrient solution when applied together.

## D. Discussion

Various studies (7) demonstrate the complexity of factors determining the competitive ability in nodule formation. One of the factors which may affect competition is the growth rate of the bacteria in the rhizosphere, but there is little evidence that the number of bacteria in the rhizosphere is related to the representation in the nodules (11). This is presumably due to the relatively small proportion of the rhizobial population on the root, which is directly involved with nodulation of the roots. Using mixtures of a virulent and an avirulent Rhizobium strain (8) it was found that very low numbers of the virulent strain were required to produce nodules, about 400 bacteria per nodule formed. In our experiments it was found that the Rhizobium strains used can multiply rapidly in the rhizosphere of pea cv. Afghanistan, whether they form nodules or not.

There are some reports, indicating that Rhizobium strains producing effective nodules are better competitors than those producing ineffective ones (10) but there is no general agreement (7). Such results must be viewed with caution since it may be assumed that the Rhizobium strains maintained in culture collections, are not only highly effective in nitrogen fixation, but presumably also unconsciously selected for competitive ability and are therefore successfully used for inoculation. Our results demonstrate that the competitive ability of a rhizobial strain may not be even related to the ability to form root nodules. The nature of this competition is still unknown, but it may be relevant that these so-called non-nodulating strains can enter the root system and induce changes in the roots, resulting in small swellings. Obviously, such initial changes may render the roots "resistant" to the following Rhizobium strain. More Rhizobium strains differing in their ability to induce slight changes in the roots are now being examined in our laboratory.

Besides the theoretical aspects these findings may have implications for the practice of inoculation of soils containing a native Rhizobium population, related to the inoculum introduced. The use of such plants as pea cv. Afghanistan is very convenient for the study of competition, and further experimentation, in soil where competition can be modified by many factors (5), is needed.

## E. References


1. Degenhardt, T.L., Larue, T.A. & Paul, E.A.: Investigations of a non-nodulating cultivar of Pisum sativum. Canadian Journal of Botany, 54, 1633-1636 (1976).
2. Govorov, L.I.: The peas of Afghanistan. Bulletin of Applied Botany, of Genetics and Plant Breeding (Leningrad), 19, 497-522 (1928).
3. Lie, T.A.: Symbiotic nitrogen fixation under stress conditions. Plant and Soil, Special Volume 117-127 (1971).
4. Lie, T.A., Hille Ris Lambers, D. & Houwers, A.: Symbiotic specialisation in pea plants: Some environmental effects on nodulation and nitrogen fixation. In: Symbiotic nitrogen fixation in plants. Nutman, P.S. (ed.). Cambridge: Cambridge University Press, pp. 319-333 (1975).
5. Lie, T.A., Soe-Agnie, I.E., Muller, G.J.L. & Göktan, D.: Environmental control of symbiotic nitrogen fixation: Limitations and flexibility of the legume/Rhizobium system. In: Proceedings of Symposium on Soil Microbiology and Plant Nutrition. Broughton, W.J. (ed.). Kuala Lumpur: Malayan University Press, in press (1977).
6. Lie, T.A.: Symbiotic specialisation in pea plants: The requirement of specific Rhizobium strains for peas from Afghanistan. Annals of Applied Biology, in press (1978).
7. Parker, C.A., Trinick, M.J. & Chatel, D.L.: Rhizobia as soil and rhizosphere inhabitants. In: A treatise on dinitrogen fixation. Hardy, R.W.F. & Gibson, A.H. (eds.) New York: John Wiley and Sons, pp. 311-352 (1977).
8. Purchase, H.F. & Nutman, P.S.: Studies on the physiology of nodule formation. VI. The influence of bacterial numbers in the rhizosphere on nodule iniation. Annals of Botany, N.S., 21, 439-454 (1957).
9. Obaton, M.: Influence de la composition chimique du sol sur l'utilite de l'inoculation des graines de lucerne avec Rhizobium meliloti. Plant & Soil, Special Volume, 273-285 (1971).
10. Robinson, A.C.: Competition between effective and ineffective strains of Rhizobium trifolii in the nodulation of Trifolium subterraneum. Australian Journal of Agricultural Research, 20, 827-841 (1969).
11. Vincent, J.M. & Waters, L.M.: The influence of the host on competition amongst clover root-nodule bacteria. Journal of General Microbiology, 9, 357-370 (1953).

# Rhizobia Associated with Indigenous Legumes of New Zealand and Lord Howe Island

R.M. GREENWOOD

## A. Introduction

New Zealand and Lord Howe Island have few indigenous legumes, and these can be grouped botanically into: (a) Carmichaelia and three small related genera, which are confined to New Zealand, except for one species on Lord Howe I: (b) Clianthus and Swainsona, one species each of these otherwise Australian genera: (c) Sophora, three species in New Zealand, and one on Lord Howe I: (d) Vigna and Canavalia, (one species each) which are strand legumes widely distributed on tropical Pacific island sandy beaches, and extend as far south as Lord Howe I. The rhizobia associated with these legumes have been investigated.

## B. Materials and Methods

Rhizobial isolates were obtained either direct from nodules, or indirectly from soil samples, using suitable trap hosts.

Effectiveness (nitrogen-fixing ability) of isolates was determined by visual assessment of plant growth, and measurement of shoot weights where appropriate. The seedlings were grown in large test tubes containing sterile pumice rooting medium supplied with N-free nutrient solution.

For detecting differences between isolates, amino acid patterns from hydrolysed extracts of nodules were determined (1).

## C. Results and Discussion

Isolates (177) of indigenous rhizobia were obtained from 47 localities throughout New Zealand, and from the Chatham Islands and Lord Howe Island. Rhizobia were not detected in soil samples taken from unmodified natural habitats where legumes were not present. The rhizobia associated with the strand legumes were quite different from all other isolates, and were of the cowpea type, as they did not produce acid in culture, and formed effective nodules with siratro (Macroptilium atropurpureum). They were the only rhizobia isolated from samples of coral sand from under the strand legumes on Lord Howe I., but were not detected in samples of lowland or mountain soil collected near roots of Lord Howe Sophora or Carmichaelia plants. The rhizobia associated with all other indigenous legumes produced acid in culture. They were unable to nodulate typical hosts of the named Rhizobium species, but nodulated all indigenous species other than the strand legumes.

Isolates from Sophora were effective with Sophora species but ineffective with all other indigenous species. Isolates from Carmichaelia, Clianthus and Swainsona all formed one group on the basis of effectiveness. Some of these isolates also fixed nitrogen with Sophora plants. However this may not indicate a close relationship with the Sophora rhizobia, as many overseas rhizobial strains including those from Leucaena, Onobrychis and Caragana can fix nitrogen with New Zealand Sophora species.

Some strains could be distinguished from each other on the basis of effectiveness with different host species, but a method we have found very useful in this rhizobial group for showing which isolates were distinct strains was the comparison of amino acid patterns from hydrolysed extracts of nodules formed with the isolates. These patterns were determined largely by the rhizobial strain rather than the host species. A great diversity of patterns was found (2), particularly among the isolates effective with Carmichaelia and Clianthus. Individual strains were shown to be very restricted in distribution. In 19 out of 22 instances where two or more isolates from one locality gave the same pattern, these isolates had come from nodules on the same plant or from the same soil sample. With isolates from different localities the patterns all appeared to be different. Thus the total number of different strains must be very great indeed.

The acid-producing rhizobia associated with New Zealand legumes are very different from those nodulating most of the numerous Australian legumes, which are of the non-acid-producing cowpea type. In more inland Australia, however, acid-producing rhizobia are associated with certain Swainsona species, and some effective cross-nodulation between Carmichaelia spp. and these Swainsona spp. can occur. Also many New Zealand isolates can nodulate and a few can fix some nitrogen with Onobrychis viciifolia from the Northern Hemisphere. Thus the rhizobia associated with New Zealand indigenous legumes can be grouped broadly with other acid-producing, fast-growing rhizobia not included in the four named fast-growing Rhizobium species. This is in agreement with recent findings on the basis of phenetic similarity and DNA base sequence homology (3).

## D. References


1. Greenwood, R.M., Bathurst, N.O.: The effect of Rhizobium strain on the amino acids of lotus root nodules. N.Z.J. Sci. 11, 280-283 (1968).
2. Greenwood, R.M., Bathurst, N.O.: Effect of rhizobial strain and host on the amino acid patterns in legume root nodules. Submitted for publication to N.Z. J. Sci. (1977).
3. Jarvis, B.D.W., MacLean, T.S., Robertson, I.G.C., Fanning, G.R.: Phenetic similarity and DNA base sequence homology of root nodule bacteria from New Zealand native legumes and Rhizobium strains from agricultural plants. N.Z.J. Agric. Res. 20, 235-248 (1977).

# The Ecology of *Rhizobium*

P.S. NUTMAN, M. DYE, and P.E. DAVIS

## A. Introduction

Definitions of habitats, taxa and counting methods, which lie at the centre of ecology present special problems for Rhizobium because the habitat is partly in the soil and partly within the host plant, and because the counting method depends on host nodulation, which defines species of Rhizobium.

In the soil rhizobia live heterotrophically (Fig. 1.I) multiplying on and in the root surface and surroundings, particularly of legumes, but not necessarily to the greatest extent on plants they are able to infect. They are stimulated by exudation and plasmoptysis of nutrients and growth factors and are subject to bacteriostasis, antibiosis, lysis, parasitism and predation, and to deleterious physical or chemical factors. They migrate slowly, either passively, or actively in wet soil, and probably exchange genetic material amongst themselves and possibly with other kinds of bacteria. Bacteriophages seem to be always present wherever rhizobia are found (13).

Rhizobia are usually polarly attached to the root surfaces (Fig. 1, II), but of these very few infect the legume, the mode of infection differing between hosts. Often it is through deformed root hairs by way of an infection thread formed by the invagination of the young hair wall (Fig. 1, III) so that the contained bacteria are intercellular (8, 12). The number of infections leading to nodulation is related to the size of the rhizosphere population, but not in a simple manner; in general the rhizosphere population is vastly in excess of that required for maximum nodulation (10).

Time-lapse microphotography shows that only bacteria near the thread tip multiply during its penetration into the host (14). Unless more than one initial thread enters the nodule, which happens uncommonly, the rhizobia in the nodule are a single clone having undergone a process analagous to line selection within the thread system.

Different strains can often be distinguished among the isolates taken from different nodules on a single plant (5, 19).

Even in clovers which normally have threads, infection can be initiated intercellularly between neighbouring epidermal cells (9). Simple intercellular infection without threads seems to be the rule in some hosts (Fig. I, IIIa). In Arachis and Stylosanthes this always occurs at points where lateral roots emerge and root cells are already somewhat separated (Chandler, person communication).

The bacteria are eventually released into the host cell from the ramifying threads or from intercellular spaces (Fig. 1, IV and IVa). They become enfolded in a plant unit membrane, multiplying rapidly as short rods in the enlarging polyploid host cell and become transformed into the swollen bacteroid forms. There seem to be two general modes of dispersion: either from the branching system of threads and intercellular infections, or by the division of already infected cells. Where there are infection threads the host cells usually complete their division before the bacteria are released, but not always. During transformation into bacteroids the bacterial nucleoid degenerates and the bacteroid's capacity for independent

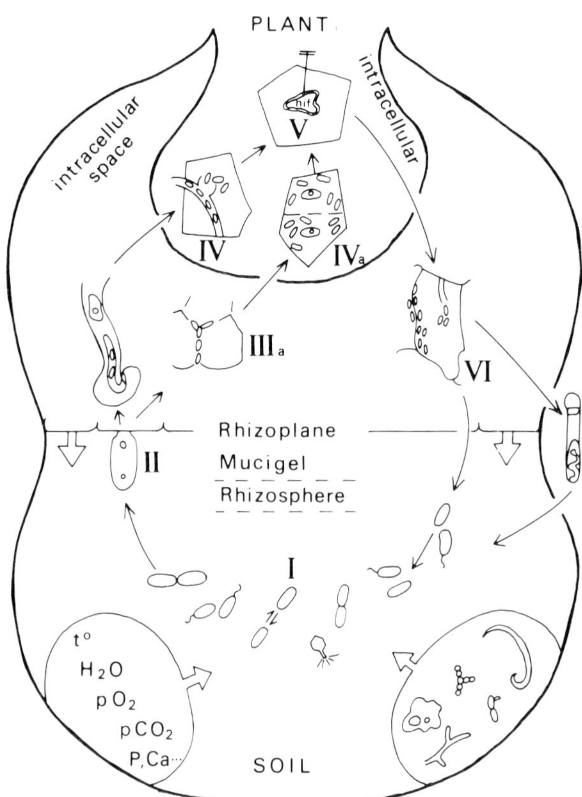


Fig. 1. Diagram showing the habitat of rhizobia in soil and in plants

multiplication is lost; at the same time conditions within the nodule enhance their power to fix nitrogen. Ecologically and genetically the stages marked IV, IVa, and V. in Fig. 1 form a biological cul-de-sac.

In the older nodule the bacteroids and their host cells break down and the nodule becomes recolonised by rod-shaped rhizobia from the infection threads and intercellular spaces and they escape into the soil to complete the cycle.

A strain isolated from a nodule therefore comprises cells descended from ones that have not entered into full symbiosis or fixed nitrogen as bacteroids in the nodule. The success of artificial inoculation suggests that relying in this way on collateral inheritance seems not to impose practical difficulties.

## B. Methods

The definition of taxa and counting methods are related because the numbers of rhizobia are estimated by Most Probable Number (MPN) techniques by recording the nodulation of a chosen host at each dilution of the soil sample; only strains of rhizobia able to nodulate the test hosts are thus counted.

Of some practical importance is the validity of the assumption for calculating the MPN, namely that a single cell of the appropriate rhizobium will multiply and form nodules. Recovery counts (3, 6) show that this is generally true. The commonest aberration is the appearance of negative results more frequently at lower dilutions than might be expected from a Poisson distribution. It has been shown

(15, 17) that such negative results are more frequent in counts of R. meliloti and R. lupini than R. trifolii and R. leguminosarum, but that cropping has little effect on their incidence providing soil pH is not acid. Negative tubes seem to be equally frequent in samples from forage, cereal or root crops soils and are not noticeably affected by mineral fertilisers, but they are increased by farmyard manure.

Shortcomings in the MPN method are not, however, so great as to obscure the very considerable effects of the natural vegetation, cropping, fertilisers and management upon soil populations of Rhizobium. Of management factors artificial inoculation can most drastically affect numbers of rhizobia and will be discussed in other papers. This paper is confined to cataloguing population changes in uninoculated soils treated in different ways.

## C. Occurrence of Rhizobia in Uninoculated Soils

Knowledge of numbers of Rhizobium in natural habitats is very incomplete and depends mostly on records of nodulation. Only about 10% of the 12,000 known species of legume have been examined for root nodules. Of these, relatively few (mostly among the Caesalpinioideae) are not naturally nodulated (1, 2).

Rhizobia are usually absent where their hosts are not found. Exceptions based on actual counts are mostly of very small populations thought to have been introduced on dust, implements or by animals. Some workers maintain that soils can contain indigenous populations of rhizobia for hosts not normally present, which then adapt and increase on the introduction of a legume host; a thesis difficult to substantiate. Work (4, 16) on the distribution of clover and medic bacteria in natural pastures in Australia shows the presence of the host to be the most important factor. Similar results were found in the naturally-seeded hill pastures of the North and West of the U.K. (18) and in Irish hill areas(11); strains isolated from higher altitudes tended to be poorly effective in fixing nitrogen.

Populations of rhizobia in certain of the differently manured plots of the Park Grass Experiment at Rothamsted which have for more than a century been undisturbed meadowland cut for hay showed a striking correspondence between the occurrence of the different Rhizobium species in the soil of each plot and the presence of its host legume in the herbage (15). R. trifolii, R. leguminosarum and R. lupinii were present in most plots, R. trifolii being most abundant, especially in limed plots. Medicago and Melilotus species were absent from the entire field and in no plot was R. meliloti found.

Abundance of each Rhizobium species corresponded to frequency of host occurrence in the different plots. Rhizobia were fewer in acid plots and absent in soil below pH about 4.

The symbiotic effectiveness on red clover strains of R. trifolii isolated from some of these plots showed that nitrogenous fertilisers per se, even after more than a century of use, did not usually affect the intrinsic nitrogen-fixing capacity of the strains of clover nodule bacteria in the soil. But where such treatments increased soil acidity they either eliminated the rhizobia or affected their capacity to fix nitrogen.

## D. Effects of Heavy Rates of Nitrogenous Fertilisers on Rhizobia and Soil Horizon

The Rothamsted results refer to experiments of medium or of very long duration in which the highest rates of nitrogen application were 20 kg N $ha^{-1}$, whereas very much larger amounts are used in modern forage production.

Table 1 shows soil counts of R. trifolii in permanent pasture at the Grasslands Research Institute, Hurley and at Jealotts Hill (I.C.I.) where different rates of N fertiliser were applied with liming. At the time of sampling, 7, 6 and 2 years after fertiliser application respectively, no clover remained except on the 0 plots and pathways between blocks. The numbers of residual rhizobia were inversely related to the amount of fertiliser used, but not in simple proportion. It was also observed that the effect of fertiliser in depressing numbers of rhizobia was most evident in the surface horizon. The effect of 879 kg N in the Jealotts Hill experiment was quite drastic even after only 2 years.

Table 2 shows similar trials at Jealotts Hill on sown swards containing red and white clover and on grass-clover mixtures on experimental husbandry farms in the U.K. Numbers were again reduced by fertiliser especially in the superficial layers of the soil. In the Jealotts Hill experiments, however, the presence of the host plant to a small degree counteracted the effect of fertiliser.

These results may do no more than quantify effects that could have been anticipated, including the greater sensitivity of the rhizobial population in the surface soil, probably caused by a temporary build up of acidity between applications of lime. The response of the test plants in the MPN tests indicated no drastic change towards lowered effectiveness.

Quantitative information of this kind is useful. It shows that although numbers may be strikingly reduced, the reduction is probably not sufficient to prevent the nodulation of any clover crop that might be sown in these areas. The reduction in numbers, however, which may be up to 1,000 -fold may be large enough to enable elite strains of Rhizobium to be introduced by methods of heavy inoculation. Nearly all attempts to improve clover growth in the lowland areas of the U.K. by inoculation have failed because of competition from the vastly larger numbers of indigenous R. trifolii present, which are generally reasonably effective.

Table 1. Effect of fertiliser N on numbers of R. trifolii in grassland Log MPN g dry soil$^{-1}$

| | kg N ha$^{-1}$ year$^{-1}$ | surface 0.5 cm | deep 15-20 cm |
|---|---|---|---|
| Jealotts Hill, Expt.C/76/A., | on mid-brown loam over London clay, N for 6 years, all plots limed. | | |
| grazed | 251 | 5.44 | 4.52 |
| " | 753 | 3.53 | 3.76 |
| cut | 251 | 5.08 | 5.21 |
| " | 753 | 2.61 | 4.31 |
| Jealotts Hill, Expt. H.V.l., | N for 2 years, all plots limed. | | |
| cut | 377 | 4.10 | 4.79 |
| " | 628 | 3.40 | 4.28 |
| " | 879 | 1.61 | 4.35 |
| Grassland Research Institute, | Broad Oak Field. Flood plain alluvium, 7 years permanent pasture, all plots limed. | | |
| cut | No fertiliser | 5.52 | 3.27 |
| " | 168 | 4.38 | 3.70 |
| " | 336 | 3.89 | 3.93 |
| " | 673 | 3.51 | 3.40 |

Table 2. Effect of fertiliser N on numbers of *R. trifolii*. Log MPN g dry soil$^{-1}$

| | Kg N ha$^{-1}$ year$^{-1}$ | surface 0-5 cm | deep 15-20 cm |
|---|---|---|---|
| Jealotts Hill, Expt. TC1A white clover/grass ley cut for hay for 2 years. | | | |
| | 189 | 5.44 | 4.30 |
| | 567 | 2.61 | 3.33 |
| Jealotts Hill, Expt. TC2. Hurgaropoly red clover/grass ley cut for hay for two years | | | |
| | 0 | 5.66 | 5.42 |
| | 189 | 5.28 | 5.13 |
| | 567 | 3.91 | 5.41 |
| Bridgets E.H.F.Winchester, Clay with flints. grazed pasture for 9 years | | | |
| permanent pasture | 112-168 | 4.76 | 2.77 |
| " " | 168-224 | 2.76 | 3.76 |
| sown pasture | 280-336 | 3.76 | 3.76 |
| Boxworth E.H.F.Cambridge,Chalky boulder clay. grazed pasture given 280-336 kg N ha$^{-1}$ | | | |
| Field | age of expt. | 0.5 cm | 15-20 cm |
| Taylers | 2 years | 4.30 | 4.30 |
| Bigfield | 7 years | 3.29 | 2.75 |
| Overhall | 10 years | 3.30 | 2.76 |
| Cooks | 12 years | 3.26 | 3.28 |
| Samsons | 17 years | 3.26 | 2.77 |

Table 3. Numbers of Rhizobia of the Cowpea miscellany at Samaru, Northern Nigeria

| | log MPN g dry soil$^{-1}$ Test Host | | |
|---|---|---|---|
| | *Stylosanthes gracilis* | *Desmodium intortum* | *Microptilium atropurpureum* |
| Site A | | | |
| 0-5 cm | <0.62 | <0.62 | 1.61 |
| 10-15 cm | <0.59 | 0.50 | 0.50 |
| 20-25 cm | 3.78 | 4.26 | 4.20 |
| Site B | | | |
| 0-5 cm | <0.76 | <0.76 | 0.76 |
| 10-15 cm | 1.45 | 3.85 | 0.86 |
| 20-25 cm | 2.75 | 3.20 | 3.20 |

An example of the influence of soil horizon on rhizobial distribution is shown in Table 3 which gives counts for two Northern Nigerian soils in an area where the groundnut is widely grown and its bacteria thought to be ubiquitous (7). Prior to sowing the crop the bare soil is exposed to severe insolation leading to very high temperatures in the upper layers of the soil and to partial sterilisation.

At both sites the soils were sampled at 0-5 cm, 10-15 cm and 20-25 cm. The three different species of host used in the MPN test gave similar results. Rhizobia were virtually absent in the upper layers and most abundant in the 20-25 cm layer. In these soils also the possibility of using heavy inoculation by outstandingly good strains is worth examining.

## E. General Conclusions

In most systems of agriculture the farmer has to be content with the strains of rhizobia nature provides. Only when an exotic legume is introduced into an area containing very few or none of its particular rhizobia can he be assured of a good response to inoculation. Not that this is so rare an event, and indeed such plant introductions combined with inoculation provide some of the most outstandingly successful crops in agriculture. There is no reason to suppose that this colonising phase will not continue. A better understanding of *Rhizobium* ecology and the development of new inoculation procedures may enable the benefits of greater biological nitrogen fixation to be spread more widely in soils containing indigenous populations of less than full effectiveness.

## F. Acknowledgements

We thank Mr. D.W. Cowling, Grassland Research Institute, Mr. A.E.M. Hood, Jealotts Hill Research Station, I.C.I. Dr. R. Dunham, Ahmadu Bello University, Zaria, Nigeria, and Directors of the Experimental Husbandry Farms for their cooperation.

## G. References

1. Allen, E.K., Allen, O.N.: The scope of nodulation in the leguminosae. Recent Advan. Bot. 1, 585-588 (1961).
2. Allen, E.K., Allen, O.N.: The nodulation profile of the genus *Cassia*. *In*: Symbiotic nitrogen fixation in plants. Nutman, P.S. (ed.). Vol. 7, IBP, Cambridge University Press, pp. 113-121. (1976).
3. Brockwell, J., Diatloff, A., Grassia, A., Robinson, A.C.: Use of wild soybean (*Glycine ussuriensis*) as a test plant in dilution nodulation frequency tests for counting *Rhizobium japonicum*. Soil. Biol. Biochem. 7, 305-311 (1975).
4. Brockwell, J., Hely, F.W., Neal-Smith, C.A.: Some symbiotic characteristics of rhizobia responsible for spontaneous effective nodulation of *Lotus hispidus* Desf. Aust. J. Exp. Agr. Anim. Husb., 6, 365-370, (1966).
5. Caldwell, B.E., Hartwig, E.E.: Serological Distribution of soybean root nodule bacteria in soils of south eastern USA. Agron, J. 62, 621-622 (1970).
6. Date, R.A., Vincent, J.M.: Determination of, the number of root nodule bacteria in the presence of other organisms. Aust. J. Exp. Agr. Anim. Husb. 2, 5-19, (1962).
7. Day, J.M., Roughley, J.R.: Effects of temperature stress on nodulation and nitrogen fixation in cowpea. Ann. Appl. Biol. (in press).

8. Higashi, S.: Electron microscopic studies on the infection threat developing in the root hair of Trifolium repens L infected with Rhizobium trifolii. J. Gen. Appl. Microbiol. 12, 147-156.

9. Kumarasinghe, R.M.K.: Ph.D. thesis. Univ. Lond. (1976).

10. Lim, G.: Studies on the physiology of nodule formation VIII. The influence of the size of the rhizosphere population of nodule bacteria on root-hair infection in clover. Ann. Bot. NS 27, 55-67 (1963).

11. Masterson, C.L.: Clover nodule bacteria survey. An Foras Taluntais Soils Div. Res. Depart. Dublin, pp. 78-79. (1961).

12. Nutman, P.S.: The influence of the legume in root nodule symbiosis. Biol. Rev. 31, 109-151 (1965).

13. Nutman, P.S.: Rhizobium in the soil. In: Soil Microbiology. Walker, N. (ed.) London: Butterworth and Co., pp. 111-131 (1975).

14. Nutman, P.S., Doncaster, C.C., Dart, P.J.: Film: The infection of clover root hairs by nodule bacteria. British Film Institute, London. (1973).

15. Nutman, P.S., Ross, G.J.S.: Rhizobium in the roils of the Rothamsted and Woburn Farms. Report Rothamsted Exp. Stn, Pt. 2. 141-167 (1969).

16. Roughley, R.J., Walker, M.H.: A study of inoculation and sowing methods for Trifolium subterraneum in New South Wales. Aust. J. Exp. Agr. Anim. Husb. 13, 284-291 (1973).

17. Thompson, J.A., Vincent, J.M.: Methods of detection and estimation of rhizobia in soil. Plant Soil, 26, 72-84 (1967).

18. Thornton, H.G.: Distribution of strains of clover nodule bacteria in Great Britain. Report Rothamsted Exp. Stn., p.75 (1939-1945).

19. Vincent, J.M.: Strains of rhizobia in relation to clover establishment. Proc. Internat. Grassl. Congress Palmerston North, New Zealand. Paper No. 16 1-11 (1956).

# Environmental Problems Management and Control

# Effect of Cadmium on Microbes *in vitro* and *in vivo*: Influence of Clay Minerals

H. BABICH and G. STOTZKY

Industrial processes (e.g., smelting), home-related activities (e.g., incineration), and extensive use of the automobile have resulted in the release of a variety of pollutants, both particulate and gaseous, into the atmosphere. Of the several particulate heavy metal contaminants currently being emitted, cadmium (Cd), an element with no known biological function, is of prime concern. Increased utilization of Cd-containing ores and other materials has lead to the rapid mobilization, transport, and subsequent deposition and accumulation of Cd in natural ecosystems. Thus, high levels of Cd have been reported in soils around smelters and major highways, and plants and animals collected in these regions have elevated concentrations of Cd.

Although there is considerable information on the effects of Cd on plants and animals (6, 7, 8), there is little information on the response of microorganisms to Cd, and, preliminary studies in this laboratory have evaluated the sensitivity to Cd of a broad spectrum of microorganisms. Bacteria were inoculated into broth amended with varying concentrations of Cd (as $CdCl_2$), and growth was determined turbidimetrically after 24 hours' incubation. For eubacteria, complete inhibition of growth was evident at 50 ppm Cd for *Brevibacterium linens*, *Bacillus megaterium*, *Rhizobium meliloti*, *Agrobacterium tumefaciens*, and *Chromobacterium orangum*; at 100 ppm Cd for *Corynebacterium* sp.; and at 500 ppm Cd for *Alcaligenes faecalis*, *Enterobacter aerogenes*, and *Bacillus cereus*. Growth of *Proteus vulgaris* was still evident in broth amended with 500 ppm Cd, the highest concentration employed. In general, the Gram negative bacteria were more tolerant of Cd than were the Gram positive organisms. For actinomycetes, total inhibition of growth occurred in broth amended with 1 ppm Cd for *Micromonospora chalcea* and with 100 ppm Cd for *Nocardia corallina*; some growth of *Streptomyces flavovirens* and *Nocardia paraffinae* was evident in broth amended with 500 ppm Cd, the highest concentration employed. Cadmium also prolonged the exponential growth of bacteria. When yeasts were inoculated into broth amended with varying concentrations of Cd and assayed turbidimetrically after 24 hours' incubation, total inhibition of growth was noted for *Schizosaccharomyces octosporus* at 500 ppm Cd, the highest concentration employed, whereas there was some growth of *Saccharomyces cerevisiae* and *S. cerevisiae* var. *ellipsoides* and more abundant growth of *Rhodotorula* sp. On the basis of these studies, yeasts appear to tolerate higher concentrations of Cd than eubacteria and actinomycetes (2).

To evaluate the response of filamentous fungi to Cd, mycelial plugs were placed on a nutrient agar medium supplemented with varying concentrations of Cd (as $CdCl_2$), and growth was assayed as mycelial extension. Based on their tolerance to Cd, the fungi studied were broadly grouped into 3 categories: (a) those capable of growth on 10 ppm but inhibited by 100 ppm Cd -- *Botrytis allii*, *B. cinerea*, *Penicillium vermiculatum*, *Aspergillus fischeri*, *A. janus*, *A. giganteus*, *Thielaviopsis paradoxa*, *Fomes annosus*, and *Pyncidiophora dispersa*; (b) those capable of growth on 100 ppm but inhibited by 1000 ppm Cd -- *A. niger*, *A. flavipes*, *Scopulariopsis brevicaulis*, *Pholiota marginata*, *Schizophyllum* sp., *Phycomyces blakesleeanus*, *Fusarium oxysporum*

f. conglutinans, and Chaetomium sp.; and (c) those capable of growth on 1000 ppm Cd, the highest concentration employed -- Rhizopus stolonifer, Trichoderma viride, Sphaerostilbe repens, P. asperum, and Cunninghamella echinulata. There was no correlation between the class of fungus and sensitivity to Cd (2).

Studies were also conducted to evaluate the influence of Cd on spore production in A. niger, R. stolonifer, and T. viride. In all instances, spore production was initially inhibited at concentrations of Cd that were not inhibitory for mycelial extension. For example, when A. niger was grown on agar amended with 1 ppm Cd, spore production was decreased to 35% of the control (i.e., 0 ppm Cd), whereas mycelial extension was unaffected by this amendment and approximately 8 ppm Cd was necessary to achieve a similar reduction in mycelial extension (2).

The differential sensitivities to Cd noted for the various bacteria and fungi in synthetic media may be indicative of similar differential tolerances in soil and other microbial habitats. Thus, deposition of Cd into a natural environment may result in the preferential growth and establishment of microbes tolerant to high concentrations of Cd, with the gradual elimination of more sensitive species. In addition, the adverse impact of Cd on microbial reproductive potential, e.g., extension of the generation times of bacteria and the retardation of sporulation by filamentous fungi, could also influence the establishment, population dynamics, interactions, and general ecology of microbes in natural habitats.

These experiments were extended to study the influence of an environmental variable - the type and concentration of clay minerals - on the toxicity of Cd to microbes. Plugs of fungal mycelium were transferred to the center of nutrient agar unamended or amended with 10, 100, or 1000 ppm Cd and containing either no clay, kaolinite (1 to 20%), or montmorillonite (1 to 5%). Montmorillonite and, to a lesser extent, kaolinite protected all the fungi studied (i.e., F. annosus, T. paradoxa, B. cinerea, A. niger, P. marginata, T. viride, S. brevicaulis, P. blakesleeanus, Schizophyllum sp., Chaetomium sp.) against inhibitory or toxic concentrations of Cd. For example, in the presence of 100 ppm Cd, growth of S. brevicaulis was 4% of the control (i.e., no Cd + no clay) on agar not amended with clay, 11% of control (i.e., no Cd + 2% kaolinite) on agar amended with 2% kaolinite, and 48% of control (i.e., no Cd + 2% montmorillonite) on agar amended with 2% montmorillonite. A similar protective effect of the clay minerals was noted with the bacteria, B. megaterium, A. tumefaciens, and N. corallina. Increasing the concentration of either clay lessened the inhibitory effects of Cd, with montmorillonite providing greater protection than equivalent or even high concentrations of kaolinite; e.g., in the presence of 1000 ppm Cd, growth of Schizophyllum sp. occurred on agar amended with 3 or 5% montmorillonite but not on agar amended with no clay or with up to 20% kaolinite (3).

The greater protection provided by montmorillonite, as compared to kaolinite, was correlated with the higher cation exchange capacity (C.E.C.) of montmorillonite. The C.E.C. reflects the capacity of the clay to exchange cations on the exchange complex for cations in the ambient environment, and, by this exchange mechanism, both kaolinite and montmorillonite were apparently able to exchange Cd in the medium for cations on the clay surface. Montmorillonite, the clay with the higher C.E.C., apparently removed more exogenous Cd than did kaolinite at equivalent concentrations. The Cd adsorbed was presumably less available for microbial uptake than Cd in solution, thus, accounting for the protective effect (3). Studies by others have also shown that kaolinite (1, 10) and montmorillonite (9) adsorb exogenous Cd, with montmorillonite being the more efficient exchanger (18). Clays homoionic to Cd did not protect fungi (e.g., A. niger, P. marginata) against exogenous Cd, as the exchange complex was already saturated with Cd. Furthermore, montmorillonite homoionic to

Cd was more detrimental to fungal growth than were equivalent concentrations of kaolinite homoionic to Cd, inasmuch as more Cd was present on, and apparently desorbed from, montmorillonite (3).

In studies designed to evaluate the effects of Cd on growth of fungi in soil, R. stolonifer, T. viride, F. oxysporum f. conglutinans, C. echinulata, and several species of Penicillium and Aspergillus tolerated higher concentrations of Cd in soil than when grown on synthetic medium (4). Several soil factors may have been responsible for lessening the toxicity of Cd: the Cd ions may have complexed with organic chelating agents, such as humic acids (17), and they may have been adsorbed to the clay minerals and organic matter in soil (1, 13). Studies with plants have also shown that the uptake of Cd was inversely related to the organic matter and clay mineral contents of soil (11, 12, 14), and studies with a green alga, Chlorella sp.,have shown that the toxicity of Cd was lessened by addition of chelators such as EDTA (20).

When the soil was amended with several concentrations (3, 6, 9, or 12%) of clay minerals, montmorillonite provided partial or total protection against the fungistatic effects of Cd, whereas kaolinite afforded little or no protection. Incorporation of montmorillonite significantly increased the C.E.C. of the soil, and increased protection provided by these clay-amended soils was apparently related to their capacity to adsorb, by cation exchange, greater quantities of exogenous Cd. Additions of kaolinite, however, did not appreciably affect the C.E.C. of the soil, thus, accounting for the lesser ability of these soils to protect against fungistatic concentrations of Cd (4).

The toxicity of Cd to the microbiota is apparently mediated by the clay mineralogy of the environment into which the pollutant is deposited. Other studies have shown that various environmental factors, such as pH (2, 4) and the levels of zinc (15, 16), magnesium (15), iron, manganese (20), cysteine (19), and chelating agents (20), also influence the toxicity of Cd to microbes. As the toxicity of Cd is apparently dependent on the physicochemical characteristics of the environment into which it is deposited, the influence of the abiotic physicochemical factors on the impact of Cd and other pollutants on the biosphere (5) should be evaluated.

## A. References


1. Andersson, A., Nilsson, K.O.: Influence of lime and soil pH on Cd availability to plants. Ambio 3, 198-200 (1974).
2. Babich, H., Stotzky, G.: Sensitivity of various bacteria, including actinomycetes, and fungi to cadmium and the influence of pH on sensitivity. Appl. Environ. Microbiol. 33, 681-695 (1977).
3. Babich, H., Stotzky, G.: Reductions in the toxicity of cadmium to microorganisms by clay minerals. Appl. Environ. Microbiol. 33, 696-705 (1977).
4. Babich, H., Stotzky, G.: Effect of cadmium on fungi and on interactions between fungi and bacteria in soil: influence of clay minerals and pH. Appl. Environ. Microbiol. 33, 1059-1066 (1977).
5. Babich, H., Stotzky, G.: Effects of cadmium on the biota: influence of environmental factors. In: Advances in Applied Microbiology, Vol. 22, Perlman, D. (ed.). New York: Academic Press, (1978). In press.
6. Fleisher, M., Sarofim, F., Fassett, D.W., Hammond, P., Shacklette, H., Nisbet, I.C.T., Epstein, S.: Environmental impact of cadmium. In: Low Level Lead Toxicity and the Environmental Impact of Cadmium. Environmental Health Perspectives, Experimental Issue No.7 Washington, D.C.: U.S. Gov't. Printing Office, pp. 253-323 (1974).

7. Flock, D.F., Kraybill, H.F., Dimitroff, J.M.: Toxic effects of cadmium: A review. Environ. Res. 4, 71-85 (1971).

8. Friberg, L., Piscator, M., Nordberg, G.F., Kjellstrom, T.: Cadmium in the Environment. Cleveland, Ohio: CRC Press.

9. Garcia-Miragaya, J., Page, A.L.: Influence of ionic strength and inorganic complex formation on the sorption of trace amounts of Cd by montmorillonite. Soil Sci. Soc. Amer. J. 40, 658-663 (1976).

10. Gardiner, J.: The chemistry of cadmium in natural water. II. The adsorption of cadmium on river muds and naturally occurring solids. Water Res. 8, 157-164 (1974).

11. Haghiri, F.: Plant uptake of cadmium as influenced by cation exchange capacity, organic matter, zinc, and soil temperature. J. Environ. Qual. 3, 180-182 (1974).

12. John, M.K.: Influence of soil characteristics on adsorption and desorption of cadmium. Environ. Lett. 2, 173-179 (1971).

13. John, M.K.: Cadmium adsorption maxima of soils as measured by the Langmuir isotherm. Can. J. Soil Sci. 52, 343-350 (1972).

14. John, M.K., Van Laerhoven, C.J., Chuah, H.H.: Factors affecting plant uptake and phytotoxicity of cadmium added to soils. Environ. Sci. Technol. 6, 1005-1009 (1972).

15. Laborey, F., Lavollay, J.: Sur la nature des antagonismes responsables de l'interaction des ions $Mg^{++}$, $Cd^{++}$, et $Zn^{++}$ dans la croissance d'Aspergillus niger. Comptes Rendus Seances Ser. D 276, 529-532 (1973).

16. Mitra, R.S., Gray, R.H., Chin, B.,Bernstein, I.A.: Molecular mechanisms of accommodation in Escherichia coli to toxic levels of $Cd^{2+}$. J. Bacteriol. 121, 1180-1188 (1975).

17. Stevenson, F.J.: Stability constants of $Cu^{2+}$, $Pb^{2+}$, and $Cd^{2+}$ complexes with humic acids. Soil Sci. Soc. Amer. J. 40, 665-672 (1976).

18. Sweeton, F.H., Tamura, T.: Adsorption of $Cd^{2+}$ from $Ca^{2+}$ solutions by clay minerals. Agron. Abstr. p. 34 (1975).

19. Tynecka, Z., Zylinska, W.: Plasmid borne resistance to some inorganic ions in Staphylococcus aureus. Microbiol. Polonica Ser. A7, 11-20 (1974).

20. Upitis, V.V., Pakalne, D.S., Nollendorf, A.F.: The dosage of trace elements in the nutrient medium as a factor in increasing the resistance of Chlorella to unfavorable conditions of culturing. Microbiology 42, 758-762 (1973).

# *In situ* Methylation of Mercury by Estuarine Sediments

B.H. OLSON

## A. Introduction

Jensen and Jernelov (8) provided conclusive evidence that mercury (Hg) could be biologically converted in sediments to the more toxic organic form methylmercury ($CH_3Hg$). Methylation of mercury has been shown to occur in Mobile Bay, Alabama, and in San Francisco Bay sediments (1, 9). Methylation of mercuric chloride in situ in estuarine sediments occurs under both aerobic and anaerobic conditions (9). Organic content and particle size distribution have been shown to be important factors in determining the ability of estuarine sediments to methylate mercury under laboratory conditions (10, 11).

The purpose of the present investigation was to examine the relationship of the methylation process to the biogeochemical cycling of mercury in the estuarine environment.

## B. Materials and Methods

In situ methylation of Hg was studied from July 18, 1975 to July 18, 1976 in sediment of an intertidal zone of Upper Newport Bay, Newport Beach, California. Wet sediment was collected, taken to the laboratory and put through a 2 mm sieve, and mercury as mercuric chloride ($HgCl_2$) dissolved in deionized water, added to each 1500 g lot to give a final mercury concentrations of 0, 10, 50, and 100 $\mu g$ Hg $g^{-1}$ respectively. Each sediment lot was then placed into a cylindrical glass ring 14.3 cm in diameter and 7.5 cm in height and implanted in the Bay intertidal zone within 12 hours of collection. When total mercury values had returned to background levels for all three concentrations of Hg added, a new experiment was initiated. Five consecutive experiments were conducted.

Sediment samples for analysis for total and methylmercury were collected at 3-7 day intervals throughout the study except during October 1975 and April 1976; a total of 75 samplings. All sediment samples were analyzed, for total Hg using the Isotope Shift Zeeman Effect Atomic Absorption Spectrometer (IZAA) (7), and for $CH_3Hg$ by a modification of the Westöö method (18). The mean extraction efficiency was 83.5% and the limit of detection was .01 ng of Hg.

At the beginning and end of each of the five experiments 500 g of additional sediment was collected and stored at -4$^0$C and these were analyzed for volatile solids, chemical oxygen demand, chloride content, organic and ammonia nitrogen and water content using standard methods (16). Particle size distribution was determined using a Hiac Model PC-320 particle size analyzer. Redox potential was determined in situ, 6 mm, 45 mm and 85 mm below the surface using a platinum electrode and Orion Specific Ion Meter (Model 401). The temperature of the sediment was recorded during sample collection.

## C. Results

I. Physical and chemical properties of the sediment. The physical and chemical properties of the sediment are shown in Table 1. There was little variation in the volatile solids component of the sediment during the course of the year with a total increase in volatile material of 1.6%. The total chlorides in the sediment showed an increase through experiments 1-4 and then decreased during experiment 5 because of increased irrigation of the land surrounding the Bay which drains into the Upper Newport Bay. The percent of water within the sediments varied by a maximum of 10% over the course of the year. The background concentration of Hg in the sediments also fluctuated during the year by a factor of 0.3μg Hg $g^{-1}$.

The carbon and nitrogen content of the sediment was extremely high, suggesting that the sediment is undergoing rapid decomposition, with a consequent high rate of nitrogen release and high oxygen demand. The latter suggestion was confirmed by the redox potential of the sediment.

The particle size distribution of the sediment over the course of the study for the diameter range of 2.5 - 125 μ is shown in Table 2. These data indicated that greater than 99% of the particles were less than 75 μ in diameter. The clay content or the percentage of particles < 2.5 μ in diameter was 76.1% by weight. The redox potential (Eh) of the sediment was -170 mv at a depth of 6 mm and -200 mv at depths of 45 and 75 mm. The temperature of the sediment varied from 14°C to 25°C.

II. Behavior of mercury added to the sediment. Over the year a total of 300 sediment samples were analyzed for methyl and total mercury. The concentration of total Hg within the sediment during experiment 1 is shown in Figure 1. For all three concentrations of $HgCl_2$ added (10, 50, and 100 μg Hg $g^{-1}$) a rapid decrease in mercury concentration occurred during the first week after mercury addition, followed by a slow but continual decrease in Hg concentration until background levels were reached.

The rate at which Hg dissipated from sediments varied over the course of this study (Table 3). The data indicated that although the amount of Hg added varied the time required for the Hg to return to background levels was similar and in some instances the same for all three concentrations added.

Table 1. Physical and chemical properties of Upper Newport Bay sediment

| | July 18-Sept. 12, 1975 | Nov.6-Dec. 12, 1975 | Dec.15,1975- Jan.14,1976 | Jan.23-March 29, 1976 | April 30- July 18,1975 |
|---|---|---|---|---|---|
| Experiment | (1) | (2) | (3) | (4) | (5) |
| No. of collection properties | 19 | 9 | 8 | 16 | 23 |
| Volatile solids (%) | 5.0 | 5.7 | 5.2 | 5.1 | 6.7 |
| Chemical oxygen demand $g^{-1}$ | 41.7 | 45.7 | 40.4 | 45.1 | 52.2 |
| Chloride mg $g^{-1}$ | 4.6 | 6.8 | 8.3 | 8.8 | 7.1 |
| Total kjeldahl nitrogen mg $g^{-1}$ | 870 | 470 | 520 | 480 | 810 |
| Water (%) | 43.4 | 45.2 | 53.9 | 50.7 | 49.2 |
| Total mercury μg $g^{-1}$ | 0.3 | 0.5 | 0.2 | 0.4 | 0.3 |
| Methylmercury ng $g^{-1}$ | 0.2 | 0.3 | 1.6 | 0.8 | 2.0 |

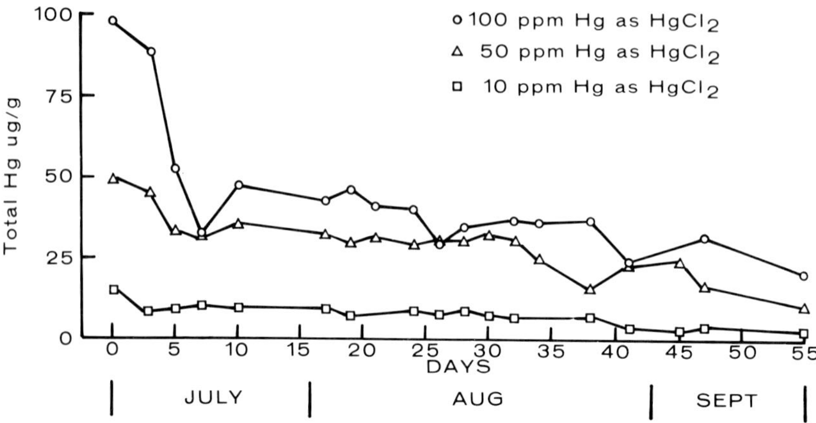


Fig. 1. Release of total mercury from sediment (July 18, 1975 to Sept. 12, 1975) to which 10, 50, and 100 μg Hg (as $HgCl_2$) $g^{-1}$ was added

Table 2. Particle size distribution of Upper Newport Bay sediment

| Experiment | Distribution (%) Particle Diameter | | | |
|---|---|---|---|---|
| | 2.5-10μ | 11-25μ | 26-75μ | 76-125μ |
| 1 | 93.47 | 5.94 | 0.59 | 0.00 |
| 2 | 89.25 | 9.72 | 0.95 | 0.07 |
| 3 | 93.62 | 5.30 | 1.05 | 0.03 |
| 4 | 91.48 | 7.34 | 1.26 | 0.03 |
| 5 | 91.61 | 6.87 | 1.35 | 0.17 |

Table 3. Time (in days) taken to return to background concentrations of total Hg following addition of mercuric chloride

| Experiment | Sediment | | |
|---|---|---|---|
| | 10 μg Hg $g^{-1}$ | 50 μg Hg $g^{-1}$ | 100 μg Hg $g^{-1}$ |
| 1 | 76 | 76 | 76 |
| 2 | 39 | 39 | 39 |
| 3 | 30 | 39 | 39 |
| 4 | 54 | 65 | 65 |
| 5 | 76 | 79 | 79 |

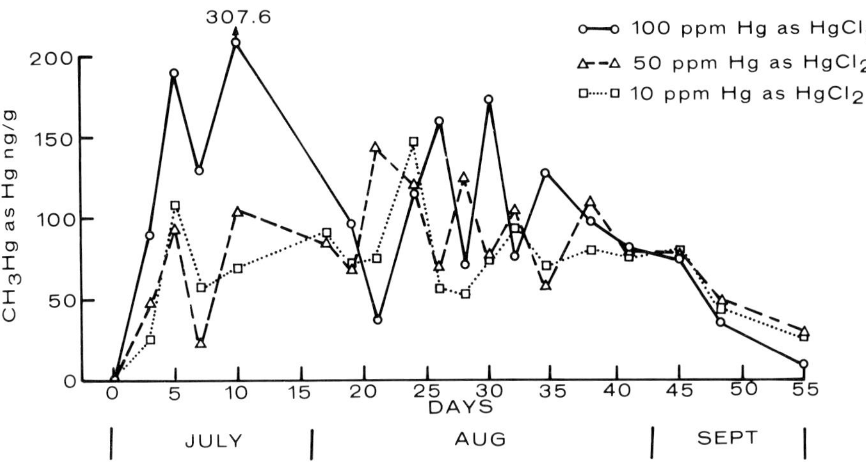


Fig. 2. Net methylmercury production from July 18, 1975 through Sept. 12, 1975 by estuarine sediment

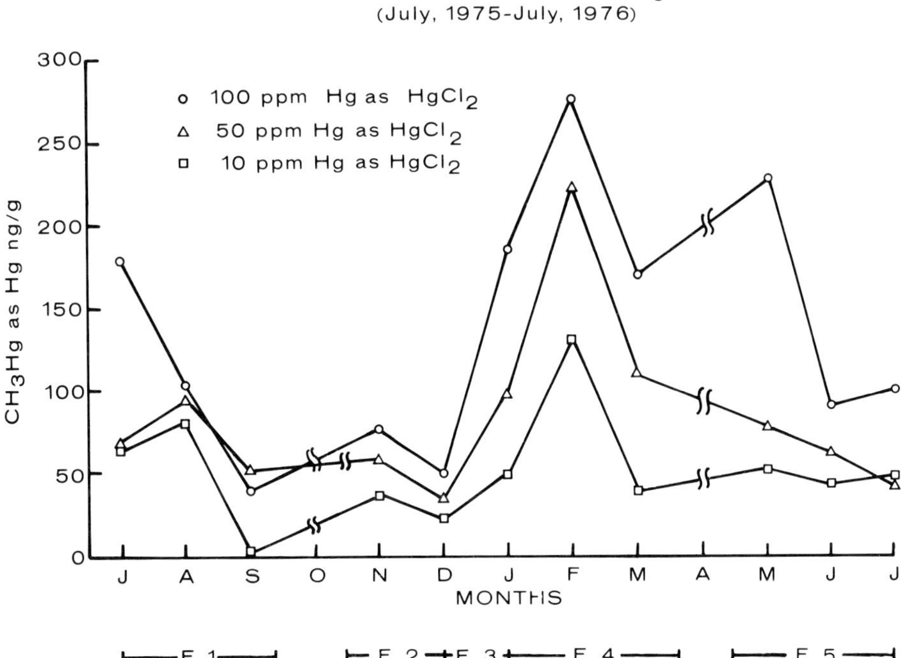


Fig. 3. Comparison of seasonal variation in net methylmercury production by sediment spiked with 10, 50 or 100 μg Hg (as $HgCl_2$) $g^{-1}$

III. Methylation properties of the sediment. As is shown in Figure 2 the production of $CH_3Hg$ in experiment 1 for all levels of Hg added, was cyclical and the pattern was similar for all five experiments. Peak net $CH_3Hg$ production occurred during day 10 of experiment 1 for the sample to which 100 $\mu g\ Hg\ g^{-1}$ added, and day 23 with 10 $\mu g\ Hg\ g^{-1}$ added. The day of peak $CH_3Hg$ production varied throughout the remaining experiments from a minimum of 8 days after the addition of $HgCl_2$ to a maximum of 76 days.

The mean monthly net $CH_3Hg$ production by sediment to which 10 $\mu g\ Hg\ g^{-1}$ was added indicated a seasonal variation in net $CH_3Hg$ production with peaks in August 1975 and February 1976 (Fig. 3). The same trend was shown in sediment to which 50 $\mu g\ Hg\ g^{-1}$ had been added. In sediment to which 100 $\mu g\ Hg\ g^{-1}$ was added the pattern of net $CH_3Hg$ production over the year was slightly different with peak production immediately after the introduction of mercury in July 1975, a second peak in February 1976, with a third peak in May, 1976. The greatest production of $CH_3Hg$ occurred in February 1976 in all treatments.

## D. Discussion

I. Behavior of total mercury in sediment. The only other work investigating the *in situ* capabilities of sediments to methylate mercury was 28 days in duration (9). The pattern of Hg loss from the sediment (Table 3) was similar for the two studies.

Studies on Rhine River sediment (4) showed that the best correlations between mercury concentration and particle size was up to a diameter size of 16 μ, but other reports suggest that higher concentrations of mercury were present on clay particles (< 2μ) than on silt particles (2-50 μ) (2). Greater than 90% of the particles examined in the present study could be placed in these size ranges which would concentrate mercury (Table 2). The data presented have indicated however that rate of release may be more a function of time of year than particle diameter size.

II. Naturally occurring methylmercury in sediment. Comparison of the data from this study with other *in situ* work (1, 9) indicates that background levels of $CH_3Hg$ were similar.

III. Methylmercury production. The amount of $CH_3Hg$ produced in this study was much greater than that reported in laboratory studies on other estuarine sediments and was most likely due to a combination of factors; the high chemical oxygen demand and total kjeldahl nitrogen in the sediment. The values for these two factors were three orders of magnitude greater than those for intertidal San Francisco Bay sediment (10).

1. Fluctuation in methylmercury production. The fluctuations in $CH_3Hg$ production in the sediment under natural conditions (Fig.2) indicated a close parallel to those shown in laboratory studies using fresh water and estuarine sediments (10, 11, 14). It is still not clear whether these fluctuations in the amount of $CH_3Hg$ over short periods are due to release of $CH_3Hg$ from the sediment at different rates, evolution of $Hg^0$ from demethylated mercury, or from decreased rates of $CH_3Hg$ production by micro-organisms.

Recent work concerning each of these mechanisms has produced interesting findings. *Proteus* sp., isolated from Upper Newport Bay sediment - grown under aerobic and anaerobic conditions showed net $CH_3Hg$ fluctuation over a period of days (12). Similar results were shown in experiments with *Enterobacter aerogenes* (6). The evolution of elemental mercury from estuarine sediments has been measured *in situ* at the mudwater interface (Colwell, personal communication) and the results should prove of interest. That demethylation of mercury is carried out by a number of bacteria that are frequently isolated from soil and sediments is now established (3, 15, 17). These data indicate that in natural environments it is likely that all three mechanisms can operate and that fluctuations in measured $CH_3Hg$ are due to one or a combination of these factors.

2. Seasonal effects. The data in Figure 3 indicates that the addition of mercury to the sediments at various times during the year produced markedly different responses in terms of $CH_3Hg$ production. This is evident from the large response in net $CH_3Hg$ production exhibited during July for sediment spiked with a 100 μg $Hg g^{-1}$ when compared with $CH_3Hg$ data after the addition of mercury in November and December. Similar results were found in sediments to which 10 μg $Hg g^{-1}$ or 50 μg $Hg g^{-1}$ had been added. A similar yearly pattern in net $CH_3Hg$ production by sediment was found at all three mercury concentrations.

Seasonal changes in the chemical and physical factors under study did not seem to be correlated with changes in methylmercury production. The highest level of methylmercury production occurred during February when both chemical oxygen demand and total kjeldahl nitrogen were low. Temperature which has been shown to be a factor in methylation rates in the laboratory (10, 13) did not seem to be a governing factor during in situ mercury methylation processes. The location of the study site in an area which has mild winters accounted for the similarity of temperature over the year.

3. Concentration of mercury in the sediment and methylmercury production. The data indicate that the concentration of Hg present is not correlated with the production of $CH_3Hg$. An example of this is shown comparing Figures 1 and 2, $CH_3Hg$ production ranges in August between 72.4 ng Hg $g^{-1}$ and 174.8 ng Hg $g^{-1}$ while total mercury fluctuates between 34.6 and 36.8 μg Hg $g^{-1}$ in sediment to which 100 μg Hg $g^{-1}$ had been added. As mentioned previously, the fluctuations in $CH_3Hg$ production most likely occur due to several environmental factors. Perhaps seasonal selection of microbial populations has a larger role in determining mercury methylation rates than does the concentration of mercury added. Maximum production of $CH_3Hg$ occurred however, at the highest Hg concentration. This agrees with previous findings (10, 11).

4. Redox potential. It can be concluded that the methylation of Hg was primarily anaerobic because only the uppermost level of the sediment (0-3 mm) was aerobic. In highly reduced medium under anaerobic conditions $CH_3Hg$ will be produced in the absence of microorganisms, while there are only trace quantities of $CH_3Hg$ produced with the same medium under anaerobic conditions (12) and may suggest in an anaerobic environment such as Upper Newport Bay mercury methylation may be partly abiotic.

This study has shown that there are seasonal fluctuations in the methylation of mercury in estuarine sediments. These fluctuations seem to be independent of chemical oxygen demand and kjeldahl nitrogen content of the sediment. In the sediment under study the methylation process was anaerobic. Further in situ investigations are needed to elucidate the cause(s) of the cycling of $CH_3Hg$ production in sediments, and in situ work should be extended to other metals such as arsenic and selenium which have been shown to be methylated in sediments.

## E. References


1. Andren, A.W. & Harriss, R.C.: Methylmercury in estuarine sediments. Nature (Lond.) 245, 156-257 (1973).
2. Armstrong, F.A.J., Metner, D. & Capel, M.J.: Mercury in sediments and water of Clay Lake, Northwestern Ontario. Fisheries Research Board of Canada (1972).
3. Billen, G., Joiris C., & Wollast, R.: A bacterial methylmercurymineralizing activity in river sediments. Water Res. 8, 219-225 (1974).
4. de Groot, A.J., de Goeij, J.J.M. & Zegers, C.: Contents and behavior of mercury as compared with other heavy metals in sediments from the Rivers Rhine and Ems. Geologie en Miujbouw 50, 393-398 (1971).
5. Fagerström, T. & Jernelov, A.: Some aspects of the quantitative ecology of mercury. Water Res. 6, 1193-1202 (1972).
6. Hamdy, M.K. & Noyes, O.R.: Formation of methylmercury by bacteria. Appl. Microbio. 30, 424-432 (1975).
7. Hadeishi, T. & McLaughlin, R.I.: Hyperfine zeeman effect atomic absorption spectrometer for mercury. Science 174, 404-407 (1971).
8. Jensen, S. & Jernelov, A.: Biological methylation of mercury in aquatic organisms. Nature (Lond.) 223, 753-754 (1969).
9. Olson, B.H. & Cooper, R.C.: In situ methylation of mercury by estuarine sediment. Nature (Lond.) 252, 682-683 (1974).
10. Olson, B.H. & Cooper, R.C.: Comparison of aerobic and anaerobic methylation of mercuric chloride by San Francisco Bay sediments. Water Res. 10, 113-116 (1976).
11. Pelo, M., Rissanen, K., Sundman, V., & Miettinen, J.K.: Microbial methylation of inorganic ionic mercury in bottom muds. First National Days of Biophysics and Biotechnique (Helsinke) (1973).
12. Reich, K. & Olson, B.H.: Mercury methylation by a bacteria isolated from estuarine sediment. Bact. Proceedings, p. 241 (1977).
13. Shin, L. & Krenkel, P.A.: Mercury uptake by fish and biomethylation mechanisms. J.W.P.C.F. 48, 473-501 (1976).
14. Spangler, W.J., Spigarelli, J.L., Rose, J.M., & Miller, H.M.: Methylmercury: bacterial degradation in lake sediments. Science (N.Y.) 180, 192-193 (1973a).
15. Spangler, W.J., Spigarelli, J.L., Rose, J.M., Flippin, R.S. & Miller, H.M.: Degradation of methylmercury by bacteria isolated from environmental sample. Appl. Microbio. 25, 448-493 (1973b).
16. Standard Methods for the Examination of Water and Waste Water. A.P.H.A., A.W.W.A., W.P.C.F., 13th edition, American Public Health Association (1971).
17. Tonomura, K., Nakagami, T., Futai, F., & Maeda, K.: Studies on the action of mercury-resistant microorganisms on mercurials (1) The isolation of mercury resistant bacterium and the binding of mercurials to the cells. J. Ferment. Tech. 46, 506-512 (1968).
18. Westöö, G.: Determination of methylmercury salts in various kinds of biological material. Acta Chem. Scand. 22, 2277-2280 (1968).

# Microbiological Observations and Chemical Analyses of Tile Line Drainage Waters and Deposits in Imperial Valley, California

J.P. MARTIN, D.S. DUMKE, B. MEEK, J.O. ERVIN, and L. GRASS

## A. Introduction

To maintain high crop yields in Imperial Valley, California it is necessary to install drainage lines to control a high water table and to prevent excessive salinity. After installation the deposition of iron and manganese oxides in and near the tile joints and in the lines may reduce the efficiency of the drainage systems (2). There is also the possibility that heterotrophic organisms growing in the joints and lines or adsorbed on the deposits and utilizing dissolved organic substances in the waters could contribute to the clogging. A microbiological and chemical study was, therefore, undertaken in order to gain a better understanding of the nature of these deposits.

## B. Methods

Tile line deposit and water samples were collected in sterile bottles after digging down to the lines with a back hoe and breaking a hole in the tile or by inserting the bottle 4 ft into the lines at sump outlets. Saturation extracts of soil samples were also made at 1 ft intervals from the surface to the drainage line. Heterotrophic bacterial and fungal populations were estimated by dilution plating. Deposits were analysed for organic and carbonate C, Fe, Mn, Ca and Mg by conventional procedures. Concentrated and desalted water samples were analysed for total carbohydrates, amino acids, organic N, phenols, organic acids, and total organic C. IR spectra were run on freeze-dried samples. Deposits were examined under oil immersion and by scanning electron micrography.

## C. Results and Discussion

The drainage line water samples usually contained <1 to 4 ppm Fe, but after addition of organic residues to the surface the Fe increased up to 15 ppm. Mn values normally ranged from <1 to 10 ppm, but occasionally exceeded 25 ppm. Organic C varied from 2 to 20 ppm but increased up to 143 ppm after organic residue applications to the soil surface. Soluble C in the soil profile was generally higher, especially in the top soil.

The black deposits contained 15 to 57% Mn, <1 to 8% Fe, 2 to 27% Ca and only 0.1 to 0.4% organic C. The red deposits contained 9 to 41% Fe, <1 to 9% Mn, 2 to 20% Ca and 0.8 to 2.5% organic C.

The tile line waters and soil solutions contained only traces of alipathic organic acids, amino acids, and phenols. The C/N ratio varied from 6-13. Polysaccharides accounted for 7 to 34% of the soluble C as estimated by the anthrone method after removal of interfering substances. On the basis of movement through biogel columns the molecular size of the organic constituents was about 10,000 or less. After organic additions to the soil surface the size increased to 15,000. The

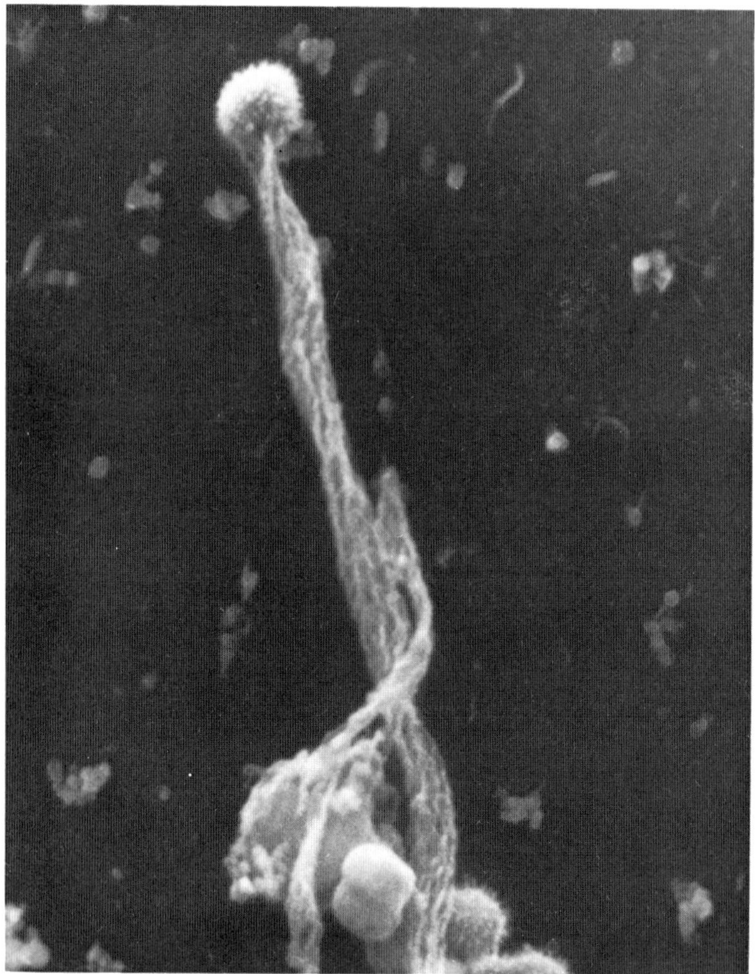

Fig. 1. Scanning electron micrograph of Gallionella sp. in red (iron oxide) deposit in tile drainage line of Imperial Valley, Calif. X 10,000

polysaccharide contents of the higher molecular weight fractions ranged up to 50%. Uronic acids estimated by the carbazole method varied from 1 to 6%.

IR spectra of freeze-dried soluble organic matter from the tile line waters and soil profile samples were characteristic of complex polysaccharides and soil humic or fulvic acids (1, 6). The bulk of the soluble C, therefore, appeared to be in the form of polysaccharides and soil fulvic acid type molecules and small molecular weight polysaccharides represent the most likely energy source for heterotrophic microorganisms.

The soils of the valley contained from 21 to 220 thousand fungal propagules per gram, the black deposits 0 to 99,000, the red deposits 0 to 79,000 and no fungi were isolated from the tile line waters. The dominant fungi from both soils and lines were: Aspergillus niger, A. terreus, A. nidulans, A. glaucus, A. carneus, A. Sydowi, A. ochraceus, Penicillium nigricans, P. decumbens, P. funiculosum,

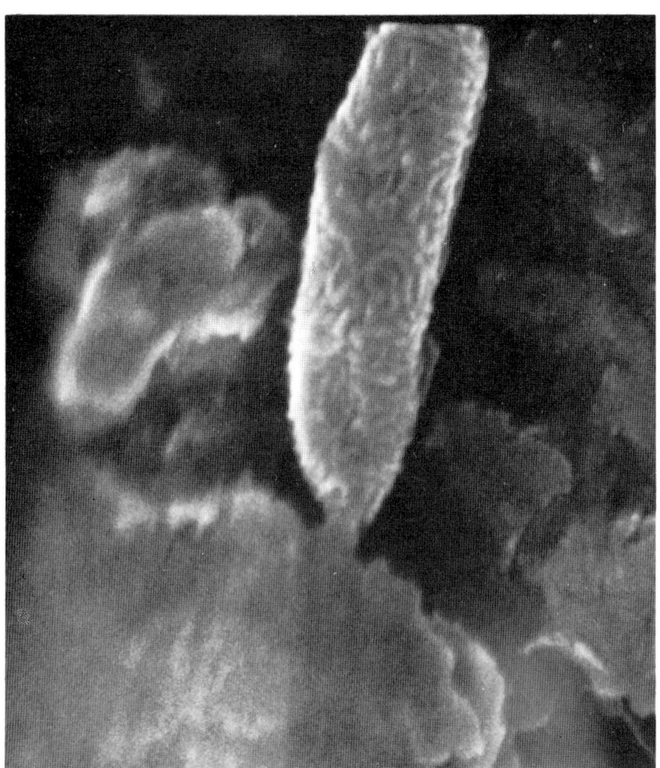

Fig. 2. Scanning electron micrograph of a bacterium attached to a $MnO_2$ deposit in tile drainage line of Imperial Valley, Calif.

Fusarium solani, Fusarium sp., Chaetonium globosum, Acremonium roseum, Cladosporium sp., Verticillium sp., Mucor sp., Stachybotrys atra and Alternaria sp. The fungus species found in the deposits were the same species as noted in the top soils. In the soils, however, 5 to 10 or more species were dominant while in the line deposits only one and rarely two or three species were isolated from any single sample. The soils contained from 1 to 61 million bacteria plus actinomycetes, the black deposits 0 to 44,000; the red deposits 0 to 9,000,000 and the waters 0 to 1,000 per gram.

Microscopic and electron micrograph examinations of the Mn deposits showed numerous bacteria present within the precipitate and attached to the surfaces at one end of rod shaped cells. A few were suggestive of the hyphomicrobia as observed by Tyler and Marshall (7). The red deposits consisted almost entirely of the twisted ferric hydrate ribbons of Galionella species (Fig. 1). Only occasionally were fungal mycelia noted.

The low organic C contents (0.1 to 2.5%), the microscopic examinations of the tile line deposits, the low numbers of bacteria developing on dilution plates, and a previous study of Mn oxidation under simulated tile line conditions (4) suggest that Mn and Fe oxidizing bacteria are more important than common heterotrophic organisms utilizing dissolved organic substrates in clogging the lines. They also suggest an autotrophic oxidation. Gallionella species are generally considered to be autotrophs, but the evidence for an autotrophic metabolism for Mn oxidizing bacteria is not definite (3, 5). Some cells attached to small Mn particles appeared to be motile and moved pulling the Mn material with them (Fig. 2). It is not known, however, whether the organisms were heterotrophs using the particle as a solid surface for attachment or whether they were involved in the precipitiation of the $MnO_2$.

Although the Fe deposits contained more organic C than the Mn the numbers of bacteria were much lower. This may suggest that the Fe complexed with some of the organic compounds and made them resistant to decomposition.

## D. References


1. Filip, Z. Haider, K., Beutelspacher and Martin, J. P. : Comparison of IR spectra from melanins of microscopic soil fungi, humic acids and model polymers. Geoderma 11: 37-52 (1974).
2. MacKenzie, A. J. : Chemical treatment of mineral deposits in drain tile. J. Soil Water Conserv. 17: 124-125.(1962).
3. Marshall, K. C. : Manganese minerals. Chapter 5 In Biological factors in mineral cycling. P. A. Trudinger and D. J. Swaine (eds.) Elsevier, Amsterdam (In press) (1977).
4. Meek, B. D., Page, A. L. and Martin, J. P. The oxidation of divalent manganese under conditions present in tile lines as related to temperature, solid surfaces, microorganisms and solution chemical composition. Soil. Sci. Amer. Proc. 37: 542-548 (1973).
5. Mulder, E. G. Le cycle biologique tellurique et a quantique du fer et du manganese. Reveu D'ecologie et de Biologie du sol. 9: 321-348 (1972).
6. Stevenson, F. J. and Goh, K. M. Infrared spectra of humic acids and related substances. Geochimica et cosmochimica Acta. 35: 471-483 (1971).
7. Tyler, P. A. and Marshall, K. C. Form and function in manganese-oxidizing bacteria. Archiv. für Mikrobiologie 56: 344-353 (1967).

# Effects of Condensed Tannins on the Growth of Microorganisms

W. D. GRANT and C. M. McMURTRY

## A. Introduction

Conifer barks are rich in condensed tannins which are readily leached out by rainfall. As tannins can inhibit the growth of microorganisms (1) their presence could upset the balance of microbial populations and hence the overall ecology of soil and natural waters near waste bark heaps. We have begun an investigation on the effects of tannins on the growth and morphology of several algae, protozoa and bacteria commonly found in soil or water.

## B. Materials and Methods

Condensed tannins were extracted and purified (2) from the bark of *Pinus radiata*, the predominant forestry species in New Zealand.

Each organism was grown under controlled conditions in defined mineral salts media (plus glucose, in the case of the bacteria) and the culture divided into several equal portions to which were added solutions of sterile condensed tannin to give final concentrations from 0.0005 to 0.05%. Growth was measured by direct cell counts (algae, protozoa) or turbidimetry (bacteria).

Algae were shaken at 56 rpm at 16$^{o}$C in light intensities of 1400 lux for *Chlamydomonas* and *Scenedesmus* or 2,500 - 2,600 lux for *Anabaena* and *Cryptomonas*. Bacteria were shaken at 80 rpm at 17$^{o}$C. Protozoan cells from broth cultures were inoculated into a suspension of *Escherichia coli* NCTC 5934 in sterile phosphate buffer (1.4 x $10^8$ cells $l^{-1}$), and incubated at 17$^{o}$C with shaking at 80 rpm. The culture was divided and tannins added as described.

## C. Results and Discussion

I. *Algae*. The growth patterns of four common freshwater algae in 0.0005 - 0.05% condensed tannins are shown in Fig. 1. Tannins at 0.005% completely inhibited *Chlamydomonas reinhardtii* and *Cryptomonas* sp. whereas this concentration only slightly affected the growth of *Scenedesmus* sp. and the blue-green alga (bacteria) *Anabaena flos-aquae*. A concentration of 0.05% tannins was required for complete inhibition of these species. The morphology of the three eukaryotes was severely distorted by tannins, even at concentrations which permitted growth to occur, but in *Anabaena* a shortening of filament length was the only change observed.

II. *Protozoa*. Growth of the ciliate *Tetrahymena pyriformis* on *E. coli* was only slightly affected by condensed tannins (Fig. 2) at 0.05%. At this or lower concentrations of tannins the cells were indistinguishable in shape and motility from those of the control culture. Preliminary results with *Acanthamoeba castellanii* grown on *E. coli* indicated that this amoeba also was unaffected by the lower concentrations of tannins.

III. *Bacteria*. The Gram-negative bacteria tested were only slightly affected by even the highest tannin concentration (0.05%). No lag periods were

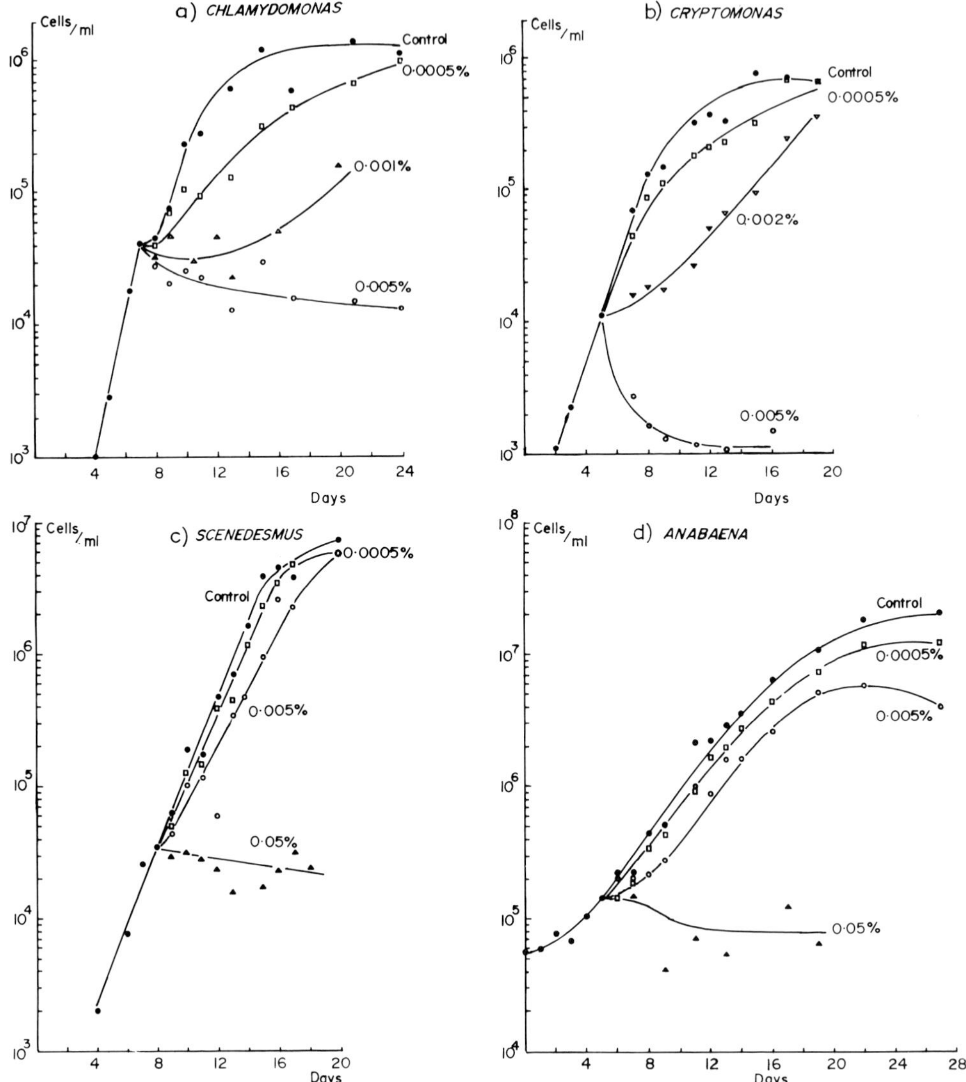


Fig. 1 a-d. Growth of (a) Chlamydomonas reinhardtii, (b) Cryptomonas sp., (c) Scenedesmus sp., (d) Anabaena flos-aquae in the presence of condensed tannins at concentrations from 0.0005% to 0.05%

observed after tannin addition. Typical growth curves are shown in Fig. 3a. In contrast, growth of three Gram-positive organisms ceased immediately upon addition of the tannins, and resumed after long lag periods, usually at altered rates. The three species gave different responses (Table 1) but in general the higher the tannin concentration the longer was the lag period, as can be seen in Fig. 3b. In every case in which the bacteria recovered, they were shown by microscopy and culturing on agar to be uncontaminated, and identical to the original organism.

In the presence of tannins, the bacteria tended to form clumps or tangled chains, although this was more marked in the Gram-negative than in the Gram-positive species,

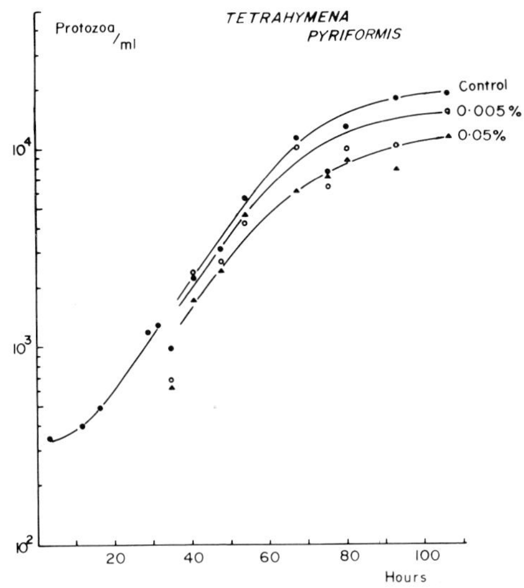


Fig. 2. Growth of Tetrahymena pyriformis on E. coli in the presence of condensed tannins. Growth curves obtained for cultures in 0.001% and 0.01% tannins were very similar to the 0.005% tannins curves, and have been omitted for clarity

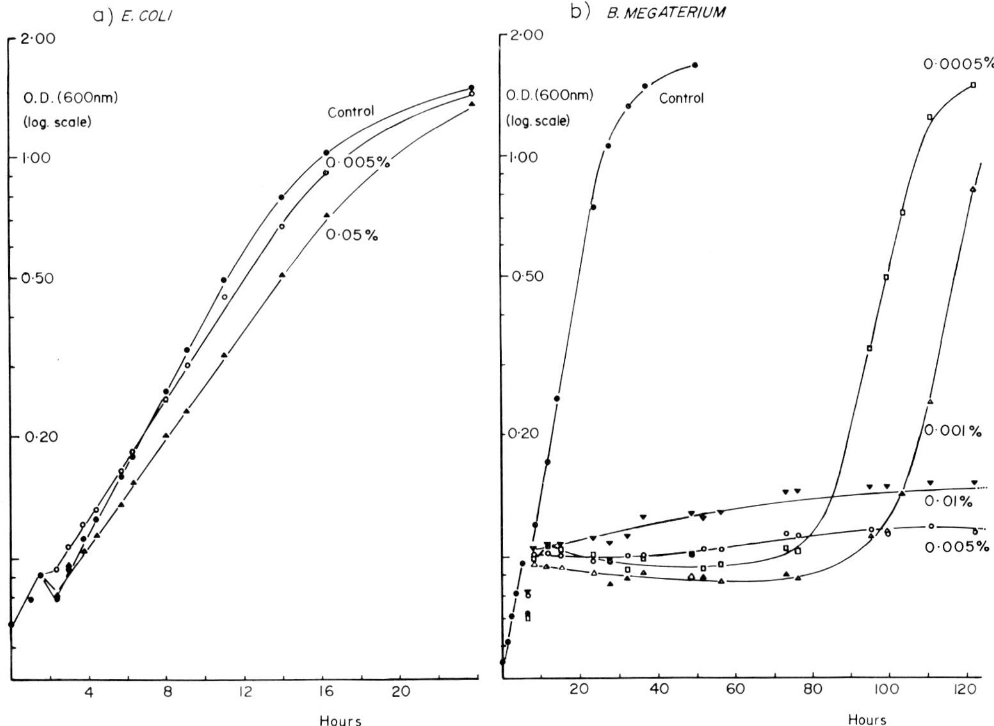


Fig. 3 a and b. Growth of (a) E. coli and (b) Bacillus megaterium in the presence of condensed tannins at concentrations from 0.0005% to 0.05%. Experimental points for a culture of E. coli in 0.01% tannins were virtually identical with those of the 0.005% tannins culture, and have been omitted for clarity. The 0.01% and 0.005% tannin-containing cultures of B. megaterium recovered after very long lag periods. (See Table 1)

Table 1. Growth of bacteria in presence of condensed tannins at several concentrations. M.G.T. is Mean Generation Time. "Lag" is the time elapsed between addition of tannins and re-commencement of growth (this occurred with the Gram-positive species only)

| Bacteria | | Tannin Concentration (%) | | | | | |
|---|---|---|---|---|---|---|---|
| | | 0 | 0.0005 | 0.001 | 0.005 | 0.01 | 0.05 |
| Escherichia coli NCTC 5934 | MGT (h) | 3.4 | - | - | 4.1 | 4.1 | 4.5 |
| Klebsiella pneumoniae NCIB 9939 | MGT (h) | 2.8 | - | - | 2.8 | 2.8 | 2.8 |
| Pseudomonas fluorescens NCTC 10036 | MGT (h) | 2.8 | - | 3.3 | 3.4 | 3.3 | 3.9 |
| Pseudomonas putida NCIB 9941 | MGT (h) | 4.4 | - | - | 4.5 | 4.5 | 5.0 |
| Arthrobacter globiformis NCIB 8602 | Lag (h) | 0 | 70 | 110 | >390 | >390 | - |
| | MGT (h) | 15.0 | 11.5 | 11.5 | ? | ? | - |
| Bacillus megaterium NCTC 10342 | Lag (h) | 0 | 70 | 85 | 170 | 235 | - |
| | MGT (h) | 5.2 | 7.5 | 7.5 | 9.0 | 9.5 | - |
| Bacillus subtilis ATCC 6633 | Lag (h) | 0 | 0 | 15 | 180 | 250 | - |
| | MGT (h) | 10.5 | 14.5 | 14.5 | 14.5 | 12.5 | - |

and was probably the cause of a loss of motility by the two Pseudomonas species. Tannin-inhibited Bacillus subtilis was immotile during the lag periods, but regained motility when the cultures recovered.

Although it seems likely that the different reactions to tannins shown by the two groups of bacteria are explicable in terms of cell wall structure, further work with other Gram-positive species is required before this can be substantiated.

This initial survey has shown that there is a surprising range in sensitivity and types of response to condensed tannins among common soil and water microorganisms.

## D. References

1. Haris, Y., Tagari, H., Volcani, R.: Effect of water extracts of carob pods, tannic acid, and their derivates on the morphology and growth of microorganisms. Appl. Microbiol. 12, 204-209 (1964).

2. Grant, W.D.: Microbial degradation of condensed tannins. Science 193, 1137-1139 (1976).

# Microbiology of Food

# Foodborne Pathogens and Indicator Tests in Ready-to-Serve and Raw Foods

M. SOLBERG, B.A. PREVOST, and D.K. MISKIMIN

## A. Introduction

Indicator tests as guides for the microbial safety of food were most likely adopted after the apparently successful application of the concept to drinking water and pasteurized milk. Reviews on the subject have been published (2, 4). The hypotheses upon which water safety is based are: (a) the coli-aerogenes group of bacteria is derived from the same source as the pathogen *Salmonella typhi*, the intestinal tract; (b) the coli-aerogenes organisms are always present in the intestinal tract and will be present whenever the pathogen is present; (c) neither the pathogen nor the coli-aerogenes organisms will multiply in water and (d) the *S. typhi* will lose viability as fast or faster than the coli-aerogenes organisms in water. These criteria applied to water supplies provide assurance of the absence of *S. typhi* but no assurance that viruses, microbial toxins or other illness-causing vectors of modern society are absent.

The use of the coli-aerogenes group as an indicator for milk safety is based upon the hypotheses that: (a) milk is adequately pasteurized destroying all pathogenic and indicator organisms; (b) milk is packaged and protected from recontamination and (c) the coli-aerogenes organisms grow as rapidly as the pathogens. The coli-aerogenes group in pasteurized milk indicates improper pasteurization, excessive post-pasteurization contamination, or a small post-pasteurizing contamination followed by extended storage at improper temperatures. These factors provide some safety assurance but tell us nothing of the presence or absence of spore-formers, toxins and other disease vectors.

The use of indicator tests for raw and ready-to-serve foods has been promulgated by official bodies in the USA (5) and Canada (6) and are included in industrial specifications (1). The commonly employed indicator tests include the aerobic plate count (APC), the coliform count and the *Escherichia coli* count.

This study evaluates the relationships between the three common microbial vectors of food-borne illness, *Staphylococcus aureus*, *Salmonella* and *Clostridium perfringens*, and indicator tests in raw and ready-to-serve (RTS) foods in a multi-unit mass-feeding system. The study also provides some insight into those *Enterobacteriaceae* which produce typical colonies on Levine's Eosin Methylene Blue (EMB) Agar but which are not *E. coli*.

## B. Materials and Methods

Samples came from a mass-feeding system in a university where 13 to 15 facilities provide approximately 150,000 meals daily (8). Sample collection, treatment

*This research was supported in part by the U.S. Department of Agriculture, Northeast Regional Project NE-83 in Food Safety and Quality and in part by contract with Dining Services Division, Rutgers University, New Brunswick, N.J. U.S.A.

and analyses were conducted as described (3). Samples were evaluated from 1972 to 1977.

Unsatisfactory samples were all of those containing >10 $g^{-1}$ of S. aureus or C. perfringens and >3 $g^{-1}$ of Salmonella. Raw foods containing >$10^6$ $g^{-1}$ APC, >$10^3$ $g^{-1}$ coliforms or >10 $g^{-1}$ E. coli were unsatisfactory. RTS foods containing >$10^5$ $g^{-1}$ APC, >$10^2$ $g^{-1}$ coliforms or >3 $g^{-1}$ E. coli were unsatisfactory.

Enterobacteriaceae identification was made using the 11 test Enterotube system (Hoffman Laroche) (9) on colonies demonstrating green sheen or dark centers on EMB Agar plates.

## C. Results and Discussion

Raw ground beef: Table 1 shows that raw ground beef in bulk form had a tendency to contain unsatisfactory levels of food-poisoning organisms and indicator organisms with greater frequency than ground beef patties. This may be due to the frozen state of the patties and the unfrozen state of the bulk ground beef during distribution. Table 1 also indicates that 48% of all the raw ground beef samples contained S. aureus. Only 15 samples contained >100 S. aureus colony forming units (cfu) $g^{-1}$. Among those only 5 contained >1000 cfu $g^{-1}$ and the maximum level detected was 3 x $10^4$ cfu $g^{-1}$. None of the samples approached the $10^6$ cfu $g^{-1}$ or more which seem necessary to cause illness. Internal temperatures were recorded in a few samples of cooked hamburger. C. perfringens survival was observed at internal temperatures of 57-59°C whereas the highest internal temperature permitting recovery of S. aureus or Salmonella was 49°C.

It is evident from Table 1 that >10 $g^{-1}$ S. aureus, >1000 $g^{-1}$ coliforms and >100 $g^{-1}$ E. coli are commonly found in ground beef. It seems that cooking ground beef to internal temperatures above 60°C will reduce the frequency of unsatisfactory product significantly. The same cooking does not alter the frequency of C. perfringens detection markedly thus demanding that cooked ground beef patties not be held at ideal C. perfringens incubation temperatures (35-46°C) for long periods after initial preparation.

The efficiency of the indicator tests in detecting food poisoning microorganisms in ground beef is shown in Table 2. The data show that <10 $g^{-1}$ E. coli was

Table 1. Number of ground beef samples containing unacceptable levels of food-poisoning or indicator organisms

| Ground Beef Type | Number of samples | CP[a] | SA[b] | SAL[c] | APC[d] | CF[e] | EC[f] |
|---|---|---|---|---|---|---|---|
| Raw bulk | 75 | 18 | 38 | 6 | 17[g] | 28 | 36 |
| Raw patties | 73 | 11 | 33 | 2 | 9[h] | 27 | 31 |
| Cooked patties | 80 | 8 | 5 | 1 | 10 | 12 | 15 |

a. CP = Clostridium perfringens; b. SA = Staphylococcus aureus;
c. SAL = Salmonella; d. APC = Aerobic Plate Count (37°C, 24 h);
e. CF = Coliform Count (35°C, MPN); f. EC = Escherichia coli (EMB, Enterotube, MPN);
g. 5 samples over 5 x $10^6$ cfu $g^{-1}$, 4 samples over $10^7$ cfu $g^{-1}$;
h. 0 samples over 5 x $10^6$ cfu $g^{-1}$.

Table 2. Success of various indicator tests at rejecting samples containing specific food-poisoning microorganisms

| Ground Beef Type | C. perfringens | | | S. aureus | | | Salmonella | | |
|---|---|---|---|---|---|---|---|---|---|
| | APC[a] | CF[b] | EC[c] | APC | CF | EC | APC | CF | EC |
| Raw, Bulk | 12/18[d] | 15/18 | 16/18 | 13/38 | 22/38 | 28/38 | 1/6 | 5/6 | 4/6 |
| Raw, patties | 2/11 | 6/11 | 8/11 | 4/33 | 17/33 | 18/33 | 1/2 | 1/2 | 1/2 |
| Cooked patties | 2/8 | 5/8 | 6/8 | 4/5 | 5/5 | 5/5 | 1/1 | 0/1 | 1/1 |

a. APC = Aerobic plate count; b. CF = coliforms; c. EC = E. coli;

d. The lower value in the ratio indicates the number of samples containing the food-poisoning organism. The upper number in the ratio indicates the number of times the indicator test would have rejected the sample.

the best index for raw ground beef and <3 $g^{-1}$ E. coli for the cooked ground beef. This observation for raw ground beef supports the conclusion previously published (4).

E. coli was the most successful indicator, and acceptance and rejection of raw ground beef samples free of and containing food poisoning microorganisms based on use of E. coli was determined. Success was achieved in 61% of the 157 samples tested. Thus 39% of all raw ground beef samples would have been either rejected when they were free of food poisoning microorganisms or accepted when the poisoning organisms were present using a criterion of <10 $g^{-1}$ E. coli for acceptability. Similar values have been reported (3) for raw foods using criteria of <3 and <100 $g^{-1}$ E. coli. Mossell (personal communication) presented the hypothesis that presence of an indicator organism and failure to detect a pathogen should be considered a correct decision due to sampling difficulties and only the absence of an indicator in the presence of a pathogen should be considered as a false response. Using this approach 20% incorrect decisions of the most dangerous type would have resulted. Thus the validity of E. coli as an indicator of safety ground beef is highly questionable.

The RTS foods selected for analysis included protein products requiring minimal and considerable post-heating handling and carbohydrate products requiring considerable preparation handling with or without prior thermal treatment. Table 3 shows that the number of samples containing food-poisoning microorganisms is very low except in precooked roast beef. This is a convenience item prepared primarily for cold slicing. It usually is only partially cooked in the center and is a pink to reddish color. The limited sampling is due to discontinued purchase resulting from the high contamination frequency. In comparison, the sliced roast beef is a product cooked to an internal temperature of 70°C which leaves no pink color in the center and seldom contains food poisoning organisms. The non-nitrite and low salt containing roast beef and turkey stand out in their frequency of contamination with 14% for coliforms and 7% for E. coli, while the cured meats have frequencies of only 4% for coliforms and 0.6% for E. coli.

A comparison of Table 4 with Table 3 shows the effect of post-thermal treatment handling. The frequency of S. aureus contamination increases from 1.7% to 7.0%

Table 3. Number of cooked meat and cooked cured meat samples containing unacceptable levels of food poisoning or indicator organisms

| Item | Number of samples | CP[a] | SA[b] | SAL[c] | APC[d] | CF[e] | EC[f] |
|---|---|---|---|---|---|---|---|
| Roast beef (precooked) | 9 | 2 | 0 | 1 | 8 | 3[g] | 3[g] |
| Roast beef (sliced) | 66 | 2 | 2 | 0 | 4 | 5 | 3 |
| Turkey (sliced) | 68 | 2 | 1 | 0 | 14 | 12 | 4 |
| Corned beef (sliced) | 16 | 0 | 0 | 0 | 6 | 1 | 0 |
| Ham (sliced) | 65 | 0 | 1 | 0 | 7 | 4 | 1 |
| Bologna | 23 | 0 | 0 | 0 | 4 | 1 | 0 |
| Salami | 21 | 1 | 1 | 0 | 0 | 1 | 0 |
| Frankfurters | 33 | 2 | 0 | 0 | 1 | 0 | 0 |

a. CP = C. perfringens; b. SA = S. aureus; c. SAL = Salmonella;
d. APC = Aerobic Plate Count; e. CF = Coliform Count; f. EC = E. coli;
g. Coliform test and E. coli test run on only 5 samples.

Table 4. Number of protein salad samples containing unacceptable levels of food poisoning or indicator organisms

| | Number of samples | CP[a] | SA[b] | SAL[c] | APC[d] | CF[e] | EC[f] |
|---|---|---|---|---|---|---|---|
| Turkey Salad | 17 | 0 | 2 | 1 | 6 | 7 | 7 |
| Chicken Salad | 12 | 0 | 1 | 0 | 3 | 3 | 1 |
| Tuna Fish Salad | 152 | 0 | 9 | 2 | 15 | 37 | 11 |
| Egg Salad | 44 | 1 | 3 | 1 | 5 | 17 | 3 |
| Ham Salad | 17 | 0 | 2 | 1 | 4 | 4 | 2 |

a. CP = C. perfringens; b. SA = S. aureus; c. SAL = Salmonella;
d. APC = Aerobic Plate Count; e. CF = Coliform Count; f. EC = E. coli.

while the Salmonella frequency changes from 0.3% to 2.1%. The elevated frequency of coliforms and E. coli in turkey salad carries over from the lesser handled sliced turkey.

The carbohydrate type salad data in Table 5 demonstrate low frequency of food-poisoning microorganism detection. The high coliform frequency in the carrot salad sets it apart from the other products.

The data in Tables 3, 4 and 5 show a higher frequency of indicator than of food poisoning microorganism detection. The reliability of the indicators in detecting the specific food poisoning microorganisms was generally poor as may be seen in Table 6. All of the indicator successes for C. perfringens were in roast beef or turkey. The indicator successes for S. aureus and Salmonella were scattered among the sliced meats and salads. The data in Table 6 demonstrate lesser indicator success in the thermally treated food than was seen in Table 2 for the raw ground beef. The APC and coliform index tests were performed identically overall but there was disagreement between the two tests in 8 samples and agreement in 31 samples. In terms of approximately $10^3$ samples of RTS foods in the complete study, 15% would

Table 5. Number of carbohydrate salad samples containing unacceptable levels of food poisoning or indicator organisms

| | Number of samples | CP[a] | SA[b] | SAL[c] | APC[d] | CF[e] | EC[f] |
|---|---|---|---|---|---|---|---|
| Macaroni salad | 35 | 1 | 0 | 0 | 4 | 3 | 1 |
| Potato salad | 42 | 1 | 0 | 1 | 4 | 1 | 7 |
| Cole slaw | 19 | 0 | 0 | 0 | 2 | 0 | 0 |
| Carrot salad | 4 | 0 | 0 | 0 | 3 | 3 | 1 |

a. CP = *C. perfringens*; b. SA = *S. aureus*; c. SAL = *Salmonella*; d. APC = Aerobic Plate Count; e. CF = Coliform Count; f. EC = *E. coli*.

Table 6. Success of various indicator tests at rejecting samples of thermally treated foods containing specific food-poisoning microorganisms

| | APC | Coliforms | *E. coli* |
|---|---|---|---|
| *C. perfringens* | 3/12[a] | 3/12 | 3/12 |
| *S. aureus* | 8/22 | 8/22 | 1/21 |
| *Salmonella* | 4/7 | 4/7 | 0/7 |

a. The lower value in the ratio indicates the number of samples containing the food-poisoning organism. The upper number indicates the number of times the indicator test would have rejected the sample.

Table 7. Incidence frequency of specific *Enterobacteriaceae* in 131 raw and 230 ready-to-serve foods

| Organism | Incidence (%) Raw | Ready-to-serve |
|---|---|---|
| *E. coli* | 88 | 39 |
| *Klebsiella* | 16 | 35 |
| *Citrobacter* | 28 | 24 |
| *E. cloacae* | 17 | 31 |
| *E. aerogenes* | 15 | 14 |
| *Erwinia* | 9 | 11 |
| *Serratia* | 27 | 28 |
| *Hafnia* | 15 | 7 |
| *Proteus* (4 species) | 16 | 16 |
| *Providencia* | 11 | 7 |
| *Edwardsiella* | 11 | 2 |
| *Shigella* | 8 | 13 |

have been judged incorrectly by either APC or coliforms. The acceptance of pathogen containing samples would have been 4% for coliforms and 4.3% for APC.

The opportunist genera, Escherichia, Enterobacter, Citrobacter, Klebsiella and Erwinia appeared 2.0 times as often as the frank pathogens, Proteus, Providencia, Hafnia, Serratia, Salmonella, Shigella and Edwardsiella, in the raw foods and 2.1 times as often in the RTS foods. There were 131 raw and 230 RTS foods which yielded typical colonies on EMB Agar. Table 7 shows that among raw foods, 88% contained E. coli while only 39% of the RTS foods yielded E. coli. Klebsiella and E. cloacae organisms were identified in 35% and 31% of the RTS foods but in only 16% and 17% of the raw foods. Hafnia organisms were identified in 15% of the raw and in 7% of the RTS foods. There were differences between Proteus species but the number of foods involved was too small to permit conclusions to be drawn. While E. coli is the only dominant isolate from raw foods, the RTS foods contained approximately equivalent numbers of E. coli, Klebsiella, E. cloacae, Serratia and Citrobacter. This observation may lend support to the suggested use of total Enterobacteriaceae as an index (4) in RTS food monitoring. Since foods requiring minimum post-heating handling showed 22% and 16% incidences of Klebsiella and Serratia compared to 38% and 30% incidences in foods requiring considerable post-heating handling, a significant amount of the contamination must be due to human unsanitary practices. Klebsiella and Serratia are the organisms of most concern. Both have been subject to classification changes with many of the Klebsiella formerly being Aerobacter and a significant portion of the Serratia formerly being E. liquefaciens. Colonization of the intestine following ingestion of Klebsiella contaminated food has been suggested (7). Reported isolation frequencies are 17% and 19% for E. coli and Klebsiella in RTS foods. Consideration must be given to the potential pathogenicity of food-borne Klebsiella and possibly Serratia.

## D. References


1. Goldenberg, N., Elliott, D.W.: The value of agreed nonlegal specifications. In: Microbiological Safety of Foods. Hobbs, B.C., Christian, J.H.B. (ed.). London: Academic Press, 1973 pp. 359-368. (1973).
2. Levine, M.: Facts and fancies of bacterial indices in standards for water and foods. Fd. Technol. 15, 29-38, Nov. (1961).
3. Miskimin, D.K., Berkowitz, K.A., Solberg, M., Riha, W.E., Franke, W.C., Buchanan, R.L., O'Leary, V.: Relationships between indicator organisms and specific pathogens in potentially hazardous foods. J. Fd.Sci. 41, 1001-1006 (1976).
4. Mossel, D.A.A.: Occurrence, prevention, and monitoring of microbial quality loss of foods and dairy products. Crit. Revs. Env. Control 5, 1-139 (1975).
5. Pace, P.J.: Bacteriological quality of delicatessen foods: are standards needed? J. Milk Fd. Technol. 38, 347-353 (1975).
6. Proposed microbiological guidelines for ground beef. Canadian Dept. of Health and Welfare Information Letter, I.L. No. 453,(Dec. 17, 1975).
7. Shooter, R.A., Faiers, M.C., Cooke, E.M., Breaden, A.L., O'Farrell, S.M.: Isolation of Escherichia coli, Pseudomonas aeruginosa and Klebsiella from foods in hospitals, canteens and schools. Lancet 275, 390-392 (1971).
8. Solberg, M., Miskimin, D.K., Kramer, R., Riha, W.E., Franke, W.C., Buchanan, R. L., O'Leary, V., Berkowitz, K.: Assurance of microbiological safety in a university feeding system. J. Milk Fd. Technol. 39, 200-205 (1976).
9. Tomfohrde, K.M., Rhoden, D.L., Smith, P.B., Balows, A.: Evaluation of the redesigned enterotube - a system for the identification of Enterobacteriaceae. Appl. Microbiol. 25, 301-304 (1973).

# The Origin of Bacteriophages in Cheese Factories

A. W. JARVIS, H. A. HEAP, and R. C. LAWRENCE

## A. Introduction

Basically cheesemaking consists of the production of coagulum in milk and the expulsion of moisture from this coagulum by the combined effects of heat, acid and salt. In Cheddar cheesemaking the acid is produced by lactic streptococci, termed starter bacteria, which form lactic acid by the metabolism of lactose. The activity of the starter may be retarded by a number of factors, the most important of which is bacteriophage. In New Zealand, cheese factories are closed down for two months in the winter and the new season begins with a factory apparently free of phage. Nevertheless, when a newly isolated starter strain is introduced it is usually only a matter of time before a phage which attacks a starter can be detected. The origin and differentiation of these phages is of considerable importance to the dairy industry. Morphological, bacterial host range and serological data have been obtained for phages isolated from factories throughout New Zealand to determine the relationships between these phages and to throw some light on the origin of bacteriophages in cheese factories. A number of possible origins have been postulated.

The most favoured hypothesis was that the source of phages was external to the cheese plant. Raw milk was considered to be the most likely origin, since it almost invariably contains wild-type lactic streptococci which could propagate phage and the phage would not be destroyed by pasteurization.

More recently however, an important finding was made, at about the same time in our laboratory and at the Universities of Minnesota and Oregon, that many, possibly the majority of strains of lactic streptococci are lysogenic. Since lysogenic bacteria in other genera have been found to release phage spontaneously in low numbers it can be assumed lactic streptococci do the same to a greater or lesser extent.

Therefore, lysogenic starter bacteria might release infective phage which could multiply if other strains in the starter rotation were sensitive to that particular phage.

An investigation into the origin of phages concerned, firstly, a survey of the types of phages found throughout New Zealand, secondly, a factory trial and, thirdly, induction experiments which support the findings made in the factory.

## B. Experimental

I. Phage survey. Lactic streptococcal phages have been grouped morphologically (3) according to head size and shape, tail length, and the presence or absence of a collar. Table 1 shows the morphology and incidence of the six phage types of 188 phages. The majority (77%) belong to group 'a' or 'b', which differ only in the presence or absence of a collar. Antisera were prepared against purified preparations of six phages (2) representing these six morphological groups (Fig. 1)

Sixty phages, isolated from factory wheys, were tested for neutralization by the six antisera (Table 2). It can be seen that there is good correlation between

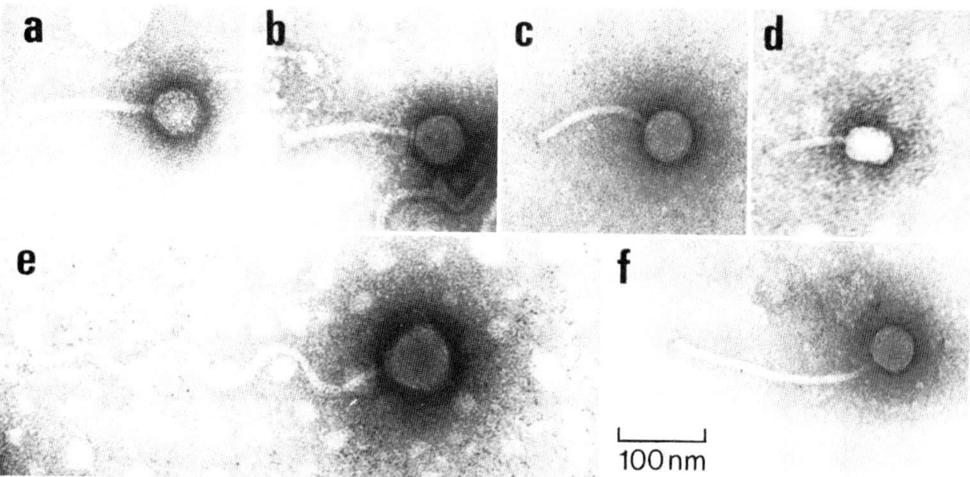


Fig. 1 a-f. Representatives of the six morphological types of phages against which antisera were prepared, a-f as in Table 1

morphological and serological grouping. With few exceptions phages reacted always and only with antiserum prepared against phage of a similar morphology, with some degree of cross-neutralization between types 'a' and 'b', which differ only in the presence or absence of a collar. However, although phages of a particular morphological type reacted with antiserum prepared against a similar phage, there was a wide variation in the rate of neutralization of the phages by the antiserum (Table 3). Rates of neutralization are reported as K values (1), K being 100% for homologous phage. Phages which grow on a particular host have similar rates of neutralization. Similar phages were isolated from widely separated factories over a period of up to three years, suggesting a common source of phage that does not depend on the factory location. Comparable results were obtained with other antisera and other phage types. The phages studied could therefore be divided into six groups by morphological and serological data and further divided serologically into subgroups which correlated with host range.

Table 1. Phages against which antisera were prepared

| Phage type | Phage | Morphology Head (nm) | Tail (nm) | Incidence[a] of Phage type % |
|---|---|---|---|---|
| a. Isometric | Ø936(158) | 58 | 137 | 47 |
| b. Collared isometric | Ø853($AM_1$) | 56 | 140 | 30 |
| c. Short-tailed isometric | Ø876($ML_8$) | 55 | 132 | 8 |
| d. Prolate | Ø923(112) | 63x47 | 86 | 11 |
| e. Large isometric | Ø949($ML_8$) | 88 | 450 | 3 |
| f. Long-tailed isometric | Ø$T_{BK_5}$ ($H_2$) | 58 | 233 | 1 |

[a] Incidence of this morphological phage type in 188 phages isolated from factory wheys during 1973-77.

Table 2. Neutralization of 60 phages by antisera prepared against phages representing six morphological types

| Phages | | Number of phages neutralized by each antiserum | | | | | | |
|---|---|---|---|---|---|---|---|---|
| Type | Number Tested | a[a] | b[a] | a & b[b] | c[a] | d[a] | e[a] | f[a] |
| a | 27 | 26 | 0 | 5 | 0 | 0 | 0 | 0 |
| b | 12 | 0 | 12 | 7 | 0 | 0 | 0 | 0 |
| c | 8 | 3 | 1 | 1 | 4 | 0 | 0 | 0 |
| d | 9 | 0 | 0 | 0 | 0 | 9 | 0 | 0 |
| e | 2 | 0 | 0 | 0 | 0 | 0 | 2 | 0 |
| f | 2 | 0 | 0 | 0 | 0 | 0 | 0 | 1 |

[a] Type of phage against which antiserum was prepared.
[b] Phages reacted with both antisera 'a' and 'b'.

Table 3. Rate of neutralization (K) of phages by antiserum prepared against type 'a' phage Ø936 (158)

| Host | Phage | Phage Type | %K [a] | Factory | Date |
|---|---|---|---|---|---|
| 158 | 936 | a | 100 | Kaikoura | November 1975 |
| 158 | 953 | a | 100 | NZDRI | January 1976 |
| 158 | 859 | a | 79 | Collingwood | October 1973 |
| 158 | 874 | a | 99 | Woodville | November 1974 |
| $P_2$ | 937 | a | 6 | Kaikoura | November 1975 |
| $P_2$ | 952 | a | 3 | Kiwi | January 1976 |
| 166 | 925 | a | 1 | Waihi | November 1975 |
| 166 | 938 | a | 1 | Kaikoura | November 1975 |
| 240 | 896 | b | 15 | Kiwi | September 1975 |
| 240 | 934 | b | 15 | Kaikoura | November 1975 |

[a] K value for homologous phage = 100%.

II. Factory trial. A trial was carried out in a commercial cheese factory in an attempt to determine the origin of phage. Each day samples were taken of the whey at milling and it was tested for phage. When phages appeared they could be eliminated in one of two ways, firstly, by removing the strain on which they were growing i.e. the indicator strain, and secondly, by replacing the strain which was believed to be the source of a temperate phage which grew on another strain in the rotation. When this strain was re-introduced the phage reappeared. In this way it was possible to select a rotation of single strain starters such that phage could not be detected in the factory.

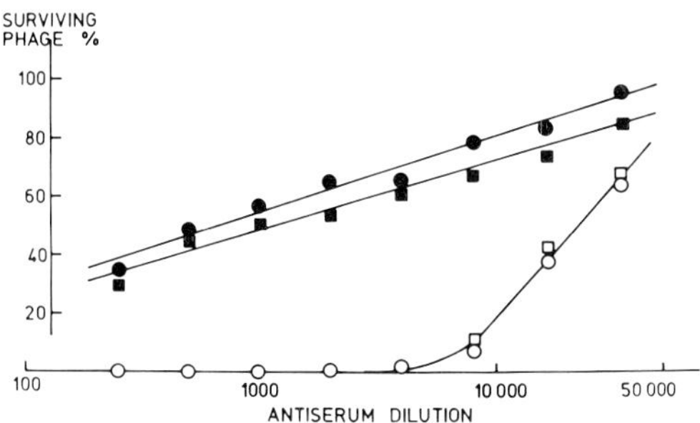


Fig. 2. Neutralization curves for phages $T_{104}$ (266) □ ; ϕ1003 (266) o ; $T_{104}$ (120) ■ ; ϕ1011 (120) ● ; with antiserum ϕ936 (158). Phages were incubated for 15 min at 37°C with antiserum dilutions, and the surviving phages expressed as percent of phages in control experiments with normal serum (2)

III. Induction experiments. We tried to reproduce this important observation in the laboratory. Each of the six strains being used at that time in the rotation was induced under ultra-violet light, the induced phage concentrated, and tested on the other five strains to check if any were sensitive to these released phages. After several attempts we obtained a UV lysate of strain 104 that contained phages which grew on strains 266 and 120. The temperate phage released from strain 104 in the laboratory and the lytic phage isolated from the cheese plant that attacked strain 266 were of the same morphological type, i.e. each had a small isometric head with no evident collar. Serological experiments showed that the two phages were identical. Similarly, the temperate phage released from strain 104 that attacked strain 120 was identical morphologically and serologically to the lytic 120 phage isolated from the cheese plant (Fig. 2). Although the induced phages which grew on strains 120 and 266 were the same morphologically, serological experiments showed that they were different, i.e. strain 104 was doubly lysogenic and carried at least two temperate phages. In similar experiments a phage was induced from strain 134 which grew on strain 168, and was similar to the 168 factory phage.

On the other hand, it was not possible to show in the laboratory that the 114 lytic phage came from a temperate phage induced from one of the six strains. It was observed however, that a large proportion of the raw milk flora consisted of lactic streptococci. These strains were isolated, grown-up in sterile broth, and induced under ultra-violet light. One of these induced lysates was shown to contain a phage that attacked the 114 strain. These findings indicated that phages found in the cheese factory originated as temperate phages in starter strains or in wild-type lactic streptococci in the raw milk.

## C. Summary and Conclusions

Starter cultures used in cheesemaking are subject to infection by bacteriophage (phages). Apparently similar phages are found throughout New Zealand and morphologically similar phages have been reported from many other countries. The origin and differentiation of these phages is of considerable importance to the dairy industry. Phages have been differentiated by morphology and serology into six main groups and further differentiated serologically into sub-groups which correlate with host range. The temperate phages induced in the laboratory from starter strains or wild-type lactic streptococci found in the raw milk were similar in morphology, serology and host range to the virulent phages isolated during cheesemaking.

## D. References

1. Adams, M.H.: Bacteriophages. Interscience. New York. p.463 (1959).
2. Jarvis, A.W.: The serological differentiation of lactic streptococcal bacteriophages. N.Z. Jl. Dairy Sci. Technol. 12, 176-181 (1977).
3. Terzaghi, B.E.: Morphologies and host sensitivities of lactic streptococcal phages from cheese factories. N.Z. Jl. Dairy Sci. Technol. 11, 155-163 (1976).

# Ecology of Meat Spoilage at Chill Temperatures

C.O. GILL and K.G. NEWTON

## A. Introduction

Most raw meat is stored at chill temperatures (-2 to $5^{o}C$) at some time before consumption. At these temperatures only a few types of bacteria are present in significant numbers in the spoilage flora of meat. These are all psychrotrophs derived from the animals' environment via the hide (3). During development of the spoilage flora bacteria are confined to the meat surface (2). Under both aerobic and anaerobic conditions, growth on meat is logarithmic after the initial lag phase. Aerobically, when the bacterial density exceeds $10^{8}$ cells $cm^{-2}$ a slime develops and off-odours and flavours can be detected, but growth continues until a maximum cell density in excess of $10^{9}$ $cm^{-2}$ is attained. Under anaerobic conditions, the maximum cell density is about $10^{8}$ $cm^{-2}$ and there is no detectable spoilage until some weeks after maximum numbers have been reached.

The growth on meat of representative strains of five genera of bacteria was examined to determine the factors which influence the development of spoilage flora of chilled meat.

## B. Methods

Strains of the following organisms were isolated from spoiling chilled meat; fluorescent *Pseudomonas*, non-fluorescent *Pseudomonas*, *Acinetobacter*, *Enterobacter*, *Microbacterium thermosphactum* and *Lactobacillus*. Bacteria were grown in a meat juice medium and on blocks of meat inoculated on one surface only which were cut into thin slices before analysis to demonstrate concentration gradients. Preparation of the medium and methods of analysis for glucose, glucose-6-phosphate, lactate, amino acids and ammonia have been reported (1).

## C. Results and Discussion

In competition between spoilage bacteria, the species with the fastest growth rate will tend to become dominant. Of the organisms which grew aerobically, the pseudomonads had a marked advantage in growth rate, and this advantage tended to increase with decreasing temperature. The *Lactobacillus* strains had a similar growth rate advantage under anaerobic conditions (Table 1).

The variable growth rate shown by *Acinetobacter* was due to variation in pH. The minimum pH found in meat is 5.5, but this can rise to 6.5 or higher. *Acinetobacter* was greatly affected by small changes of pH below 6.0, but the growth rates of the other bacteria were unaffected by the pH of the meat.

During growth the bacteria utilize the low molecular weight soluble components of meat and utilization of these substrates by the bacteria studied is shown in Table 2.

Since growth occurs only at the meat surface, the rate of diffusion to the surface of preferentially utilized substrates present only in low concentration could be too low to meet the bacterial demand at high cell densities. The limited availability of some substrates could affect the development of the spoilage flora. With a pseudomonad, a concentration gradient of glucose developed and became more pronounced with increasing cell density, until at 3 x $10^8$ cells/$cm^2$ the glucose concentration at the surface was zero. The surface concentration of ammonia then started to rise. Growth ceased when the pH reached 7, although a final pH of 8 was attained. This showed that the appearance of detectable spoilage in aerobic cultures is not due simply to the increased bacterial numbers, but is a result of the change

Table 1. Growth rates of bacteria on meat at 5°C

| Genus | Generation time (h) Aerobic | Anaerobic |
|---|---|---|
| Pseudomonas | 5.4 | - |
| Acinetobacter | 6.3 - 10.7 | - |
| Enterobacter | 7.8 | 23.2 |
| Microbacterium thermosphactum | 7.3 | 20.1 |
| Lactobacillus | - | 6.5 |

Table 2. Utilization of low molecular weight soluble components of meat by spoilage bacteria

| Genus | Substrates in order of utilization | | | |
|---|---|---|---|---|
| | Glucose | Glucose-6-phosphate | Amino acids | Lactate |
| Aerobic | | | | |
| Pseudomonas | ++ | - | ++[a] | + |
| Acinetobacter | - | - | ++[a] | + |
| Microbacterium thermosphactum | ++ | - | ++[b] | - |
| Enterobacter | ++ | ++ | +[a] | + |
| Anaerobic | | | | |
| Lactobacillus | ++ | - | ++[c] | - |
| Microbacterium thermosphactum | ++ | - | - | - |
| Enterobacter | ++ | ++ | - | - |

(++) Maximum growth rate (+) reduced growth rate (-) no growth

(a) glutamate, aspartate, serine, proline, alanine

(b) glutamate only

(c) arginine only

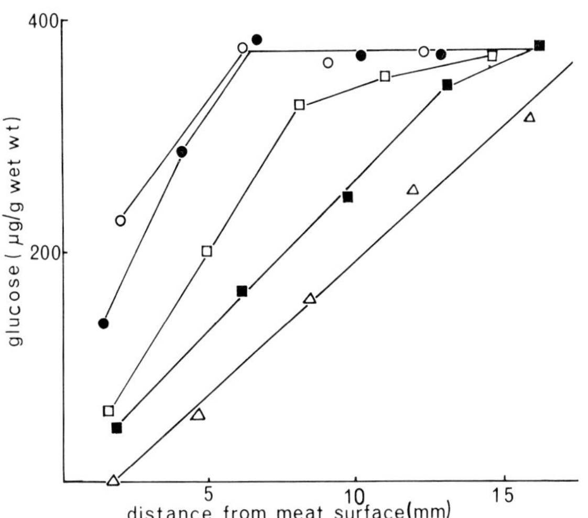


Fig. 1. Concentration gradients of glucose in meat with *Pseudomonas* on the surface at cell densities $cm^{-2}$ of 2.7 x $10^7$ (o), 6.3 x $10^7$ (●), 3.2 x $10^8$ (□), 1.1 x $10^9$ (■) and 2.8 x $10^9$ (Δ)

from glucose to amino acids as the main substrate. *Enterobacter* at near maximum numbers produced a concentration gradient of G-6-P, but amino acids and lactic acid remained abundant at the meat surface when bacterial growth had ceased.

Anaerobically, concentration gradients of the three utilized substrates could be demonstrated with the appropriate bacteria when they approached their maximum cell densities (Fig. 2). In this case it appears that the maximum cell density is determined by the availability of fermentable substrates.

To determine what factor limits aerobic growth of bacteria, mixed cultures of the test organisms were grown on meat. When two organisms were inoculated in similar initial numbers, both grew at their maximum rates until the maximum cell density was approached. When a pseudomonad however, was allowed to grow to near

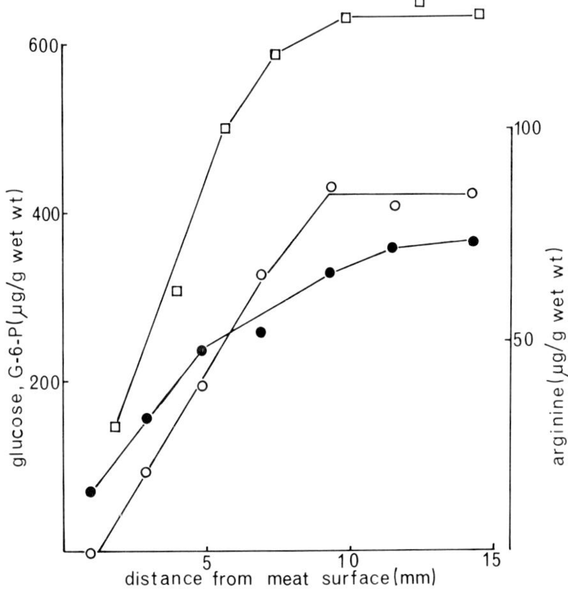


Fig. 2. Concentration gradients produced in meat by anaerobic bacteria. Glucose (o), arginine (●); *Lactobacillus* at 4.0 x $10^7$ $cm^{-2}$. Glucose-6-phosphate (□); *Enterobacter* at 2.7 x $10^7$ $cm^{-2}$

its maximum density before inoculation of the minor species, the growth rates of Enterobacter and M. thermosphactum were reduced and they attained a cell density typical of anaerobic growth. Enterobacter had a similar effect on the growth of M. thermosphactum. In experiments with a pseudomonad as the minor species, the growth rate of the pseudomonad was not affected and the bacteria reached a cell density of the same order as the major species (Table 3). These results suggest that cessation of aerobic growth is due to limited availability of oxygen and that the pseudomonads can reduce the oxygen concentration to a level which forces the facultative organisms into fermentative metabolism.

In experiments with anaerobic mixed cultures, the growth of M. thermosphactum was inhibited by Enterobacter when the latter approached its maximum cell density ($<10^7$ $cm^{-2}$). Both M. thermosphactum and Enterobacter however, were inhibited by Lactobacillus when it reached a cell density in excess of $10^6$ $cm^{-2}$ (Fig. 3).

At growth-limiting substrate concentrations, the substrate affinity of the organism, rather than the maximum growth rate, may be decisive in microbial competition. When mixed cultures grow in a chemostat under glucose limitation at sub-maximum growth rates, the organism with the greatest affinity for glucose will displace those with less affinity. In such experiments it was found that Enterobacter displaced M. thermosphactum, and both displaced Lactobacillus. On meat, maximum numbers of Enterobacter limited growth of M. thermosphactum, presumably by reducing the glucose concentration below that required for growth, but inhibition of the other species by Lactobacillus on solid test media appeared to be due to the production of an antimicrobial agent. This is a frequently observed phenomenon with Lactobacillus species (4).

We are therefore dealing with very different systems under aerobic and anaerobic conditions. Aerobically, substrate limitation does not affect the composition of developing spoilage cultures, and individual bacteria do not interfere with each other's growth until maximum numbers are attained, when oxygen is probably the limiting factor. Pseudomonads dominate aerobic spoilage cultures because they can outgrow competitors and at high cell densities suppress their growth, probably by

Table 3. Growth of bacteria on Meat at $10^{\circ}C$ in air in the presence of a species which had attained the maximum cell density

| Species at maximum cell density | Species at low initial cell density | Generation time of minor species (h) | Maximum cell density of minor species (h) |
|---|---|---|---|
| fluorescent Pseudomonas | Enterobacter | 10.4 | $6.2 \times 10^7$ |
| fluorescent Pseudomonas | Microbacterium thermosphactum | 7.2 | $7.1 \times 10^7$ |
| Enterobacter | fluorescent Pseudomonas | 3.2 | $1.1 \times 10^9$ |
| Enterobacter | Microbacterium thermosphactum | 9.1 | $2.7 \times 10^7$ |
| Microbacterium thermosphactum | fluorescent Pseudomonas | 3.2 | $1.6 \times 10^9$ |
| Microbacterium thermosphactum | Enterobacter | 6.0 | $7.6 \times 10^8$ |

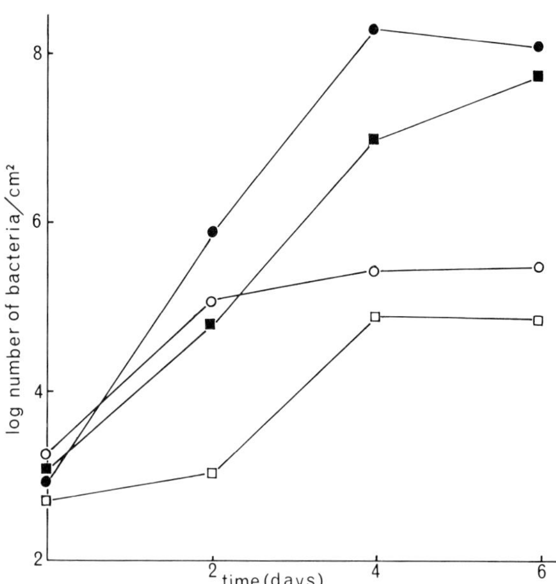


Fig. 3. Anaerobic growth on meat at 10°C of Lactobacillus (●) with Enterobacter (○) and Enterobacter (■) with M. thermosphactum (□)

preventing the other bacteria obtaining oxygen. Anaerobically, substrate limitation determines the maximum cell density, and inability to compete for a common substrate, such as glucose, may result in the suppressed growth of some spoilage types. However, lactobacilli usually predominate, because they outgrow competitors and produce an agent which inhibits growth of other species.

## D. References

1. Gill, C.O.: Substrate limitation of bacterial growth at meat surfaces. J. Appl. Bacteriol. 41, 401-410 (1976).
2. Gill, C.O. & Penney, N.: Penetration of bacteria into meat. Appl. Environ. Microbiol. 33, 1284-1286, (1977).
3. Grau, F.H.: In: Advances in Meat Science and Technology 1. C.S.I.R.O., Australia (1974).
4. Hurst, A.: Microbial antagonism in foods. Can. Inst. Food Sci. Technol. J. 6, 80-90 (1973).

# Subject Index

## Microbial Ecology of a Brackish Water Environment

Editor: G. Rheinheimer
With contributions by M. Bölter, K. Gocke, H.-G. Hoppe, J. Lenz, L.-A. Meyer-Reil, B. Probst, G. Rheinheimer, J. Schneider, H. Szwerinski, R. Zimmermann
1977. 77 figures, 87 tables. XI, 291 pages
(Ecological Studies, Volume 25)
ISBN 3-540-08492-4

This book describes the results of a 15-month investigation in the Kiel Bight (Baltic Sea), in which more than 50 hydrographical, chemical, and microbiological parameters were measured. Using methods such as scanning electron microscopy and autoradiography, a comparative analysis of most of the measured data on the bacterial colonization of detritus and the relations between pollution, production, and remineralization led to the construction of a model for the ecosystem, indicating the energy fluxes.
Together with Volume 24 by Kremer and Nixon on ecological studies conducted at Naragansett Bay, and system-simulation by a computer model, the work represents a comprehensive study of coastal marine ecosystems.

## Antibiotics

Volume 1: **Mechanism of Action**
Editors: D. Gottlieb, P. D. Shaw
1967. 197 figures. XII, 785 pages
ISBN 3-540-03724-1

Volume 2: **Biosynthesis**
Editors: D. Gottlieb, P. D. Shaw
1967. 115 figures. XII, 466 pages
ISBN 3-540-03725-X

Volume 3: **Mechanism of Action of Antimicrobial and Antitumor Agents**
Editors: J. W. Corcoran, F. E. Hahn
1975. 193 figures. XII, 742 pages
ISBN 3-540-06653-5

"This review is to be commended. It is deliberately selective and concise, has litle redundancy, is well edited, and quite complete... These books should provide excellent reference volumes and, in addition, pleasurable reading. They will be extremely useful to all investigators using the techniques of modern molecular biology, handy for graduate students, and will provide an important adjunct to the physicians' library."

*Annals of Internal Medicine*

Springer-Verlag
Berlin
Heidelberg
New York